Anaerobic Degradation of Chlorinated Solvents

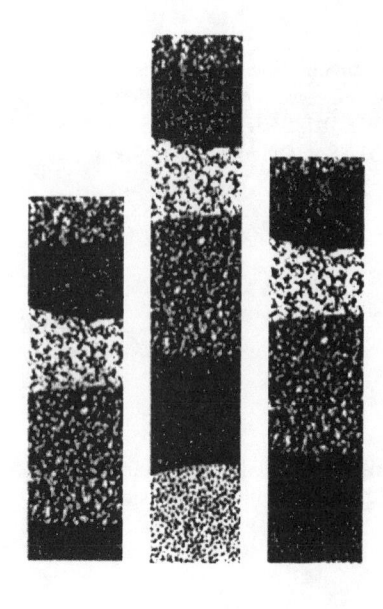

Editors

Victor S. Magar, Donna E. Fennell, Jeffrey J. Morse, Bruce C. Alleman, and Andrea Leeson

The Sixth International In Situ and On-Site Bioremediation Symposium

San Diego, California, June 4–7, 2001

BATTELLE PRESS
Columbus • Richland

Library of Congress Cataloging-in-Publication Data

International In Situ and On-Site Bioremediation Symposium (6th : 2001 : San Diego, Calif.)
   Anaerobic degradation of chlorinated solvents : the Sixth International In Situ and On-Site Bioremediation Symposium : San Diego, California, June 4-7, 2001 / editors, V. Magar ... [et al.].
   p. cm. -- (The Sixth International In Situ and On-Site Bioremediation Symposium ; 7)
   Includes bibliographical references and index.
   ISBN 1-57477-117-5 (hc. : alk. paper)
   1. Solvents--Biodegradation--Congresses. 2. Organochlorine compounds--Biodegradation--Congresses. 3. Bioremediation--Congresses. 4. Anaerobic bacteria--Congresses. I. Magar, V. (Victor), 1964-   . II. Title. III. Series: International In Situ and On-Site Bioremediation Symposium (6th : 2001 : San Diego, Calif.). Sixth International In Situ and On-Site Bioremediation Symposium ; 7.
   TD192.5.I56 2001 vol. 7
   [TD196.S54]
   628.5 s--dc21
   [628.5'2]

2001044130

Printed in the United States of America

Copyright © 2001 Battelle Memorial Institute. All rights reserved. This document, or parts thereof, may not be reproduced in any form without the written permission of Battelle Memorial Institute.

Battelle Press
505 King Avenue
Columbus, Ohio 43201, USA
614-424-6393 or 1-800-451-3543
Fax: 1-614-424-3819
Internet: press@battelle.org
Website: www.battelle.org/bookstore

For information on future environmental conferences, write to:
   Battelle
   Environmental Restoration Department, Room 10-123B
   505 King Avenue
   Columbus, Ohio 43201-2693
   Phone: 614-424-7604
   Fax: 614-424-3667
   Website: www.battelle.org/conferences

CONTENTS

Foreword   vii

Source Zone Remediation

Effects of Potassium Permanganate Oxidation on Subsurface Microbial Activity.
M.A. Rowland, G.R. Brubaker, K. Kohler, M. Westray, and D. Morris   1

Sequestration and Degradation of TCE Following DNAPL Source Remediation.
A. Gavaskar, E. Foote, J. Butler, and S. Yoon   13

Two Novel Methods for Enhancing Source Zone Bioremediation: Direct Hydrogen Addition and Electron Acceptor Diversion. *C.J. Newell, C.E. Aziz, P.E. Haas, J.B. Hughes, and T.A. Khan*   19

Enhanced Bioremediation of High Contaminant Concentrations in Source Residual Area. *V.B. Dick, N.L. Case, and S.L. Boyle*   27

Technologies Competitive with Enhanced Bioremediation of Source Zones.
R.D. Norris   35

A Comparison of Chemical Oxidation and Biological Reductive Dechlorination Technologies for the Treatment of Chlorinated Solvents. *R.J. Fiacco, G. Demers, R.A. Brown, G. Skladany, and D. Robinson*   45

Mobilization of Sorbed-Phase Chlorinated Alkenes in Enhanced Reductive Dechlorination. *F.C. Payne, S.S. Suthersan, F.C. Lenzo, and J.S. Burdick*   53

Laboratory Studies of Reductive Processes

Biodegradation of Chlorinated Ethenes Under Various Redox Conditions. *X. Lu, G. Li, S. Tao, T.N.P. Bosma, and J. Geritse*   61

Differential Stimulation of Haloreduction by Carbon Addition to Subsurface Soils. *M.A. Panciera, O. Zelennikova, B.F. Smets, and G.M. Dobbs*   69

Interaction of Microbes and Uranium During Enrichment for TCE Reductive Dechlorination. *B.D. Lee, M.R. Walton, and J.L. Meigio*   77

Enhanced Reductive Dechlorination on an Industrial Site in Belgium. *D. Nuyens, M. Meyer, D. Wanty, V. Miles, and V. Dries*   87

Coupling of Toluene Oxidation with PCE Dechlorination Under Sulfidogenic Conditions. *T.P. Hoelen, J.A. Cunningham, C.A. LeBron, and M. Reinhard*   95

Comparison of Reducing Agents for Dechlorination in a Simulated Aquifer.
W.A. Farone and T. Palmer 103

Potential for Bioremediation of Groundwater Contaminated with Landfill
Leachate. *D.L. Freedman, J. Cox, L. Baiden, C. Carvalho, C.K. Gunsch, J. Hunt,
and R.L. Brigmon* 109

Varying Substrate Concentration to Enhance TCE Degradation in Dual-Species
Bioreactors. *J. Komlos, A. Cunningham, A. Camper, and R. Sharp* 117

Bioremediation Field Case Studies

The Federal Integrated Biotreatment Research Consortium: Flask to Field.
J.W. Talley, L. Hansen, D. Felt, J. Spain, H. Pritchard, G. Sewell, and J. Tiedje 125

Design of an Enhanced Anaerobic Bioremediation System for a Low
Permeability Aquifer. *S.M. Dean, L. Wiseman, A.W. Bourquin, J.V. Accashian,
M. Edgar, and J. Callendar* 133

Monitoring Stimulated Reductive Dechlorination at the Rademarkt in
Groningen, The Netherlands. *A.A.M. Langenhoff, A.A.M. Nipshagen, C. Bakker,
J. Krooneman, and G. Visscher* 141

Complete PCE Degradation and Site Closure Using Enhanced Reductive
Dechlorination. *M.S. Maierle and J.L. Cota* 149

Anaerobic Bioremediation of Trichloroethene Near Duluth International
Airport. *R. Semer and P. Banerjee* 157

Enhanced CAH Reductive Dehalogenation at a Former Wastewater Treatment
Facility. *I.R. Schaffner, A.T. Doherty, J.M. Wieck, and S.R. Lamb* 165

Enhanced Reductive Dechlorination of Ethenes Large-Scale Pilot Testing.
J.A. Peeples, J.M. Warburton, I. Al-Fayyomi, and J. Haff 173

In Situ Enhanced Reductive Dechlorination of PCE. *S.A. Koch and J.M. Rice* 181

Effective Enhancement of Biological Degradation of Tetrachlorothene (PCE)
in Groundwater. *R.W. North, S.E. Burkett, and M.J. Sincock* 189

Enhanced Bioremediation of Chlorinated Solvents. *W. Murray, M. Dooley, and
S. Koenigsberg* 197

Accelerating the Reductive Dechlorination Process in Groundwater. *D. South,
J. Seracuse, K. Garrett, and D. Li* 205

Biologically-Enhanced Reductive Dechlorination. *G.J. Skladany, D. Brown,
D.A. Burns, M. Bell, and M.D. Lee* 213

Enhanced Bioremediation in Clay Soils. *J.C.Bensch and Z.M. Zahiraleslamzadeh* 221

Aerobic and Anaerobic Bioremediation of 1,1-DCE and Vinyl Chloride in Groundwater. *J.R. Larson and V.J. Voegeli* 229

Electron Donor Injection Strategies

Technical Protocol for Enhanced Reductive Dechlorination Via Vegetable Oil Injection. *T.H. Wiedemeier, B.M. Henry, and P.E. Haas* 241

Effective Distribution of Edible Oils – Results From Five Field Applications. *M.D. Lee, B. Borden, M.T. Lieberman, W. Beckwith, T. Crotwell, and P.E. Haas* 249

Time-Release Electron Donor Technology: Results of Forty-Two Field Applications. *S. Koenigsberg, C. Sandefur, and K. Lapus* 257

Favoring Efficient In Situ TCE Degradation Through Amendment Injection Strategy. *J.P. Martin, K.S. Sorenson, and L.N. Peterson* 265

Design of a Novel Injection Scheme for Enhanced Anaerobic Bioremediation. *M.E. Miller and J.T. Drake* 273

Enhanced Bioremediation Under Difficult Geologic Conditions — Case Studies. *N.L. Case, S.L. Boyle, and V.B. Dick* 281

Enhanced Microbial Dechlorination of Chloroethenes and Chloroethanes: From Laboratory Tests to Application in a Bioscreen. *H. Slenders, J. Ter Meer, M. Van Eekert, T. Verheij, P. Verhaagen, and J. Theeuwen* 289

Liner — A New Concept for the Stimulation of Reductive Dechlorination. *E.C.L. Marnette, H. Tonnaer, A. Alphenaar, and G.J. Groenendijk* 297

Enhanced CAH Dechlorination Using Slow and Fast Releasing Polylactate Esters. *P.K. Sharma, H.T. Voscott, and B.M. Swann* 305

Application of In Situ Remediation Technologies by Subsurface Injection. *B.E. Duffy* 313

Groundwater Circulation Wells for Introduction of Vitamin B_{12}

Rehabilitation of a Biofouled Recirculation Well Using Innovative Techniques. *S.R. Forman, T. Llewellyn, S. Morgan, C. Mowder, S. Lesage, K. Millar, S. Brown, G. DeLong, D.J. Green, and H. McIntosh* 321

Biocide Application Prevents Bifouling of a Chemical Injection/Recirculation Well. *K. Millar, S. Lesage, S. Brown, C.S. Mowder, T. Llewellyn, S. Forman, D. Peters, G. DeLong, D.J. Green, and H. McIntosh* 333

Use of a Recirculation Well for the Delivery of Vitamin B_{12} for the In-Situ Remediation of Chlorinated Solvents in Groundwater. *S. Lesage, S. Brown, K. Millar, C.S. Mowder, T. Llewellyn, S. Forman, G. DeLong, and D. Green*   341

Author Index   349

Keyword Index   377

FOREWORD

The papers in this volume correspond to presentations made at the Sixth International In Situ and On-Site Bioremediation Symposium (San Diego, California, June 4-7 2001). The program included approximately 600 presentations in 50 sessions on a variety of bioremediation and supporting technologies used for a wide range of contaminants.

This volume focuses on *Anaerobic Degradation of Chlorinated Solvents*. Several case studies presented in this volume testify to the fact that research conducted over the past several years has advanced this technology to the point where it has become a common, full-scale method for treating groundwater plumes containing dissolved-phase chlorinated solvents. Research continues to improve the technology by optimizing electron donor injection and recirculation strategies. Other research is aimed at broadening the application of this technology to treat source zones, where remediation continues to pose a significant challenge.

The author of each presentation accepted for the symposium program was invited to prepare an eight-page paper. According to its topic, each paper received was tentatively assigned to one of ten volumes and subsequently was reviewed by the editors of that volume and by the Symposium chairs. We appreciate the significant commitment of time by each of the volume editors, each of whom reviewed as many as 40 papers. The result of the review was that 352 papers were accepted for publication and assembled into the following ten volumes:

Bioremediation of MTBE, Alcohols, and Ethers — 6(1). Eds: Victor S. Magar, James T. Gibbs, Kirk T. O'Reilly, Michael R. Hyman, and Andrea Leeson.

Natural Attenuation of Environmental Contaminants — 6(2). Eds: Andrea Leeson, Mark E. Kelley, Hanadi S. Rifai, and Victor S. Magar.

Bioremediation of Energetics, Phenolics, and Polycyclic Aromatic Hydrocarbons — 6(3). Eds: Victor S. Magar, Glenn Johnson, Say Kee Ong, and Andrea Leeson.

Innovative Methods in Support of Bioremediation — 6(4). Eds: Victor S. Magar, Timothy M. Vogel, C. Marjorie Aelion, and Andrea Leeson.

Phytoremediation, Wetlands, and Sediments — 6(5). Eds: Andrea Leeson, Eric A. Foote, M. Katherine Banks, and Victor S. Magar.

Ex Situ Biological Treatment Technologies — 6(6). Eds: Victor S. Magar, F. Michael von Fahnestock, and Andrea Leeson.

Anaerobic Degradation of Chlorinated Solvents— 6(7). Eds: Victor S. Magar, Donna E. Fennell, Jeffrey J. Morse, Bruce C. Alleman, and Andrea Leeson.

Bioaugmentation, Biobarriers, and Biogeochemistry — 6(8). Eds: Andrea Leeson, Bruce C. Alleman, Pedro J. Alvarez, and Victor S. Magar.

Bioremediation of Inorganic Compounds — 6(9). Eds: Andrea Leeson, Brent M. Peyton, Jeffrey L. Means, and Victor S. Magar.

In Situ Aeration and Aerobic Remediation — 6(10). Eds: Andrea Leeson, Paul C. Johnson, Robert E. Hinchee, Lewis Semprini, and Victor S. Magar.

In addition to the volume editors, we would like to thank the Battelle staff who assembled the ten volumes and prepared them for printing: Lori Helsel, Carol Young, Loretta Bahn, Regina Lynch, and Gina Melaragno. Joseph Sheldrick, manager of Battelle Press, provided valuable production-planning advice and coordinated with the printer; he and Gar Dingess designed the covers.

The Bioremediation Symposium is sponsored and organized by Battelle Memorial Institute, with the assistance of a number of environmental remediation organizations. In 2001, the following co-sponsors made financial contributions toward the Symposium:

Geomatrix Consultants, Inc.
The IT Group, Inc.
Parsons
Regenesis
U.S. Air Force Center for Environmental Excellence (AFCEE)
U.S. Naval Facilities Engineering Command (NAVFAC)

Additional participating organizations assisted with distribution of information about the Symposium:

Ajou University, College of Engineering
American Petroleum Institute
Asian Institute of Technology
National Center for Integrated Bioremediation Research & Development (University of Michigan)
U.S. Air Force Research Laboratory, Air Expeditionary Forces Technologies Division
U.S. Environmental Protection Agency
Western Region Hazardous Substance Research Center (Stanford University and Oregon State University)

Although the technical review provided guidance to the authors to help clarify their presentations, the materials in these volumes ultimately represent the authors' results and interpretations. The support provided to the Symposium by Battelle, the co-sponsors, and the participating organizations should not be construed as their endorsement of the content of these volumes.

Andrea Leeson & Victor Magar, Battelle
2001 Bioremediation Symposium Co-Chairs

EFFECTS OF POTASSIUM PERMANGANATE OXIDATION ON SUBSURFACE MICROBIAL ACTIVITY

Martin A. Rowland, Ph.D., P.E., **Lockheed-Martin Michoud Space Systems, New Orleans, LA**
Gaylen R. Brubaker, Ph.D., ThermoRetec Corporation, Durham, NC
Keisha Kohler, ThermoRetec Corporation, Durham, NC
Mark Westray, ThermoRetec Corporation, Durham, NC
Damon Morris, ThermoRetec Corporation, Seattle, WA

INTRODUCTION

Biological Reductive Dechlorination. Under anaerobic conditions, chlorinated ethenes can be biologically remediated through the process of reductive dechlorination. In this process, chlorine atoms are replaced by hydrogen atoms through nucleophilic substitution. For example, trichloroethene (TCE) can degrade into 1,2-dichloroethene (DCE), and vinyl chloride (VC), then ethene:

$$C_2Cl_3H \rightarrow C_2Cl_2H_2 \rightarrow C_2ClH_3 \rightarrow C_2H_4$$
$$\text{TCE} \quad\quad \text{cis/trans-1,2-DCE} \quad\quad \text{VC} \quad\quad \text{Ethene}$$

The rate of this reaction typically decreases as the number of chlorine substituents decreases. Therefore, the biological reductive dechlorination of TCE to cis-1,2-DCE (the predominant DCE isomer resulting from anaerobic TCE biodegradation) occurs much more rapidly than the biotransformation of cis-1,2-DCE to VC, and VC to ethene. Consequently, cis-1,2-DCE and VC often tend to accumulate at sites where historical releases of TCE to the subsurface have occurred (Freedman and Gossett, 1989; Mohn and Tiedje, 1992; Kao and Prosser, 1999). Conversely, the rate of chloroethene biodegradation under aerobic conditions typically increases as the number of chlorine substituents decreases. As a result, the complete biodegradation of chlorinated solvents can require the combined activity of both anaerobic and aerobic microorganisms.

The study site, referred to as the Building 190 Tank Farm area, is located at the NASA Michoud Assembly Facility (MAF), 16 miles east of New Orleans, Louisiana. In 1966, a spill of TCE estimated at 16,000 gallons occurred at the facility due to the failure of a piping connector, although additional undocumented releases from chemical sewers, drains and sumps may have also occurred at various times. Impacts to subsurface soils and groundwater were first discovered in 1982. In 1984, extensive subsurface investigations and groundwater corrective action, consisting of groundwater recovery and above-ground treatment, were initiated and continue to the present. Despite aggressive remedial efforts, residual dense non-aqueous phase liquids (DNAPL) and dissolved-phase concentrations of TCE, DCE and VC in excess of 100 mg/L still persist in some areas of the site. As a result, additional remedial technologies are being evaluated to enhance the overall effectiveness of the site remediation program.

The documented presence of TCE and its breakdown products, including ethene and ethane, in groundwater provide strong evidence that biological reductive dechlorination is occurring at the site. In addition, subsurface environmental conditions are especially conducive to microbial degradation of TCE, including:
- High concentrations of naturally-occurring organic carbon (averaging ~30 mg/L, with individual measurements as high as 100 mg/L) in the site groundwater
- Highly reducing groundwater conditions, as indicated by low oxidation/reduction potentials (ORP), generally non-detectable levels of nitrate, and high concentrations of dissolved methane
- Neutral pH values and high alkalinity levels (>1,000 mg/L)
- Relatively high groundwater temperature (>20° C) (RETEC, 1997)

The site data strongly indicate that natural biodegradation of TCE is occurring in the Building 190 Tank Farm area, and therefore suggest that it may be possible to enhance this natural attenuation process to improve the effectiveness of the overall site remediation program. The process of reductive dechlorination involves the utilization of the chlorinated solvents as electron acceptors by the anaerobic microbial populations, with molecular hydrogen serving as the electron donor. However, the halo-respiring microorganisms compete for available hydrogen with other anaerobes, such as methanogens. The literature indicates that the efficiency of chloroethene dechlorination may be related to the rate of hydrogen production and its steady-state concentration (Fennell et al., 1997). Studies have also shown that a variety of simple organic substrates can stimulate reductive dechlorination by producing relatively low steady-state concentrations of hydrogen when these substrates are metabolized anaerobically. The low steady-state hydrogen concentrations appear to provide dechlorinating microbial populations with a competitive advantage compared to the methanogens.

Previous microcosm studies, conducted with TCE-impacted aquifer material from the Building 190 Tank Farm area, evaluated ethanol, propionic acid, and a proprietary poly-lactate ester (HRC™) for their ability to enhance microbial dechlorination of TCE (ThermoRetec, 2000). Degradation of TCE and daughter products was optimized in the ethanol and propionic acid treated microcosms. Figures 1 and 2 show the results of the ethanol-amended study for aquifer material collected from the surficial aquifer and the upper shallow aquifer, respectively.

The results of this microcosm study indicated that ethanol can be an effective and economical amendment for enhancing the anaerobic biodegradation of chloroethenes at the Building 190 Tank Farm site. Additionally, the study demonstrated biological dechlorination activity at initial TCE concentrations of up to 55 mg/L. The literature suggests that TCE can be toxic to microbes at concentrations ranging from 10 to 100 mg/L (Baley and Daniels, 1987). Concentrations in some parts of the Building 190 Tank Farm area exceed 100 mg/L, and DNAPL TCE is known to occur in some of these areas. Therefore, a

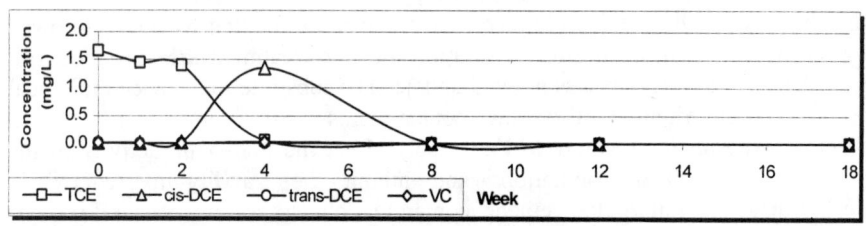

FIGURE 1. Concentrations of TCE and dechlorination products in ethanol amended surficial aquifer soils.

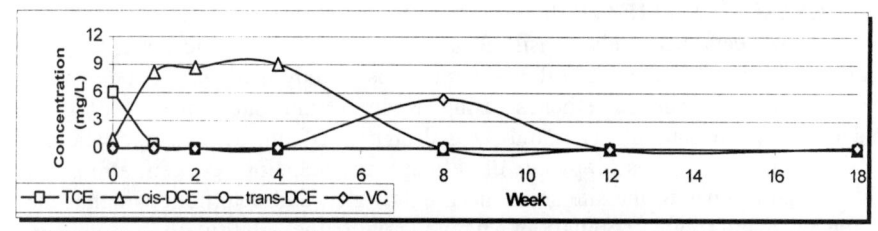

FIGURE 2. Concentrations of TCE and dechlorination products in ethanol amended upper shallow aquifer soils.

more aggressive remediation scheme may be necessary to reduce the mass of contaminants in highly-impacted areas within a relatively short time period, which might then be followed with enhanced natural attenuation to manage residual levels of contamination.

In Situ **Chemical Oxidation.** A number of field and laboratory studies have exhibited that potassium permanganate ($KMnO_4$) is capable of significantly reducing levels of chlorinated solvents in contaminated soils (Yin and Allen, 1999; Yan and Schwartz, 1999; Schnarr et al., 1998; Hood et al., 2000; Lunn et al., 1994). The basic reaction between TCE and $KMnO_4$ is:

$$C_2Cl_3H + 2MnO_4^- \longrightarrow 2CO_2 + 2MnO_2(s) + 3Cl^- + H^+$$

These studies have shown that $KMnO_4$ has the potential to be an effective chemical for *in situ* oxidation; however, the effectiveness of this remediation approach is primarily related to the ability to achieve transport of the oxidant through the contaminated regions of the subsurface. Consequently, the application of *in situ* oxidation at sites with low permeability and geologic heterogeneity is likely to meet with limited success. Since these conditions are found at the Building 190 Tank Farm area, it is not expected that *in situ* chemical oxidation alone will be capable of achieving complete removal of TCE in highly contaminated subsurface soils. However, it is conceivable that *in situ* chemical oxidation using $KMnO_4$ could rapidly remove a substantial portion of the chlorinated solvent contamination and thereby improve the long-term effectiveness of enhanced microbial degradation.

Based upon the completed studies, a conceptual remediation strategy for the site is to oxidize TCE in highly contaminated areas using $KMnO_4$, followed by the introduction of an ethanol solution into the treatment area to react with (and thus "neutralize") excess oxidant and act as a substrate to enhance biological reductive dechlorination of residual contaminants. However, it is not known whether exposure to $KMnO_4$ will adversely affect the metabolic activity of the indigenous microbial populations and inhibit their ability to reductively dechlorinate the chlorinated ethene contamination at the site. Therefore, the purpose of the study described in this paper was to evaluate the effects of $KMnO_4$ concentrations and exposure duration on microbial declorination in aquifer material from the Building 190 Tank Farm area.

MATERIALS AND METHODS

The treatablity study consisted of initial screening tests to determine the oxidant "demand" of the site soil, and a microcosm study that evaluated chemical oxidation of chlorinated ethenes using $KMnO_4$ and subsequent biological reductive dechlorination. The study was designed to both demonstrate whether $KMnO_4$ oxidation was capable of effectively reducing concentrations of chlorinated ethenes in the site soils and groundwater, and determine whether the indigenous microbial populations would retain the ability to reductively dechlorinate the chlorinated ethenes following $KMnO_4$ treatment.

Screening Studies to Determine Appropriate $KMnO_4$ Concentrations. A series of preliminary screening tests were conducted to determine the concentration of $KMnO_4$ required to completely oxidize the reduced constituents in the site soil matrix. Screening tests were established in 160-ml serum bottles with 75 ml of site groundwater and 15 gm (dry weight) of site soil. Amended conditions (supplemented with nutrients and ethanol) and unamended (control) conditions were established at each $KMnO_4$ concentration. The concentrations of $KMnO_4$ used for the screening study were 0.001, 0.005, 0.01, 0.05, 0.1, and 0.5 percent ($KMnO_4$/soil dry weight). Following a 7-day exposure period, the absence of the purple color characteristic of $KMnO_4$ solutions indicated that all of the $KMnO_4$ had been consumed in each of the microcosms.

The results of the initial screening test indicated that the $KMnO_4$ demand in the test microcosms exceeded the highest $KMnO_4$ concentration; therefore, a second series of screening tests at higher $KMnO_4$ concentrations (0.5, 1.0, 1.5, 2.0, and 2.5 percent) was conducted. Following the 7-day exposure period, only the microcosms receiving the 2.0 and 2.5 percent $KMnO_4$ treatments retained the purple color. These results suggest that complete oxidation of reduced constituents in the microcosm soil matrix was achieved at a $KMnO_4$ concentration between 1.5 percent and 2.0 percent. Based on these results, $KMnO_4$ concentrations of 0.8 and 2.4 percent (1,600 and 4,800 mg/L final aqueous concentration) were selected to examine the effects of $KMnO_4$ oxidation on biological reductive dechlorination. The 0.8 percent $KMnO_4$ concentration was expected to be sufficient to oxidize TCE and, potentially, some of the naturally-occurring soil humic material. The 2.4 percent KMnO$_4$ concentration was selected to provide an excess of KMnO$_4$ throughout the exposure period.

Treatablilty Study. Microcosms consisted of 160-ml Wheaton bottles containing 15 gm of soil (dry weight) and 75 ml of site groundwater. Microcosms were spiked with 2 ml of a TCE stock solution, which resulted in a final aqueous TCE concentration of approximately 20 mg/L (taking into account the mass of TCE initially present in the site soil).

The study consisted of two parallel sets of microcosms, each with five treatments. Treatments consisted of four combinations of KMnO$_4$ exposure times and concentrations (designated Treatments A, B, C, and D), and one set of vials that was not treated with KMnO$_4$ (Treatment E). The first set of microcosms (Set 1) were amended with ethanol and nutrients prior to KMnO$_4$ addition to enhance microbial activity. Ethanol and nutrient concentrations were established in a previous microbial degradation study (ThermoRetec, 2000). The second set of microcosms (Set 2) were not amended with ethanol and nutrients. Table 1 outlines the composition of the two sets of microcosms.

TABLE 1. Composition of KMnO$_4$ oxidation study microcosms.

Contents	Set 1 Amended	Set 2 Unamended
Soil	15 grams dry wt.	15 grams dry wt.
Water	75 ml groundwater	75 ml groundwater
TCE	20 mg/L microcosm liquid-phase	20 mg/L microcosm liquid-phase
Ethanol[1]	100 mg/L	None
Nutrients[2]	200 mg/L nitrogen 100 mg/L phosphorous	None

Notes:
1 – Ethanol was added to site groundwater immediately prior to placement in microcosms by mixing 0.1g ethanol per liter of groundwater.
2 – Inorganic nutrients included 200 mg/L nitrogen as (NH$_4$Cl) and 100 mg/L phosphorus as an equal molar mixture KH$_2$PO$_4$ and K$_2$HPO$_4$. These nutrients were added to the amended treatments by dissolving 0.76g NH$_4$Cl, 0.22g KH$_2$PO$_4$, and 0.28g K$_2$HPO$_4$ per liter of site groundwater.

Individual microcosms were sacrificed at appropriate time points for analysis of aqueous VOC concentrations using EPA Method 8260. At each time point, samples of the liquid phase were collected for VOC analysis by cooling the microcosms to 4°C, centrifuging to separate the soil and water matrices, and transferring the liquid phase to two 20-ml VOC vials. The initial dissolved VOC concentrations were determined from three random vials to establish baseline concentrations. VOC analyses were repeated after 3 weeks to identify any changes in the VOC concentrations prior to the addition of KMnO$_4$.

Immediately following the Week 3 VOC analysis, the microcosms in Treatments A and B were treated with 0.8 percent KMnO$_4$. Treatments C and D were treated with 2.4 percent KMnO$_4$. After 1 week of exposure to KMnO$_4$ (Week 4), appropriate microcosms were sacrificed for VOC analysis, and ethanol was added to the A and C treatments to react with any residual $KMnO_4$ and re-establish conditions suitable for anaerobic biodegradation. A 0.01 ml aliquot of ethanol was added to each of the A-treatment vials (treated with 0.8 percent $KMnO_4$) and a 2 ml aliquot was added to the C-treatment vials (treated with 2.4 percent $KMnO_4$). The amount of ethanol addition was determined using screening studies in which various aliquots of ethanol were added to microcosms until all $KMnO_4$ was consumed. The reaction was shown to proceed slowly requiring a full 24-hour period to reach completion. As a result, the unamended C-treatment vials received an excessive addition of ethanol, which apparently resulted in toxicity to the microbial populations. The procedure was modified for the B- and D-treatment vials after 3 weeks of $KMnO_4$ exposure (Week 6). A 0.01 ml aliquot of ethanol was added to each of the B-treatment vials (treated with 0.8 percent $KMnO_4$) and a 0.175 ml aliquot of ethanol was added to each of the D-treatment vials (treated with 2.4 percent $KMnO_4$). This ethanol addition was successful in removing residual $KMnO_4$ from the higher concentration treatment, but did not adversely affect the viability of the microbial populations.

Following neutralization of the residual $KMnO_4$ with ethanol, additional TCE was added to the vials to establish a TCE concentration of approximately 10 mg/L. VOC analyses were conducted on duplicate vials from each treatment following TCE addition. Concentrations of VOCs were then determined at various time periods to monitor biological TCE degradation and the appearance of daughter products. The A and C treatments were sampled and analyzed for VOCs at Weeks 6, 8, and 12 of the study. The B and D treatments were analyzed at Weeks 8 and 10 of the study. The control vials (Treatment E) were also analyzed for VOCs during this period at Weeks 6, 8 10, and 12 to monitor biological reductive dechlorination in microcosms which did not receive the $KMnO_4$ treatment. Neither ethanol or ethene were monitored during the study.

RESULTS

Effects of $KMnO_4$ Addition on Chloroethene Concentrations. The results of the VOC analyses at each time point are summarized in Table 2 below. The results show little evidence of TCE degradation during the initial three weeks of the treatability study. Initial TCE concentrations for the nutrient-amended and unamended treatments were 9,133 µg/L and 15,000 µg/L, respectively, and were essentially unchanged after three weeks. Likewise, there was essentially no change in concentrations of cis-DCE or VC during this period. However, TCE concentrations in both treatments were reduced by an average of 99 percent following $KMnO_4$ addition. The reduction of chlorinated ethenes was as effective at the 0.8 percent addition concentration of $KMnO_4$ as it was at the 2.4 percent concentration, suggesting that a considerable excess of $KMnO_4$ was present following oxidation of the chlorinated ethenes when the 2.4 percent concentration was used. In contrast, relatively little change in TCE concentrations occurred during the course of the study in both the nutrient-amended and unamended microcosms that were not treated with $KMnO_4$, although a slight increase in cis-DCE was observed by the conclusion of the study. These data demonstrate that

Source Zone Remediation

TABLE 2. Average concentrations of chloroethenes in nutrient amended and unamended microcosms resulting from KMnO₄ treatment.

Treatment	Time (wks) relative to TCE Spike	Nutrient-Amended					Unamended				
		PCE (ug/L)	TCE (ug/L)	c-DCE (ug/L)	t-DCE (ug/L)	VC (ug/L)	PCE (ug/L)	TCE (ug/L)	c-DCE (ug/L)	t-DCE (ug/L)	VC (ug/L)
A	-4	<225	9133	983	<225	403	<225	15000	1333	<225	<225
	-1	<75	9233	737	<75	227	<75	16333	2600	<75	218
	0	<0.5	74	5	<1	2	<2	230	35	<2	4
	0	<75	9900	<75	<75	<75	<75	9250	133	<75	<75
	2	<50	<50	8950	<50	<50	<100	4550	4293	<100	<100
	4	<31	<28	4780	<31	1100	<50	53	6050	<50	840
	8	<50	<50	8750	<50	100	25	31	3324	26	1040
B	-6	<225	9133	983	<225	403	<225	15000	1333	<225	<225
	-3	<75	9233	737	<75	227	<75	16333	2600	<75	218
	0	<3	320	238	3	4	<2	335	128	2	3
	0	<50	10000	130	<50	<50	<75	10150	150	<75	<75
	2	<50	575	9500	<50	<50	<50	2041	8050	<50	<50
	4	<38	2025	6950	<38	138	<50	87	8850	<50	<50
C	-4	<225	9133	983	<225	403	<225	15000	1333	<225	<225
	-1	<75	9233	737	<75	227	<75	16333	2600	<75	218
	0	<1	<1	<1	<1	<1	<1	<1	<1	<1	<1
	0	<75	9900	<75	<75	<75	<75	10450	<75	<75	<75
	2	<38	6400	665	<38	71	<100	9500	<100	<100	<100
	4	<50	2450	6200	<50	<50	<50	11000	95	<50	<50
	8	50	150	7900	50	50	50	9700	1655	50	50
D	-6	<225	9133	983	<225	403	<225	15000	1333	<225	<225
	-3	<75	9233	737	<75	227	<75	16333	2600	<75	218
	0	<2	155	51	<2	2	<1	45	42	<1	<1
	0	<50	9650	<50	<50	<50	<50	9950	<50	<50	<50
	2	<50	3385	5900	<50	<50	<50	2045	8150	<50	<50
	4	<50	175	8050	<50	215	<38	408	6500	<38	<38
E	0	<225	9133	983	<225	403	<225	15000	1333	<225	<225
	3	<75	9233	737	<75	227	<75	16333	2600	<75	218
	6	<33	5800	1025	<33	110	<33	11500	3350	<33	<50
	8	<50	6600	2150	<50	78	<50	17000	2650	<50	<50
	10	<50	11700	2500	<50	<85	<50	13350	3150	<50	<50
	14	50	9800	4500	50	51	50	7650	3660	50	56

Notes:
The first 2 time points are the average of triplicate microcosms, all other time points are the average of duplicates.
Bold values indicate concentrations above the analytical detection limit.
Treatments -   A - One week exposure to 0.8% KMnO₄
B - Three week exposure to 0.8% KMnO₄
C - One week exposure to 2.4% KMnO₄
D - Three week exposure to 2.4% KMnO₄
E - No KMnO₄

KMnO₄ is highly effective in reducing concentrations of chlorinated ethenes in soils and groundwater from the Building 190 Tank Farm site, under controlled laboratory conditions.

Effects of KMnO₄ Concentrations on Microbial Dechlorination Capacity. Figures 3 and 4 show the concentrations of chlorinated ethenes following a one-week exposure to 0.8 percent and 2.4 percent KMnO₄, respectively, for the nutrient-amended study group. It should be noted that, since there was very little difference in the results obtained for the nutrient-amended group, compared to the unamended group, only the data from the nutrient-amended treatment are summarized in these graphs. The shaded area on the graphs represent the period during which the microcosms were exposed to KMnO₄. Following the exposure period, ethanol was added to neutralize any residual KMnO₄, after which TCE was added to achieve an aqueous phase concentration of about 10 mg/L.

The figures show that, following the TCE spike, reductions in TCE concentrations occurred rapidly, with a simultaneous production of cis-DCE. The apparent rates of TCE degradation and cis-DCE production were somewhat lower in the microcosms that were exposed to 2.4 percent KMnO$_4$, which may indicate that the microbial populations were stressed by the exposure to the higher KMnO$_4$ concentrations. However, similar TCE and cis-DCE concentrations were achieved by the end of the study under both KMnO$_4$ treatments.

It is noted that, unlike the previous microcosm study described by Figures 1 and 2, significant reduction of cis-DCE and production of VC was not observed in either the microcosms treated with KMnO$_4$, or in the control microcosms that

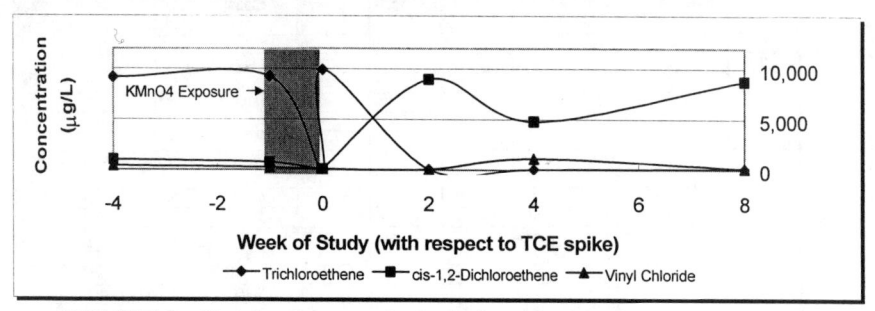

FIGURE 3. Nutrient-amended study, one-week exposure to 0.8 percent KMnO$_4$ (Treatment A).

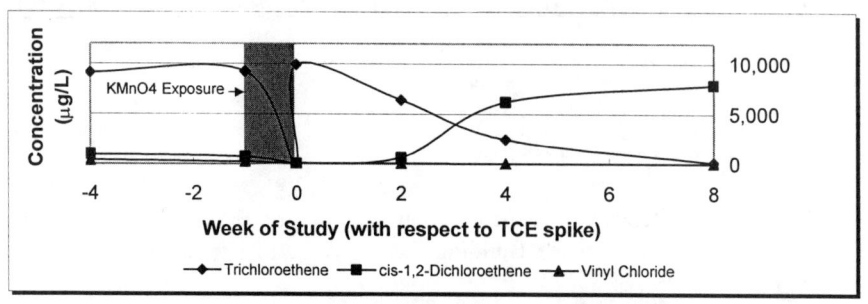

FIGURE 4. Nutrient-amended study, one-week exposure to 2.4 percent KMnO$_4$ (Treatment C).

were not treated with KMnO$_4$. It is not clear why dechlorination did not progress beyond cis-DCE. The most likely explanation is that the duration of the study was insufficient for observation of cis-DCE dechlorination. A slight increase in VC and decrease in cis-DCE concentrations is evident at the end of the study for the nutrient-amended condition that received the 3-week exposure to 0.8 percent KMnO$_4$ (Figure 5). It is also possible that the oxidation/reduction potential (Eh) within the microcosm vials was too high to permit reductive dechlorination of cis-DCE, or that the levels of available electron donors (organic substrates and/or hydrogen) were inadequate for dechlorination to proceed. Additionally, it is conceivable that the microbial populations capable of cis-DCE dechlorination were more sensitive to KMnO$_4$. Regardless, the data indicate that the ability of the indigenous microbial populations to dechlorinate TCE to cis-DCE was not impaired by additions of KMnO$_4$.

Effects of KMnO$_4$ Exposure Duration on Microbial Dechlorination Capacity. Figures 5 and 6 show the concentrations of chlorinated ethenes in the nutrient-amended microcosms following a three-week exposure to 0.8 percent and 2.4 percent KMnO$_4$, respectively. These figures show decreases in TCE concentrations and concomitant increases in cis-DCE concentrations that are similar to those observed for the one-week exposures to comparable KMnO$_4$ concentrations (Figures 3 and 4). Once again, apparent rates of TCE degradation and cis-DCE production were somewhat slower following treatment with the higher KMnO$_4$ concentration. In addition, Figure 5 and the data in Table 2 suggest that reduction of cis-DCE and production of VC may be starting by the final sampling point. However, the data show no evidence that microbial dechlorination of TCE to cis-DCE is adversely affected by the duration of exposure to KMnO$_4$.

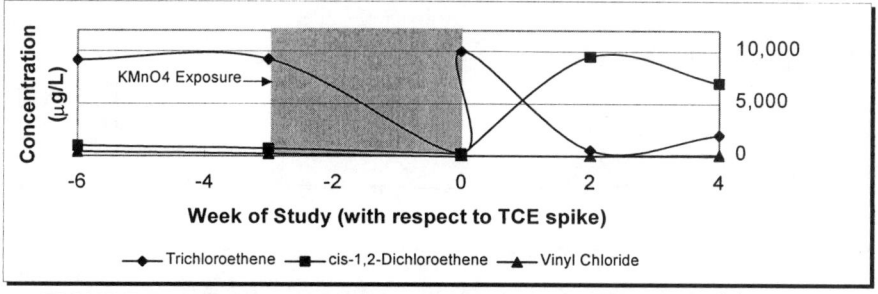

FIGURE 5. Nutrient-amended study, three-week exposure to 0.8 percent KMnO$_4$ (Treatment B).

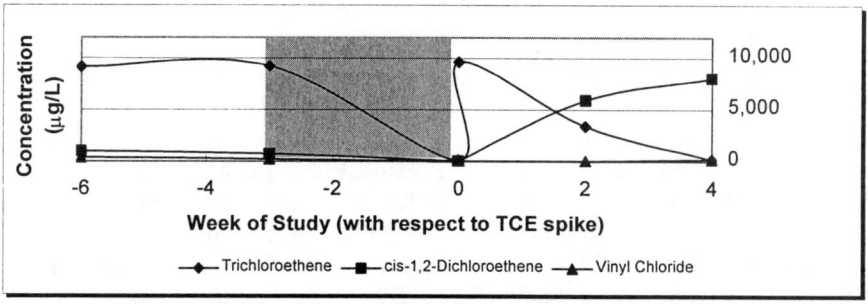

FIGURE 6. Nutrient-amended study, three-week exposure to 2.4 percent KMnO$_4$ (Treatment D).

Apparent Ethanol Toxicity in the Unamended Treatment C Microcosms. Of the eight sets of nutrient-amended and unamended microcosms that were treated with KMnO$_4$, only the unamended set that received the one-week exposure to 2.4 percent KMnO$_4$ (Treatment C) failed to exhibit significant reduction of TCE following treatment with KMnO$_4$ (Figure 7). It is believed that the absence of TCE reduction in these microcosms was due to the inadvertent addition of an excessive amount of ethanol to quench residual KMnO$_4$ at the end of the one-week exposure period. This set of microcosms was the first that was treated with ethanol. It was initially expected that the reaction of ethanol with KMnO$_4$ would occur rapidly and would be visually detectable by a quick change in the color of the microcosm water from purple-to-clear. However, the purple color persisted in the Treatment C microcosms in spite of repeated additions of ethanol. When the vials were inspected approximately eight hours later, no purple color was detected, indicating that the reaction of ethanol and KMnO$_4$ was considerably slower than anticipated. Subsequent analysis of the microcosms following re-spiking with TCE showed ethanol concentrations of approximately 13,000 mg/L. Consequently, a period of 24-hours was allowed for the complete reaction of ethanol and KMnO$_4$ for all other treatments, all of which subsequently exhibited TCE degradation and cis-DCE production. Therefore, it is likely that the high concentration of ethanol in the Treatment C microcosms was responsible for the absence of reductive dechlorination in that treatment condition.

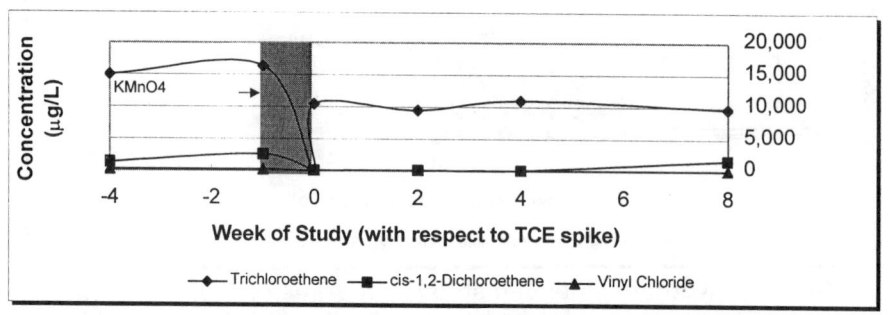

FIGURE 7. Unamended study, one-week exposure to 2.4 percent KMnO$_4$ (Treatment C) – apparent ethanol toxicity.

DISCUSSION AND CONCLUSIONS

This study has evaluated the compatibility of combined *in situ* remediation of chlorinated ethene contamination using chemical oxidation with KMnO$_4$ and biological reductive dechlorination. The results of the study demonstrated that microbial dechlorination activity in aquifer material from the Building 190 Tank Farm site was not significantly impaired at KMnO$_4$ concentrations of up to 2.4 percent and exposure durations of up to three weeks. In fact, the data indicated that TCE dechlorination did not occur to a significant degree until after the microcosms had been treated with KMnO$_4$. During the first three weeks of the study (prior to KMnO$_4$ addition), the concentrations of TCE, cis-DCE and VC in both the nutrient-amended and unamended conditions were essentially unchanged. The relative lack of dechlorination activity in the absence of KMnO$_4$ treatment is further indicated by the control microcosms, which did not receive

KMnO$_4$. Figure 8 shows the results of the nutrient-amended condition that was not treated with KMnO$_4$ (Treatment E). The data from the unamended Treatment E microcosms exhibit trends similar to those shown in Figure 8, but have not been presented graphically in this paper.

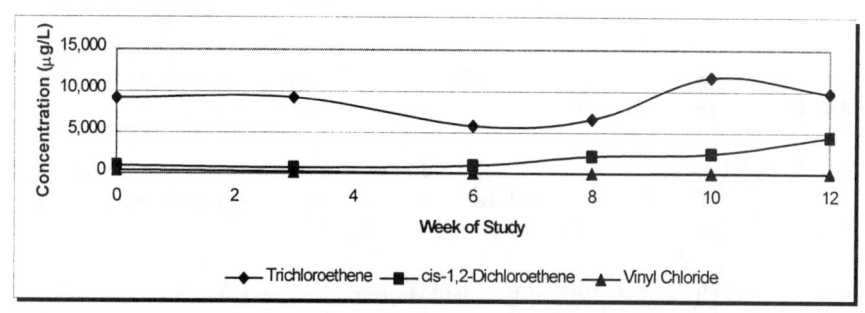

FIGURE 8. Nutrient-amended study with no exposure to KMnO$_4$ (Treatment E).

The absence of TCE dechlorination in the Treatment E microcosms during the study, and by the other treatment conditions prior to the addition of KMnO$_4$, differs from the previous microcosm study illustrated in Figures 1 and 2, in which dechlorination occurred rapidly at the outset of the study. The initial absence of TCE dechlorination in the current study suggests that the initial amendment of ethanol was an insufficient quantity of electron donor to stimulate dechlorination and satisfy other electron acceptors (i.e., oxygen, nitrate, ferric iron, etc.) that may have been present. It is possible that KMnO$_4$ treatment may have partially oxidized long-chain humic compounds in the soils from the site (which are known to contain significant amounts of peat), making them more easily biodegraded. This additional supply of electron donor, in concert with the re-amendment with ethanol following the period of KMnO$_4$ exposure, may have provided the reducing equivalents necessary to support reductive dechlorination of TCE

On the basis of this study, an integrated *in situ* remediation consisting of initial chemical oxidation using KMnO$_4$ followed by enhanced anaerobic biodegradation using ethanol amendments, is technically feasible and has the potential for effective application in highly-impacted areas of the MAF site. The data indicate that KMnO$_4$ concentrations on the order of 1-2 percent are not likely to result in microbial toxicity or inhibition of biological reductive dechlorination. It is, therefore, appropriate to initiate pilot studies at the site to evaluate this remediation approach under actual field conditions.

REFERENCES

Baley, N. and L. Daniels. 1987. "Production of Ethane, Ethylene and Acetylene from Halogenated Hydrocarbons by Methanogenic Bacteria." *Applied and Environmental Microbiology 53(7):1604-1610.*

Fennell, D.E., J.M. Gossett, and S.H. Zinder, 1997. "Comparison of Butyric Acid, Ethanol, Lactic Acid, and Propionic Acid as Hydrogen Donors for the Reductive Dechlorination of Tetrachloroethene." *Environmental Science and Technology. 31: 918-926.*

Freedman, D.L. and J.M. Gossett. 1989. "Biological Reductive Dechlorination of Tetrachloroethylene and Trichloroethylene to Ethylene under Methanogenic Conditions". *Applied and Environmental Microbiology 55:2144-2151.*

Hood, E.D., Thomson, N.R., Grossi, D. and Farquhar, G.J., 2000. "Experimental Determination of the Kinetic Rate Law for the Oxidation of Perchloroethylene by Potassium Permanganate." *Chemosphere 40: 1383-1388.*

Kao, C.M. and Prosser, J. 1999. "Intrinsic Bioremediation of Trichloroethylene and Chlorobenzene: Field and Laboratory Studies." *Journal of Hazardous Materials B69:67-79.*

Lunn, G., Sansone, E.B., De Meo, M., Laget, M. and Castegnaro, M., 1994. "Potassium Permanganate Can Be Used for Degrading Hazardous Compounds." *American Industrial Hygiene Association Journal 55(2):167-171.*

Mohn, W.W. and Tiedje, J.M. 1992. "Microbial Reductive Dehalogenation." *Microbiology and Molecular Biology Reviews 56(3):482-507.*

RETEC, 1997. *Phase II Report Evaluation, Selection, and Design of Remediation Approach; Lockheed Martin Manned Space Systems, Michoud Assembly Facility, Building 190 Tank Farm*, Remediation Technologies, Inc." February 1997.

Schnarr, M., Truax, C., Farquhar, G., Hood, E., Gonullu, T. and Stickney, B., 1998. "Laboratory and Controlled Field Experiments Using Potassium Permanganate to Remediate Trichloroethylene and Perchloroethylene DNAPLs in Porous Media." *Journal of Contaminant Hydrology 29:205-224.*

ThermoRetec, 2000. *Report for the Microcosm Study to Evaluate Natural Attenuation and In Situ Bioremediation of Chlorinated Ethylenes Within the Building 190 Tank Farm Area*, Remediation Technologies, Inc. February 2000.

Yan, Y.E. and F.W. Schwartz. 1999. "Oxidative Degradation and Kinetics of Chlorinated Ethylenes by Potassium Permanganate." *Journal of Contaminant Hydrology 37:343-365.*

Yin, Y. and H.E. Allen. 1999. *Technology Evaluation Report: In Situ Chemical Treatment* (Publication Document Number: TE-99-01). Ground-Water Remediation Technologies Analysis Center.

SEQUESTRATION AND DEGRADATION OF TCE FOLLOWING DNAPL SOURCE REMEDIATION

Arun Gavaskar, *Eric Foote*, Jenny Butler, and Sam Yoon
Battelle, Columbus, Ohio, U.S.A.

BACKGROUND

One concern that site owners and regulators have with accepting DNAPL source remediation is that DNAPL mass removal efficiency often is not 100%. Technologies, such as resistive heating, chemical oxidation, steam injection, and surfactant flushing, have reportedly removed around 60 to 95% DNAPL mass at various sites, with the removal efficiency depending on the site geology and the treatment technology applied. Although dissolved PCE/TCE concentrations in the source zone can decline significantly after remediation in the short term, in the long term, the remaining DNAPL can cause PCE/TCE concentrations to rebound. In this case, the site owner is often left with a plume (albeit, perhaps, a weaker one) that may linger for many more years or decades and require continued plume control actions.

Subsurface application of vegetable oil as a *secondary* treatment of the source zone is the focus of this work. The application of vegetable oil as a secondary treatment may prove to be a cost effective mechanism for preventing rebound in source zones that have previously undergone physical-chemical remediation (primary treatment). Laboratory tests have shown that the vegetable oil acts as a weak solvent for sequestering some residual TCE. In addition to sequestration, vegetable oil acts to replenish organic carbon lost during primary remediation. Primary treatments, such as, permanganate oxidation, deplete the native organic matter/carbon sources in the treated aquifer. If the carbon source in the source zone is replenished, biodegradation of residual TCE can be enhanced. Sequestration and biodegradation are the two mechanisms through which vegetable oil addition can minimize the rebound in PCE/TCE concentrations and eliminate or reduce subsequent plume control requirements.

EXPERIMENTAL APPROACH

This study investigated the applicability of vegetable oil as a secondary treatment option for source zone removal of TCE. Two preliminary bench-scale experiments were conducted to demonstrate the secondary treatment concept with soil and groundwater obtained from a TCE-contaminated site in Florida. At this site, thermal (resistive heating) and chemical oxidation (permanganate) treatments had been applied in two test plots in the source zone to determine the efficiency of DNAPL removal. The laboratory experiments focused on the ability of vegetable oil to sequester the TCE present in groundwater. The ability of vegetable oil to provide a slow-release carbon source into the groundwater and to promote microbial growth and anaerobic dechlorination of TCE was also examined in this study.

TCE Sequestration Experiments. The ability of vegetable oil to sequester TCE was examined by using an untreated-contaminated groundwater from the Florida site. Table 1 shows the experimental conditions. The site groundwater contained approximately 6.2 mg/L TCE. Approximately 155 mL each of this groundwater was transferred into a series of 160-ml serum bottles and 0.5 mL of crude canola oil was added. The bottles were immediately sealed using Mininert™ valves and shaken by hand for a period of 0.5 minutes each. The experimental matrix included control bottles without vegetable oil to determine any loss of TCE from the bottles due to volatilization. After shaking by hand for a few seconds, the bottles were allowed to stand undisturbed at room temperature (22°C) for 30 minutes. After 30 minutes the groundwater was sampled and analyzed for dissolved organic carbon (DOC) and TCE concentrations.

Table 1. Verifying Dissolution of Vegetable Oil in Groundwater and Sequestration of TCE in the Oil.

Sample Description	DOC (mg/L)	Avg. DOC (mg/L)	TCE (µg/L)	Avg. TCE (µg/L)
Millipore Water (Blank)	1	1	ND	ND
Groundwater Containing TCE (Before Shaking)	169	169	6,182	6,182
Groundwater Containing TCE + Veg. Oil (After Shaking) [a]	189 / 192	190	888 / 832	860
Groundwater Spiked with TCE (Before Shaking Control) [b]	144 / 141	143	8,340 / 7,906	8,123
Groundwater Spiked with TCE (After Shaking Control) [b]	140 / 141	140	7,294 / 6,831	7,062

[a] Duplicates tested and averaged.
[b] Shaded rows are duplicate controls sampled before and after shaking.
Controls were spiked with 2,000 µg/L of TCE to raise initial TCE levels
ND = Not detected. Below minimum detection limit (25 µg/L)

Anaerobic Microcosms. Microcosms were constructed to determine microbial population densities and TCE degradation. As shown in Table 2, a series of microcosms were constructed with 160 mL serum bottles using aquifer materials collected from the Florida site. Contaminated soil and groundwater were obtained from different regions of the treated aquifer to represent both the heavily treated and fringe regions for both treatment types. This resulted in four experimental conditions. The "High-Temperature" and "Low-Temperature" soil samples represent regions of the source zone that received relatively more and less of the thermal treatment, respectively. The "High-Dose and "Low-Dose" soil samples represent regions of the source zone that received relatively more and less of the applied potassium permanganate, respectively. Each microcosm was constructed using 100 g of soil (wet wt.), 75 mL of groundwater, and 0.5 mL of crude canola oil. Because the native aquifer is anaerobic, the samples were maintained under anaerobic conditions. Each bottle was hand shaken for 0.5

minutes once per week and stored in the dark for stagnant incubation for 5 months at room temperature (~22°C). Population densities and the level of PCE in each microcosm were monitored at time T = 0 and after 5 months of anaerobic incubation. Population density was determined by aerobic and anaerobic total heterotroph enumeration. Time T = 0 enumerations gave valuable insight to the robustness of microbial populations following each aggressive primary treatment.

Table 2. Microbial Counts in Thermally Treated and Chemically Oxidized Soils Before (T=0) and After (T=5 Months) Addition of Vegetable Oil

Treatment Type/Area	Heterotrophic Enumerations (CFU/mL)[a]			
	T=0		T= 5 Months	
	Aerobic	Anaerobic	Aerobic	Anaerobic
High Temperature	4.6×10^7	2.9×10^7	3.3×10^5	1.7×10^7
Low Temperature	2.7×10^7	8.7×10^6	4.3×10^5	3.9×10^5
High KMnO$_4$	$<1.0 \times 10^1$	$<1.0 \times 10^1$	1.9×10^6	8.4×10^5
Low KMnO$_4$	1.0×10^4	7.1×10^4	1.3×10^4	7.0×10^3
Killed Control[b]	$<1.0 \times 10^1$	$<1.0 \times 10^1$	$<1.0 \times 10^1$	$<1.0 \times 10^1$

[a] Average of triplicate microcosms
[b] Chemically treated with HgCl$_2$

RESULTS AND CONCLUSIONS

The results for the TCE sequestration experiment are presented in Table 1. Dissolved organic carbon (DOC) was used as a measure of the level of oil that dissolved into the groundwater; this is important for ensuring the availability of a long-term carbon source for microorganisms. When vegetable oil was added to the groundwater, DOC levels increased slightly due to the slow release of oil into solution. TCE levels present in the groundwater declined considerably due to sequestration by the vegetable oil. The control bottles resulted in an approximate 10 to 15% loss of TCE during experimental setup and incubation, which was relatively insignificant in comparison to the decreases observed in the experimental bottles supplemented with vegetable oil, where TCE removal was approximately 86%.

Soil samples from source zone regions subjected separately to thermal and chemical oxidation treatment were analyzed for survival of microbial populations. Based on monitoring of temperatures and permanganate concentrations, the treated DNAPL source zone had been divided into "high" and "low" treatment regions as defined previously. The "high" and "low" categories indicate that not all regions of the source zone received the same degree of primary treatment due to the heterogeneous geology. The expectation was that even though some regions of the source zone that received strong treatment may not be able to sustain microbial populations, other regions that escaped the harshest treatment would have surviving populations and would be amenable to secondary (vegetable oil) treatment. Table 2 shows the results of microbial population analyses conducted at time T = 0 and after T = 5 months of anaerobic incubation. Microorganisms were found to be surprisingly resilient to these relatively harsh treatments. Most regions had significant surviving populations. The high-dose permanganate regions in the site soil were the only areas that appeared to be void of detectable microorganisms. In general, the permanganate treatment appeared to be harsher on the native microbial populations than the thermal treatment. Both aerobic and anaerobic populations were sustained over the 5-month incubation period. The high-dose permanganate conditions showed a significant rebound in population by the 5-month sampling event, indicating the potential for recovery of microorganisms after secondary treatment.

The sterilization of crude vegetable oil is both undesirable and impractical on a field scale and, therefore was not considered a parameter for investigation during these experiments. The $HgCl_2$-killed control conditions indicated no detectable growth over the 5-month incubation period.

TCE, DCE, and VC levels were monitored, as also was methane generation at T = 0 and after T = 5 months of anaerobic incubation. These results are presented in Table 3.

Table 3. TCE Levels in Thermally Treated and Chemically Oxidized Soils Before (T=0) and After (T=5 Months) Addition of Vegetable Oil

Treatment Type/Area	Sampling Event (Months)	TCE (µg/L) [a]	Cis-1,2 DCE (µg/L) [a]	VC (µg/L) [a]	Methane (ppmv) [a]
High Temperature	T = 0	329	104	13	3
	T = 5	ND	4	ND	3
Low Temperature	T = 0	325	108	94	2
	T = 5	ND	224	80	2
High KMnO₄	T = 0	1	5	33	1
	T = 5	ND	ND	NS	NS
Low KMnO₄	T = 0	80	1	190	15
	T = 5	0	29	6	162
Killed Control [b]	T = 0	89	40	265	6
	T = 5	119	41	271	3

[a] Represents median groundwater concentration values of triplicate microcosms.
[b] Chemically treated with $HgCl_2$
ND = Not detected, below the minimum detection level (10 µg/L).
NS = Not sampled.

Source Zone Remediation 17

As compared with the chemically killed control, addition of vegetable oil to the bottles stimulated TCE degradation. Production of cis-DCE was observed in some bottles, indicating anaerobic degradation of TCE.

DISCUSSION

The tests described here are preliminary, but the results appear promising for the use of vegetable oil as a secondary treatment option for partially remediated DNAPL source zones. Further testing at both bench- and field scale would be necessary to determine the effectiveness of the application. Figure 1 shows a conceptual model that describes a field application of vegetable oil for secondary treatment of a DNAPL source. Mechanisms by which vegetable oil is expected to control plume rebound are discussed below.

Sequestering Capability. Bench-scale tests have shown that the vegetable oil acts as a sequestering agent by partially solubilizing TCE. After injection, the oil physically sequesters some of the PCE/TCE that it comes into contact with. The oil smears across the aquifer thickness when injected at multiple locations and depths.

Flow Control Capability. The relatively viscous oil gets entrapped in the soil pores, reduces the permeability of the soil, and, consequently, reduces the groundwater flow through the source region. This is a desirable side effect because the potential for plume formation is reduced, as some groundwater flow is diverted around the source zone. Also, unlike bio-enhancements applied in the plume, any potentially excessive proliferation of microbial populations or biofouling in the source zone can only help by causing the groundwater to bypass the source zone. This will further reduce the

(a) Pre-Remediation Aquifer

(b) Post-Primary-Remediation Aquifer

(c) Secondary Treatment of Partially-Remediated Aquifer

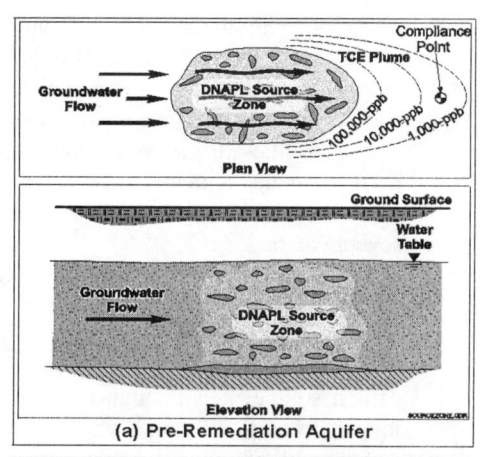

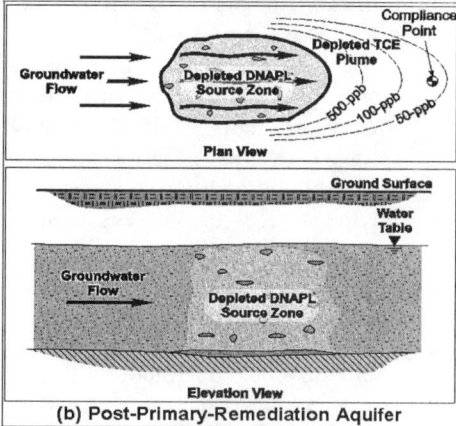

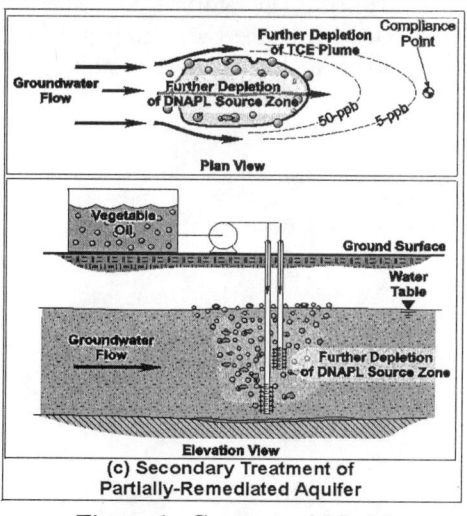

Figure 1. Conceptual Model

potential for plume formation.

Slow-Release Capability. Laboratory tests show that vegetable oil is only sparingly soluble in groundwater. Any excess oil accumulates in the aquifer and serves as a long-term, slow-release carbon source. Unlike the more soluble carbon sources or electron donors used in other studies, vegetable oil does not need frequent replenishment. This greatly improves the economics of the application. A single application of vegetable oil to the aquifer can last for several years.

Ease of Distribution. Unlike more soluble carbon sources used in the past, it is not necessary to get a thorough distribution of the oil in the target zone, as it does not solubilize and dissipate right away. Intimate contact between the oil and the remaining DNAPL also is not required. For it to serve as a long-term carbon source, it is enough that the oil remains trapped in the aquifer near the injection points. Through a combination of dissolution, advection-dispersion, and diffusion, the dissolved carbon source will eventually be available in most parts of the source zone.

Slower Treatment. Although vegetable oil could be added to the source zone as a primary treatment (instead of first applying thermal or chemical treatments), it is likely to be most effective as a secondary treatment. Once primary treatment (faster treatment) has removed the bulk of the DNAPL, vegetable oil addition (slower treatment) can control dissolved TCE/PCE levels in the source region, thus eliminating or reducing subsequent plume control requirements and costs.

ACKNOWLEDGEMENTS

This work was done under an internal research and development grant from Battelle. The authors thank William Keigley, Michael Holdren, Patricia Holowecky, Kristen Hartzell, and Christy Burton from Battelle for their assistance with the laboratory experiments. The authors also acknowledge Jacqueline Quinn from the U.S. National Aeronautic and Space Administration for providing access to the Florida site soil.

TWO NOVEL METHODS FOR ENHANCING SOURCE ZONE BIOREMEDIATION: DIRECT HYDROGEN ADDITION AND ELECTRON ACCEPTOR DIVERSION

Charles J. Newell and Carol E. Aziz, Groundwater Services, Inc., Houston, TX
Patrick E. Haas, Air Force Center for Environmental Excellence, Brooks AFB, TX
Joseph B. Hughes, Rice University Dept. Envi. Sci. and Eng., Houston, TX
Tariq A. Khan, Groundwater Services, Inc., Houston, TX

ABSTRACT: Two novel methods for enhancing bioremediation in DNAPL source zones are now being developed. The first method, direct hydrogen addition (i.e., electron donor addition without the use of fermentation substrates), is now being tested in the field by the Technology Transfer Division at the Air Force Center for Environmental Excellence (AFCEE). This approach permits the direct delivery of a relatively large mass of electron donor to DNAPL source zones undergoing biodegradation. The second method, electron acceptor diversion, is an emerging technology where competing electron acceptors are diverted around the source zone, thereby greatly increasing the naturally-occurring rate of reductive dechlorination in the source zone. This approach may have the potential to increase the naturally-occurring rate of DNAPL mass destruction at many sites without any long-term operating costs except monitoring.

INTRODUCTION

Hydrogen is now widely recognized as a key electron donor required for the biologically-mediated dechlorination of chlorinated compounds. Hydrogen acts as an *electron donor,* and halogenated compounds such as chlorinated solvents act as *electron acceptors* that are reduced in the reductive dechlorination process. To enhance beneficial anaerobic processes for the purpose of bioremediation, numerous research groups have focused on methods to increase the supply of electron donor to the dechlorinating bacteria. Most researchers and technology developers have focused on adding an indirect electron donor (such as lactate, molasses, mulch, edible oil, or other carbon source) that is fermented by one type of bacteria to produce hydrogen for the dechlorinators. However, there are two other methods to increase the effective supply of electron donor to the dechlorinating bacteria: 1) direct delivery of dissolved hydrogen to the subsurface (Hughes, et al. 1997); and 2) diversion of competing electron acceptors around the chlorinated solvent source zone (Newell et al., 2001).

DIRECT HYDROGEN DELIVERY

Direct delivery methods that have been proposed include circulation of groundwater containing dissolved hydrogen, placement of chemical agents that release dissolved hydrogen, electrolysis of water with subsurface electrodes, use of colloidal gas aphrons (foams), and low-volume pulsed biosparging (Hughes et al., 1997). Because of its simplicity and low-cost, AFCEE funded an 18-month long field trial of low-volume pulsed biosparging of hydrogen gas in the subsurface. With this approach, small volumes of hydrogen gas from cylinders was sparged directly into the contaminated zone in short intervals. In this case the sparge interval was approximately one 20 minute pulse once a week for most of the test. Small volumes are used to ensure that breakthrough to the surface will not present safety problems. The hydrogen is pulsed to allow effective dissolution of the trapped gas, thereby transferring the residual hydrogen gas to the aqueous phase.

Results from an eighteen-month low-volume pulsed hydrogen biosparging pilot test at Cape Canaveral Air Station Florida showed extensive biological dechlorination in a 30 x 30 ft (9.1 m x 9.1 m) zone located 15 to 20 ft (4.6 to 6 m) below the water table in a sandy aquifer. The test zone was in or very near a DNAPL source zone, as chlorinated ethene concentrations were very high (~300 mg/L). Hydrogen gas was pulsed into three sparge points at regular intervals (weekly for most of the test) to form residual hydrogen gas bubbles, which then dissolved to deliver electron donor directly to the test zone. Table 1 shows the observed changes in concentration in the test zone.

TABLE 1. Change in total chlorinated ethenes (CE) after 18 months of low-volume pulsed biosparging.

Location of Monitoring Well Group	Distance to Sparge Pts (ft)	Baseline CE Concentration (mg/L)	CE Concentration After 18 Months (mg/L)	Percent Change (%)
Close to Sparge Point	3 - 6	291	16	- 95%
Downgradient of Sparge	15	294	151	- 49%
Nitrogen Sparge Control	15	42	37	- 12%
Natural Attenuat. Control	20	207	165	- 20%

The change in the concentration in the test zone are presented in the following figures: TCE and cis 1,2-DCE (Figure 1); vinyl chloride and ethene (Figure 2); and methane (Figure 3). Figure 4 shows the vertical distribution of dissolved hydrogen in groundwater around the three hydrogen sparge points four days after a pulse event.

Wells in the hydrogen delivery zone showed greater reduction in chlorinated ethene concentrations compared to: 1) two wells (A-5 and B-5) located in a nitrogen control zone where nitrogen was pulsed into the subsurface at the same rate and frequency as the hydrogen sparge; and 2) two wells (A-1 and B-1) located in a natural attenuation control zone located outside the effective radius of the hydrogen sparge points. The greater reduction in chlorinated ethene concentrations in the hydrogen test monitoring wells compared to the control wells shows that reductive dechlorination resulted from the direct addition of hydrogen.

Source Zone Remediation

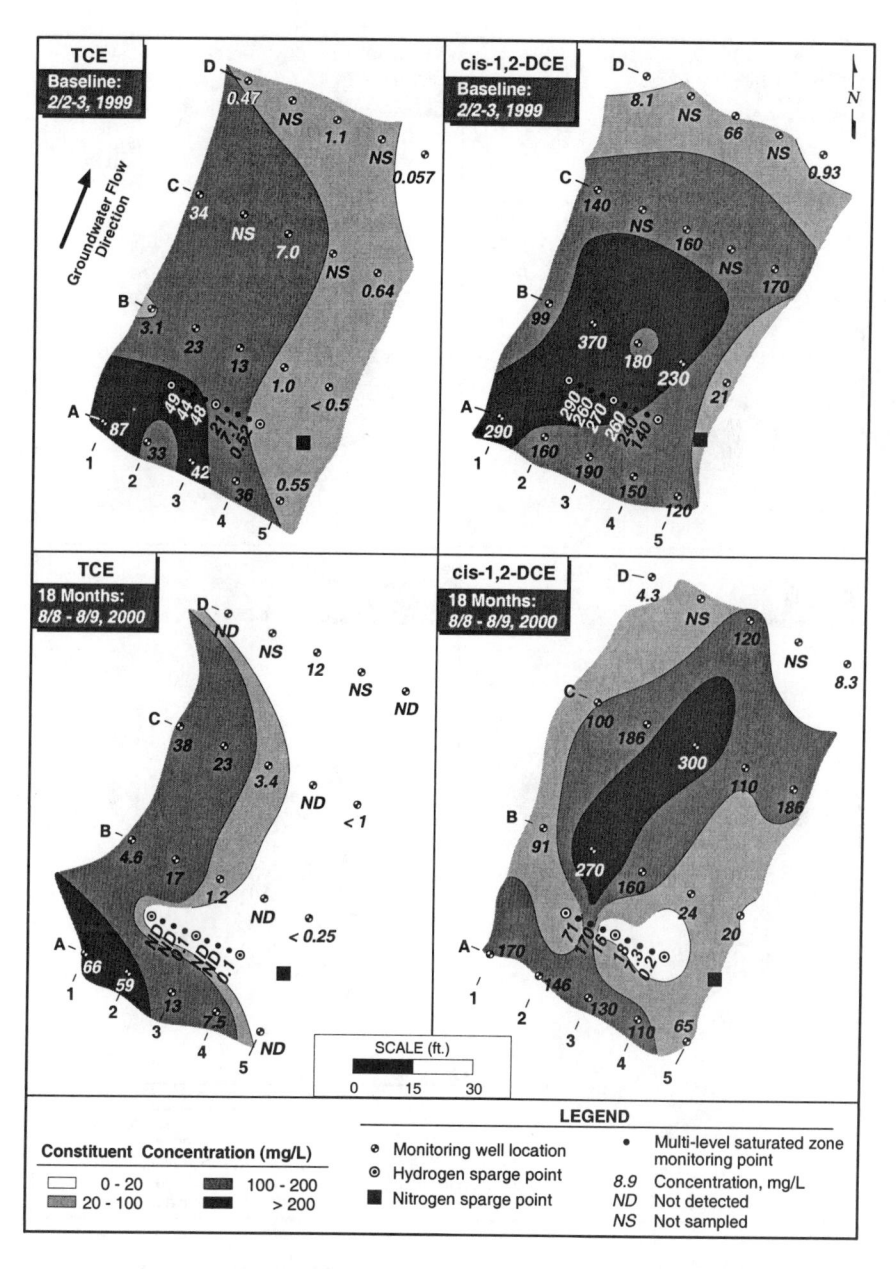

FIGURE 1. Change in TCE and cis-1,2-DCE concentration during 18-month pilot test of low-volume pulse biosparging Launch Complex 15, Cape Canaveral Air Station, Florida.

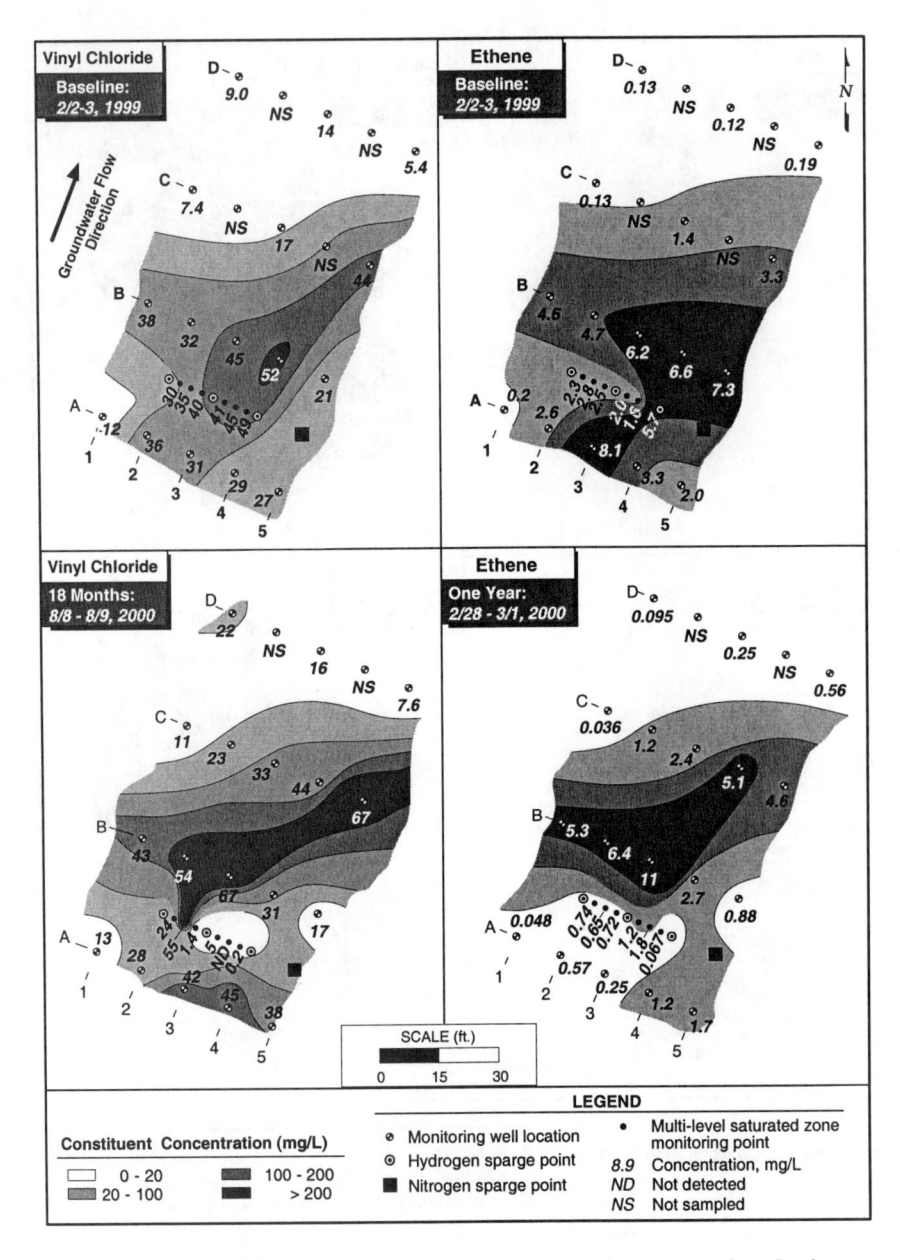

FIGURE 2. Change in vinyl chloride and ethene concentration during 18-month pilot test of low-volume pulse biosparging Launch Complex 15, Cape Canaveral Air Station, Florida

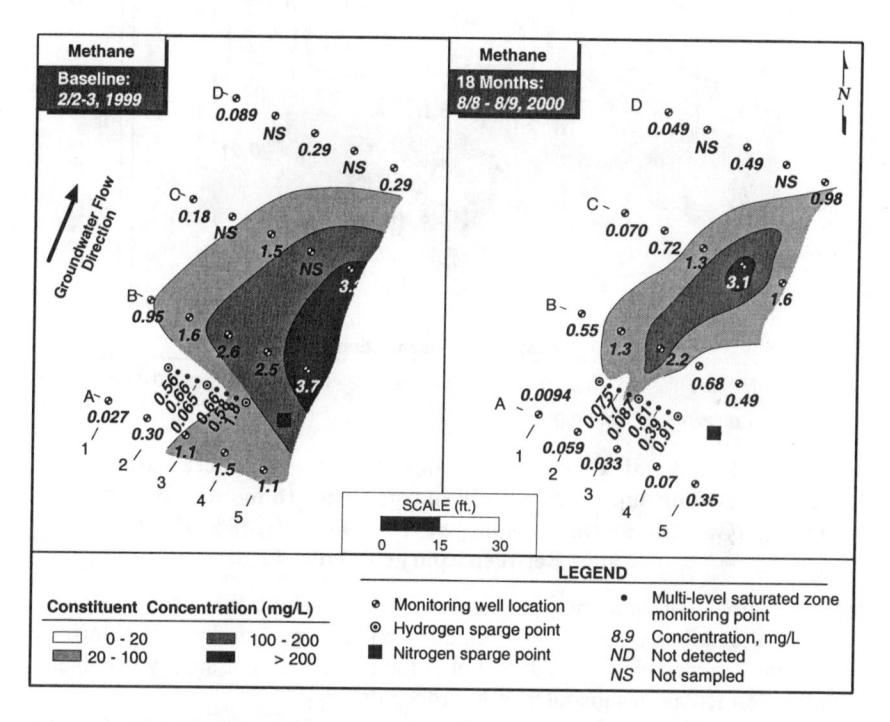

FIGURE 3. Change in methane concentration during 18-month pilot test of low-volume pulse biosparging

Evaluation of other indicators, such as the consumption of hydrogen vs. non-biodegraded tracers and increasing daughter/parent ratios, also support the conclusion that high rates of reductive dechlorination were initiated then sustained throughout the test zone by direct hydrogen delivery. No excessive methane production was observed during the test (see Figure 3), although increased concentrations of vinyl chloride were observed in the test zone after 18 months of treatment (see Figure 2).

DIVERSION OF COMPETING ELECTRON ACCEPTORS

The effective electron donor supply may be increased by reducing the transport of competing electron acceptors to a chlorinated solvent source zone (Newell et al., 2001). The presence of competing electron acceptors (primarily dissolved oxygen, nitrate, and sulfate) in a source zone will result in biodegradation reactions that compete with beneficial dechlorination reactions for

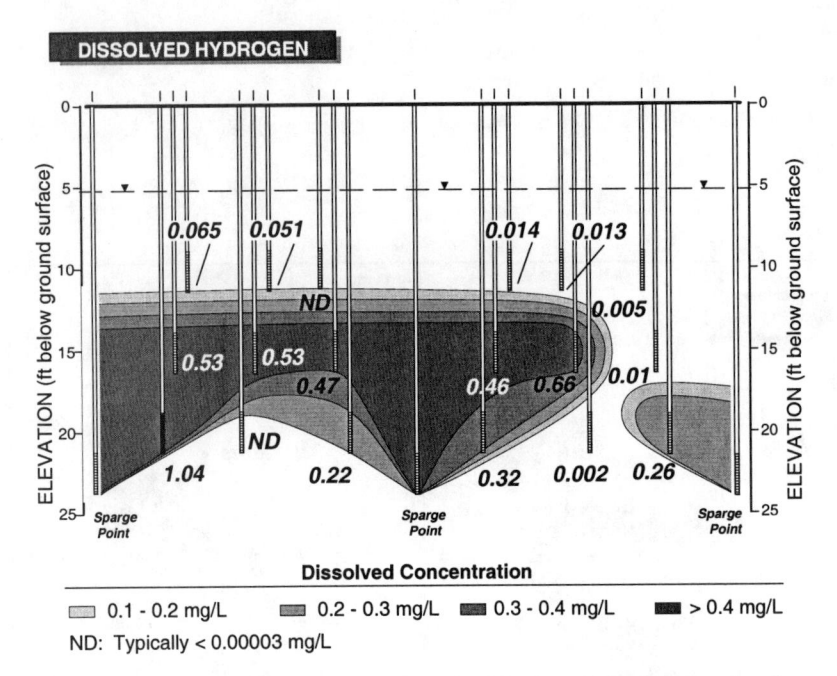

FIGURE 4. Vertical profile of dissolved hydrogen concentration in groundwater four days after weekly sparge event, 18 months after project startup. Gas mixture in last sparge event: 48% H_2, 48% He, 2% SF_6. Distance Between Sparge Points: 12 ft.

electron donor. This competition occurs in cases where the electron donor is present in the source zone prior to remediation (a Type I or Type II chlorinated solvent site; Wiedemeier et al., 1999) or if the electron donor supply is enhanced by adding fermentation substrates or hydrogen directly.

By diverting the transport of competing electron acceptors (oxygen, nitrate, and sulfate) around a contaminated groundwater zone in a Type I site (typically with anthropogenic donor in the source zone NAPL), the electron donor supply may be effectively increased. For example, a 14-site chlorinated site database in Wiedemeier et al. (1999) show the following characteristics (Table 2).

TABLE 2. Selected hydrogeologic, plume, and background groundwater characteristics from 14 chlorinated solvent sites.

	MEDIAN	STANDARD DEV.
Plume/source width	400 ft	604 ft
Seepage velocity	110 ft/yr	532 ft/yr
Saturated thickness	20 ft	19 ft
Background D.O.	8.0 mg/L	3.8 mg/L
Background NO3	5.8 mg/L	4.0 mg/L
Total Chlorin. Solvents in Source	1.5 mg/L	34 mg/L

Assuming a porosity of 0.3, a representative specific discharge through a chlorinated solvent source zone is equivalent to 15 x 10^6 L/yr of flow. Approximately 120 kg of dissolved oxygen and 87 kg of nitrate flow into a representative source zone per year, where they compete for electron donor. One method to account for the potential amount of lost reductive dechlorination to competing electron acceptors is to assume that every 16 kilograms of dissolved oxygen can consume the equivalent of 2 kilograms of dissolved hydrogen (based on the stoichiometry of water formation), and that every 50 kilograms of nitrate can consume the equivalent of 4 kilograms of dissolved hydrogen (based on the stoichiometry of nitrate reduction). Therefore the introduction of the 120 kilograms of dissolved oxygen and 87 kilograms of nitrate into the source zone per year is equivalent to the consumption of 22 kilograms of dissolved hydrogen per year (i.e., 120*2/16+87*4/50). Finally, if one uses the accepted stoichiometry where 1 kg of hydrogen has the potential to completely dechlorinate 21 kilograms of PCE, then an additional 462 kilograms of PCE could be completely dechlorinated to ethene per year assuming no loss to other mechanisms. Note this calculation does not account for sulfate as a competing electron acceptor, which would increase the potential benefits from electron acceptor diversion.

By comparison, naturally-occurring reductive dechlorination processes in a source zone at a typical chlorinated solvent site may be on the order of tens of kilograms per year. Using the BIOCHLOR natural attenuation model (Aziz et al., 2000a) with the representative site data above and a typical biodegradation rate coefficient for chlorinated solvents from the BIOCHLOR database (Aziz et al, 2000b), it is estimated that only 20 kilograms of solvents are biodegraded naturally per year in a 400 ft by 400 source zone.

Therefore diverting the competing electron acceptors away from the source zone has the potential to increase the biodegradation of chlorinated solvents in the source zone of a representative chlorinated solvent site from about 20 to 462 kilograms per year, greater than a 20-fold increase in the naturally-occurring biodegradation rate. Note that these calculations are estimates only and should be confirmed with detailed field measurements. This approach does have limitations, as it would not address vertical oxygen infiltration via diffusion or via rainfall, or remove competion from ferric iron in the source zone. The diversion of competing electron acceptors can be performed in a number of ways, but most likely can be achieved by the construction of a physical barrier upstream of the source zone using conventional geotechnical barrier techniques (slurry wall, grout curtain, etc.) (Figure 5).

In summary, by constructing a low-cost, low-permeability containment barrier upgradient of a chlorinated solvent source zone, three benefits may be realized: 1) competing electron acceptors will be diverted away from the source zone, thereby increasing the rate of naturally-occurring bioremediation; 2) the plume will shorten, greatly reducing long-term monitoring costs; and 3) the plume will be controlled without pumping. Such a barrier system will be inexpensive, reliable, and have the potential to significantly increase the rate of chlorinated solvent biodegradation at a typical chlorinated solvent site already undergoing natural attenuation.

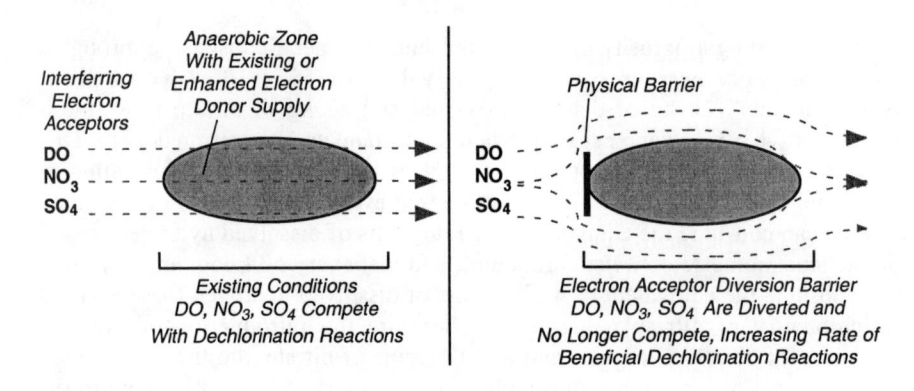

FIGURE 5. Conceptual diagram of physical barrier to divert competing electron acceptors.

REFERENCES

Aziz, C.E., C.J. Newell, and J.R. Gonzales. 2000b. *BIOCHLOR Chlorinated Solvent Plume Database Report*, Air Force Center for Environmental Excellence, Brooks AFB, Texas.

Aziz, C.E., C.J. Newell, J.R. Gonzales, P.E. Haas, T.P. Clement, and Y. Sun. 2000a. *BIOCHLOR Natural Attenuation Decision Support System, User's Manual Version 1.0*, U.S. EPA, Office of Research and Development, EPA/600/R-00/008, Washington D.C., January, 2000. www.gsi-net.com

Hughes, J.B., C.J. Newell, and R.T. Fisher. 1997. *Process for In-Situ Biodegradation of Chlorinated Aliphatic Hydrocarbons by Subsurface Hydrogen Injection*, U.S. Patent No. 5,602,296, February 11, 1997.

Newell, C.J., C.E. Aziz, and G.A. Cox. 2001. "Novel Method to Enhance Chlorinated Solvent Biodegradation by the Use of Barriers," 2001 Int. Containment & Rem. Technology Conf., June 10-13, 2001, Orlando, Florida.

Newell, C.J., P.E. Haas, J. B. Hughes, and T.A. Khan. 2000. "Results From Two Direct Hydrogen Delivery Field Tests For Enhanced Dechlorination," *Bioremediation and Phytoremediation of Chlorinated and Recalcitrant Compounds*, G. B. Wickramanayake, A. R. Gavaskar, B.C. Alleman, and V.S Magar, eds., Battelle Press, Columbus, Ohio, pg. 21-38.

Wiedemeier, T.H., H.S. Rifai, C.J. Newell, and J.T. Wilson. 1999. *Natural Attenuation of Fuel Hydrocarbons and Chlorinated Solvents*, John Wiley and Sons, New York, New York.

ENHANCED BIOREMEDIATION OF HIGH CONTAMINANT CONCENTRATIONS IN SOURCE RESIDUAL AREA

Vincent B. Dick, Nichole L. Case and Susan L. Boyle, Haley & Aldrich, Inc., Rochester, NY.

ABSTRACT: Although enhanced bioremediation has proven to be effective at many sites, typically it has been applied to relatively low contaminant concentrations and in relatively permeable media. The project site described herein is an application that potentially extends the practical limits of *in situ* bioremediation applications. An active eyewear manufacturing facility located in Western NY was found to be contaminated with 1,1,1- Trichloroethane (TCA) with groundwater concentrations as high as 400-500 mg/L in the source residue area. The facility entered into a Voluntary Cleanup Agreement (VCA) with New York State and was required to investigate site conditions and remediate the site.

Site investigations were completed in April 2000. Site soils consist of a very dense till comprised primarily of clayey silts with some sand. Due to facility constraints, many typical remedies such as soil vapor extraction, excavation, or soil heating, could not be applied. Enhanced bioremediation was evaluated through microcosm studies that concluded that site conditions were conducive for enhanced bioremediation.

Site conditions necessitated different injection scenarios at various locations of the site, comprised of varying injection grid patterns. The most significant field limitation was the need to penetrate the dense tills to accomplish effective substrate delivery from 11.6m (38 ft) bgs to approximately 1.6m (5 ft) bgs with effective loading of bioremediation substrate (HRC used as a lactate slow-release agent). Because of the depth of injection and the density of the site soils, an innovative injection/drilling program was designed. While HRC is typically introduced to the subsurface via direct-push methods, direct-push alone was found to be inadequate for site conditions. The injection design consisted of a combination air rotary and direct-push method with tooling modified to allow for rotary pilot holes followed by direct push injection.

INTRODUCTION

Enhanced bioremediation has proven to be effective at many sites, but field-scale applications typically have involved relatively low to moderate contaminant concentrations (ug/L to low mg/L range of concentrations) and have involved delivery of bioremediation nutrients or substrates into relatively permeable media. Many practitioners have been exploring ways to adapt bioremediation techniques to higher contaminant concentrations and more difficult settings (see for example Hansen, et al., 2000 and Harms, et al., 2000). This paper describes an application of *in situ* bioremediation related to the chlorinated solvent 1,1,1-trichloroethane (TCA) at relatively high groundwater concentrations in a very dense glacial till where direct injection with conventional direct push methods was not possible.

The site is an active eyewear manufacturing facility located in Western NY that was found to be contaminated with TCA. The majority of the property is occupied by active manufacturing buildings and related parking and support improvements, therefore investigation methods and any future remediation need to accommodate the facilities ongoing operational restrictions. The general site layout is shown in Figure 1.

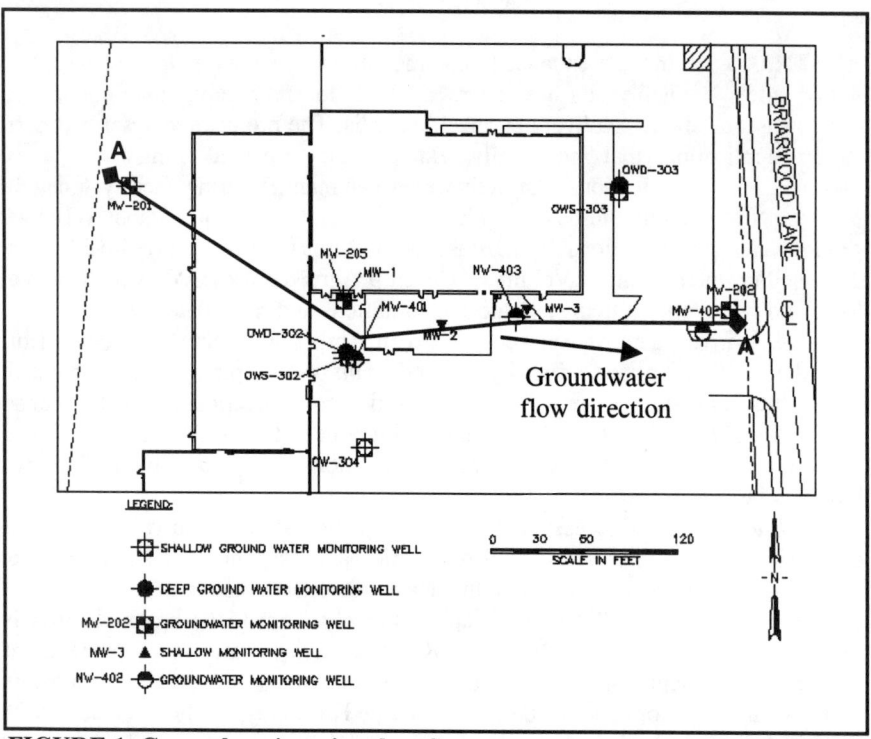

FIGURE 1. General project site plan. Source area located in and around wells MW-205 to OWD-302.

SITE SETTING

Initially, source area groundwater concentrations were found as high as 400-500 mg/L in the source area. The facility entered into a Voluntary Cleanup Agreement (VCA) with New York State and was required to investigate site conditions and remediate the site.

Site investigations were completed in April 2000 and peak total chlorinated volatile organic compound concentrations (TCA and daughter products) in the source zone were confirmed to be in the same range as the preliminary investigation had indicated (400 - 500 mg/L range). Groundwater flow direction is generally toward the east-southeast across the site (Figure 1), and multiple sampling events showed the plume to be limited to the site.

Source Zone Remediation 29

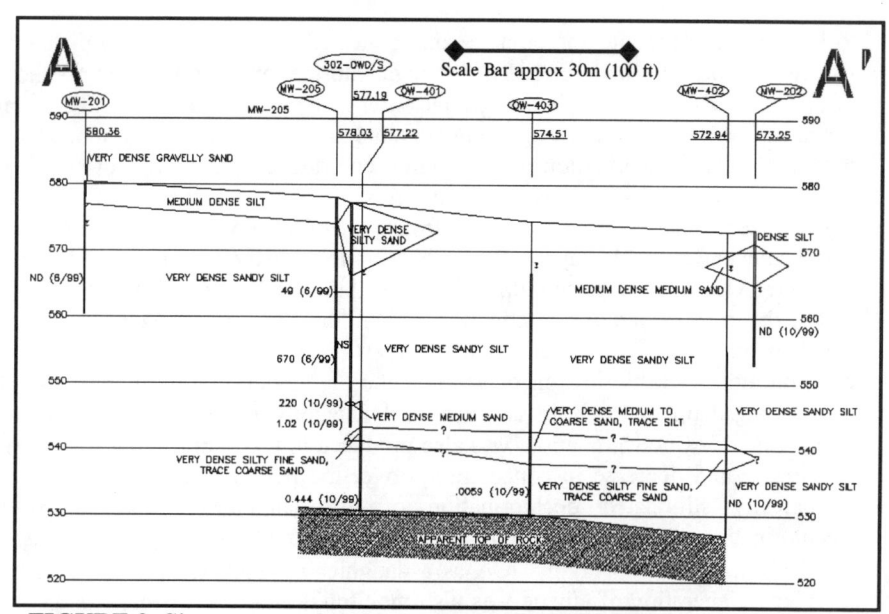

FIGURE 2. Site cross section from west to east through source area (around wells MW-205, 302-OWD and OW-401).

Impacted soils were found to be present up to 11.6m (38 ft) below ground surface (bgs) – see Figure 2. Site soils consist of a very dense till (K = 10^{-4} to 10^{-7} cm/sec, standard penetration "N" values ranging from 60 to >100 blow counts) which is comprised primarily of clayey silts with some sand. Blow counts generally increase with depth reflecting some weathering effects of the shallow till. A slightly sandier layer of the till was found at depth (see at approximately elevation 540 in Figure 2) but determined to have no significant role in plume flux. Bedrock is present at approximately 13.7m (45 ft) bgs across the area of interest.

A high-vacuum vapor/water extraction pilot test was completed at the site and provided no meaningful subsurface response to the vacuum. Additionally, due to facility constraints (existing building layout, limited available surface area for support systems), typical presumptive remedies such as soil vapor extraction and excavation could not be applied. Also, remedies that involved significant infrastructure, such as extensive wells, header or equipment networks, would seriously affect ongoing site business activities. Therefore the site remedy needed to be in-situ and "low profile."

BIOREMEDIATION EVALUATION

Enhanced bioremediation was evaluated through microcosm studies that tested the compatibility of lactate as a substrate to promote anaerobic dechlorination (in this study Hydrogen Release Compound "HRC™" was used). Samples of site soils and groundwater at various concentrations of site compounds were tested. Viable populations of microorganisms capable of chlorinated volatile organic compound (CVOC) dechlorination were determined to be present in the site source area soils collected, and could be stimulated by lactate addition. The study included testing dechlorination of both Trichloroethene (TCE) and TCA in the site's soil/groundwater samples, because some TCE was found to be present in the site contaminant suite, and it was important to determine that dechlorination past the vinyl chloride daughter product could be accomplished. The TCE was tested in site samples at concentrations up to 25 mg/L. Dechlorination of TCE ranging from 68% of 80% of initial concentrations was achieved after 28 days. Both cis-1,2-Dichloroethene (DCE) and Vinyl Chloride (VC) were produced (up to 1.2mg/L) but both declined markedly by the end of the test. Generation of ethene was monitored and was detected as an end daughter-product. TCA was tested at low (25 mg/L) and high (250 mg/L) concentrations. Dechlorination of TCA in the "low" concentration samples (25 mg/L) ranged up to 79±% in 28 days. 1,1-Dichloroethane (DCA), a daughter product of TCA biodegradation, was also produced in the samples, but was found to increase and decrease in concentration over the period of the test indicating biologically stimulated dechlorination of the daughter product as well. Dechlorination of TCA in the "high" concentration samples (250 mg/L) ranged up to 92±% in 28 days. Again, DCA as a daughter product was produced in the samples. Generation of ethane was also monitored and was detected as an end daughter-product.

BIOREMEDIATION SUBSTRATE DELIVERY

Site conditions (high density till, presence of buildings over portions of the plume, etc.) and source residues up to required different injection scenarios at various locations of the site. Groundwater concentration gradients vary from 400 to 500 mg/L at shallow depths in the source area to 0.4 – 0.6 mg/L in the hydrostratigraphic zone near the base of the till. Laterally, concentrations decrease to approximately 1 – 3 mg/L approximately 45m (150 ft) downgradient, and to non-detect near the property line (73m or 240 ft from the source area). Therefore injection was broken into several grid-zones for substrate mass load estimation (see Figure 3).

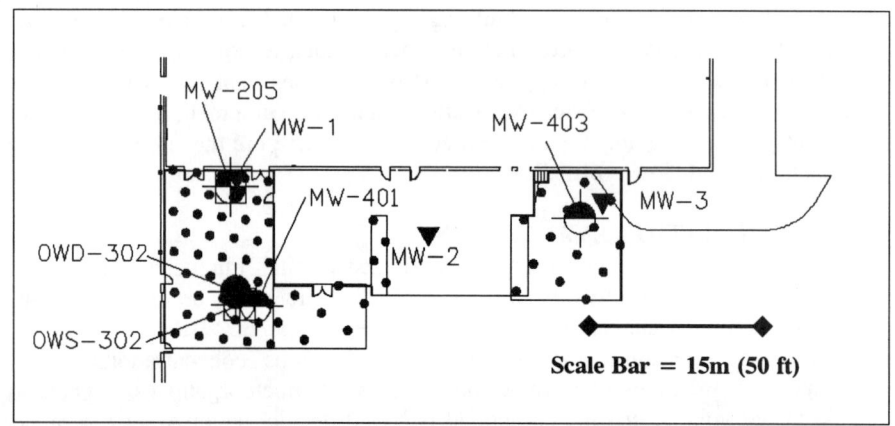

FIGURE 3. Site injection grid pattern broken into four injection grids.

The source zone injection (grid area around well OWD-302) was designed on a 2.1m (7-ft) grid, and resulted in 40 injection points, with completion to a depth of 11.6m (38 ft) bgs. The injection design spanned from 11.6m (38 ft) bgs to approximately 1.5m (5 ft) bgs with approximately 2.3kg (5 lb) of HRC injected per linear foot for a total of approximately 74.8 kg (165 lb) per hole. Other areas of the site (near well MW-2, around well MW-3, etc.) have similar injection designs, but modified based on depth and concentrations of contamination. There are a total of 65 injection points throughout the treatment areas.

Bioremediation nutrients or substrates are typically introduced to the subsurface using Geoprobe-type direct push methods or wells that allow soluble nutrients to "infuse" the affected subsurface area. However, because of the required depth of injection for this site and the high density of the soils, it was determined that a rotary injection method would likely be required to reach the desired injection depth, the deepest being approximately 11.6m (38 feet) below ground surface. A field test, with what is reported by Geoprobe to be their most powerful rig (Model 6600), took place at the site in July 2000. This field test was completed to determine if a more powerful direct-push rig was capable of reaching the desired injection depth. Because of the exceptionally dense soils, the rig was only able to penetrate to approximately 3m (10 ft) below ground surface. Based on this field test, it was determined that a combination rotary injection-direct push technique was required for the HRC injection design.

Through discussions with various drillers, a combination rotary injection-direct push technique method was developed (Haley & Aldrich, 2001). The combination rotary injection-direct push technique method involves advancing a tri-cone rotary bit or narrow diameter solid-stem auger from the ground surface to the desired depth. Once the desired depth is reached the bit/auger is removed from the hole and a direct-push tool inserted to act as the injector. HRC is then injected through the tool tip under pressure using a Rupe Pump, or equivalent, as the tip is slowly raised from the borehole base.

In a normal direct-push type injection, pressure is maintained in the borehole by a snug match of tool diameter with the diameter of the formation borehole. Pressure is measured directly at the pump. For this project, several options were developed to address potential pressure problems in the field. The initial borehole uses the following procedure: a pilot hole is completed prior to the injection using a bit/auger as described above. Injection follows by inserting an injection tip or tool whose diameter matches the pilot hole as tightly as possible. Because there will be a pilot hole, there is a potential for loss of some friction between the tool and the formation. To help mitigate this, the tool used to create the pilot hole must match the injection tool in diameter as closely as possible so as to maintain a high-friction contact against the borehole sides. Again, the pressure is measured directly at the pump.

If blowback occurs near the surface (i.e. HRC ejects from the borehole rather than going into the formation), the injection is terminated and another hole is completed adjacent to the first hole. Because blowback is observed near the surface, the tool string only needs to penetrate to a shallow depth to complete injection, therefore the second hole is advanced using direct push methods only, creating a hole with the injection tool and injecting as the tool is retracted from the hole. By doing this a greater amount of friction can be created between the injection tool and the formation, thereby creating a better seal in the hole, facilitating the injection of substrate into the formation rather than to the ground surface. If blowback is observed at depth, then the hole is repeated using a smaller bit, attempting to find the "best fit" for the hole and create a better seal between the injection tool and the formation. If neither of the initial approaches work then a pilot hole is drilled with a tri-cone bit while driving casing (see Figure 4). The injection then takes place through the casing as the casing is being retracted from the hole. Inflatable packers may also be used to create a better seal between the casing and injection tool. Injection proceeds from the target depth to approximately 1.6m (5 ft) bgs, the approximate high water mark.

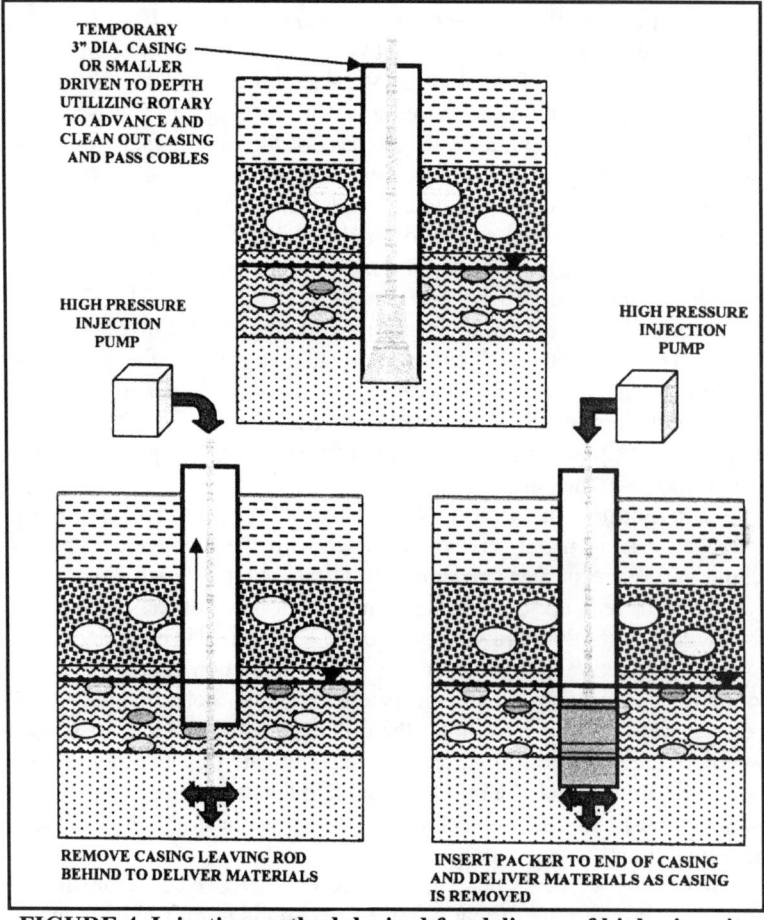

FIGURE 4. Injection method devised for delivery of high-viscosity bioremediation substrate into high-density till.

The volume of HRC injected in each hole is calculated by the driller prior to the start of injection by determining the volume of HRC delivered with each stroke of the pump. The volume is then converted into weight using a conversion table provided by Regenesis. This calculation and conversion determines the pounds of HRC delivered in each stroke of the pump.

It should be noted that the procedure above assumes that the injected substrate's density and viscosity for the various formulations of HRC. Other substrates such as veg-oil, dilute molasses, or other forms of lactate, or propionate, etc. may require alternate injection procedures.

FIELD RESULTS

The field injection for this site is scheduled for Spring 2001 and evaluation data will be shown at the time of paper presentation in June 2001. Monitoring parameters for the site include CVOC concentrations, HRC component concentrations, biodegradation indicators (e^- acceptors, endpoint gases), microbial analyses, and field parameters (DO, Eh, pH, temperature).

CONCLUSIONS

Enhanced in-situ bioremediation is an effective remediation technology that warrants increased attention by practitioners to develop delivery methods for high concentration areas and under difficult field conditions. In other words, two of the factors that have been used in the past to reject enhanced bioremediation as a viable alternative, namely source concentrations that approach dense non-aqueous phase liquid (DNAPL) levels and difficult subsurface soil conditions, should not be an automatic basis for rejection of enhanced bioremediation. The installation techniques described here may be applicable to other unconsolidated soil site settings.

REFERENCES

1. Haley & Aldrich of New York, 2001. *VCA Remediation Work Plan*

2. Harms, W.D., K.A. Taylor and B.S. Taylor. 2000. "HRC-Enhanced Reductive Dechlorination of Source Trichloroethene in an Unconfined Aquifer." *In Bioremediation and Phytoremediation of Chlorinated and Recalcitrant Compounds*, pp295-302. Battelle Press, Columbus OH

3. Hansen, M.A., J. Burdick, F.C. Lenzo and S. Suthersan. 2000 "Enhanced Reductive Dechlorination: Lessons Learned at Over Twenty Sites." *In Bioremediation and Phytoremediation of Chlorinated and Recalcitrant Compounds*, pp263-270. Battelle Press, Columbus OH.

TECHNOLOGIES COMPETITIVE WITH ENHANCED BIOREMEDIATION OF SOURCE ZONES

Robert D. Norris (Eckenfelder/Brown and Caldwell, Denver Colorado)

ABSTRACT: Chlorinated solvents are most frequently released as non-aqueous liquids and when present in sufficient quantity reach soils and groundwater as Dense Non-Aqueous Liquids (DNAPL). Dissolution is a slow process and thus DNAPL can remain as a groundwater source for very long times. The distribution of these compounds is quite complex and challenging to remediation of sites. Frequently, naturally occurring anaerobic degradation (reductive dechlorination) is sufficient to limit the size of plumes and in a limited number of cases can result in remediation of source areas. In most cases, natural attenuation will not be sufficient to address source areas in a time frame that is likely to be acceptable to regulatory agencies. In these situations, active remedies are required. Enhanced bioremediation has until recently, been considered to be more applicable to plumes than source areas. Recent studies were presented at this conference by Joe Hughes, Cathy Vogel, and by Chuck Newell, Vincent Dick, and Kent Sorenson concerning enhanced bioremediation of DNAPL impacted zones. Other source area technologies are available. The effects of these technologies on geochemistry in turn impact anaerobic bioremediation, which requires the absence of dissolved oxygen, low redox potential, and the presence of suitable electron donors. These conditions frequently exist, especially where DNAPL is present as a result of chlorinated solvents having been used as degreasers (where a solvent/grease mixture was released) or where both chlorinated solvents and petroleum hydrocarbons were released in close proximity. The impacts on these conditions should be considered during remedy selection. For example, chemical oxidation in addition to destroying contaminant mass will increase the redox potential, destroy electron donors, and alter the microbial population. Thus one needs to consider the trade off between mass reduction and changes in conditions that affect reductive dechlorination. This paper serves as a summary of available technologies and related issues without specific examples.

INTRODUCTION

In situ bioremediation, as introduced by Richard L. Raymond in 1972, was essentially a method to accelerate natural biodegradation of petroleum hydrocarbons (Norris, 1994). This process, as well as the variety of implementation techniques developed by countless others, has provided much of the basis for enhanced natural attenuation (bioremediation) of chlorinated hydrocarbons. The level of understanding of the microbiology of reductive dechlorination has evolved over a relatively recent period thanks to efforts by numerous researchers. This has provided the basis for utilizing reductive dechlorination for monitored natural attenuation, permeable barriers, passive treatment systems, and active remediation systems for mitigation of chlorinated solvents outside source areas. Recently, techniques for carrying out reductive dechlorination in close proximity to DNAPL with the intent of source area remediation has become of great interest, as evidenced by the inclusion of a dedicated session at this conference.

Like many in situ remediation processes, in-situ bioremediation processes are designed to change conditions in the aquifer to promote specific degradation or conversion reactions. For petroleum hydrocarbons an oxygen source is sufficient to achieve acceleration of the degradation process. Other electron acceptors, i.e., nitrate, sulfate, manganese, and iron play important roles in the natural attenuation of hydrocarbons. The utilization of these electron acceptors leads to conditions favorable to reductive dechlorination. The utilization of electron acceptors lowers the oxidation/reduction potential, can lead to generation of compounds that undergo fermentation reactions that yield low levels of hydrogen, and may be beneficial in other ways. We also know that if we supply electron donors, especially those that are efficient at producing hydrogen, reductive dechlorination rates will increase and oxidation/reduction potentials will decrease. This is the basis for enhanced reductive dechlorination.

CHALLENGES OF REMEDIATING DNAPL SOURCE AREAS

Remediation of DNAPL zones is challenging. First, because we are addressing a source area, there is likely a large mass of contaminant and thus a large effort is needed regardless of the technology. Furthermore, because the densities of most chlorinated compounds are greater than the density of water, their distribution in aquifers is typically more complex. DNAPLs tend to sink below the water table, leaving a complex trail of globules within the soil pore spaces. If present in sufficient quantities, DNAPLs will continue downward until reaching an impervious layer where pools of DNAPL may form. These pools are typically hard to find, let alone define in detail. Furthermore, it may be difficult to treat or remove sufficient fractions of pooled DNAPL to make real improvements to the aquifer. Frequently it may be necessary to sequence technologies including monitored natural attenuation.

The natural attenuation or reductive dechlorination processes, in which the chlorinated compounds serve as electron acceptors, require or are favored by the following:

- Microorganisms specific to the desired degradation pathway
- Near absence of dissolved oxygen
- Low redox potential
- Near neutral pH
- Presence of sufficient electron donors
- Low to moderate levels of dissolved hydrogen (higher hydrogen concentrations may favor methane formation)
- Low nitrate
- Maybe low sulfate

Enhanced bioremediation of chlorinated solvents is based on enhancing or creating these conditions. High concentrations of chlorinated ethenes can be quite toxic to microorganisms in industrial waste systems but this appears not to be the case with many microorganisms found in impacted aquifers. Observations of DNAPL droplets being biodegraded have formed the basis for the current activities devoted to developing commercially applicable techniques for remediating DNAPL containing source areas.

As discussed in the following section, application of various other remediation technologies will be dependent upon and cause changes in geochemistry. When selecting a site remedy these effects should be considered.

UNINTENTIONAL EFFECTS OF REMEDIATION TECHNOLOGIES

Aggressive remediation of source areas is intended to remove or destroy nearly all of the mass of the COCs. Thus the effectiveness of mass removal or destruction will typically be the primary factor in the evaluation process. As a result of creating conditions that destroy constituent mass, the aquifer geochemistry is altered and/or electron donors beneficial to reductive dechlorination may be removed or destroyed. Changes in geochemistry may either support or inhibit reductive dechlorination. Thus during remedy selection it is necessary not only to compare cost effectiveness of various technologies, but also whether their impact on the geochemistry will result in conditions that favor natural attenuation as a polishing step within the source zone or downgradient of the source. This has implications for life cycle costs of the project.

COMMON DNAPL ZONE REMEDIAITON TECHNOLOGIES

In order for a remediation technology to successfully address a DNAPL zone, the DNAPL must have been sufficiently delineated and the remedial design must be sufficiently robust to treat nearly all of the DNAPL. Remediation that leaves a few percent of the DNAPL behind may not have resulted in much long-term benefit to the aquifer. The impact on natural attenuation is considered critical because it is unlikely that MCLs will be reached by DNAPL treatment alone and achieving a reasonable life cycle cost is likely to include monitored natural attenuation as a component of the remedy.

Examples of technologies that have been applied or are being developed for source area treatment of DNAPL are summarized along with their impacts on the potential for natural attenuation to address the residual COCs. A more in depth summary is provided in a recent National Academy of Science publication (NRC, 1999).

Free Product Recovery. Where pooled DNAPL is found, it is common to utilize systems that remove DNAPL with as little water as practical. These systems are relatively inexpensive to operate and typically provide the lowest cost per mass of DNAPL removed or treated. These methods also have very little impact on the aquifer geochemistry nor do they have much potential to mobilize DNAPL. Because all of the pooled DNAPL is seldom found or accessible and frequently most of the DNAPL is retained in the interstitial spaces of the soil, free product recovery is expected to leave behind at least fifty percent of the total DNAPL with minimal beneficial impact on groundwater quality. However, the cost of applying other technologies can be substantially reduced by applying DNAPL removal to reduce the mass requiring treatment before applying more costly technologies.

Solvent/Surfactant Recovery. These methods use solutions that solubilize DNAPL constituents and remove them to the surface. Generally, the surfactant or solvent, typically alcohol based, is injected into the area of suspected or known DNAPL, allowed to contact the DNAPL, and then removed to the surface. Typically, this is accomplished by injecting the solvent or surfactant blend on one side of the DNAPL zone and recovering the blend containing the DNAPL constituents from the other side of the DNAPL zone. Various other schemes such as injecting from the perimeter of the DNAPL area and extracting from the center can be utilized. The recovered DNAPL constituents are then typically separated from the solvent or surfactant blend by air stripping or other extraction process. Water in excess of that injection is recovered and thus the solvent or surfactant blend must be concentrated by removal of excess water prior to reuse.

Solvent and surfactant extraction can be highly efficient for the removal of DNAPL where the injection and recovery system sweeps nearly all of the DNAPL. Efficiency is impeded by heterogeneity of the geological matrix. Use of foaming agents can mitigate this problem. Concerns expressed over the use of this technology include the potential to mobilize the DNAPL, especially vertically through an aquitard. The solvents and surfactants must be biodegradable. Thus, residual solvent or surfactant not recovered during treatment may enhance reductive dechlorination. On the other hand, the process may remove beneficial electron donors and the solvents and surfactants may be biotoxic, although one should anticipate that the microorganisms will recover.

Air sparging. In this well-known technology for treating dissolved phase constituents, air is injected through wells or injection points that are screened over a fairly short interval within the aquifer. If air sparging is applied to a DNAPL area, a very robust vapor extraction and off-gas treatment system will be required. Thus most of the cost is associated with the adjuncts to sparging. In order for air sparging to be potentially effective, the screened intervals must be either very closely spaced or located sufficiently below the DNAPL zone. Thus it is not likely to be highly effective for pooled DNAPL. Frequently expressed concerns include the usual limitations of site geology and heterogeneity as well as the potential to mobilize the DNAPL and its constituents. Furthermore, air sparging will result in increased dissolved oxygen levels and oxidation/reduction potentials and may biodegrade or remove beneficial electron donors.

Thermal Enhancements. Steam injection, Six Phase Heating, and radio frequency heating have been used to remove DNAPL (as well as LNAPL) from aquifers. These technologies rely on increased volatilization, dissolution, or steam stripping of the DNAPL and its constituents. The DNAPL constituents generally are volatilized and move upward towards the unsaturated zone or surface where they are captured by vapor extraction and the off-gas treated. Some steam stripping processes are also designed to thermally oxidize some classes of compounds. As with all technologies, the details of the site geological matrix are important. Typically, steam stripping is more limited by fine soil matrices than are the other two technologies. Where the geology and access are favorable, these methods can offer a relatively rapid method of addressing DNAPL. The processes are costly and thus present a trade off between cost and time. Concerns sometimes expressed include the potential to mobilize DNAPL. These methods have the potential to destroy both beneficial bacteria and remove or destroy electron donors that would be supportive of natural attenuation of chlorinated solvents. However, current studies indicate that microorganisms are more resilient to thermal technologies than might be anticipated.

Chemical Oxidation. These technologies are designed primarily to degrade the constituents of interest. The chemical oxidation methods commonly applied to dissolved phase organics are potassium (or sodium) permanganate, Fenton's Reagent (hydrogen peroxide, typically with an iron catalyst and pH adjustment) and ozone. Ozone and Fenton's Reagent are not very discriminating and degrade nearly all organics with which they come in contact. Permanganate is effective for a narrower range of compounds and has primarily been used for the treatment of chlorinated ethenes. Treatment of DNAPL zones requires the addition of large quantities of oxidant, in part due to overcoming reduced conditions within the aquifer. This may provide challenges to their design as well as impacting cost. Significant decreases in source area concentrations have been reported for these technologies, although, as with all technologies, the amount of residual contamination is an issue. Concerns that have been expressed include the potential for mobilization of DNAPL constituents, impact on soil permeability due to precipitation of manganese dioxide or fracturing from high pressures under certain applications of Fenton's Reagent. For ozone and Fenton's Reagent, as well as air sparging, the potential for migration of volatile compounds through utilities or into basements should be evaluated. All three of these processes will create highly oxidizing conditions, may create elevated temperatures in the soil and aquifer, are likely to degrade most electron donors, and may kill off a significant fraction of the microorganisms.

Summary of Technologies. Table 1 provides a brief summary of the mechanisms of the technologies discussed above, effects on geochemistry, and potential changes on electron donor availability. How well these technologies will work and potentially interfere with natural attenuation will depend upon the design of a particular system, the mass and types of anthropogenic constituents, the hydrogeology, location and sensitivity of receptors relative to the current extent of the plume, and site specific cleanup levels. Potential effects are not limited to the microbiology (biodegradation). Acceleration of solubilization and mobilization of constituents may occur. Where groundwater gradients are altered, transport mechanisms are affected. Seepage velocity (advection) is altered and thus so may dispersion be affected.

TABLE 1. Effects of active remediation in addition to mass reduction.

Technology	Mechanism(s) of Treatment	General Comments	Changes to Geochemistry	Potential Unintended Effects
Surfactant Flush and Solvent Extraction	Extraction	Can remove large fraction of DNAPL mass.	May decrease DO and ORP	Removes some electron donors. May add electron donors. May be toxic to microorganisms.
Air/Biosparging	Removal and Degradation of some compounds.	More appropriate for plumes.	Increase DO Increase ORP	Remove or degrade electron donors. May be toxic to microorganisms.
Chemical Oxidation	Degradation (some methods can also result in extraction).	Efficiency for DNAPL being tested.	Increase DO Increase ORP	Degrade and possibly remove electron donors. May be toxic to microorganisms.
Enhanced Microbial Dehalogenation	Degradation	A very new Field.	Decrease DO Decrease ORP	Solubilization of certain metals
Free Product Removal	Extraction	Cost effective method, but leaves large residual.	Minimal	May remove electron donors.
Thermal Treatment	Heating of soil/volatilization	Effective and fast where applicable but costly.	Oxidation	Sterilization, potential spreading of contaminants.

DOCUMENTED IMPACT ON NATURAL ATTENUATION

As discussed in an earlier paper (Norris et al. 1998), several studies have demonstrated that active aerobic treatment can destroy naturally occurring reductive dechlorination. In situ bioremediation of a fuel spill using hydrogen peroxide was documented to inhibit reductive dechlorination (J. T. Wilson et. al., 1994). Sewell and Gibson (1991) documented a site where aerobic bioremediation removed the BTEX compounds from the aquifer and thus deprived the organisms that carried out reductive dechlorination of an electron donor source.

Gates and Siegrist (1995) have observed that chemical oxidation (hydrogen peroxide or potassium permanganate) for remediation of chlorinated solvents may actually lead to increased dissolved organic carbon concentrations. Apparently, partial degradation of humic acids can release lower molecular weight and more bioavailable compounds that can serve as electron donors.

Siegrist and coworkers (Siegrist, et al., 1995) observed one to three orders of magnitude reductions in biomass following chemical oxidation of TCE. However, post treatment populations remained substantial although the viability of the microorganisms was not evaluated.

Kastner and coworkers (Kastner, et. al., 2001) recently documented the impact of Fenton's reagent on TCE cometabolism. Both the oxidant and pH effects are discussed. As with all of these studies, negative impacts on microbial populations or processes are discussed. More research is needed to distinguish between transient and permanent impacts and methods of reestablishing microbial processes.

MODIFICATION OF REMEDIAL STRATEGIES TO ALLOW SEQUENTIAL MONITORED NATURAL ATTENUATION

The application of active source area remedies will reduce the mass of COCs within the area of treatment. As a result, less mass will be introduced into groundwater within the source area and/or plume. This will reduce the amount of attenuation necessary to meet the project goals. The extent to which subsequent natural attenuation is benefited or inhibited will be site-specific depending upon the initial conditions, site goals, and the technology as well as the design and operation of the active remedy.

In general three types of modification to remedial designs are contemplated for those instances where the active remedy might impede further natural biodegradation. The first is to modify the process itself, the second is to limit the area over which the process is implemented, and the third is to implement a post treatment addition of chemicals that would stimulate the biodegradation process. Examples of these approaches include:

- Use of nitrogen for sparging within the source area or along the down gradient edge of the plume.
- Introduction of electron donors along the downgradient edge of a oxidative treatment system.
- Introduction of electron donors or hydrogen sources subsequent to active treatment.
- Bioaugmentation in treatment area (possibly, with maintained native microorganisms) including addition of carbon source.

In general, it would be best to use a remedial technology that is compatible with the specific natural attenuation/biodegradation process occurring at the site.

SUMMARY

We have available a number of technologies for addressing DNAPL. All are difficult to apply because of the complex nature of DNAPL distribution in aquifers. Selection of a remedy is based on many factors, one of which may be the impact on natural attenuation when used as a polishing step. Active remediation will affect natural attenuation. A reduction in contaminant mass will reduce the demand on natural attenuation. Other effects may be beneficial or detrimental. A careful review of all the mechanisms of active remediation and natural attenuation is recommended.

REFERENCES

Farone, W. A. 2000, Personal communication.

Gates, D.D. and R.L. Siegrist. 1995. "In Situ Chemical Oxidation of Trichloroethylene Using Hydrogen Peroxide.". *J. Environmental Engineering.* 121(9):639-641.

Knox, R. C., D. A. Sabatini, J. H. Harwell, R. E. Brown, C. C. West, F. Blaha, and C. Griffin. 1997. "Surfactant Remediation Field Demonstration Using a Vertical Circulation Well." *Ground Water* 35(6): 948-953.

Kastner, J. R., J. S. Domingo, M. Denham, M. Mariosa, and R. Brigmon. 2001. "Effect of Chemical Oxidation on subsurface Microbiology and Trichloroethene (TCE) Biodegradation" *Bioremediation Journal* Vol. 4. Issue 3. 219-236.

Kennedy, L.G., J. W. Everett, K. J. Ware, R. Parson, and V. Green. 1998. "Iron and Sulfur Mineral Analysis Methods for Natural Attenuation Assessments" *Bioremediation Journal* Vol 2.issue 3-4. 259-276.

Norris, R. D., M. C. Marley, and D. J. Wilson. 1998. "Modification of Remedial Methods to Preserve the Benefits of Natural Attenuation." *Natural Attenuation: Chlorinated and Recalcitrant Compounds.* Ed. Wickramanayake, G. B. and R. E. Hinchee. Batelle Press, Columbus, Ohio: 269-274.

Norris, R.D, et. al. 1984. *Handbook of Bioremediation,* Lewis Publishers, Bacon Raton, Florida.

Norris, R.D., D.J. Wilson, D.E. Ellis, and R. Siegrist. 1999. "Consideration of the Effects of Remediation Technologies on Natural Attenuation" *Natural Attenuation of Chlorinated Solvents, Petroleum Hydrocarbons, and Other Organic Compounds.* Alleman, B.C. and A. Leeson, eds. Battelle Press, Columbus Ohio, 5(1) 59-64.

National Research Council. 1999. *Groundwater & Soil Cleanup.* National Academy Press, Washington, D. C.: Chapter 4.

Siegrist, R.L., O.R. West, J.S. Gierke, et al. 1995. "In Situ Mixed Region Vapor Stripping of Low Permeability Media. 2. Full Scale Field Experiments. *Environmental Science & Technology.* 29(9): 2198-2207.

Sewell, G. W. and S. A. Gibson. 1991. "Stimulation of the Reductive Dechlorination of Tetrachloroethene in Anaerobic Aquifer Microcosms by the Addition of Toluene." *Environmental Science & Technology.* 25:982-984

Wilson, J. T., J. M. Armstrong, and H. S. Rafai. 1994. "A Full-Scale Field Demonstration of the Use of Hydrogen Peroxide for In Situ Bioremediation of an

Aviation Gasoline-Contaminated Aquifer." *In Bioremediation Field Experience* Ed. P. F. Flathman, D. E. Jerger, and J. H. Exner. Pp. 333-359. Lewis Publishers (1994).

REMEDIATION REFERENCES

AATDF (Advanced Applied Technology Demonstration Facility). 1997. Technologies Practice Manual for Surfactants and Cosolvents. Houston, Texas. AATDF, Rice University.

Fountain, J.C. 1998. Technologies for Dense Nonaqueous Phase Liquid Source Zone Remediation. Ground Water Remediation Technology Analysis Center, Technology Evaluation Report. 70 pp.

Ward, C. H., et. al. 1999 "Groundwater Soil Cleanup". National Academy Press, Washington D. C.

Web Sites on Related Topics

http://clu-in.org/conf/itrc/cvoc/resource.htm
Http://www.napl.net/newsletter.html
http://www.em.doe.gov/plumesfa/intech/spsh/techdes.html
http://www.epa.gov/ada/pubs/reports.html

A COMPARISON OF CHEMICAL OXIDATION AND BIOLOGICAL REDUCTIVE DECHLORINATION TECHNOLOGIES FOR THE TREATMENT OF CHLORINATED SOLVENTS

R. Joseph Fiacco, Jr., and Gregg A. Demers (ERM, Boston, Massachusetts); Richard A. Brown, George Skladany, and David Robinson (ERM, Ewing, New Jersey)

Abstract: Over the last several years, two new remediation technologies have been developed and applied to treat chlorinated solvents: chemical oxidation and biological reductive dechlorination. Chemical oxidation involves the injection of a chemical oxidant directly into soil or groundwater to destroy the chlorinated solvents present, while reductive dechlorination involves the injection of a readily biodegradable carbon amendment and inorganic nutrients. Since chemical oxidation creates oxidative subsurface conditions while reductive dechlorination creates reductive conditions, there are major differences between the two technologies in how they are selected, designed, implemented and monitored. This paper will discuss the differences between the two technologies and offer criteria for selecting one over the other.

INTRODUCTION

Widespread contamination of groundwater and soil with chlorinated solvents has led to a search for faster, more effective, and lower cost *in situ* remediation technologies. Over the last several years, *in situ* chemical oxidation (ISCO) and biologically-mediated reductive dechlorination (BMRD) have been developed to treat chlorinated solvents in soil and groundwater.

ISCO involves direct chemical oxidation of the contaminant through the direct injection of a chemical oxidant into soil or groundwater, while BMRD involves the injection of a readily biodegradable carbon amendment and inorganic nutrients. Since ISCO creates oxidative subsurface conditions while BMRD creates reductive conditions, there are major differences between the two technologies in how they are selected, designed, implemented and monitored.

This paper will focus on the remediation of three common chlorinated solvents (trichloroethene [TCE], tetrachloroethene [PCE], and 1,1,1-trichloroethane [TCA]) and their associated daughter products by comparing ISCO treatment with BMRD treatment.

***In Situ* Chemical Oxidation.** Four oxidants are commonly used: hydrogen peroxide, potassium permanganate, sodium persulfate, and ozone. A brief discussion of each oxidant follows.

Hydrogen Peroxide (H_2O_2) is typically used in conjunction with acid and an iron catalyst to generate hydroxyl radicals (Fenton's Process), which have an oxidation potential of 2.76V:

$$Fe^{+2} + H_2O_2 \rightarrow Fe^{+3} + OH^- + {}^\bullet OH \qquad (1)$$
$$2\,{}^\bullet OH + 2H^+ + 2e^- \rightarrow 2H_2O \qquad (2)$$

The hydroxyl radical formed by this process has a high oxidation potential capable of oxidizing a wide range of chlorinated solvents. In addition to hydroxyl radical formation, hydrogen peroxide also undergoes a rapid and often exothermic decomposition, which can cause thermal stripping of the contaminants:

$$2H_2O_2 \rightarrow 2H_2O + O_2 + \Delta \text{ (heat)} \tag{3}$$

Because of its high reactivity, hydrogen peroxide does not persist in soils, so it is often ineffective in treating slow-reacting contaminants at high soil concentrations.

Potassium permanganate (KMnO$_4$) is a fairly strong oxidant with an oxidation potential of 1.68 V.

$$MnO_4^- + 4H^+ + 3e^- \rightarrow MnO_2 + 2H_2O \tag{4}$$

Permanganate is stable in groundwater, will effectively oxidize chlorinated ethenes (TCE and PCE), but is ineffective against chlorinated ethanes (TCA). It will also react with naturally-occurring organics and inorganics (metals) in soil, which can lead to an excessive demand for the oxidant. Unreacted permanganate imparts a purple color to water. Permanganate is used to treat soil and groundwater and has had some success against dense non-aqueous phase liquids (DNAPL).

Sodium persulfate (Na$_2$S$_2$O$_8$) has only recently been used for ISCO (US Patent #6,019,594), and has an oxidation potential of 2.01 V. Persulfate generally reacts via a free radical pathway, catalyzed by iron:

$$S_2O_8^= \rightarrow 2^\bullet SO_4^- \tag{5}$$
$$^\bullet SO_4^- + e^- \rightarrow SO_4^= \tag{6}$$

Persulfate will oxidize chlorinated ethenes, as well as some chlorinated ethanes and methanes. Persulfate is fairly stable in groundwater, and unlike permangnate, does not readily react with naturally-occurring organics. The oxidation of organics with persulfate produces acid and increased levels of sulfate. Additionally, the stoichiometric demand of persulfate is greater than other oxidants.

Ozone (O$_3$) is a highly reactive gas with an oxidation potential of 2.1V:

$$O_3 + 2H^+ + 2e^- \rightarrow O_2 + H_2O \tag{7}$$

Ozone will readily oxidize chlorinated ethenes and some chlorinated ethanes. In additon to being oxidized, volatile organic compounds (VOCs) can also be stripped as a result of ozone gas injection.

Ozone also reacts with hydrogen peroxide to generate hydroxyl radicals that will subsequently oxidize organics:

$$2O_3 + 3H_2O_2 \rightarrow 4O_2 + 2\,^\bullet OH + 2H_2O \tag{8}$$

Ozone is typically delivered in an air or pure oxygen carrier gas stream. Ozone has a low mass-delivery rate and is relatively unstable.

Biologically Mediated Reductive Dechlorination. BMRD involves the stepwise reduction of chlorinated solvents to non-toxic end products (ethane and ethene) by removing a chlorine atom and replacing it with a hydrogen atom. Although reductive dechlorination can take place under naturally-occurring conditions, active remedial systems are often required because natural attenuation mechanisms may not lead to complete dechlorination, or the rate of reaction may not be adequate to satisfy regulatory requirements for site cleanup.

The implementation of a remedial program to promote BMRD is often accomplished in two stages: biostimulation and bioaugmentation. Biostimulation involves achieving the desired reducing conditions by injecting a carbon source (e.g., acetate, methanol, lactate, soybean oil) and inorganic nutrients into the zone of treatment. Bioaugmentation, which is not always required, involves adding bacteria that are known to be effective at complete dechlorination of chlorinated solvents. Reductive dechlorination of TCE and/or PCE using native microorganisms often stalls at dichloroethene (DCE) or vinyl chloride because of a slow reaction rate and/or the lack of appropriate dechlorinating bacteria (Weidemeier et al., 1999).

COMPARATIVE EVALUATION

There are a number of factors that need to be considerd in comparing these two remedial technologies. These factors include level of effectiveness, range and ease of implementation, timeliness, cost, contaminant type, test work required, and regulatory acceptance.

There is no clear advantage for either technology under all conditions, and relative advantages are site dependent. However, there are some fundamental differences in how these two technologies are applied and how they perform. For example, ISCO is generally a faster process. BMRD is a more flexible process, and is less sensitive to the type of chlorinated solvent or to site conditions. There is no clear advantage to one or the other on cost. BMRD typically requires more testing and more monitoring, but uses much less expensive chemicals. Table 1 summarizes the major differences between these two technologies.

REMEDY SELECTION CRITERIA

In general, both technologies have potential applicability at most sites. However, they may not be equally viable or effective. There are several factors which need to be considered in determining the viability of using ISCO or BMRD. Since the physical application of each technology is similar, the appropriate remedy selection factors would not, therefore, include the equipment and systems needed for application. The remedy selection factors are more fundamental and include site-specific factors, as well as economic and regulatory factors.

TABLE 1. Comparison of Technologies

Factor	ISCO	BMRD
Remediation Goal	- Concentration-based standards	- Monitored Natural Attenuation
Contaminants Amenable	- Ethenes: PCE, TCE, DCE & VC	- Ethanes: TCA & DCA - Ethenes: PCE, TCE, DCE & VC
Effect on DNAPL	- Some treatment possible	- Not applicable
Timeliness	- Relatively fast process (weeks to months)	- Relatively long process (months to years)
Treatability Testing	- Simple tests	- Complex studies
	- 1 to 3 months duration	- 3 to 6 months duration
	- Common equipment	- Specialized equipment (anaerobic)
	- Critical tests: soil demand and oxidation effectiveness	- Critical tests: degradation effectiveness, formation of intermediates, nutrients, carbon substrate required, and bioaugmentation
Field Pilot	- Not always required	- Essential
	- 1 to 3 months duration	- 3 to 6 months duration
	- Key parameter: distribution	- Key test parameters: distribution, microbial response, and by-product accumulation
	- Easy monitoring	- Complex monitoring
Regulatory Acceptance	- New process	- Extension of MNA
	- UIC permit often needed	- Additives benign; no UIC permit needed
	- Concern with metals	- Little concern with metals
	- Safety and handling concerns	- Few safety concerns
Costs	- Treatability tests: $3,000 to $10,000	- Treatability tests: $15,000 to $30,000
	- Pilot (if needed): $15,000 to $30,000	- Pilot: $30,000 to $100,000
	- Full-Scale: $150,000 to $400,000 per acre	- Full-Scale: $100,000 to $300,000 per acre

Generally, bioremediation (i.e., reductive dechlorination) is more appropriate for large groundwater sites, sites having low hydraulic conductivity (because it is a slower process and can make use of diffusive flow), or sites with reducing geochemical conditions and intrinsic dechlorinating bacteria. BMRD is the more appropriate technolgy if there is a high degree of regulatory concern or a potential for surface water impacts. Chemical oxidation is most appropriate for smaller sites, sites with high concentrations and/or soil contamination, and sites having low natural oxidant demand. Chemical oxidation is more appropriate for sites requiring fast response and where remediation is on a voluntary basis.

Table 2 summarizes the key factors that should be considered when choosing a technology. Site-specific factors are important in determining which technology to use. In general, ISCO is more advantageous in areas with rapid groundwater flow (advective), homogenous lithology, and with groundwater conditions not favorable to microbial activity. Conversely, on sites with heterogenous lithology, strong microbial growth and more diffuse groundwater flow, BMRD would be the preferred remedial technology. Economic and regulatory factors will also determine which technology to use. If rapid closure (i.e., <6 months) is desired, ISCO is the preferred technology.

TABLE 2. Remedy Selection Factors

Favors BMRD		Favors ISCO
Site Specific Factors		
Large (>5 acres)	**Plume Size**	Small (<1 acre)
Low (<10^{-5} cm/second)	**Permeability**	Moderate to high
Non-pervious, layered	**Lithology**	Pervious, uniform
Diffusive	**Contaminant transport characteristics**	Advective
Chloro-ethenes/ethanes	**Contaminant type**	Chloroethenes
Low (<25 mg/L or <1 mg/kg)	**Contaminant concentration**	High (>25 mg/L or >100 mg/kg)
Little (<100 yd^3)	**Soil contamination area**	Significant (>1,000 yd^3)
Many, ecological	**Surface & groundwater receptors**	Few
Usable	**Groundwater supports microbial activity**	Poor (such as pH & TDS)
Pronounced	**Natural attenuation parameters**	Low indication
High	**Matrix reactivity**	Low
Economic & Regulatory Factors		
Programmatic	**Regulatory environment**	Voluntary
MNA	**End point**	Standards
Slow (>1 year)	**Response time desired**	Fast (<6 months)
High	**Testing & Design Tolerance**	Low
High	**Concern with metals**	Low

CASE STUDIES

Site A: A historical release of TCA occurred beneath an active manufacturing facility. Average TCA concentrations of 400 mg/L were detected in the shallow bedrock aquifer in the source area. TCA degradation products, DCA and DCE, were also detected in source area groundwater at average concentrations of 10 mg/L and 20 mg/L, respectively.

Source area geology consists of sandy fill over glacial till over crystalline bedrock. Groundwater is primarily located in the bedrock aquifer at a depth of 20 to 25 feet BGS. Shallow bedrock is very competent with few significant water-bearing fractures (i.e., defined as having a sustainable yield of 0.5 gallons per minute or greater) noted in source area monitoring wells. Additional water-bearing fractures have been detected down-gradient of the source area, primarily at depths of 50 to 75 feet BGS. The degree of fracturing generally decreases with increasing depth beyond 75 feet BGS. Numerous water-bearing micro-fractures were observed in and down-gradient of the source area at varying depths. Due to aquifer characteristics, advective flow dominates in significant water-bearing fractures and diffusive flow dominates in water-bearing micro-fractures.

Because the site is not located within a drinking water protection area, state environmental regulations require reduction of average TCA concentrations in groundwater to below 50 mg/L. Due to bedrock characteristics and the lack of recharge beneath the building, the horizontal extent of the shallow bedrock source area (i.e., defined by concentrations greater than 50 mg/L) is approximately 5,000 ft^2. Migration control is not mandated, but may be considered to control potential off-site migration. Because the client was interested in rapid reduction of contaminant concentrations, both ISCO and BMRD were considered for source area remediation.

BMRD was considered potentially infeasible given the elevated source area concentrations, although it is being considered for migration control because of its longer half-life, mobility, and ability to be distributed via advective and diffusive flow. Permanganate and persulfate oxidation were deemed infeasible to facilitate rapid reduction in source area chlorinated ethane concentrations. Therefore, ISCO using hydrogen peroxide was selected as the preferred remedial alternative.

A bench-scale study indicated that hydrogen peroxide was effective at destroying approximately 99% of TCA, independent of site hydrogeology. Because of peroxide's short half-life and limited radius of influence, a pilot study was conducted. The pilot system consisted of a network of 16 injection wells installed in a general grid at 15-foot centers. The pilot study was conducted in two phases: a smaller-scale preliminary phase and a larger-scale final phase. The preliminary phase consisted of three one-week oxidant application periods resulting in injection of 1,260 gallons of hydrogen peroxide. The final phase consisted of three two-week application periods, resulting in injection of 3,200 gallons of hydrogen peroxide.

Based on a comparison of pre-remediation TCA concentration data with data collected four months after the final application, TCA concentrations within the source area decreased by approximately 40%. However, acidification of the source area aquifer resulted in the leaching of cadmium, nickel and zinc into groundwater at concentrations up to 0.5 mg/L, 26 mg/L and 57 mg/L, respectively. An evaluation of the relationship between pH and the solubility of these metals indicated that they are more soluble at pH values less than 6.

The ISCO pilot study resulted in a significant reduction in contaminant mass. The feasibility of expanding the ISCO system into deeper bedrock beneath the source area is largely dependent on our ability to deliver oxidant to the contaminants using a limited number of deep injection wells. This will require an enhancement in oxidant distribution (i.e., greater radius of influence per injection well). To enhance the treatment area per injection well, ERM is conducting bench-scale studies to evaluate the effectiveness of using sodium persulfate, in conjunction with hydrogen peroxide, to destroy chlorinated ethanes and ethenes. Persulfate has a significantly longer half-life, which facilitates the utilization of both advective and diffusive transport mechanisms to distribute the oxidants.

Simultaneously, ERM is conducting microcosm studies to evaluate biostimulation and bioaugmentation, in the event that persulfate proves to be ineffective at this site. Given that the pilot study significantly decreased source area concentrations in the shallow bedrock aquifer, we believe that bioremediation may be effective over longer periods of time if implemented initially in areas of lower contaminant concentrations.

Site B: Industrial manufacturing operations resulted in PCE (and various degradation products) contamination of a shallow coastal plain aquifer. Up to 120 mg/L total VOCs were detected in wells at the facility. Site geology consists of a ten-foot, coarsening downward sequence of silty, fine to medium sand with a thick, confining clay layer below. Impacts to groundwater are limited to the shallow water table aquifer from 5 to 10 feet BGS. There are four distinct PCE plumes at the site totaling an area of 40,000 ft^2.

Although the site is currently being used commercially and served by municipal drinking water, remediation efforts are focused on obtaining closure with few or no restrictions on land or groundwater use. Adjacent properties include residential and light industrial areas.

A risk assessment performed indicated that a critical exposure pathway from volatilization of VOCs from the groundwater to enclosed spaces existed. Therefore, remediation efforts are focused on achieving acceptable residential levels off-site and on-site levels that are acceptable to commercial workers. Monitored attenuation on site may be acceptable if off-site contamination is prevented.

Soil in the area has a low permeability and moderate levels of metals. Data collected since 1995 indicated natural attenuation is occurring throughout the site as evidenced by general reductions in PCE levels and consistent or slightly increasing levels of daughter compounds. Dissolved oxygen (DO) levels in the contaminated areas were typically below 1 mg/L and nitrate and sulfate concentrations indicated the native microbial population was using several electron acceptors in a reducing environment.

ERM evaluated the use of several remedial technologies including air sparging, natural attenuation, chemical oxidation, and enhanced bioremediation. The dense soil eliminated the possibility of air sparging and the slow rate of natural attenuation was determined to be insufficient in preventing off-site migration. Initial chemical oxidation bench-scale tests using permanganate were unable to completely remove the PCE due to a high soil demand. Based on the client's objectives and site hydrogeochemical conditions, ERM recommended BMRD.

A microcosm study was then performed to evaluate the effects of various combinations of food sources (lactate, soybean oil), nutrients (i.e., biostimulation) and non-indigenous bacteria (i.e., bioaugmentation) on samples collected from the aquifer to facilitate reductive dechlorination of PCE. Results of the microcosm study demonstrated that complete degradation of PCE occurred in all of the microcosms provided with a source of food and nutrients, including the the stoichiometrially transformation of VC to ethene and methane.

One of the main concerns with the use of BMRD is whether the microbial population is able to completely degrade the elevated vinyl chloride concentrations. The vinyl chloride concentrations were reduced to below regulatory levels during the microcosm study. Based on the microcosm study results, soybean oil was chosen as the most cost-effective food source. The indigenous microbial population was capable of completely dechlorinating the PCE.

ERM designed a BMRD pilot study for the site. The pilot test began in April 2000 and included the injection of a soybean oil mixture into the shallow aquifer on a grid pattern using a Geoprobe® and grout pump.

Six months after injecting the soybean oil solution, total chlorinated ethene concentrations have declined in all piezometers within the pilot area. PCE concentrations decreased by 57%, with some increases occurring in TCE, DCE, and VC. The concentrations of DCE and VC in some areas increased even though there was initially relatively low levels of TCE. Ethene levels increased from an initial concentration of 623 ug/l to 870 ug/l, a 40% increase.

Although select VOCs have been eliminated in several areas within the pilot test area, the pilot study has been extended from six to twelve months to determine the extent of VOC reduction in the most heavily contaminated areas.

BMRD was the appropriate choice for this site. Seven out of 11 site specific selection factors and three out of five of the economic and technology regulatory factors favored BMRD over ISCO. Of particular significance in the selection were the high matrix demand, the low permeability, and the layered geology. Also, since there was strong evidence of significant natural attenuation already occurring on the site, the regulatory environment was predisposed to accepting MNA.

CONCLUSION

Both ISCO and BMRD are effective technologies for the treatment of many chloroethenes. BMRD is also effective in treating many chloroethanes. Each technology offers specific advantages and disadvantages for its use. The proper choice between these technologies relies on understanding both the specific site factors, as well as project regulatory and economic requirements. In certain circumstances, sequential use of these technologies may provide a more timely and/or cost-effective remedy than the use of either technology alone.

REFERENCE

Weidemeier, T. H., H. S. Rifai, C. J. Newell, and J. T. Wilson. 1999. *Natural Attenuation of Fuels and Chlorinated Solvents in the Subsurface.* John Wiley & Sons, Inc., New York, NY.

MOBILIZATION OF SORBED-PHASE CHLORINATED ALKENES IN ENHANCED REDUCTIVE DECHLORINATION

Frederick C. Payne (ARCADIS Geraghty & Miller, Novi, Michigan, USA)
Suthan S. Suthersan (ARCADIS Geraghty & Miller, Langhorne, Pennsylvania, USA)
Frank C. Lenzo (ARCADIS Geraghty & Miller, Langhorne, Pennsylvania, USA)
Jeffrey S. Burdick (ARCADIS Geraghty & Miller, Langhorne, Pennsylvania, USA)

ABSTRACT: Enhanced reductive dechlorination (ERD) is a groundwater remedial technology that entails injection of a mixture of highly biodegradable organic carbon into a solvent-contaminated aquifer formation. The organic loading directs aquifer microbial consortia into low-redox behaviors such as sulfate reduction and methanogenesis. When sulfate-reducing or methanogenic conditions are established, chlorinated alkenes such as perchloroethene (PCE) and trichloroethene (TCE) can be reduced to ethene and ethane. A common collateral effect of the soluble organic carbon injections is an increase in dissolved chlorinated alkene and alkane concentrations downgradient of the injection location. It is believed that these increases are a natural consequence of four mechanisms related to soil organic carbon: organic carbon partition coefficients (K_{oc}) decrease as chlorine atoms are removed by dechlorination; biosurfactant production by microbial consortia; solvent action of fermentation products; and aqueous-phase carbon flooding. At sites where soil pH is well buffered, enhanced reductive dechlorination can be managed specifically to attack sorbed-phase contaminant that is otherwise unexposed to remedial measures. ERD studies were conducted at two chlorinated solvent sites: a low-carbonate fractured bedrock aquifer in the northeastern United States and a porous high-carbonate aquifer in the midwestern United States. Changes in contaminant composition at both sites were consistent with ERD-induced release of non-aqueous-phase contaminants and their subsequent dechlorination to ethene and ethane. At the Midwestern site, apparent degradation rate constants were 0.04 day^{-1} or greater (corresponding to half-lives of fewer than 17 days).

INTRODUCTION

Reductive dechlorination of chlorinated alkenes in aquifers can be enhanced through the injection of readily degradable carbonaceous materials such as molasses, whey or vegetable oil (Lenzo, 2000; Suthersan, 1996; Suthersan, 2000). Microbial degradation of the injected carbon consumes available electron acceptors such as oxygen and nitrates and forces the aquifer microbial communities into utilization of alternative electron acceptors such as ferric iron and sulfates. When sulfate-reducing or methanogenic conditions develop, chlorinated alkenes such as perchloroethene, trichloroethene and cis-1,2-dichloroethene can be rapidly dechlorinated.

Reductive dechlorination reactions occur primarily in the aqueous phase. For many chlorinated alkenes such as perchloroethene and trichloroethene, a large portion of the mass in an aquifer can reside in the sorbed phase, unavailable to reductive dechlorination reactions (cf. Scow and Johnson, 1997). Aqueous-phase treatment processes such as oxidation and enhanced reductive dechlorination generally require desorption of sorbed phase contaminants to achieve remedial objectives. Contaminants sorbed to "soft" forms of carbon such as poorly decomposed plant and animal tissues and non-target organic contaminants are more easily desorbed than those bound to "hard" carbon forms such as soot and charcoal that often occurs naturally in aquifer soils (Pignatello and Xing, 1996; Pignatello, 2000).

Figure 1 describes the portion of aquifer mass that resides in sorbed phase as a function of aquifer organic carbon fraction, for selected chlorinated alkenes. It was calculated from partitioning equations summarized in US EPA (1996). At moderate levels of soil organic matter, less than 15 percent of perchloroethene in an aquifer can be found in aqueous phase. Substantial mass transfer of contaminants from the sorbed phase to the aqueous phase must occur, along with reductive dechlorination reactions, to fully treat most aquifer formations. Four mechanisms associated with the enhanced reductive dechlorination process increase the availability of contaminants and are essential to the success of the technology.

Sequential K_{oc} Reductions. In the common reductive dechlorination sequence of PCETCEcis-1,2-DCEvinyl chloride, the organic carbon partition coefficients decrease from 265 to 19 L/kg (U.S. EPA, 1996). As reductive dechlorination proceeds, the products resulting from each step in the sequence are less susceptible to sorption than the previous compound in the sequence, as can be seen in Figure 1.

Dissolved Organic Carbon Flooding. Under normal conditions, dissolved-phase organic matter is dwarfed by solid-phase organic matter in the matrices of an aquifer. However, the organic carbon flooding an aquifer during enhancement of reductive dechlorination constitutes a large pool of uncontaminated organic matter. Simple equilibrium partitioning with the added carbon will drive a portion of the sorbed-phase contaminants into aqueous phase. This effect was observed by Hunchak-Kariouck, et al. (1997), even though the levels of dissolved organic matter they studied were two orders of magnitude less than those used in field applications of enhanced reductive dechlorination.

Fermentation Products. Fermentation reactions create organic acids, alcohols, ketones, fatty acids and other soluble constituents that may act as solvents, mobilizing a portion of the non-aqueous soil organic matter. As reducing conditions develop, fermentation reactions become more prominent and the treatment process reaches a critical phase. If the aquifer is poorly buffered, organic acids produced by fermentation can drive the pH downward to levels that may curtail sulfate reduction sulfate reduction or can occur (according to Wetzel [1983], bacterial sulfate reduction is restricted to pH levels greater than 5.5 when the redox potential is near

0 mV and is restricted to pH levels of 7 or greater at redox potentials of −400 mV). Further reduction of pH may also affect methanogenesis. As a consequence, the aquifer pH must be carefully monitored and controlled as it approaches or falls below 5; thus reducing the possibility that the enhancement process will become trapped in fermentation. Uncontrolled, the process can lead to mobilization of sorbed-phase alkenes without adequate reductive degradation mechanisms to degrade the released mass. This can lead to mobilization of contaminants in the aquifer. Conversely, if the aquifer is well buffered, fermentation products cannot decrease pH levels appreciably, and sulfate reduction and methanogenesis proceed. Under these conditions, reductive dechlorination reactions can rapidly destroy alkenes in aqueous phase, while the products of fermentation recruit sorbed-phase contaminant mass to feed the process.

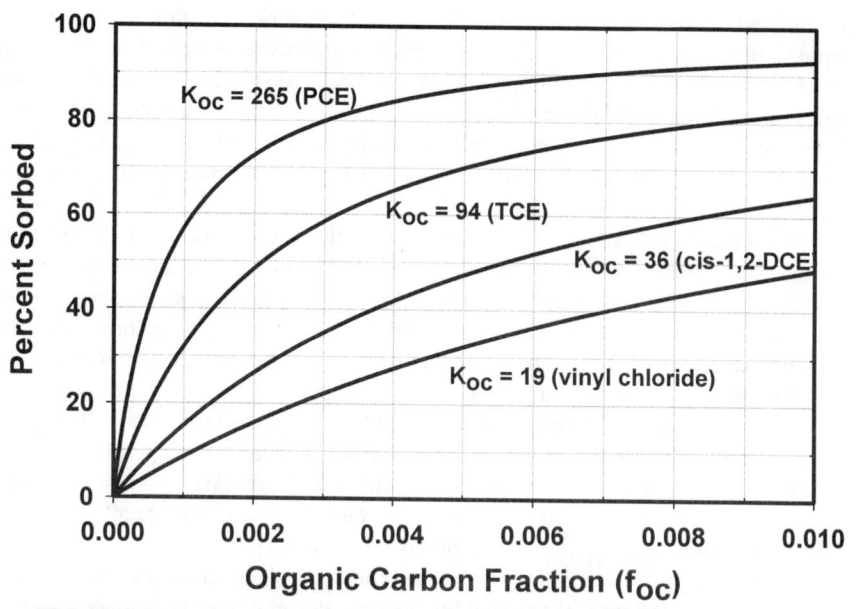

FIGURE 1. Percent of selected chlorinated alkenes predicted to reside in sorbed phase, as a function of the organic carbon fraction in the aquifer soil. In aquifer formations containing high organic carbon, only a small fraction of high-Koc compounds such as perchloroethene will be observed in aqueous phase. Koc values were reported in US EPA (1996) and are expressed as L/kg.

Biosurfactants. The microbial consortia that participate in low-redox metabolism have been shown to produce large quantities of biosurfactants, further enabling solubilization of sorbed-phase materials. Several studies have demonstrated the potential for mobilization of hydrophobic contaminants by rhamnolipid biosurfactants produced by *Pseudomonas* species in fermentation processes (Mercade and Manresa, 1994; Noordman, et al., 2000; Zhang and Miller, 1992). The bacterial genus *Rhodococcus* has also been shown to produce effective biosurfactants (Kanga, et al., 1997).

CASE STUDY RESULTS AND DISCUSSION

The two case studies discussed below demonstrate the net effects of interactions between enhanced reductive dechlorination and sorbed-phase chlorinated alkenes. It was not possible in these field studies to separate the magnitude of each process, nor to know certainly how much each of the suggested processes contributed to the observed effects. However, the patterns observed are instructive because they clearly demonstrate rapid desorption and destruction of chlorinated alkenes.

Low-Carbonate Fractured Bedrock Aquifer. A reactive zone was established in a fractured, low-permeability bedrock formation in the Northeastern U.S., to treat an aquifer that was contaminated by perchloroethene and trace amounts of its degradation products. The organic carbon fraction in this formation is approximately 0.001. A groundwater recovery system was operated in the source area for several years prior to implementation of enhanced reductive dechlorination, and a contaminant plume extended more than 500 meters from the source. Aqueous-phase perchloroethene in the bedrock fractures was observed at concentrations up to 23 umol/L (3,800 ug/L) prior to starting the enhanced reductive dechlorination process.

Molasses injections were conducted in the mid-plume area during pilot testing, then were expanded to allow treatment of the entire contaminated formation. Pilot testing was later conducted in the source area. A surge in the perchloroethene concentration was observed in a source area monitoring well 60 days after injections were started in this area; the surge exceeded 560 umol/L (94,000 ug/L). Carbon injections were stopped but the perchloroethene concentration continued to rise for a short period, peaking at more than 670 umol/L (112,000 ug/L) before returning to near-baseline levels. During this time, no significant degradation was observed. Source area molasses injections were restarted at a lower carbon feed rate on day 150 and the perchloroethene concentration rebounded only to a fraction of its previous peak.

The surge of aqueous-phase perchloroethene observed at the source area monitoring well was probably caused by the rapid desorption or non-aqueous-phase liquid mobilization that can accompany carbon source injections. Although plug flow displacement of non-aqueous-phase liquid or high-PCE groundwater may have contributed to the observed PCE increase, it seems unlikely: dissolved organic carbon concentrations at the monitoring well increased 25-fold from baseline levels, 60 days *before* the perchloroethene peak was observed. If plug flow or advective displacement had caused the PCE concentration surge, the dissolved organic carbon increase would have *followed* the PCE concentration increase.

Although the relative contribution of each mechanism could not be isolated from available data, fermentation products and carbon flooding are likely to have supported the mass transfer into aqueous phase. In addition to the dissolved organic carbon increase mentioned above, pH at the injection location fell to 3.7 S.U., suggesting that fermentation reactions were occurring at a high rate. pH levels at the source area monitoring wells, however, did not decrease from their pre-treatment levels (7.0 S.U.); so the fermentation-driven pH decrease was very localized.

Chlorinated alkene concentrations were also measured at a location approximately 30 days downgradient from the carbon injection location. These observations, shown in Figure 2, were consistent with the pattern expected to develop in ERD-driven desorption: PCE concentrations remain stable initially, while reductive dechlorination products increase. Recruitment of PCE from sorbed phase sustains aqueous-phase concentrations during this period, while the lower-K_{oc} degradation products accumulate in aqueous phase. In the second stage of the pattern, sorbed-phase PCE mass dwindles and its aqueous-phase concentration begins to decline. During this stage, degradation processes consume the TCE and cis-1,2-DCE. Ethene concentrations did not build in the monitor well shown here, although concentrations in a deeper well at the same location exceeded 18 umol/L (500 ug/L). The cause of this distribution has not been determined.

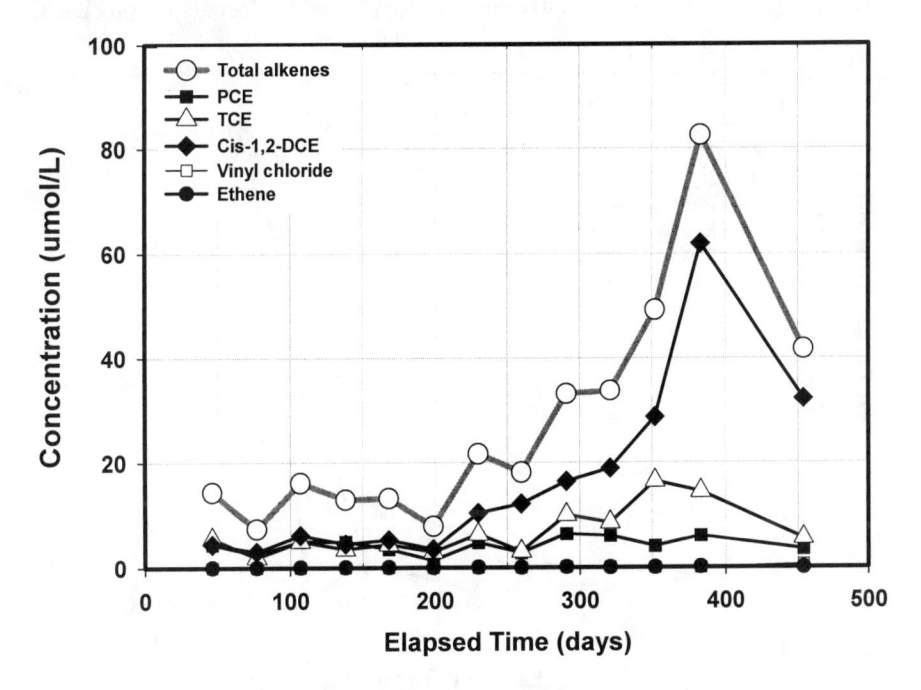

Figure 2. Results of enhanced reductive dechlorination in a fractured bedrock aquifer. Carbon source injections were initiated at zero days elapsed time.

The enhanced reductive dechlorination process was slowed at the Northeastern fractured bedrock site to control displacement of high-concentration groundwater observed in the pilot study area. Despite that restraint, significant reductions of chlorinated alkenes have been documented across the site.

High-Carbonate Porous Aquifer. A reactive zone was established in a porous, high-carbonate aquifer in the Midwestern U.S., that was contaminated by perchloroethene and trichloroethene releases prior to 1980. Darcian groundwater velocities at the site were approximately 30 cm per day. The organic carbon fraction in the aquifer ranged from 0.001 to 0.006, while aqueous-phase perchloroethene and trichloroethene concentrations were 3 and 5 umol/L (500 and 700 ug/L), respectively, prior to treatment. The sorbed-phase portions of these compounds that were not reflected in aqueous-phase measurements can be estimated from Figure 1. At a median organic carbon fraction of 0.003, 80 percent of the PCE and 58 percent of the TCE were expected to reside in sorbed phase prior to the start of carbon injections.

Enhanced reductive dechlorination was induced through injections of 5 or 10 percent molasses solution every 2 weeks over a six-month period. Chlorinated alkene concentrations were observed at a groundwater monitoring well located approximately 30 meters downgradient from the reactive zone. The results of enhanced reductive dechlorination are shown in Figure 3. Molar concentrations are shown in Figure 3 to clearly display the stoichiometry of the degradation processes.

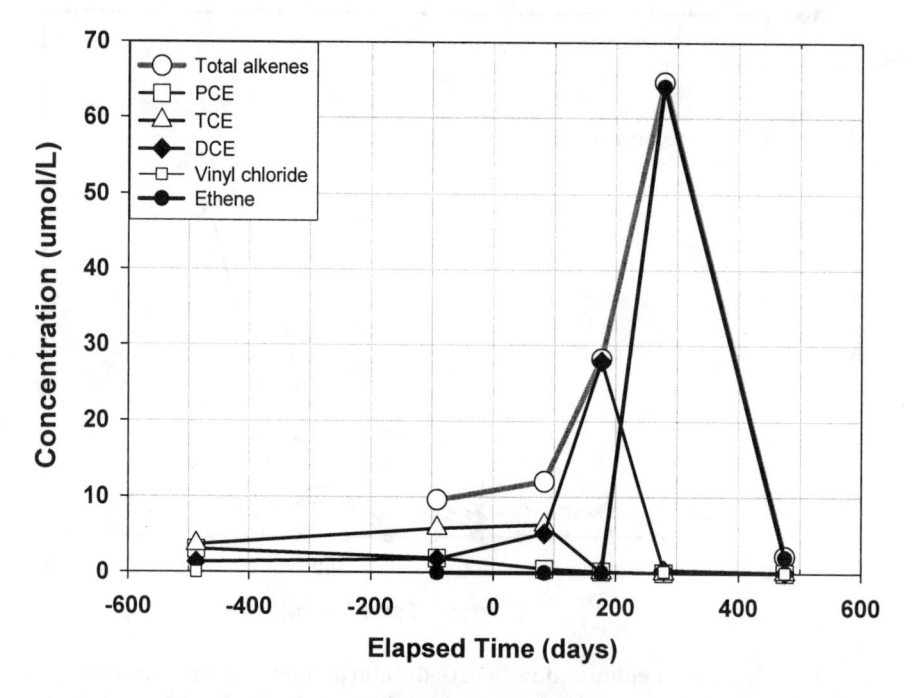

Figure 3. Results of enhanced reductive dechlorination in a high-carbonate aquifer. Carbon injections began at 0 days elapsed time. Ethene monitoring began with the baseline sampling event which occurred at –91 days.

As reductive dechlorination proceeded, a 6-fold increase in total dissolved alkenes was observed. This was consistent with the pattern shown in Figure 1, with the soil organic carbon fraction, f_{oc}, in the range of 0.006.

The observed contaminant decreases represented the combined effects of desorption and degradation. As a result, field observations could not be used to determine true degradation rate constants. Instead, "apparent" degradation rate constants were estimated from the aqueous-phase data assuming simple first-order decay. The resulting values were 0.015 day^{-1} for perchloroethene and 0.042 day^{-1} for trichloroethene and cis-1,2-dichloroethene. Because the apparent degradation was the net of both desorption releases of contaminant and reductive dechlorination reactions, the actual rate constants for perchloroethene and trichloroethene were likely 0.05 day^{-1} or greater, corresponding to a half life shorter than 14 days. It is important to note that vinyl chloride did not accumulate during the study period. The pre-treatment vinyl chloride concentrations was 0.05 umol/L (3 ug/L), and the peak observed was only 0.2 umol/L (12 ug/L) – occurring after 27 umol/L (2,700 ug/L) of cis-1,2-dichloroethene was degraded.

Fermentation reactions were documented by the appearance of ketones at the carbon injection wells, but pH at the injection wells never fell below 5.5. At the downgradient monitoring well, the pH remained above 6.9 during all observations. This is a sharp contrast with the fractured bedrock study site, where pH levels fell to 3.7 at the injection well. The dominant ketone at the study location was 2-butanone (methyl ethyl ketone, or MEK). It's concentrations exceeded 180 umol/L (13,000 ug/L) in the injection wells and more than 40 umol/L (3,000 ug/L) were observed at the downgradient monitoring well. Ketones and other fermentation products are expected to degrade very rapidly downgradient from the reductive treatment zone.

CONCLUSIONS

Mobilization of sorbed-phase chlorinated alkenes (mass transfer from sorbed to aqueous phase) was clearly demonstrated in the case studies. The site with a well-buffered aquifer withstood heavy carbon loadings without an adverse pH excursion, permitting development of fermentation. The resulting releases of sorbed-phase alkenes, along with contaminant destruction under strongly reducing conditions, combined to generate net half-lives of 15 days or less and did so without the accumulation of vinyl chloride.

In the poorly buffered fractured bedrock aquifer, there was a mobilization of sorbed-phase contaminant without their destruction, over a limited area. This required limitation of the carbon injection rate and led to a more restrained remedial program than was possible at the well-buffered, porous aquifer site.

The *fermentation trap*, in which fermentation-induced pH decreases may block development of the sulfate-reducing or methanogenic conditions critical to the reductive dechlorination processes, was not widespread. The well-buffered site showed no pH excursions, while the poorly-buffered aquifer fell to low pH only at the injection well.

The speed of the desorption process was consistent with partitioning reactions of "soft" carbon adsorption (e.g., Pignatello, 2000; Luthy, et al., 1997), rather than the slower desorption expected to result from "hard" carbon adsorption described by Pignatello and Xing (1996). Longer-term observations will be required to determine whether desorption continues at acceptable levels.

REFERENCES

Hunchak-Kariouk, K., L. Schweitzer, and I.H. Suffet. 1997. "Partitioning of 2,2',4,4'-Tetrachlorobiphenyl by the Dissolved Organic Matter in Oxic and Anoxic Porewaters." *Environ. Sci. Technol.* 31:639-645.

Kanga, S.A., J.S. Bonner, C.A. Page, M.A. Mills, and R.L. Auntenrieth. 1997. "Solubilization of Naphthalene and Methyl-Substituted Naphthalenes from Crude Oil Using Biosurfactants." *Environ. Sci. Technol.* 31(2): 556-561.

Lenzo, F. 2000. "Reactive Zone Remediation." In: *In Situ Treatment Technology, 2nd ed.*. Nyer, E.K., P.L. Palmer, E.P. Carmen, G. Boettcher, J.M. Bedessem, F. Lenzo, T.L. Crossman, G.J. Rorech, and D.F. Kidd. Lewis Publishers, New York.

Luthy, R.G., G.R. Aiken, M.L. Brusseau, S.D. Cunningham, P.M. Gschwend, J.J. Pignatello, M. Reinhard, S.J. Traina, W.J. Weber, Jr., and J.C. Westall. 1997. "Sequestration of Hydrophobic Organic Contaminants by Geosorbents." *Environ. Sci. Technol.* 31(12):3341-3347.

Mercade, M.E., and M.A. Manresa. 1994. "The Use of Agroindustrial By-Products For Biosurfactant Production." *J. Amer. Oil Chemists Soc.* 71(1): 61-64.

Noordman, W.H., J.-W. Bruining, P. Wietzes, and D.B. Janssen. 2000. "Facilitated Transport of a PAH Mixture by a Rhamnolipid Biosurfactant in Porous Silica Matrices." *J. Contam. Hydrol.* 44(2):119-140.

Pignatello, J.J. 2000. "The Measurement and Interpretation of Sorption and Desorption Rates for Organic Compounds in Soil Media." *Adv. Agron.* 69:1-73.

Pignatello, J.J. and B. Xing. 1996. "Mechanisms of Slow Sorption of Organic Chemicals to Natural Particles." *Environ. Sci. Technol.* 30:1-11.

Scow, K.M., and C.R. Johnson. 1997. "Effect of Sorption on Biodegradation of Soil Pollutants." *Adv. Agron.* 58:1-56.

Suthersan, S.S. 1996. *Insitu Anaerobic Reactive Zone for Insitu metals precipitation and to achieve microbial denitrification.* U.S. Patent 5,554,290.

Suthersan, S.S. 2000. *Engineered In Situ Anaerobic Reactive Zones.* U.S. Patent 6,143,177.

U.S. EPA. 1996. *Soil Screening Guidance: Technical Background Document.* EPA/540/R95/128. May, 1996.

Wetzel, R.G. 1983. *Limnology (2nd edition).* Saunders College Publishing.

Zhang, Y. and R.M. Miller. 1992. "Enhanced Octadecane Dispersion and Biodegradation by a Psuedomonas Rhamnolipid Surfactant (Biosurfactant)." *Appl. Environ. Microbiol.* 58(10):3276-3282.

BIODEGRADATION OF CHLORINATED ETHENES UNDER VARIOUS REDOX CONDITIONS

Xiaoxia Lu and Guanghe Li (Tsinghua University, Beijing, P.R.China)
Shu Tao (Peking University, Beijing, P.R.China)
Tom N.P. Bosma and Jan Gerritse (TNO Environment,
Energy and Process Innovation, Apeldoorn, The Netherlands)

ABSTRACT: Batch experiments were performed in the laboratory to investigate the biodegradation of chlorinated ethenes under various redox conditions and to determine the H_2 threshold concentrations corresponding to different terminal electron-accepting processes (TEAPs). Results from fifteen months of operation indicated that chlorinated ethenes could be reductively dechlorinated to less or non-chlorinated compounds under favourable circumstances, such as methanogenic conditions. Reductive dechlorination was stimulated by the addition of volatile fatty acids. TEAPs with nitrate, manganese oxide and ferric iron as the terminal electron acceptors exhibited H_2 threshold values similar to those observed in PCE/TCE dechlorinating cultures, in contrast, cis-DCE and VC dechlorinations exhibited H_2 thresholds in the range of sulfate reduction and methanogenesis, respectively. These observations may be one explanation why on certain locations reductive dechlorination is impeded at the level of cis-DCE/VC, whereas at other locations complete dechlorination to ethene readily occurs.

INTRODUCTION
The contamination of soil and groundwater with chlorinated ethenes is a widespread and serious environmental problem. Studies from the past decade indicated that chlorinated ethenes can be degraded biologically through their use by micro-organisms as electron donors and/or electron acceptors, depending on the oxidation form of the chlorinated compound and the redox condition of the environment (Wiedemeier et al., 1998). In an anaerobic subsurface environment, reductive dechlorination, where chlorinated ethenes act as electron acceptors, is an important process. However, this process often has been observed to be incomplete, resulting in accumulation of cis-DCE and/or VC (Vogel et al, 1985; Bagley and Gossett, 1990; Chu and Jewell, 1994; Garant and Lynd, 1996; Christiansen et al., 1997).

Recent studies suggested that the dissolved hydrogen concentration can be a useful indicator for in-situ redox condition and reductive dechlorination (Lovley et al.; 1988, 1994; Chapelle et al., 1996, 1997; Yang and McCarty, 1998). Based on experimental data as well as on microbiological and thermodynamic theory, a terminal electron-accepting process (TEAP) at steady state is characterized by a specific H_2 level due to competitive exclusion of other microbiologically mediated TEAPs (Jakobsen, et al. 1998). However, knowledge on the relationships among dissolved H_2 concentrations, redox conditions, and reductive dechlorinations is still limited.

Objective. The objective of this study is to reveal the biodegradation potentials of chlorinated ethenes under various redox conditions and to evaluate their relationship to the dissolved H_2 concentrations. The studied chlorinated ethenes included tetrachlorothene (PCE), cis-dichloroethene (cis-DCE) and vinyl chloride (VC). Nitrate (NO_3), oxidized iron ($Fe(OH)_3$) and manganese (MnO_2), sulfate (SO_4), carbon dioxide (CO_2), or a mixture of volatile fatty acids (VFAs) were used as electron acceptors or donors, respectively.

MATERIALS AND METHODS

Chemicals. Liquid PCE($\geq$99.5%) and cis-DCE(>98%) were obtained from Fluka Chemika, Germany. Gas VC (>99.9%) was bought from Praxair, Alltech Associates, USA. Electron acceptor stocks, prepared under a N_2 atmsphere, included 1M $NaHCO_3$, 1M Na_2SO_4, 1M $NaNO_3$, 0.5M MnO_2 and 0.4M $Fe(OH)_3$. Electron donor stock was 1M volatile fatty acids (VFAs) prepared by mixing acetic, propionic, valeric, isobutyric, isovaleric, and butyric acids in concentration ratio of 18:1.5:1.2:1.2:1:1.

Microcosm Setup. Batch microcosms were prepared under a N_2 atmosphere in 600 ml bottles filled with 250 ml basal medium. Chloroethenes including PCE, cis-DCE and VC were respectively supplied to the cultures. For each chloroethene, a series of bottles was prepared with the addition of different electron acceptors or donors including $NaNO_3$, MnO_2, $Fe(OH)_3$, Na_2SO_4, $NaHCO_3$, or VFAs, respectively. A sterilized bottle was used as a control. To maintain anaerobic conditions, viton septa (for PCE and cis-DCE) or rubber stoppers (for VC) were used to seal the bottles. The basal medium contained: NH_4Cl (1.0 g/L), $MgSO_4 \cdot 7H_2O$ (0.1 g/L), $CaCl_2 \cdot 2H_2O$ (0.05 g/L), yeast extract (0.1 g/L), trace elements solution (1.0 ml/L), vitamin solution (1.0 ml/L), 1M phosphate buffer (20 ml/L, pH 7), and 0.1% resazurin (1.0 ml/L). Pure PCE, cis-DCE or VC was injected to the corresponding bottles at nominal concentrations of 100 µM. $NaNO_3$, Na_2SO_4, $NaHCO_3$, MnO_2, and VFAs were added to final concentrations of 5 mM, while $Fe(OH)_3$ was added to 10 mM. Initial molar concentration of each VFA constituent in the batch enrichments was as follows: acetic 3.77 mM, propionic 0.31 mM, valeric 0.25 mM, isobutyric 0.25 mM, isovaleric 0.21 mM, and butyric 0.21 mM.

The PCE and VC bottles were inoculated (3.0 ml) with a mixture of a PCE to ethene dechlorinating pre-enrichment cultures isolated from chloroethene contaminated aquifer material, from Maassluis (Ter Meer and Sinke, 1998) and Arnhem (Borger et al.1998). The inoculum of the cis-DCE bottles was 3.0 ml from a pre-enrichment batch culture obtained from PCE-contaminated aquifer materials from Maassluis. All the bottles were continuously mixed on an orbital shaker (100 rpm) in a temperature controlled dark room (20 °C).

Analytic Methods. PCE, TCE, cis-DCE, VC, ethene, ethane and methane were determined on a Varian Genesis GC equipped with a Porabondcolumn (i.d. 0.32mm, length 25m, film thickness 5 µm) and a flame ionization detector (FID;

detector temperature 300 °C). The temperature of the GC column oven was programmed as follows: 35 °C for 3 minutes; increased at 10 °C/min to 250 °C; kept at 250 °C for 5.5 minutes. Helium served as the carrier gas with a flow of 14 ml/min. For each sample, 100 µl gas in the sampling vial was used.

H_2 was measured on a RGA3 Reduction Gas Analyzer (Trace Analytical, USA) with a 60/80 Unibeads pre-column and a 60/80 molecular sieve 5A column and a reduction gas detector (RGD; detector temperature 265 °C; column temperature 105 °C). The carrier gas was N_2 with a flow rate of 20 ml/min.

Nitrate and sulfate were analyzed by a DX-100 Dionex ion-chromatograph with an Ionpac AS9-SC analytical Column. Manganese, iron, sulfide and nitrite were determined by colorimetric methods. The VFAs were measured on a CP9001 gas chromatograph with a Chrompack FFAP column (i.d. 0.53 mm, length 25 m, film thickness 2 µm) and an electron capture detector (ECD; detector temperature 275 °C).

Calculations. Nominal concentrations (M), which ignored headspace-liquid partitioning equilibrium within bottles, were used to quantify the chlorinated ethenes and their degrading products. According to mass balance and Henry's law, the nominal concentrations were determined as: $M=(C_wV_w+C_gV_g)/V_w= C_w(1+H_cV_g/V_w)$, where C_w, C_g are concentrations of the substance in the water and in the gas (mol/L), respectively; V_w, V_g are volume of water in the bottle and volume of headspace in the bottle (ml), respectively; and H_c is Henry's law constant (dimensionless).

H_2 measured in the gas phase (ppm) was converted to molar units (nM) using a conversion factor derived from the ideal gas law and Henry's law (20 °C): $C_w=C_g/52.4$, where C_w and C_g are the H_2-concentrations in the aqueous phase and in the gas phase, respectively.

RESULTS AND DISCUSSION

Transformation of chlorinated ethenes under different redox conditions. Over the experiment, redox species, H_2 concentrations, and chlorinated ethenes and their degradation products were determined periodically in various enrichment cultures. Table 1 shows transformation results of the studied chlorinated ethenes within fifteen months of operation.

Under denitrifying and manganese reducing conditions, no degradation of the studied chlorinated ethenes was observed. Although theoretically cis-DCE and VC can act as electron donors and have been shown to be oxidized to CO_2 in anaerobic aquifer (Bradley and Chapelle, 1998), this process of anaerobic oxidation was not observed in the current batch experiments.

Complete transformation of PCE to ethene, cis-DCE and TCE occurred in the presence of VFAs, carbon dioxide, and ferric iron, respectively, whereas PCE was slightly transformed to TCE in the presence of sulfate. Similarly, complete transformation of cis-DCE and VC to ethene occurred in the VFAs-enriched cultures, whereas cis-DCE was only partially transformed in the CO_2-enriched and SO_4-enriched cultures. The mechanism involved in all the observed transformations was reductive dechlorination, as indicated by the near stoichiometric formations of reductive products.

TABLE 1. Transformations of chlorinated ethenes in various enrichment cultures (Temperature 20°C)

Electron Acceptors/ Eelectron Donors	PCE	cis-DCE	VC
VFAs	+	+	+
CO_2	+	+/-	-
SO_4	+/-	+/-	-
$Fe(OH)_3$	+	-	-
MnO_2	-	-	-
NO_3	-	-	-
Control	-	-	-

+: complete transformation   +/-: partial transformation   -: no transformation

With VFAs as the electron donors, and under methanogenic condition, complete dechlorination of all chlorinated ethenes to ethene was observed. In the PCE/VFAs culture, sequential dechlorination of PCE via cis-DCE and VC to ethene proceeded, with each step of the dechlorination characterised by a specific H_2 level. Figure 1(a) and 1(b) display the changes of chlorinated ethenes over time and the changes of VFAs and H_2 over time, respectively. There was a succession of preferential use of the individual VFAs, i.e., propionic acid was the first to be used, followed by acetic acid, butyric acid, isobutyric acid and valeric acid. However, isovaleric acid was not degraded till the end of this experiment (data not shown).

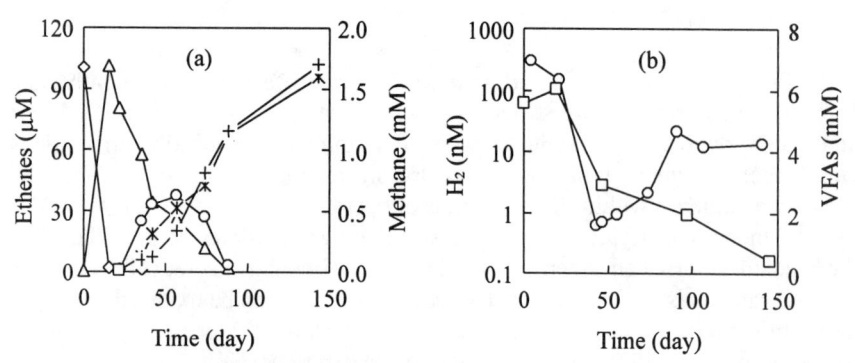

Figure 1. Biodegradation in the PCE/VFAs enrichment culture. (a) Changes of chlorinated ethenes over time. Symbols: ◇-PCE; □-TCE; △-cis-DCE; ○-VC; ✕-ethene; +- methane.   (b) Changes of dissolved H_2 and VFAs over time. Symbols: ○-H_2; □-VFAs.

In the PCE/CO$_2$ and PCE/Fe(OH)$_3$ cultures, dechlorination of PCE stopped at cis-DCE and TCE, respectively. In both cases, the dissolved H$_2$ concentrations remained below 2 nM (data not shown). Iron reduction proceeded actively in the PCE/Fe(OH)$_3$ culture, corresponding to a steady-state H$_2$ level of 0.4 nM. As high as 8.0 mM Fe (II) was detected on day 70. Clearly, the cis-DCE dechlorinators might be out-competed for H$_2$ by iron-reducing bacteria. In contrast, no significant methane production occurred in the PCE/CO$_2$ culture, probably due to electron donor limitation of the methanogenic bacteria. The H$_2$ in the PCE/NO$_3$ and PCE/SO$_4$ cultures stabilized at about 0.3 nM and 4.0 nM, respectively. The processes of NO$_3$-reduction and SO$_4$-reduction were confirmed by the detections of nitrate (0.31 mM) and sulfide (0.62 mM). Nevertheless, no obvious steady-state H$_2$ was observed in the PCE/MnO$_2$ culture. The measured 0.46 mM Mn (II) on day 90 and 0.48 mM Mn(II) on day 160 suggested that Mn-reduction occurred during the early stage and was inhibited by other TEAPs thereafter.

Dechlorination Kinetics. Assuming all biological reactions follow first-order kinetics, the dechlorination rate constants (K-values) and half-lives ($t_{0.5}$) of chlorinated ethenes in various enrichment cultures were calculated, as summarized in Table 2. Lag phases (T) of the dechlorinations were also listed. Dechlorination kinetics for the PCE/SO$_4$ and cis-DCE/SO$_4$ enrichment cultures were not calculated, for only slight transformations of PCE and cis-DCE were observed and the data used for the calculation were limited.

TABLE 2. Dechlorination rates of chlorinated ethenes in various enrichment cultures (Temperature 20°C)

Enrichment Culture	Dechlorination	T (day)	K (/day)	$t_{0.5}$ (day)
PCE/VFAs	PCE→cis-DCE	-	0.571	1.2
	cis-DCE→VC	-	0.033	21.3
	VC→ethene	-	0.023	30.5
PCE/CO$_2$	PCE→cis-DCE	~14	0.308	2.3
PCE/Fe(OH)$_3$	PCE→TCE	-	0.261	2.7
cis-DCE/VFAs	cis-DCE→VC	~42	0.008	85.0
	VC→ethene	-	0.100	6.9
cis-DCE/CO$_2$	cis-DCE→VC	~138	0.005	142.6
VC/VFAs	VC→ethene	~186	0.024	29.3

-: dechlorination occurred with no lag phase

Two features can be seen from Table 2. First, the less chlorinated ethenes like cis-DCE and VC exhibited dechlorination rates approximately ten times lower than those of PCE; and second, the dechlorination rate of PCE in the presence of VFAs was significantly higher than that in the absence of VFAs.

Clearly, under similar circumstances, the dechlorination rates of chlorinated ethenes were greatly affected by their oxidation forms. PCE is most susceptible to reductive dechlorination process because it is in a highly oxidized form. In contrast, cis-DCE and VC are less oxidized and less readily subject to reduction. The fact that the dechlorination rates of PCE in the PCE/CO$_2$ and PCE/Fe(OH)$_3$ cultures were half of those in the PCE/VFAs batch indicates that electron donors stimulated the dechlorination process. The cessation of dechlorination at cis-DCE and TCE in the PCE/CO$_2$ and PCE/Fe(OH)$_3$ cultures, respectively, is likely due to the lack of electron donors.

H$_2$ Threshold Concentrations. The H$_2$ measurements from all the enrichment cultures, are summarized in Figure 2 in a Box & Whisker plot of apparent steady-state H$_2$ concentrations obtained under different TEAPs.

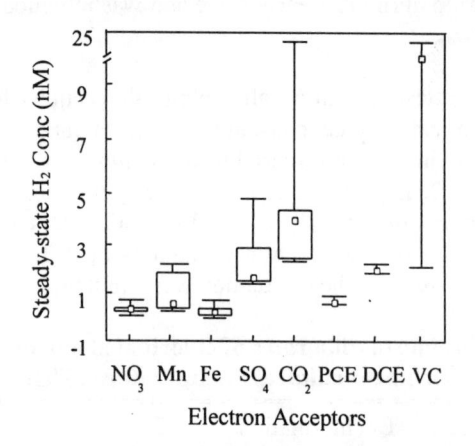

FIGURE 2. Box & Whisker plot of the steady-state H$_2$ concentrations for various TEAPs
Symbols: ⊤ Min-Max   ▢ 25%-75%   ▫ Median value

The H$_2$ threshold concentrations obtained in this study were similar to the reported values of 5-30 nM for methanogensis, 1-4 nM for sulfate reduction, and 0.2-0.8 nM for iron reduction (Chapelle et al., 1997). However, for denitrification and manganese reduction, the observed H$_2$ levels were higher than the reported values of less than 0.1 nM. In the current study, TEAPs with ferric iron, manganese oxide and nitrate as the terminal electron acceptors exhibited H$_2$ threshold values close to those of PCE/TCE dechlorination. Similar results were recently obtained by Loffler et al. (1999). In contrast, cis-DCE and VC dechlorinations exhibited significantly higher H$_2$ threshold values, in the range of sulfate reduction and methanogenesis. Indeed, VC dechlorination only proceeded under methanogenic conditions, whereas PCE dechlorination also occurred under less reduced conditions like iron-reducing batches. These latter results differ with the findings of Loffler et al. (1999), but may well explain why on many locations PCE dechlorination ceases at cis-DCE and VC.

CONCLUSIONS

Batch experiments indicated that chlorinated ethenes were reductively dechlorinated to less or non-chlorinated products, particularly under methanogenic conditions. The addition of electron donors such as volatile fatty acids stimulated the complete dechlorination process to ethene. Lack of electron donors restrain dechlorination, resulting in accumulation of TCE, cis-DCE or VC. The dechlorination rates of cis-DCE and VC were ten times lower than those of PCE. TEAPs with nitrate, manganese oxide, and ferric iron as the terminal electron acceptors exhibited H_2 threshold values similar to those observed in PCE/TCE dechlorination, whereas cis-DCE and VC dechlorinations exhibited H_2 thresholds in the range of sulfate reduction and methanogenesis, respectively.

ACKNOWLEDGEMENTS

This project was a co-operation between Peking University and TNO (Netherlands Organisation for Applied Scientific Research) and was financially supported by the Royal Netherlands Academy of Arts and Science, and Chinese Major State Basic Research Development Program: Study on the Environmental Pollution Mechanisms and Control Theory (G1999045711).

REFERENCES

Bagley, D. M., and J. M. Gossett. 1990. "Tetrachloroethene Transformation to Trichloroethene and Cis-1,2-dichloroethene by Sulfate-reducing Enrichment Cultures." *Appl. Enviorn. Microbiol.* 56(8): 2511-2516.

Borger, A. R., J. Gerritse, R. F. W. Baartmans, H. Slenders, and T. N. P. Bosma. 1998. *Stimulation of PCE Dechlorination in NEPROMA Batch Cultures.* TNO-report, R98/354.

Bradley, P. M., and F. H. Chapelle. 1998. "Microbial Mineralization of VC and DCE under Different Terminal Electron Accepting Conditions." *Anaerobe* 4(2): 81-87.

Chapelle, F. H., .K. Haack, P. Adriaens, M. A. Henry, and P. M. Bradley. 1996. "Comparison of E_h and H_2 Measurements for Delineating Redox Processes in a Contaminant Aquifer." *Environ. Sci .Technol.*, 30(12): 3565-3569.

Chapelle, F. H., P. M. Vroblesky, J. C., Woodward, and D. R., Lovley. 1997. "Practical Consideration for Measuring Hydrogen Concentrations in Ground-water." Environ. Sci. Technol, 31(10): 2873-2877.

Christiansen N., S. R. Christensen, E. Arvin, and B. K. Ahring. 1997. "Transformation of tetrachloroethene in an upflow anaerobic sludge blanket reactor." *Appl. Microbiol. Biotechnol.*, 47 (1): 91-94

Chu K. H., and W. J. Jewell. 1994. "Treatment of tetrachloroethylene with anaerobic attached film process." *J.Environ.Eng.*, 120 (1): 58-71.

Garant H., and L. R. Lynd. 1996. "Perchloroethylene utilization by methanogenic fed-batch cultures: Acclimation and degradation." *Appl. Biochem. Biotechnol.*, 57-8: 895-904.

Jakobsen, R., H. J. Albrechtsen, M. Rasmussen, H. Bay, P. L. Bjerg, and T. H. Christensen. 1998. "H_2 Concentrations in a Landfill Leachate Plume (Grindsted, Denmark): In situ Energetics of Terminal Electron Acceptor Processes." *Environ. Sci.Technol.* 32(14): 2141-2148.

Loffler, F. E., J. M., Tiedje, and R. A. Sanford. 1999. "Fraction of Electrons Consumed in Electron Acceptor Reduction and Hydrogen Thresholds as Indicators of Halorespiratory Physiology." *Appl. Enviorn. Microbiol.*, 65(9): 4049-4056.

Lovley, D. R., and S. Goodwin. 1988. "Hydrogen Concentrations as an Indicator of the Predominant Terminal Electron-Accepting Reactions in Aquatic Sediments." Geochimica et cosmochimica acta. 52: 2993-3003.

Lovley, D. R, F. H., Chappelle, and J. C. Woodward. 1994. "Use of Dissolved H_2 Concentrations to Determine Distribution of Microbially Catalyzed Redox Reactions in Anoxic Groundwater." Environ. Sci. Technol. 28(7): 1205-1210.

Meer, J. T., and A. J. C. Sinke. 1998. *The Importance of Geochemistry in Biodegradation of Chloroethenes.* TNO-report. TNO-MEP-R 98/251.

Wiedemeier, T. H., M. A. Swanson, D. E. Moutoux, E. K. Gordon, J. T. Wilson, B. H. Wilson, D. H. Kampbell, P. E. Haas, R. N. Miller, J. E. Hansen, and F. H. Chapelle. 1998. *Technical Protocol for Evaluating Natural Attenuation of Chlorinated Solvents in ground water.* EPA/600/R-98/128.

Vogel, T. M., and P. L. McMarty. 1985. "Biotransformation of Tetrachloro-ethylene to Trichlrorethylene, Dichloroethylene, Vinyl chloride, and Carbon Dioxide under Methanogenic Conditions." *Appl. Enviorn. Microbiol.* 49(5): 1080-1083.

Yang, Y., and P. L. McCarty. 1998. "Competition for Hydrogen within a Chlorinated Solvent Dehalogenating Anaerobic Mixed Culture." Environ. Sci. Technol. 32(22): 3591-3597.

DIFFERENTIAL STIMULATION OF HALOREDUCTION BY CARBON ADDITION TO SUBSURFACE SOILS

Matthew A. Panciera[*], Olga Zelennikova[#], Barth F. Smets[*,#], Gregory M. Dobbs[&]
[*]Environmental Engineering Program, [#]Microbiology Program
University of Connecticut, Storrs, Connecticut
[&]United Technologies Research Center, East Hartford, Connecticut

ABSTRACT: A study was performed to evaluate whether soil from a bulk aerobic aquifer, historically contaminated with chlorinated ethenes (CEs), could be made anaerobic by carbon source addition to enhance reductive dehalogenation. We hypothesized that sulfate-reducing bacteria (SRBs) were responsible for dechlorination in this soil. Microcosms were constructed to examine the effects of various electron donors on reductive dehalogenation. Six electron donor treatments were compared to include acetate, propionate, lactate, ethanol, vegetable oil (soy oil), and hydrogen release compound (HRC™, Regenesis). Dehalogenation was observed in all microcosms regardless of the carbon source used. Perchloroethene (PCE) and trichloroethene (TCE) were quickly reduced to *cis*-1,2-dichloroethene (*cis*-1,2-DCE) when acetate, propionate, ethanol, or HRC™ was the carbon source. Microbial community structure profiles based on 16s rDNA indicated the stimulation of different microbial communities depending upon the carbon source added. Results from chemical analysis revealed the importance of SRBs to dehalogenation when ethanol, lactate, vegetable oil, or HRC™ was the electron donor. When acetate or propionate was the carbon source, dehalogenation occurred irrespective of whether SRBs were present.

INTRODUCTION

CEs are ubiquitous contaminants in aquifers at sites with historic industrial use. The half-lives of CEs in the environment depend on local biogeochemistry, but are often very long. While lower substituted ethenes can be oxidized under aerobic conditions, more highly substituted ethenes such as PCE and TCE are susceptible to microbial reductive dehalogenation and halorespiration under anaerobic conditions. Dehalogenating bacteria require access to some type of electron donor and acceptor. The electron donor is typically formed through the fermentation of organic substrates. The electron acceptor can be the halogenated compound (halorespiration) or another physiological terminal electron acceptor (cometabolic dehalogenation). For example, the ability to dechlorinate CEs cometabolically is possible by methanogenic or sulfate reducing bacteria that contain the acetyl-CoA pathway [1].

In order to understand the key microbial groups required for reductive dehalogenation, community DNA profiles based on Terminal Restriction Fragment Length Polymorphism (T-RFLP) analysis can be compared [2]. In this method, the total community 16S rDNA gene is amplified with universal oligonucleotide primers. After subsequent restriction digest and resolution of 16S rDNA fragments, the resulting electropherogram depicts the distribution and abundance of restriction fragments as a function of size. Fragments are consistent among a phylogenetically close group of organisms and some inferences regarding community structure can be made from this analysis.

The objective of this study was to evaluate whether soil from a bulk aerobic aquifer, historically contaminated with CEs, could be made anaerobic by carbon source addition to enhance dehalogenation. The aquifer is adjacent to a tidal estuary and wetland. In marine or brackish aquifer systems, sulfate is typically a dominant terminal electron acceptor. We therefore hypothesized that SRBs were responsible for dechlorination in this soil. Using soil and water from the aquifer, microcosms were constructed to examine the effects of various electron donors on reductive dehalogenation. Six electron donor treatments were compared; they included acetate, propionate, lactate, ethanol, vegetable oil (soy oil), and hydrogen release compound (HRC™, Regenesis). An SRB-inhibited control (Na$_2$MoO$_4$) and an abiotic control (NaN$_3$) were included with each treatment. Concentrations of CEs in the aqueous phase were measured via headspace analysis and GC/MS detection. T-RFLP analysis was employed using a primer set targeted for the total bacterial community in order to compare microbial community differences between carbon source treatments. The effect of SRBs on microbial community structure was examined by comparing T-RFLP profiles from SRB-inhibited and non-inhibited microcosms. Considering that each set of microcosms was fed a different carbon source, we expected a difference in microbial community structure. Combined with the data on CE degradation, we sought a relationship between microbial community structure and its dechlorinating ability. Results of CE dehalogenation are shown and discussed for a period of one year of operation.

METHODS AND MATERIALS

Chemicals. The following chlorinated compounds were used in this study: perchloroethene (PCE), trichloroethene (TCE), *cis*-1,2-dichloroethene (*cis*-1,2-DCE), *trans*-1,2-dichlorothene (*trans*-1,2-DCE), 1,1-dichloroethene (1,1-DCE), and vinyl chloride (VC). As carbon sources we used sodium acetate, sodium propionate, sodium lactate, ethanol, vegetable oil (soy oil), and Hydrogen Release Compound™ (HRC™). The vegetable (soy) oil was a commercial product (Wesson 100% Pure Vegetable Oil). HRC™ was supplied by Regenesis (Cinnaminson, NJ).

Soil and Water. Soil and water for this study were obtained from Geoprobe™ sampling at two test locations in a sandy/gravely overburden aquifer adjacent to a tidal wetland and estuary. The two sampling locations were located directly next to previously installed monitoring wells. The two test locations are thus referred to as monitoring well 89 (MW89) and monitoring well 101 (MW101). A soil core was taken from MW89 at a depth of 48 to 52 ft. The soil core taken from MW101 was taken at a depth of 28 to 32 ft. Three gallons of groundwater were collected from depths of 60 and 31 ft. at MW89 and MW101, respectively. Before use, aqueous samples were refrigerated at 4°C under a nitrogen headspace in the sampling bags.

Determination of Carbon Source Additions. Given the concentration of PCE, TCE, and *cis*-1,2-DCE in groundwater from MW101, the total milli-equivilents of electrons per liter (meq e$^-$/L) needed to totally reduce all CEs was determined at 0.602 meq e$^-$/L. This value was multiplied by 100 as a safety factor, and then converted to the grams carbon source needed (based on e$^-$ eq/g carbon source). For water from MW101 the various carbon source concentrations were calculated to be 0.444 g/L acetate, 0.314 g/L propionate, 0.447 g/L lactate, 0.231 g/L ethanol, 0.167 g/L vegetable oil, and 0.389 g/L HRC™.

Besides CEs, oxygen and sulfate are two other major electron acceptors in this aquifer. The calculated total milli-equivilents of each carbon source that was to be added to the microcosms was compared to the oxygen and sulfate concentrations in MW101 groundwater. Analysis on water from MW101 revealed the sulfate concentration to be 28.6 mg/L. Since this is an aerobic aquifer, the oxygen concentration was assumed to be approximately 8 mg/L. When these two other electron acceptors are considered, it is evident that they are negligible when compared to the 100-fold excess of electron donor added. Sulfate concentrations in the microcosms were not measured during this study.

Microcosm Set-up. The set of microcosms for each carbon source treatment consisted of three replicates, an abiotic control, and a sulfate reduction inhibited control. The microcosms consisted of 100 ml nominal volume serum bottles sealed with a teflon-lined butyl rubber stopper and an aluminum crimp seal. To each bottle in a treatment set, the following ingredients were added at the beginning of the study: soil from the MW101 location, the appropriate carbon source at the required concentration, sodium phosphate for a desired final concentration of 5 mM to provide pH buffering, and 90.0 ml of ground water from MW101. To each abiotic control, sodium azide (NaN$_3$) was added to achieve a final concentration of 0.005 g/ml and then autoclaved twice. For the no-sulfate-reducing bacteria (NSRB) control, sodium molybdate (Na$_2$MoO$_4$) was added at a final concentration of 20 mM. The soil mass added to each microcosm was approximately 20 g. The soil mass was slightly adjusted so the final total (soil plus liquid) volume in the serum bottle was 100 ml. The mean headspace volume for the microcosms was 20.93 (±0.43) ml. The serum bottles were crimp sealed.

Carbon source was added only once at the beginning of the study. Since it was added in a 100-fold excess, it was assumed that residual carbon source would remain in the microcosms for the duration of the study. Electron donor concentrations were not measured at any time during the duration of the study.

All bottles were incubated upside down on a shaker table at 150 rpm allowing water to septum contact and minimizing loss of VOCs from the headspace [3] and incubated in a 25°C constant temperature room. Initial volatile organic compound (VOC) measurements in the headspace were taken after 24 hrs. of incubation. After 4 months of operation, yeast extract was added to duplicate microcosms of each carbon addition treatment and to the controls (replicate 2 and 3, abiotic control, and SRB inhibited control) to a final concentration of 10 mg/L in order to try and stimulate any relevant fastidious microorganisms that were limited by some growth factors.

Measurements by Gas Chromatograph/Mass Spectrometer (GC/MS). The sampling method for the microcosms [3] was performed by first purging a gas-tight valved syringe (Hamilton, Reno, NV) with high purity nitrogen (N$_2$) three times and taking a 500 µl gas sample. The sample was immediately injected in the manual sampling port of the GC/MS system for analysis (Varian Saturn 3 GC/MS, Walnut Creek, CA). The approximate detection limit for all compounds was 1 µg/L aqueous.

T-RFLP Analysis. DNA from 1ml aqueous samples was extracted using the UltraClean™ Soil DNA Kit (Mo Bio Laboratories, Inc., Solana Beach, CA). An aliquote of 1 µl was used as a template for subsequent PCR reactions as described previously [4]. In total, three 100 µl PCR reactions were set for each DNA sample. Final triplicate PCR products were combined and purified with Wizard® PCR Preps DNA Purification System (Promega, Madison, WI) and eluted in 50 μl of milli-Q water. All samples were dried and resuspended in 10 μl of milli-Q water prior the digest. Digest mixtures (10 μl DNA sample, 5 U/μl *MspI*, 1x buffer, milli-Q water) were incubated for 3.5 hrs at 37°C followed by 10 min at 65°C for enzyme inactivation. The resulting digests were run in an ABI sequencer in Gene Scan mode under specified conditions modified for ABI377XL [2].

RESULTS

Water and soil from both MW89 and MW101 were analyzed for VOCs. Because groundwater from MW101 already contained some evidence of reductive dehalogenation (traces of *cis*-1,2-DCE), soil and groundwater from MW101 was used in subsequent microcosm studies. Initial CE concentrations in MW101 water were: 579 μg/L PCE, 2957 μg/L TCE, and 127 μg/L *cis*-1,2-DCE.

Results from microcosm studies showed varying profiles for individual CEs as a percentage of the total CEs present. A mass balance at the conclusion of the study accounted for an average of 84% of the initial CE mass. Reduction of PCE and TCE to *cis*-1,2-DCE was observed with the addition of all carbon sources. Notably, when acetate, propionate (Figure 1), ethanol, or HRC™ were added, PCE and TCE concentrations quickly dropped (~20 days) and *cis*-1,2-dichloroethene (*cis*-1,2-DCE) was formed. With the lactate (Figure 3) and vegetable oil treatments, *cis*-1,2-DCE was formed only after a considerable lag period (~60 days). Time from onset of reduction to completion (where *cis*-1,2-DCE essentially reached 100% of the CEs in the microcosm) ranged from about 20 days for ethanol to about 82 days for lactate. The set of microcosms using vegetable oil as a carbon source did not reach 100% conversion to *cis*-1,2-DCE by 120 days. Upon examination of the microcosms one year later, 100% conversion to *cis*-1,2-DCE was achieved in the vegetable oil microcosms as well.

Throughout the study, distribution of CEs in all abiotic controls remained constant indicating that PCE and TCE were not abiotically dehalogenated to *cis*-1,2-DCE. CE distributions in the NSRB controls remained constant in the lactate (Figure 4), ethanol, vegetable oil, and HRC™ treatment bottles but changed in the acetate and propionate (Figure 2) bottles.

In the acetate NRSB control bottle, dehalogenation began at about 5 days and continued until day 20 when all the CEs were converted to *cis*-1,2-DCE. In the propionate NRSB control bottle (Figure 2), dehalogenation started at day 5 and continued until day 40 when all the CEs were converted to *cis*-1,2-DCE. For both the acetate and propionate treatments, dehalogenation was faster in the NSRB control bottle than in the three uninhibited replicates.

For all carbon sources, PCE and TCE concentrations dropped to below 5 μg/L and 10 μg/L respectively while *cis*-1,2-DCE concentrations increased to about 2,000 μg/L. Depending on the carbon source added, *cis*-1,2- DCE concentrations increased and surpassed TCE concentrations at times. Where transformation had occurred, for all treatments there was no reduction past *cis*-1,2-DCE, evidenced by its concentration remaining near steady until the end of the study (ethene, ethane, and methane were not measured in this study).

At the end of one year, low, yet detectable, levels of vinyl chloride were evident in the vegetable oil and HRC™ treatments. Some low level traces of *trans*-

Laboratory Studies of Reductive Processes

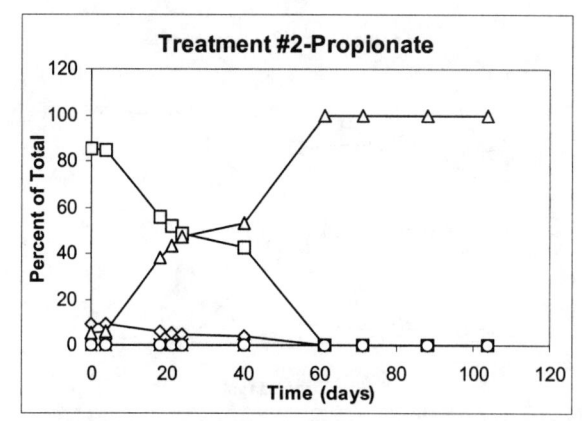

FIGURE 1. Average % CEs in three Non-inhibited Replicates

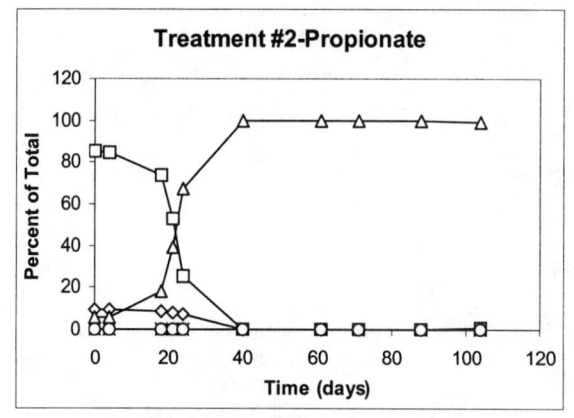

FIGURE 2. % CEs in SRB-inhibited Microcosm

- ◇ Perchloroethene
- □ Trichloroethene
- △ cis-1,2-
- ✕ trans-1,2-
- ✻ 1,1-Dichloroethene
- ○ Vinyl Chloride

1,2-DCE and 1,1-DCE were also detected in some of the microcosms. After one year, the microcosm vials were sacrificed. CEs were extracted from the soil with methanol and mass values obtained were compiled with CE masses found in the aqueous and gaseous phases. A mass balance was performed and an average of 84% of the initial CE mass from the beginning of the study could be accounted for. There were no significant differences between replicates and inhibited controls.

T-RFLP profiles of total bacterial communities were similar for the acetate and propionate treatments (Figure 5) and for the lactate and HRC^{TM} treatments. T-RFLP profiles from microcosms not treated with yeast extract were markedly different from those treated with yeast extract. T-RFLP profiles obtained for the lactate, vegetable oil, and HRC^{TM} treatments indicate that significantly less peaks are present in microcosms not treated with yeast extract.

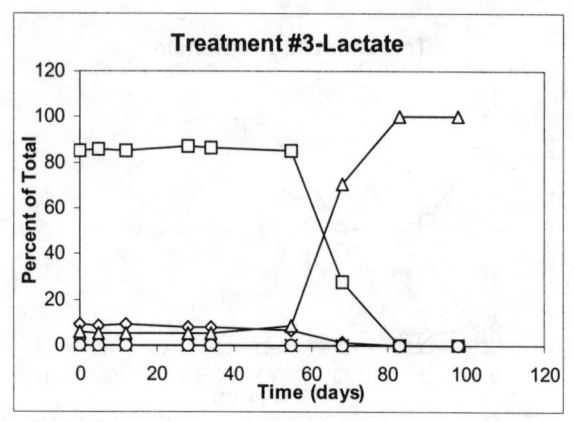

FIGURE 3. Average % CEs in three Non-inhibited Replicates

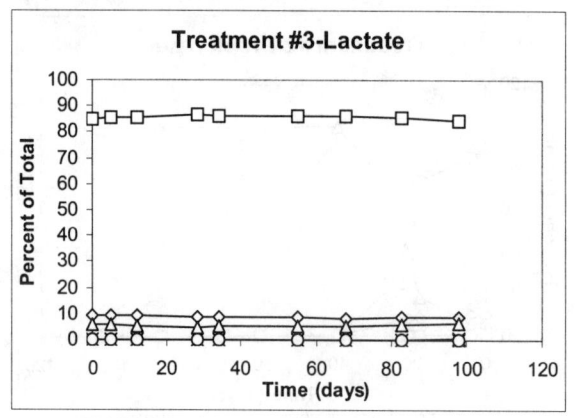

FIGURE 4. % CEs in SRB-inhibited Microcosm

Symbol	Compound	Symbol	Compound
◇	Perchloroethene	✕	trans-1,2-
□	Trichloroethene	✱	1,1-Dichloroethene
△	cis-1,2-	○	Vinyl Chloride

T-RFLP profiles for microcosms after 6 month of incubation (data not shown) indicate that inhibition of sulfate reduction causes a dramatic shift in the structure of the HRC™ spiked bacterial community. New phylogenetic groups appear in the absence of sulfate reduction.

Although significant amounts of DNA were extracted at the end of experiment (9 month of incubation) in the NSRB controls, no 16S rDNA amplicon was obtained after PCR with eubacterial primers.

DISCUSSION

It is apparent from the chemical profiles that reduction of PCE and TCE to *cis*-1,2-DCE occurred in the microcosms and is consistent with microbial reductive dehalogenation [5]. In some instances the reduction to *cis*-1,2-DCE began relatively

Laboratory Studies of Reductive Processes 75

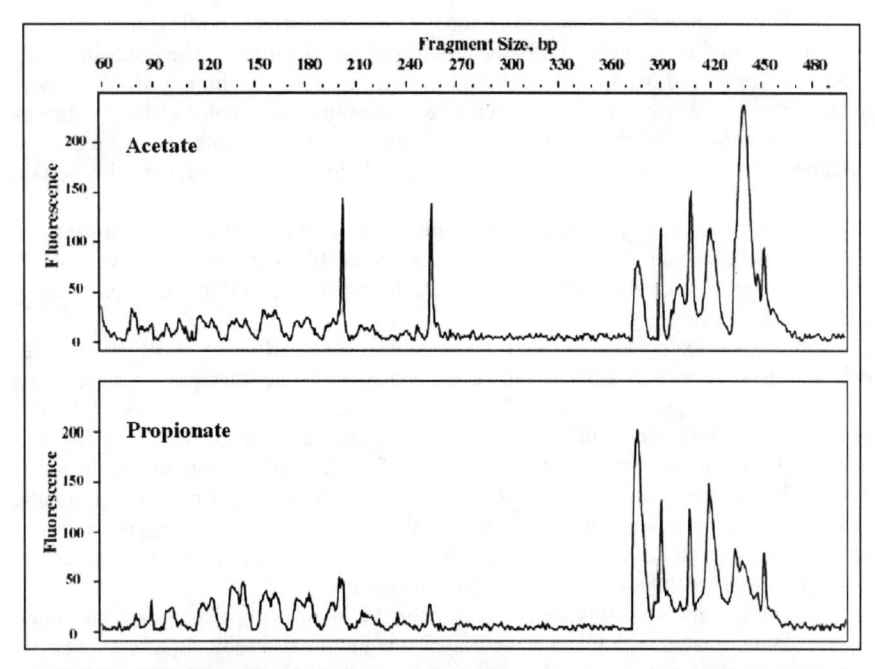

FIGURE 5. T-RFLP profiles of microbial communities grown on acetate and propionate after a 9 month of incubation. Yeast extract was added after 4 months of incubation.

quickly. In the case of acetate, propionate, ethanol and HRC™, reduction occurred within approximately 10 days. Reduction of PCE and TCE in the lactate and vegetable oil treatments started only after a lag period (~55 days). This may indicate some sort of mass transfer limitation of the carbon source or a biochemical adaptation that was required within the microbial community in order to use lactate or vegetable oil as a source of energy and reducing equivalents.

CE distributions for the NSRB control bottles revealed that reductive dehalogenation was absent for the lactate, ethanol, vegetable oil, and HRC™ treatments. CE distributions remained constant over the duration of the study in the NSRB controls, while they changed in the uninhibited treatments indicating that sulfate-reducing microorganisms are required for reduction of CEs with lactate, ethanol, vegetable oil, and HRC™ as the carbon source. Reductive dehalogenation occurred in the NSRB control bottles supplemented with acetate and propionate (Figure 2) indicating that reductive dehalogenation using acetate or propionate as a carbon source is not contingent on the presence of sulfate-reducing microorganisms. In fact, when SRBs are inhibited, more electrons may be shuttled to dechlorination, thus speeding up dehalogenation.

One must also consider that since sodium molybdate is only an inhibitor of sulfate reduction at the initial steps where sulfate is reduced to sulfite, it is possible that the SRBs can switch form sulfate reduction to a fermentative lifestyle possible utilizing propionate or lactate when sulfate is depleted. At this time, the correlation between sulfate and dehalogenation in this aquifer is not fully understood and further research is required.

A black precipitate formed in most of the microcosm bottles indicating the reduction of sulfate to sulfide and precipitation of metal sulfides. The contents of the serum bottles turned from an initial light brown color at the beginning of the study to a dark gray or black color over the period of 2 months. This color change indicates sulfate reduction in the absence of oxygen strengthening our hypothesis that reductive dehalogenation was cometabolically performed by sulfate reducers in certain treatments.

The constancy of the CE distribution in the abiotic controls indicates that no reductive dehalogenation occurred. The absence of black precipitate in the NSRB controls suggests that the sodium molybdate treatment was effective at prohibiting sulfate-reducing activity.

As we expected, bacterial communities had different T-RFLP profiles depending on source of carbon. However, some of the peaks representing distinct phylogenetic groups were shared by two or more sets of microcosms. Similar profiles were observed with acetate and propionate, as well as with HRC™ and lactate. This is consistent with the fact that HRC™ is a poly-lactate molecule and is eventually converted to lactate. Inhibition of sulfate reduction caused a dramatic change in the structure of the HRC™ bacterial community. New phylogenetic groups appeared in the absence of sulfate reduction, although chemical analysis suggests that those groups do not mediate reductive dehalogenation.

The results from this study are conclusive in showing that a carbon-limited aquifer contaminated with CEs can be transformed to an anaerobic aquifer expressing reductive dehalogenation by carbon source addition. Of the six carbon sources used, acetate, propionate, ethanol, and HRC™ seem to stimulate dehalogenation quickly while lactate and vegetable oil require a substantial lag period. Acetate, propionate, and HRC™ appear to facilitate slow and steady dechlorination while ethanol allowed the quick transformation of PCE and TCE. The indigenous microorganisms in this soil exhibited the ability to reductively dechlorinate PCE and TCE to *cis*-1,2-DCE and it was found that SRBs play an important role in dechlorination when lactate, ethanol, vegetable oil, or HRC™ is the carbon source. Future work is required to further define the role of SRBs in contaminated anaerobic environments where reductive dehalogenation is occurring.

REFERENCES
1. Mohn, W. W. and J. M. Tiedje. 1992. "Microbial Reductive Dehalogenation." *Microbiol. Rev.* 56: 482-507.
2. Liu, W. T., Marsh, T. L., Cheng, H. and Forney, L. J. 1997. "Characterization of microbial diversity by determining terminal restriction fragment length polymorphisms of genes encoding 16S rRNA." *Appl. Environ. Microbiol.* 63(11): 4516-22.
3. Gossett, J. M. 1987. "Measurement of Henry's Law Constants for C_1 and C_2 Chlorinated Hydrocarbons." *Environ. Sci. Technol.* 21(2): 202-208.
4. More, M. I., Herrick ,J. B. , Silva M.C., Ghiorse W.C., Madsen E. L. 1994. "Quantitative Cell Lysis of Indigenous Microorganisms and Rapid Extraction of Microbial DNA from Sediments." *Appl. Environ. Microbiol.* 60(5): 1572-1580.
5. Freedman, D. L. and J. M. Gossett. 1989. "Biological Reductive Dechlorination of Tetrachloroethylene and Trichloroethylene to Ethylene under Methanogenic Conditions." *Appl. Environ. Microbiol.* 55(9): 2144-2151.

INTERACTION OF MICROBES AND URANIUM DURING ENRICHMENT FOR TCE REDUCTIVE DECHLORINATION

Brady D. Lee, Michelle R. Walton and Jodette L. Meigio
Idaho National Engineering and Environmental Laboratory, Idaho Falls, Idaho

ABSTRACT: Laboratory research was performed to understand the effect of anaerobic enrichments on the mobility of uranium in basaltic aquifers. Trichloroethylene (TCE) was used as a co-contaminant in the research. Interaction of uranium and hydrous iron oxide moieties on the mineral oxide surfaces was maximized by using iron oxide coated quartz from Oyster, VA. Results showed that adsorption and desorption of U(VI) to iron oxide coated sand appears to be affected by anaerobic enrichments and is culture dependent. Laboratory grown consortium enriched with or without sulfate caused the release of low concentrations of adsorbed uranium into the growth medium, followed by a steady decrease in the soluble fraction with a concomitant increase in particles filterable by a 0.2 µm filter. No chemical reduction of U(VI) was noted using the laboratory consortium. Bacteria from groundwater taken from the Test Area North (TAN) site at the Idaho National Engineering and Environmental Laboratory (INEEL) enriched with or without sulfate were able to chemically reduce the iron on the surface of the sand which may decrease the reactive surface area on mineral surfaces significantly affecting adsorption of uranium and subsequent mobility in the groundwater. Enrichments from the TAN groundwater demonstrated reduction of U(VI) to less reduced forms whether enriched with or without sulfate.

INTRODUCTION

Weapons production and energy research activities have resulted in widespread contamination of subsurface sediments and groundwater with chlorinated organics, metals and radionuclides. More than 5,700 known waste plumes have contaminated over 600 billion gallons of water and 50 million cubic meters of soils and sediments at DOE sites. Some of the more prevalent contaminants that have been co-disposed at DOE sites are chlorinated solvents (e.g. perchloroethylene (PCE) and TCE), metals (e.g. lead, chromium and mercury) and radionuclides (e.g. uranium, cesium and strontium) (Riley, et al., 1992). These contaminants, especially the metals and radionuclides, can persist in the environment for extremely long periods. Migration of contaminants within these plumes threatens local and regional water sources, and in some cases has already adversely impacted off-site receptors (DOE, August 1996).

Over the past several years, biological treatment of chlorinated organic contaminants has gained popularity as a treatment technology (Ellis et al. 2000; Harkness et al. 1999). When considering biotreatment of the chlorinated organic component of a mixed contaminant plume in saturated porous media, a number of related chemical and biological interactions must be considered. First, and most obvious, is the interaction of the microbial population of interest with the contaminant to be remediated. A large amount of fundamental and applied knowledge has been gained regarding reductive dechlorination of chlorinated organic contaminants (Adriaens and Vogel, 1995; Fetzner, 1998). Second, is the interaction of codisposed inorganic contaminants, such as metals and actinides within the contaminant plume. A wide variety of laboratory research has gone into understanding the partition coefficients and surface complexation of compounds such as uranium on the surface of mineral oxides (Waite et al. 1994; Kohler et al. 1996; Reich et al. 1998; Rosentreter et al. 1998; Bargar et al. 1999). Finally, the interaction of the microbial population enriched for degrading the chlorinated organic contaminant with actinide contamination, as well as the mineral oxides making up the growth environment, must be determined.

Microorganisms may affect the mobility of the actinide directly by changing the speciation, or indirectly by transforming the mineral matrix to which the metal has complexed. Once the actinide in its ionic form has been released it is available for complexation to counter ions in solution, allowing for mobilization of the actinide. During the reduction of iron oxides and liberation of soluble ferrous iron, potential sites for adsorption may also be lost.

The purpose of this research was to determine the effect of anaerobic microbial populations on uranium adsorption to and desorption from mineral oxides. TCE was used as a co-contaminant for the research.

MATERIALS AND METHODS

Culture Selection. Since the project was designed to look at the effect of stimulating microbial populations for the purpose of TCE degradation on uranium adsorption and desorption, a known TCE degrading consortium was sought for the research. The TAN site at the INEEL is an evaluation site for the enhanced biodegradation of chloroethenes in deep fractured basalt. The point source of the plume is currently being remediated by injecting sodium lactate to stimulate the indigenous microbial population to degrade the contaminants.

There were two sources of inoculum from TAN that were used for the experiments that were performed for the current research. Initial experiments were run using a microbial culture enriched from TAN groundwater that had been used to determine TCE removal kinetics. For presentation of results this culture will be called the "Laboratory Consortium." The second culture used was enriched from groundwater obtained from sampling well TAN-37. A 1-Liter sample of groundwater was filtered and the filter was added to growth medium containing lactate, sulfate, and iron oxide coated sand for enrichment. This culture was designated as the "Field Consortium" for presentation of results.

Iron Oxide Coated Quartz. The iron oxide coated sand used for testing was taken from the Oyster site, which is a sand borrow pit on the Delmarva Peninsula in Virginia, near the village of Oyster. While the surface area, extractable iron and aluminum were quite variable, the average values were 1.17 m^2/g, 7.93 µmol Fe/g and 21 µmol Al/g (Rosentreter et al. 1998). The average partition coefficient of the sand for uranium at pH 5 was 1.47 ml/g. While there was no correlation between uranium sorption and bulk mineralogy, there was a significant correlation noted between the uranium sorptive capabilities and the surface area and extractable metals. The ratio of sand used in each experiment was 0.5 g of sand per 25 ml of growth medium.

Uranium Adsorption/Desorption Testing. Testing was performed in 250 ml amber vials closed to the atmosphere using Mininert™ valves. Approximately 4.8 g of iron oxide coated quartz was added to each vial. For each leg of the experiment, five treatments were set up; one blank, one blank plus TCE, a killed control, an enrichment containing sulfate (40 mg/L) and an enrichment without sulfate. A nitrogen phosphorus (N/P) growth medium containing 1 mM ammonium nitrate and 1mM phosphate was used as the testing matrix. Resazurin was used as a redox indicator in the growth medium. Lactate at a final concentration of 1,000 mg/L was used as the carbon and energy source for growth and was included in all legs of the experiment except the blanks. Uranium at a concentration of 1 mg/L and TCE at a concentration of 5 mg/L were used as co-contaminants during testing. Enrichment vials were inoculated with the cultures described above, while the killed control was inoculated with the same cultures that had been autoclaved. The initial pH for each experiment was approximately 7.0.

Sampling and Analysis. U(VI) and total uranium were measured using kinetic phosphorescence analysis (KPA) and inductively coupled plasma atomic absorption spectrometry (ICP-AA), respectively. Concentrations of total U(VI) were determined following acidification of an untreated sample while soluble U(VI) was determined by filtering an aliquot through a 0.2 µm syringe filter prior to acidification and analysis on the KPA. TCE and methane concentrations were measured using gas chromatography (GC). TCE was sampled using solid phase microextraction and was analyzed using a flame ionization detector (FID) on the GC while methane was sampled by taking headspace samples followed by analysis using the FID. Organic acids (i.e., lactate, acetate, propionate, butyrate and formate) were measured using high pressure liquid chromatography (HPLC). Ferrous iron was measured using the ferrozine dye assay with the adsorbance being measured using a spectrophotometer set at a wavelength of 562 nm.

RESULTS AND DISCUSSION
Adsorption/Desorption Testing – Laboratory Consortium. Enrichments of the Laboratory Consortium caused the initial desorption of uranium followed by a longer-term loss of uranium from the growth medium in batch experiments in which iron oxide coated sand had been exposed to uranium prior to inoculation. Figure 1 shows U(VI) concentration over time for enrichments with and without sulfate. In both enrichments, the soluble U(VI) was less than the mass that was filterable through 0.2 µm filters. In the enrichments without sulfate, the soluble U(VI) decreased steadily over the duration of the experiment. The filterable fraction of U(VI) initially decreased, but then began to increase as the experiment neared completion. This disparity in results between the filterable and soluble

fraction may be due to the formation of fine particles or colloids rather than readsorption to the surface of the iron oxide. Formation of fine particles or

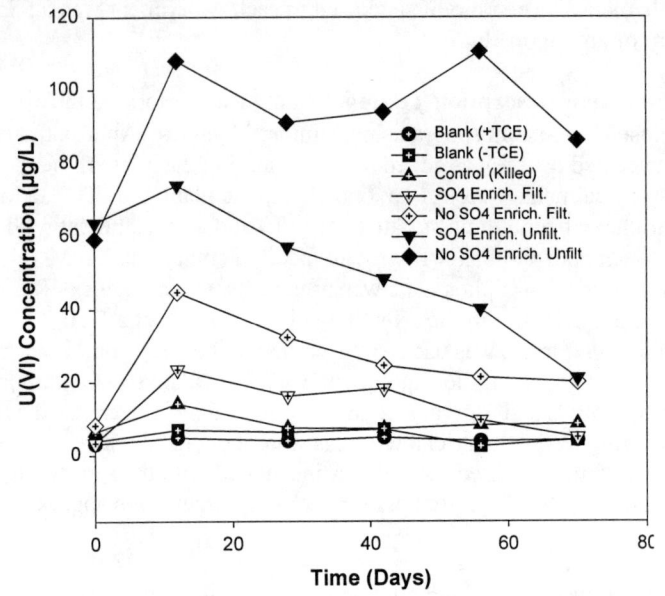

FIGURE 1. Desorption of uranium from iron oxide coated quartz in enrichments of the Laboratory Consortium with and without sulfate.

colloids may lead to the mobility of U(VI) in groundwater of microbially impacted aquifers with mixed contamination (Beak and Pitt 1996).

U(VI) in the enrichments with sulfate showed a similar decrease as the experiment continued, but there was little difference in the rate of U(VI) disappearance between the filtered and unfiltered fractions. The amount of U(VI) desorbed initially was less for the enrichment containing sulfate than for the enrichment without sulfate. The difference may have been caused by the chemical precipitation of uranium by the presence of sulfides produced by the culture. Precipitation of black iron sulfide on the surface of the iron oxide was the primary indicator of sulfate reduction in the test vials. In contrast, no build up of the black encrustation of the iron oxide coated sand was noted in the sulfate-free enrichments.

Enrichments with sulfate degraded the lactate (initial concentration ~500 mg/L) within the first ten days, producing approximately 100 mg/L acetate and small amounts of propionate. Complete lactate removal in the sulfate-free enrichments required nearly 30 days.

Analysis of the growth medium using ICP-AA indicated no significant difference between the total uranium and the U(VI) measured using the KPA. Neither of the enrichments were able to reductively dechlorinate the TCE that was added to the assay vials as a cocontaminant with the U(VI).

Adsorption/Desorption Testing – Field Consortium. Due to the inability of the laboratory consortium to reductively dechlorinate TCE, measures were taken to obtain a culture that was demonstrating reductive dechlorination of TCE in the field. A 1-L sample of groundwater from TAN (TAN 37) was filtered through a 0.2 μm filter membrane using a sterile filter apparatus. The filter was then submerged in nitrogen/phosphorus growth medium containing lactate, sulfate and iron oxide coated sand. Following growth the culture was transferred to fresh growth medium and then a culture was set up for use in the adsorption/desorption testing. The culture was fed lactate at the start of the experiment.

Enrichments with and without sulfate significantly affected the initial adsorption of U(VI) in the assay vials compared to the killed control and the blanks with and without TCE (Figure 2). Within the first 24 hours, the U(VI)

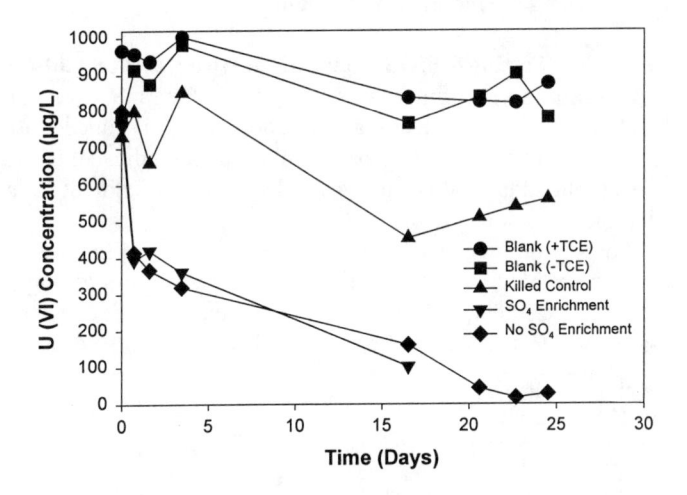

FIGURE 2. Effect of enrichments of the Field Culture with and without sulfate on uranium adsorption to iron oxide coated quartz.

concentration had decreased to near 400 ppb and steadily decreased to near zero by the completion of the experiment. The initial decrease in U(VI) noted in the killed control may be due to transfer of an exudate from the inoculation culture causing a chemical reaction with the U(VI) in solution or adsorption to biomass. The blanks for the experiment showed a minimal decrease in uranium compared to the enrichments. On day 17 the vial containing the sulfate enrichment broke, ending this leg of the experiment.

ICP-AA analysis of nitric acid digests of the iron oxide coated sand from the enrichments without sulfate yielded evidence that the microbial culture was able to chemically reduce the U(VI) to lower oxidation states. At the completion of the experiment KPA and ICP-AA analysis of the sand and the growth medium indicated a total uranium mass of 148.4 μg and a U(VI) mass of 99.2 μg. A similar analysis for the sulfate enrichment was not performed due to the breaking of the flask.

Lactate in both sulfate and sulfate-free enrichments was converted to small amounts of acetate and propionate. Complete removal of lactate occurred within 20 days of incubation in both enrichments.

Solubility changes of the U(VI) due to pH are not likely since the pH remained stable between 6.5 and 7 during the entire experiment. As with previous experiments using the Laboratory Consortium, there was no reductive dechlorination of TCE noted in either enrichment. Activity of iron reducing bacteria (IRB) in the sulfate-free enrichment may have diminished the number of iron oxide sites on the surface of the sand as indicated by the presence of ferrous iron (0.019 mM) in the growth medium at the end of the test. A loss of iron oxide in an aquifer impacted by IRB may decrease the number of uranium adsorption sites, potentially increasing the mobility of uranium.

Adsorption/Desorption Testing – Field Consortium with Lactate Addition. A second adsorption/desorption experiment was set up using the Field Consortium to determine the effect of lactate depletion and addition on U(VI) concentration in the enrichments. Vials were set up in a manner similar to that indicated in the previous experiment, but additional lactate was added to the enrichment vials after 20 days of incubation.

Figure 3 shows uranium concentration in adsorption/desorption assay vials set up to determine the effect of lactate addition on U(VI) adsorption/desorption. As with the previous experiment, U(VI) in the enrichments decreased to near zero. Interestingly, the U(VI) concentration in the killed control also dropped to

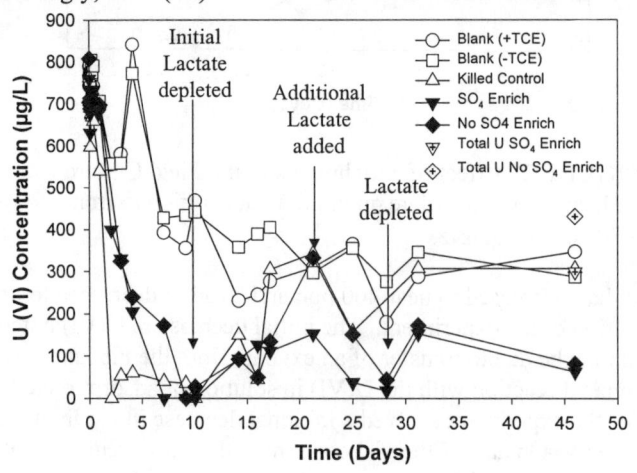

FIGURE 3. Effect of lactate addition to enrichments of the Field Culture with and without sulfate on uranium adsorption to iron oxide coated quartz.

zero, indicating contamination or significant chemical precipitation of uranium in solution. By day 10 the initial supply of lactate was depleted in both the sulfate and sulfate-free enrichments. Following this time the amount of propionate and acetate produced by fermentation of the lactate in the culture began to decrease. During this time period (between days 10 and 20) the U(VI) concentration began to increase. Upon addition of lactate on day 20, the U(VI) concentration again dropped to near zero. Upon depletion of lactate on day 28 the U(VI) again began to increase.

Metal reducing microorganisms appeared to be dominant in both enrichments, especially the enrichment without sulfate. Samples analyzed for total uranium after 46 days of incubation indicated high concentrations of reduced forms of uranium compared to U(VI) in the samples (Figure 3). The ferrous iron concentration in both enrichments, as well as the killed control and blank can be seen in Figure 4. Bacteria from the enrichments with and without sulfate showed

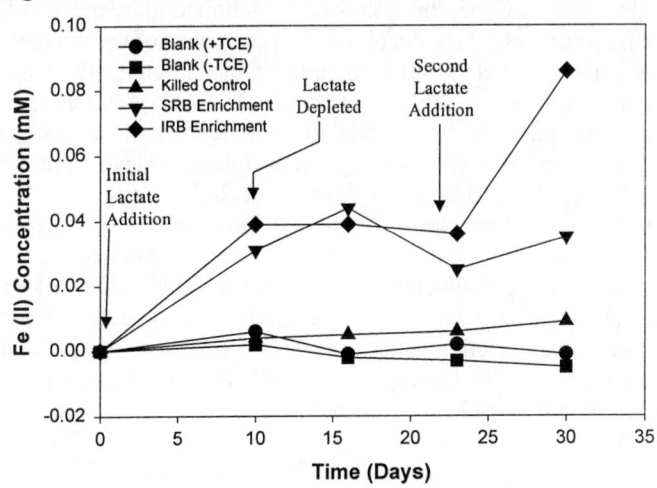

FIGURE 4. Ferrous iron concentration in enrichments of the Field Culture with and without sulfate on uranium adsorption to iron oxide coated quartz. Experiments to determine the effect of lactate addition on metal reduction.

the ability to reduce the iron present on the iron oxide coated quartz. When lactate was depleted the ferrous iron concentration in the enrichment without sulfate stabilized. Upon addition of lactate, the ferrous iron concentration in solution increased. In contrast, when lactate was depleted in the enrichment with sulfate, the ferrous iron concentration began decreasing, but did come back when lactate was added. The loss of ferrous iron in the enrichments containing sulfate may have been due to the formation of ferrous sulfide as indicated by the formation of a dark precipitate on the iron oxide surface of the sand. The ferrous iron concentration in the killed controls and blanks remained near zero for the duration of the experiment.

As with previous experiments, the enrichments were not able to reductively dechlorinate TCE. Enrichment with and without sulfate may be selecting for metal reducing microbes in the culture, shifting the community away from TCE degraders that were prevalent in the field. The loss in some TCE dechlorination capacity appears to be inherent with the culture from TAN. Microcosms set up specifically to monitor reductive dechlorination of TCE have encountered similar problems, including lags in rates of dechlorination and initiation of dechlorination (Kent Sorenson, personal communication).

CONCLUSIONS

Results generated from the above research showed that microbes in cultures originally utilized for reductive dechlorination of TCE were able to reduce metals such as iron and U(VI). The ability to reduce U(VI) also appeared to be culture dependent. The Laboratory Consortium was not capable of U(VI) reduction, while the Field Consortium showed reduction to more reduced forms of uranium. During testing, the TCE dechlorination capability of the microbes was diminished resulting in no net loss of TCE added to the enrichments. This loss in dechlorination activity was more than likely due to the form of enrichment used during testing. During the adsorption and desorption testing, the microbes were exposed to a growth environment with ample reducible metals which may have been preferentially used as electron acceptors over TCE.

While TCE was not dechlorinated, findings from these experiments indicated that establishment of conditions to stimulate microbial reductive dechlorination of chlorinated solvents may also have a significant effect on radionuclides and metals co-disposed with the chlorinated solvent. The primary effects shown were: the reduction of surface sites for complexation with the uranium, formation of colloids leading to the mobilization of uranium in groundwater, and finally, the chemical reduction of contaminants such as uranium leading to immobilization.

ACKNOWLEDGMENTS

Work was supported by the U.S. Department of Energy Office of Environmental Management, INEEL Environmental Systems Research and Application Program Under DOE Idaho Operations Office Contract DE-AC07-99ID13727. The authors would also like the thank Dr. Robert Smith and Jonathan Ferris for supplying the iron oxide coated quartz and Byron White for performing the ICP-AA analysis for total uranium.

REFERENCES

Adriaens, P. and T. M. Vogel, "Biological Treatment of Organics." In Microbial Transformation and Degradation of Toxic Organic Chemicals," L. Young and C. Cerniglia, eds. John Wiley & Sons, Inc. New York. 1995 pp. 435-486.

Bargar, J. R., R. Reitmeyer, and J. A. Davis. 1999. "Spectroscopic Confirmation of Uranium(VI)-Carbonato Adsorption Complexes on Hematite." *Environ. Sci. Technol.* 33: 2481-2484.

Beak, I., and W. W. Pitt, Jr., "Colloid-Facilitated Radionuclide Transport in Fractured Porous Rock," *Waste Manag.*, 1996, 16(4): 313-325.

Ellis, D. E., E. J. Lutz, J. M. Odom, R. J. Buchanan, Jr., C. L. Bartlett, M. D. Lee, M. R. Harkness, and K. A. Deweerd. 2000. "Bioaugmentation for Accelerated In Situ Anaerobic Bioremediation." *Environ. Sci. Technol.* 34: 2254-2260.

Fetzner, S., "Bacterial Dehalogenation," *Appl. Microbiol. Biotechnol.* 1998, 50: 633-657.

Harkness, M. R., A. A. Bracco, M. J. Brennan, Jr., K. A. Deweerd, and J. L. Spivack. 1999. "Use of Bioaugmentation to Stimulate Complete Reductive Dechlorination of Trichloroethene in Dover Soil Columns." *Environ. Sci. Technol.* 33: 1100-1109

Kohler, M., G. P. Curtis, D. B. Kent, and J. A. Davis. 1996. "Experimental Investigation and Modeling of Uranium(VI) Transport under Variable Chemical Conditions." *Wat. Resources Res.* 32: 3539-3551.

Reich, T., H. Moll, T. Arnold, M. A. Denecke, C. Henning, G. Geipel, G. Bernhard, H. Nitsche, P. G. Allen, J. J. Bucher, N. M. Edelstein, and D. K. Shuh. 1998. "An EXAFS Study of Uranium(VI) Sorption onto Silica Gel and Ferrihydrite." *J. Electron Spec. Related Phenomena* 96: 237-243.

Riley, R. G., J. M. Zachara, and F. J. Wobber, *"Chemical Contaminants on DOE Lands and Selection of Contaminant Mixtures for Subsurface Science Research,"* DOE/ER—0547T. April, 1992.

Rosentreter, J. J., H. S. Quarder, R. W. Smith, and T. McLing. 1998. "Uranium Sorption onto Natural Sands as a Function of Sediment Characteristics and Solution pH. In Adsorption of Metals by Geological Media." E. A. Jenne (Ed.) Academic Press, pp. 181-192.

U. S. Department of Energy, 1996. Subsurface Contaminants Focus Area, Summary Report. Washington, D. C., August, 1996.

Waite, T. D., J. A. Davis, T. E. Payne, G. A. Waychunas, and N. Xu. 1994. "Uranium(VI) Adsorption to Ferrihydrite: Application of a Surface Complexation Model." *Geochem. Cosmochem. Acta* 58: 5465-5478.

ENHANCED REDUCTIVE DECHLORINATION ON AN INDUSTRIAL SITE IN BELGIUM

Dirk Nuyens, Michael Meyer (Environmental Resources Management, ERM, Brussels, Belgium), Duane Wanty (Gillette, Boston, Massachusetts), Victor Miles (Duracell, Bethel, Connecticut), and Victor Dries (Flemish Public Waste Agency, OVAM, Mechelen, Belgium)

ABSTRACT: A partnership was developed with the Flemish Public Waste Agency (OVAM) to test, design and install the first enhanced reductive dechlorination bioremediation application in Belgium. Distinct historical releases of chlorinated solvents including trichloroethene (TCE), trichloroethane (TCA), and dichloroethane have extended from these respective on/off-site sources into downgradient areas. Natural attenuation has been observed, though complete dechlorination of the parent compounds TCE/TCA appears to have been limited. More recently, a significant groundwater mound and rising temperatures near the source areas appears to have accelerated biodegradation. The goal of this project is characterize the historical groundwater plume, and recent changes in chlorinated solvent concentrations and biodegradation/attenuation due to groundwater mounding and increasing temperature, and to evaluate the possible remedial application of enhanced reductive dechlorination for the observed chlorinated solvents. Bench-scale tests are currently running to evaluate possible substrates (lactate, vegetable oil, etc.) for application in the field. This paper addresses the feasibility testing, design, and construction considerations of the practical field-application of enhanced reductive dechlorination in this complex setting with the goals of 1) controlling off-site migration, 2) mitigating source areas, and 3) enhancing bioremediation in combination with monitored natural attenuation of off-site contamination.

INTRODUCTION

Chlorinated solvents have been widely used in industrial and commercial activities (U.S. EPA, 1996). To reduce human and environmental exposure to these contaminants, efforts are under way to develop innovative and cost-efficient remediation technologies as alternatives to traditional technologies such as pump & treat. One of these innovative technologies is anaerobic bioremediation using native bacteria to facilitate the destruction of chlorinated compounds (Pankow & Cherry, 1996, Fountain, 1998).

Effective solvent degradation of even higher chlorinated species such as tetrachloroethene has been observed using anaerobic bacteria. The feasibility and field application of chlorinated solvent remediation by (enhanced) bioremediation and monitored attenuation are being studied (Van Cauwenberge & Roote, 1998, U.S. EPA, 1998, Boulicault et al., 2000, U.S. EPA, 2000). It is important to obtain and maintain optimal growing conditions (pH, redox, organic substrates, nutrients, temperature, etc.) for remedial implementation (Daly et al., 2000; U.S. EPA, 2000).

Site characterization, including evaluation, biodegradation, and natural attenuation processes, are critical in the feasibility evaluation and remedial design. Bench-scale tests are required to determine the need for amendments (bio-augmentation, nutrients, and substrates) and to allow an efficient and cost-effective field application.

Objective. The goal of this study is to characterize the historical groundwater plume, recent changes in chlorinated solvent concentrations, and biodegradation and/or attenuation due to groundwater mounding and increasing temperature caused by a leaking underground steam/sewer pipe. It is thought that this leaking pipe has locally elevated groundwater temperatures and introduced organic matter, thereby inadvertently stimulating biological activity and contaminant reductive dechlorination. The possible remedial application of enhanced reductive dechlorination in combination with natural attenuation, within this complex geological setting, was investigated. The future remedial strategy will focus on 1) controlling further off-site migration, 2) mitigating source areas, and 3) enhancing bioremediation in combination with monitored natural attenuation of off-site contamination.

Site Description. The test site has a forty-year manufacturing history and lies in an area of complex geology including Quaternary silty/peat material overlying Tertiary glauconitic, sands of marine origin. A protective clay layer at a depth of 45 m bgs (meters below ground surface) separates the first aquifer (Diestian) from the main drinking water aquifer (Brussellian) for the region. The phreatic groundwater table has been detected at 1.5-2.0 m bgs. The leaking steam/sewer pipe(s) intersects the VOC plume underneath the plant building.

MATERIALS AND METHODS

Groundwater Sampling and Analysis. Samples were collected for initial bioremediation and natural evaluation testing. The wells to be sampled were selected based on the results of the site characterization and delineation studies. Additional chemical analysis (pH, oxidation-reduction potential or ORP, conductivity, dissolved oxygen, temperature, chlorinated solvents or VOC's) was carried out. Water level measurements were collected from all sampling wells in the targeted zone.

Bench-scale Sampling and Biodegradation Testing. Representative aquifer samples (soil and groundwater from both Quaternary and shallow Diestian) were collected for biodegradation testing. Samples were collected under a nitrogen blanket to maintain anaerobic conditions. The soil samples were analyzed for pH, ORP, dry matter, organic matter, total organic carbon, N-Kjehldal, N-NO$_3$, VOC's, BTEX, Fe, Mn, sulphide, phenol index, organic acids, Ca, Mg, K, and B), and the groundwater samples were analyzed for pH, Eh, conductivity, dissolved oxygen, ORP, chloride, alkalinity, ammonia, nitrate, nitrite, phosphate, sulphate, Fe, Mn, dissolved organic carbon, dissolved inorganic carbon, carbon dioxide, VOC, BTEX, ethane, ethene, methane, phenol index, C_2-C_5 organic acids, hydrogen, Ca, Mg, K, and B. A number of electron donor and nutrient combinations will be selected and tested in an anaerobic laboratory microcosm and column test. This will be followed by monitoring over time to evaluate each amendment for its effectiveness in supporting chloroethene dechlorination. Lactate, butyrate, and vegetable oil are considered as the main testing substrates based on currently available published bench-scale and pilot-test results for other chlorinated solvent contaminations. The bench scale bioremediation tests were initiated at the moment of drafting this paper.

RESULTS AND DISCUSSION

Groundwater data. Strong variations in pH, conductivity, ORP, and dissolved oxygen were observed in the groundwater samples used for bioremediation and natural attenuation screening (Table 1) in the elevated groundwater table zone influenced by the leaking steam/sewer pipe.

Table 1. Groundwater results target area (March, 2001).

Sampling well	pH	Temp °C	Conductivity µS/cm	ORP mV	Dissolved Oxygen mg/L
P1	6.37	9.8	743	37	0.08
P3	6.43	7.9	293	34	0.08
P4R	6.60	9.0	1,263	23	1.05
P5	6.50	9.2	409	29	0.32
P58	6.16	12.5	1,825	47	0.34
B1	5.78	11.2	1,126	70	0.11
B5	6.57	11.8	1,348	26	0.23
6	6.14	13.2	912	50	0.40
17	6.00	12.4	234	58	0.18
103	6.45	15.4	1,212	32	0.17
GP30	9.55	18.7	3,150	143	0.17
HA39	6.56	10.5	698	25	0.34
HA43	6.76	21.3	4,430	15	0.15
MGP01	6.89	21.6	4,310	7	0.03
MGP19	6.85	21.8	2,010	10	0.07
MGP22	7.83	19.8	1,838	-46	0.02
MGP24	7.10	21.4	4,980	-4	0.29
MGP26	6.50	20.6	2,590	29	0.11
MGP46	6.20	19.5	973	46	0.17

Increased groundwater temperatures up to 22°C were measured; the background groundwater temperature values ranged between 8-12°C. Increased conductivity values were measured in the high temperature wells. Observed dissolved oxygen concentrations (< 0.5 mg/L) and ORP-values (< 50 mV) indicate possible favorable conditions for reductive dechlorination and natural attenuation in the aquifer.

The groundwater level data clearly indicates the presence of a groundwater recharge mound under the main production area. The groundwater recharge mound coincides with the high temperature mound (Figures 1 and 2).

The exact source for the recharge mound is currently unclear, but is probably related to a leaking steam/sewer pipe.

Reduction in VOC Concentrations. As a result of the groundwater mounding and heating, reductions in VOC concentrations have been observed both in the source area and along the downgradient perimeter of the property. The chlorinated solvent results for well HA43, in one of the source areas located in the temperature and groundwater recharge mound, show elimination of PCE, TCE and a significant reduction in cDCE. This may be explained by a combination of dilution and degradation. Over the same time period, observations show increased concentrations of vinyl chloride (VC), indicating significant reductive dechlorination occurring in the anaerobic aquifer (Table 2) prior to any nutrient or substrate addition.

Laboratory Studies of Reductive Processes

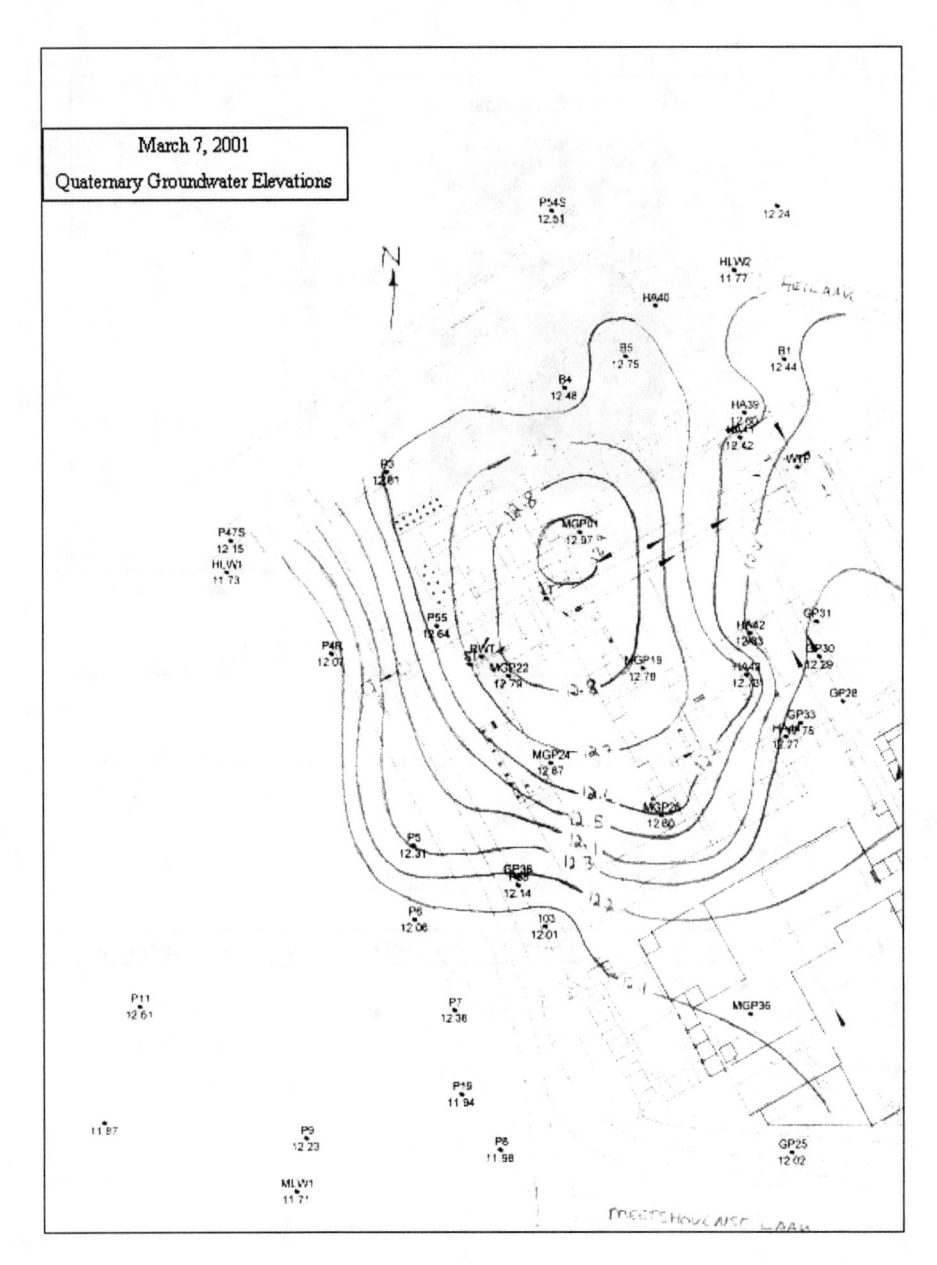

FIGURE 1: Groundwater elevation map – Phreatic aquifer (March, 2001)

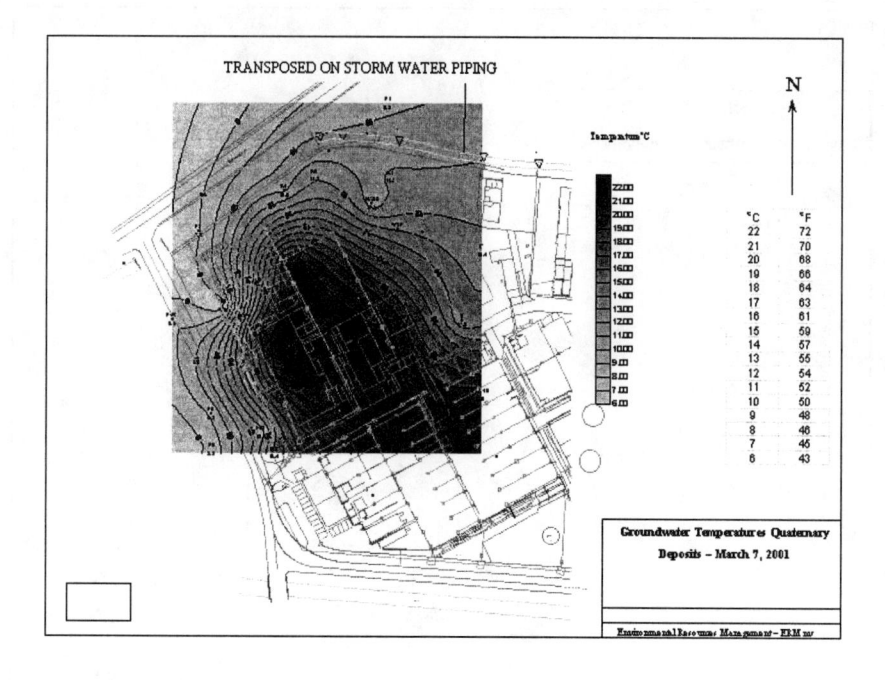

Figure 2. Temperature map for the phreatic aquifer (March, 2001)

Table 2. VOC-data for Well HA 43 (µg/L).

VOC	HA 43 – July 1999	HA 43 – March 2001
Tetrachloroethene	1,700	<1,000
Trichloroethene	11,000	<1,000
1,2-Cis-dichloroethene	430,000	210,000
Vinyl chloride	37,000	57,000

Along the perimeter of the site, a similar reduction in the concentration of chlorinated species is evident from 1999 sampling data to the most recent 2001 data. For example, at Well 3, a Quaternary (shallow) well, 1,1,1-TCA concentrations decreased from 12,000 to 1,500 µg/L, 1,1-DCA concentrations decreased from 31,000 to 6,400 µg/L, and cDCE concentrations decreased from 6,600 to 880 µg/L.

The reductive dechlorination process is not only supported by the inflow of warm and nutrient-rich water, it is also stimulated by the presence of a peat layer in the impacted aquifer interval. VOC-degradation rates are temperature, redox and substrate dependent (Van Cauwenberghe and Roote, 1998; Middeldorp et al., 1999; Suarez and Rifai, 1999; Yang and McCarty, 2000).

Further characterization work is currently running to evaluate the observed natural biodegradation/attenuation, to isolate the dechlorinating bacterial activity for the scheduled microcosm studies (microcosm and column tests), and to design and run a field pilot test.

REFERENCES

Boulicault, K. J., Hinchee, R. E., Wiedemeier, T. H., Hoxworth, Swingle, T. P., Carver, E., and Haas, P. E., 2000. "Vegoil: A novel approach for stimulating reductive dechlorination". In Remediation of Chlorinated and Recalcitrant Compounds: Vol C2-4. Papers from the 2nd International Conference on Remediation of Chlorinated and Recalcitrant Compounds. Monterey, CA, May 22-25, 2000. p. 1, Battelle Press, Columbia, OH.

Fiacco, R. J., Daly, M. H., Demers, G., Lee, M., and Wanty, D., 2000. "Anaerobic bioremediation of chlorinated VOCs in a fractured bedrock aquifer – preparations for an in-situ pilot study". In Remediation of Chlorinated and Recalcitrant Compounds: Vol C2-4. Papers from the 2nd International Conference on Remediation of Chlorinated and Recalcitrant Compounds. Monterey, CA, May 22-25, 2000. p. 405, Battelle Press, Columbia, OH.

Fountain, J. C., 1998. Technologies for dense non-aqueous phase liquid source zone, GWRTAC Technology Evaluation Report, TE-98-02, Pittsburgh, PA.

Middeldorp, P. J., Luijten, M. L., van de Pas, B. A., van Eekert, M. H., Kengen, S. W., Schraa, G., and A. J. Stams, 1999. "Anaerobic microbial reductive dehalogenation of chlorinated ethenes.", Bioremediation Journal 3 (3): 151-169

Pankow, J. F., and Cherry, J. A., Dense chlorinated solvents and other DNAPLs in groundwater. Waterloo Press, Portland, OR.

Suarez, M. P., and Rifai, H. S., 1999. "Biodegradation rates for fuel hydrocarbons and chlorinated solvents in groundwater". Bioremediation Journal 3 (4): 337-362.

U.S. EPA, 1996. Recent developments for in-situ-treatment of metal contaminated soils. U.S. Environmental Protection Agency Report, EPA contract number 68-W5-0055, Washington DC.

U.S. EPA, 1998. Technical protocol for evaluation of natural attenuation of chloirinated solvents in groundwater. U.S. Environmental Protection Agency Report, EPA 600-R-98-128, Washington DC.

U.S. EPA, 1999. Field applications of in-situ remediation technologies : permeable reactive barriers. U.S. Environmental Protection Agency Report, EPA 542-R-99-002, Washington DC.

Van Cauwenberghe, L., and Roote, D. S., 1998. In-situ bioremediation. GWRTAC Technology Evaluation Report, TE-98-01, Pittsburgh, PA.

Yang, Y., and McCarty, P. L., 2000. "Biomass, oleate, and other possible substrates for chloroethene reductive dehalogenation". Bioremediation Journal 4 (2) : 125-133.

COUPLING OF TOLUENE OXIDATION WITH PCE DECHLORINATION UNDER SULFIDOGENIC CONDITIONS

Thomas P. Hoelen (Stanford University, Stanford, CA, USA)
Jeff Cunningham (Stanford University, Stanford, CA, USA)
Carmen A. LeBron (NFESC, Port Hueneme, CA, USA)
Martin Reinhard (Stanford University, Stanford, CA, USA)

ABSTRACT: Toluene was evaluated as an electron donor for the reductive dehalogenation of PCE (perchloroethylene) under sulfate reducing conditions. Microcosms were constructed using aquifer material and groundwater from a sulfidogenic site contaminated with fuel hydrocarbons and chlorinated solvents. Amended toluene was oxidized with sulfate as the electron acceptor. The amendment stimulated slow dehalogenation of PCE, forming vinyl chloride (VC) and ethylene. Details of the stimulating process are unknown but may involve the fermentation of biomass formed during toluene degradation.

INTRODUCTION
Chlorinated solvents and aromatic hydrocarbons (including benzene, toluene, ethylbenzene, and xylenes (BTEX)) are among the most common groundwater contaminants in the U.S. and often occur in commingled plumes. In fuel hydrocarbon plumes, oxygen and nitrate are often used rapidly, creating sulfidogenic or methanogenic conditions (Cunningham et al., 2001). Methanogenic conditions promote biological dechlorination of contaminants, such as chlorinated ethylenes (e.g. Bouwer and McCarthy, 1983; De Bruin et al., 1992). The link between BTEX degradation and PCE or trichloroethylene (TCE) dehalogenation has been studied in the absence of sulfate (Sewell and Gibson, 1991; Liang and Grbic-Galic, 1992; Johnston et al., 1996; Harkness et al, 1999). These authors observed that dechlorination stopped at cis-dichloroethylene (cis-DCE), while VC and ethylene (ETH) were not formed. PCE or TCE dehalogenation under sulfate reducing conditions with various electron donors has been shown to produce cis-DCE (Pavlostathis and Zhuang, 1993; Semprini et al., 1995; Cabirol et al., 1998; Harkness et al., 1999). In some cases, extremely slow formation of VC was observed, but conversion to ethylene has never been observed before now. These findings led to the conclusion that complete dechlorination of PCE to ethylene does not occur under sulfidogenic conditions (Holliger, 1995). The frequently observed accumulation of VC is a concern because it is more toxic than its chlorinated precursors.

Microorganisms that couple reductive dechlorination to growth have been isolated and studied intensively (Holliger et al., 1993). All strains transform PCE and TCE to cis-DCE except for strain 195 (Maymo-Gatell et al., 1997), which can completely dehalogenate chlorinated ethylenes. To date, no pure cultures that grow with cis-DCE or VC as electron acceptor have been isolated. Various organic electron donors sustain the transformation of cis-DCE and VC to ethylene, but the process appears to depend on the availability of hydrogen produced through fermentation of organic donors by other members of the microbial consortium (Fennell et al., 1997; Yang and McCarty, 1997). In natural systems, dehalogenating organisms have to compete for hydrogen with other microorganisms such as homoacetogens, hydrogenophilic methanogens, and sulfidogens. Lovley and Goodwin (1988) have shown that various terminal electron-accepting reactions are associated with defined hydrogen threshold levels required for growth. Typically, organisms using an electron acceptor with a higher electrochemical potential are associated with a lower hydrogen threshold. Hydrogen threshold levels for methanogens (5 – 95 nM) are higher than those for dehalogenating organisms (< 2 nM) (Smatlak et al, 1992; Yang and McCarty, 1998; Löffler et al, 1999). Therefore, methanogens can be outcompeted by dehalogenators at hydrogen concentrations below the threshold levels for methanogens.

Competition of dehalogenating organisms with sulfate reducers is poorly understood. While hydrogen threshold levels for dehalogenators are expected to be close to those of nitrate reducers based on free energy considerations, i.e., around 0.05 nM (Lovley and Goodwin, 1988), experimental results show levels varying from less than 0.3 nM (Löffler et al., 1999) to 2 nM (Yang and McCarty, 1998). This range partly overlaps with the hydrogen threshold levels for sulfate reducers, ranging between 1 and 15 nM. Outcompetition of sulfate reducers by dehalogenaters may therefore only be possible at very low hydrogen levels. Low hydrogen concentrations may be expected when the fermentation process is slow. In this paper, we report data that demonstrate that addition of toluene as an electron donor can promote the dechlorination of PCE to ethylene under sulfate reducing conditions.

MATERIALS AND METHODS

Chemicals. All compounds used were 99+% pure unless noted, and used as received without further purification. Liquid toluene, PCE, TCE, and *cis*-DCE (97%) (Aldrich Chemical Co., Milwaukee, WI) were used as chemical amendments and to prepare analytical standards. Neat toluene, PCE, TCE, and *cis*-DCE were added to microcosms with a 10 µL Hamilton syringe if the volumes of the solvents were larger than 1 µL; otherwise a saturated solution was prepared and added to the microcosm with a 1 ml or 2.5 ml Hamilton syringe. Hydrogen, VC, ETH, and methane gases (Scott Specialty Gases, Alltech Associates, Inc., Deerfield, IL) were used as analytical standards. Final concentrations of toluene, chlorinated solvents, and VC were determined with GC/FID analysis (see analytical procedures). Sulfate (Baker, sodium salt) was used as electron acceptor and analytical standard.

Batch Experiments. Aquifer sediment and ground water were collected from a site adjacent to the San Francisco Bay, CA. The ground water was degassed under vacuum and purged with nitrogen gas to remove the oxygen. Microcosms were constructed in an anaerobic glove box by weighing 20 g aquifer sediment in

160 mL serum bottles and adding 110 mL groundwater (sulfate concentration approximately 2 mM (190 mg/L)). The vials were sealed with butyl rubber stoppers and aluminum crimp seals, and the headspace (about 40 ml) was removed and replaced with nitrogen. Microcosms were amended with either PCE (Experiment 1), toluene (Experiment 2), or PCE and toluene (Experiment 3). Experiment 4 is an autoclaved control experiment amended with toluene and PCE. Toluene and PCE were added at 200 µM (18.4 mg/L) and 150 µM (24.9 mg/L), respectively. Microcosms used in Experiments 1, 2, and 3 were amended with an additional 0.5 mM sulfate (48 mg/L) to compensate for sulfate removal that may occur as a result of toluene oxidation.

Analytical Methods. Headspace samples (250 µl) were withdrawn from microcosms with a 500 µL valved precision gas-tight syringe equipped with a side-port needle, and submitted to gas chromatochraphy (GC) analysis to quantify toluene, PCE, TCE, cis-DCE, trans-DCE, 1,1-DCE, VC, ethylene, ethane, and methane concentrations. Samples were injected splitless at 225 °C onto a HP 5890 GC equipped with a 30-m megabore GSQ-PLOT column (J&W) and a FID detector. The GC was operated at 60 °C for 1 minute, increased to 150 °C at 30 °C/min, and hold at 150 °C for 30 minutes. Known volatile compounds in the microcosms were identified and quantified through comparison of retention times and peak surface areas with those of external standards. The total amount of volatile compounds in microcosms and external standards was calculated using Henry's constant determined by Gossett (1987). Sulfate analysis was performed with a Dionex series 4000i ion chromatograph (IC) equipped with a conductivity detector, a Dionex IonPac AS4A column (4 x 250 mm), and an AG4A guard column (4 x 50 mm), operated with sodium bicarbonate as eluent. 200 µL liquid samples were withdrawn from the microcosms, centrifuged, and 50 times diluted prior to analysis. Sulfate concentrations were identified and quantified through comparison of retention times and peak surface areas with those of external sulfate standards. Hydrogen was not measured.

RESULTS AND DISCUSSION

Toluene only. The transformation of toluene (162.2 µM) in the presence of sulfate (2.44 mM) is shown in Figure 1A. After a lag time of approximately 100 days, toluene removal was complete within 50 days. Toluene removal was accompanied by removal of 442 µM sulfate indicating toluene oxidation by sulfate reducers. The ratio of toluene to sulfate utilized was 1:2.73, less than that expected for the energy reaction (i.e., complete mineralization), which is 1:4.5. Apparently, a significant fraction of the toluene is used for biomass synthesis. Comparison with an autoclaved control (Figure 1E) indicates that abiotic reactions are insignificant. The initial concentrations of cis-DCE, VC, and ethylene in the sediment were at 2.0 µM, 0.95 µM, and 0.39 µM, respectively (Figure 1B) and originated from the site. Shortly after the onset of toluene degradation (after 120 days), cis-DCE transformation began and was complete in 35 days. Conversion of VC to ethylene commenced after 456 days. It is

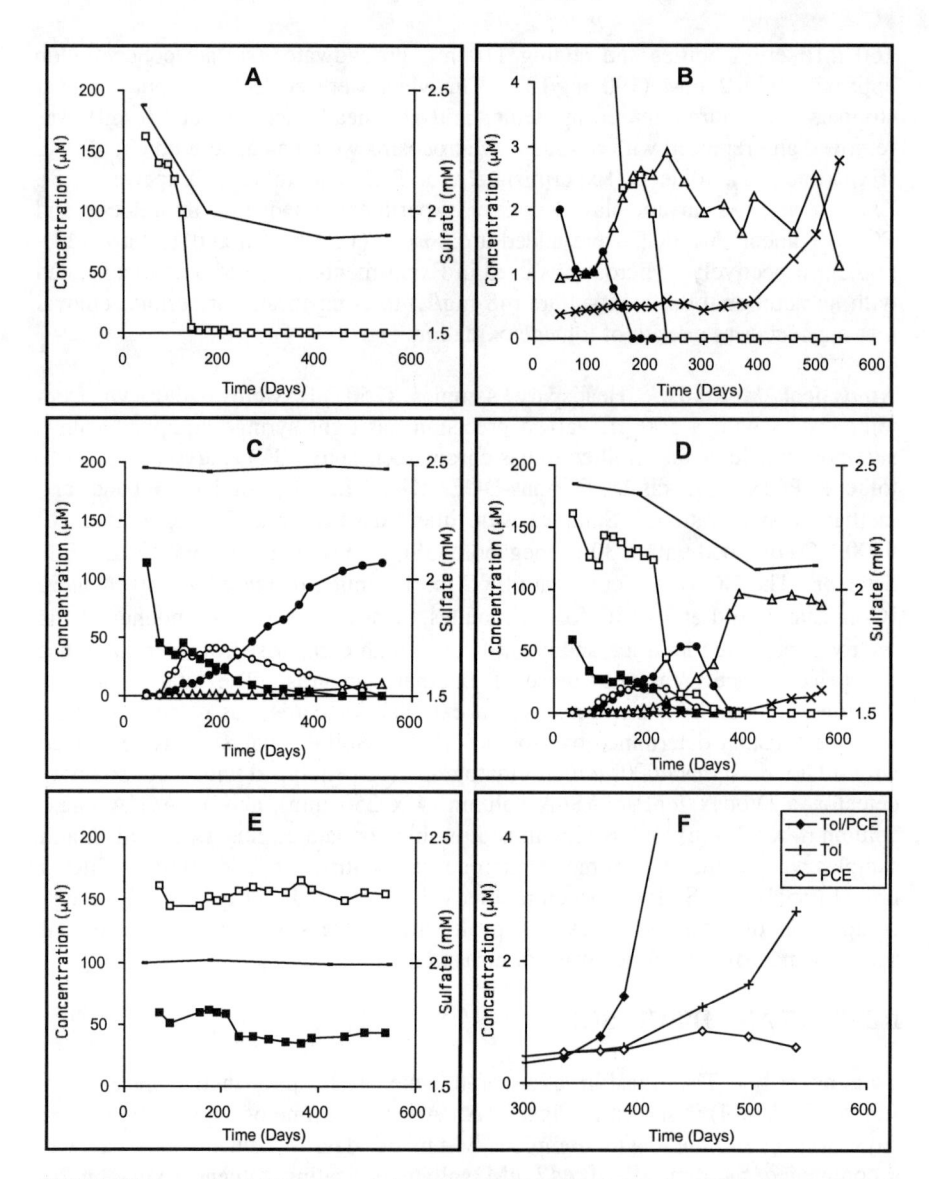

FIGURE 1. A-E: Toluene (□), PCE (■), TCE (○), cis-DCE (●), VC (Δ), Ethylene (×), and Sulfate (-). A) Experiment 1: Toluene and Sulfate. B) Experiment 1: Toluene, cis-DCE, VC, and ETH. C) Experiment 2: PCE, TCE, cis-DCE, VC, and Sulfate. D) Experiment 3: Toluene, PCE, TCE, cis-DCE, VC, ETH, and Sulfate. E) Experiment 4: Toluene, PCE, and Sulfate (Autoclaved control). F) Ethylene concentrations in experiments 1 through 3: Toluene (Expt. 1), PCE (Expt. 2), Toluene + PCE (Expt. 3).

noteworthy that conversion of cis-DCE to ethylene occurred in the presence of 1.9 mM sulfate (182 mg/L).

PCE only. Figure 1C indicates the transformation of PCE (113.5 μM) in a microcosm that was not amended with toluene (Experiment 2). The initial sulfate concentration (2.48 mM or 240 mg/L) remained constant during the course of this experiment. The initial concentrations of cis-DCE, VC, and ethylene (originating from the field) were 2.2 μM, 0.92 μM, and 0.33 μM, respectively. Dehalogenation of PCE commenced after 80 days and led to increases in the TCE and cis-DCE concentrations. Formation of cis-DCE continued until PCE and TCE were completely removed at Day 538. VC production started after Day 366 and appeared to be much slower than TCE or cis-DCE formation. Ethylene was not formed within 538 days. The origin of the donor used for dehalogenation of PCE to cis-DCE and VC is uncertain. Perhaps it was present in the sediment but biologically unavailable, and shaking mobilized unavailable donor compounds. Some introduction of hydrogen into the sample may have occurred during preparation of the microcosm in the glove box.

PCE and toluene. Figure 1D shows the stimulating effect of toluene (161.0 μM) on PCE (59.0 μM) transformation in the presence of sulfate (2.42 mM). The lag time for toluene degradation lasted to Day 209 and was about twice as long as observed in Experiment 1. The prolonged lag phase may have been due to PCE toxicity. As in Experiment 1, PCE transformation began at the time toluene degradation commenced. Sulfate utilization occurred mainly between day 172 and 419, and was presumably linked to toluene degradation. Until Day 209, transformation rates of PCE to TCE and cis-DCE were similar to Experiment 2 (Figure 1C). Thereafter, dehalogenation reactions increased as indicated by rapid PCE and TCE removal, increased formation and subsequent removal of cis-DCE, and the formation of VC. At Day 387, PCE, TCE, and cis-DCE were depleted and stoichiometrically converted to VC. VC was slowly converted to ethylene. These observations are in accordance with the results of Experiment 1 (Figure 1B). The sudden increase in dehalogenation rates compared to Experiment 2 coincides with the onset of toluene degradation, indicating a stimulating effect of toluene on the dehalogenation reactions.

Comparison with an autoclaved control (Figure 1E) indicates that abiotic reactions are insignificant. Methane and ethylene concentrations remained approximately constant (10 and 0.5 μM, respectively) during the course of the experiment. Figure 1F depicts the formation of ethylene in Experiments 1, 2, and 3. Ethylene formation is directly coupled to the dechlorination of VC, the slowest step in the dehalogenation of chlorinated ethylenes. Ethylene formation is most rapid in the toluene-amended microcosms, consistent with toluene stimulating the dehalogenation reaction.

CONCLUSIONS

Dechlorination of PCE to vinyl chloride and ethylene was demonstrated in sulfate reducing microcosms amended with toluene. The microcosms were constructed with sediments from a sulfate-reducing site that was contaminated with fuel hydrocarbon compounds and chlorinated ethylenes. Dehalogenation occurred with and without toluene added, but was more extensive with toluene present, i.e., proceeded to vinyl chloride and ethylene. Enhanced dehalogenation reactions continued for more than a year after toluene was transformed and occurred without significant sulfate reduction, thus excluding toluene as the direct electron donor. Simple metabolic intermediates of toluene oxidation, such as benzoate or acetate, can be excluded as the source of reducing power for dehalogenation since these intermediates are expected to be degraded rapidly by sulfate reducing organisms. Details of the stimulating process are unknown. Perhaps the biomass formed during toluene degradation is slowly fermented, releasing hydrogen. Dechlorination activity in the presence of sulfate is expected if the low levels of hydrogen can be used by dechlorinators but not by sulfate reducers.

Results imply that, at contaminated field sites, dehalogenation of chlorinated ethylene metabolites can continue long (many months) after BTEX compounds are transformed. This finding has important implications for assessing the fate of chlorinated ethylenes in high-sulfate groundwater. In this study, the process was observed to occur in the presence of sulfate (> 150 mg/L). Previously, methanogenic conditions were thought to be necessary for complete dehalogenation. The creation of methanogenic conditions, and therefore the removal of sulfate, may be unnecessary at field sites. Complete sulfate removal would require large quantities of an electron donor and would significantly increase remediation cost. Also, complete sulfate reduction may lead to excessively high sulfide concentrations that may inhibit dechlorination. Furthermore, at sulfate-reducing sites where cis-DCE is the apparent end product, VC may form slowly over time causing an unanticipated problem.

ACKNOWLEDGMENTS

This work was partially sponsored by the Navy through the Naval Facilities Engineering Service Center.

REFERENCES

Bouwer, E. J., and P. L. McCarty. 1983. "Transformation of 1-carbon and 2-carbon Halogenated Aliphatic Organic Compounds under Methanogenic Conditions." *Appl. Environ. Micr.* *45*: 1286-1294.

Cabirol, N., F. Jacob, J. Perrier, B. Fouillet, and P. Chambon. 1998. "Interaction between Methanogenic and Sulfate-reducing Microorganisms during Dechlorination of a High Concentration of Tetrachloroethylene." *J. Gen. Appl. Micr.* *44*: 297-301.

De Bruin, W. P., M. J. J. Kotterman, M. A. Posthumus, G. Schraa, and A. J. B. Zehnder. 1992. "Complete Biological Reductive Transformation of Tetrachloroethene to Ethane." *Appl. Environ. Micr.* *58*: 1996-2000.

Cunningham, J. A., H. Rahme, G. D. Hopkins, C. A. LeBron, and M. Reinhard. 2001. "Enhanced In-Situ Bioremediation of BTEX-Contaminated Groundwater by Combined Injection of Nitrate and Sulfate." *Environ. Sci. Techn.* (in press).

Fennel, D. E., J. M. Gossett, and S. H. Zinder. 1997. "Comparison of Butyric Acid, Ethanol, Lactic Acid, and Propionic Acid as Hydrogen Donors for the Reductive Dechlorination of Tetrachloroethene." *Environ. Sci. Techn.* *31*: 918-926.

Gossett, J. M. 1987. "Measurement of Henry's Law Constant for C1 and C2 Chlorinated Hydrocarbons." *Environ. Sci. Techn.* *21*: 202-208.

Harkness, M. R., A. A. Bracco, M. J. Brennan, jr., K. A. Deweerd, and J. L. Spivack. 1999. "Use of Bioaugmentation to Stimulate Complete Reductive Dechlorination of Trichloroethene in Dover Soil Columns." *Environ. Sci. Techn.* *33*: 1100-1109.

Holliger, C., G. Schraa, A. J. M. Stams, and A. J. B. Zehnder. 1993. "A Highly Purified Enrichment Culture Couples the Reductive Dechlorination of Tetrachloroethene to Growth." *Appl. Environ. Micr.* *59*: 2991-2997.

Holliger, C. 1995. "The Anaerobic Microbiology and Biotreatment of Chlorinated Ethenes." *Curr. Opin. Biotechn.* *6*: 347-351.

Johnston, J. J., R. C. Borden, and M. A. Barlaz. 1996. "Anaerobic Biodegradation of Alkylbenzenes and Trichloroethylene in Aquifer Sediment Down Gradient of a Sanitary Landfill." *J. Cont. Hydrol.* *23*: 263-283.

Liang, L-N., and D. Grbic-Galic. 1992. "Biotransformation of Chlorinated Aliphatic Solvents in the Presence of Aromatic Compounds under Methanogenic Conditions." *Environ. Tox. Chem.* *12*: 1377-1393.

Löffler, F. E., J. M. Tiedje, and R. A. Sanford. 1999. "Fraction of Electrons Consumed in Electron Acceptor Reduction and Hydrogen Thresholds as Indicators of Halorespiratory Physiology." *Appl. Environ. Microbiol.* *65*: 4049-4056.

Lovley, D. R., and S. Goodwin. 1988. "Hydrogen Concentrations as an Indicator of the Predominant Terminal Electron-accepting Reactions in Aquatic Sediments." *Geochim. Cosmochim. Acta 52*: 2993-3003.

Maymo-Gatell, X., Y. Chien, J. M. Gossett, and S. H. Zinder. 1997. "Isolation of a Bacterium That Reductively Dechlorinates Tetracholroethene to Ethene." *Science.* *276*: 1568-1571.

Pavlostathis, S. G., and P. Zhuang. 1993. "Reductive Dechlorination of Chloroalkenes in Microcosms Developed with a Field Contaminated Soil." *Chemosphere, 27*: 585-595.

Semprini, L., P. K. Kitanidis, D. H. Kampbell, and J. T. Wilson. 1995. "Anaerobic Transformation of Chlorinated Aliphatic Hydrocarbons in a Sand Aquifer Based on Spatial Chemical Distributions." *Water Resour. Res. 31*: 1051-1062.

Sewell, G. M., and S. A. Gibson. 1991. "Stimulation of the Reductive Dechlorination of Tetrachloroethene in Anaerobic Aquifer Microcosms by the Addition of Toluene." *Environ. Sci. Techn. 25*: 982-984.

Smatlak, C. R., J. M. Gossett, and S. H. Zinder. 1996. "Comparative Kinetics of Hydrogen Utilization for Reductive Dechlorination of Tetrachloroethene and Methanogenesis in an Anaerobic Enrichment Culture." *Environ. Sci. Techn. 30*: 2850-2858.

Yang, Y., and P. L. McCarty. 1998. "Competition for Hydrogen within a Chlorinated Solvent Dehalogenating Anaerobic Mixed Culture." *Environ. Sci. Techn. 32*: 3591-3597.

COMPARISON OF REDUCING AGENTS FOR DECLORINATION IN A SIMULATED AQUIFER

William A. Farone, Applied Power Concepts, Inc., Anaheim, CA
Tracy Palmer, Applied Power Concepts, Inc., Anaheim, CA

ABSTRACT: A simulated aquifer system (Aquifer Simulation Vessel – ASV) was used to study the efficacy of various reducing agents used in one-time applications to assist microbial dechlorination. The soil samples were homogenized and the soil was packed into 6.0-foot long tubes with an internal diameter of 5.75 inches. Flow rates can be calibrated for flow at 0.1 to 2.0 feet/day. The system has a volume of 2025.44 in^3. A solution of TCE and water is held in a reservoir at the beginning of the tube and connected to a nitrogen tank under slight positive pressure to retard volatilization of TCE. The TCE concentration in the reservoir is maintained at 20 ppm.

This system has been used for several years and found to be quite reproducible in determining gradient profiles of contaminants, as they are bioremediated with an active soil and reducing agent. Lactate released from a polylactate ester was compared to vegetable oil and molasses for efficacy of bioremediation of the TCE. Vegetable oil and molasses were not able to maintain a profile of reducing agent along the tube. Controlling the location of the reducing agent appears critical in order to obtain consistent results. The efficacy of a reducing agent as well as the economics of use should consider the total amount of material that is need to maintain contact with the bioremediation zone as well as the cost and timing of application. With the polylactate ester the source of the agent remained stationary and only the released lactate acid migrated with the water flow while with oil and molasses the entire reducing agent plume moved away from the treatment zone over time.

INTRODUCTION

Over the last several years, microcosm tests have been conducted in our laboratory using soil from sites contaminated with various chlorinated compounds. Recently the rates in the microcosm tests were compared to rates at some filed remediation sites (Farone, 2000) when TCE was the contaminant. The purpose of these microcosm tests were to determine whether the soil contained the appropriate microbiological population to accomplish dechlorination when augmented over a long time with slow release source of lactic acid and other nutrients released from a polylactate ester (HRC™).

Kinetic studies on the rates of degradation in these systems indicated that the most important factor is the number and type of microbes (Farone et al., 1999). The microcosm tests do not take into account the dynamic nature of an aquifer. The movement of contaminated water through the aquifer can displace the reducing agent providing only momentary assistance to the biological remediation process. The tests were designed to see whether a single injection of material could cause significant remediation with out the need for continuous applications of the molasses, vegetable oil or polylactate ester.

In the current study a model simulated aquifer was used. Plastic (PVC) tubes that are 6.0 feet long and 5.75 inches inside diameter were packed with a homogenized mixture of sand and soil. The tubes are connected to a reservoir of TCE contaminated water through a pump that can maintain a steady flow of liquid through the soil in the range of 0.1 to 2.0 feet per day. Ports every 6 inches along the bottom of the tubes are used to sample the water flowing through the tube periodically. The sampling is performed to minimize the disturbance to the flow

Objective. The objective of this study was to model the characteristics of dynamic flow of the water in an aquifer as it may affect a molasses solution, a vegetable oil mixture and a polylactate ester when they are used to assist bioremediation. The ability of these materials to continue to provide appropriate substrate for bioremediation in a moving water column was to be determined. These electron donor substrates were to be evaluated for their ability to assist remediation in a single injection avoiding the need for continuous injection. They have all been suggested for their use in this manner.

EXPERIMENTAL METHODS

The aquifer simulation vessels (ASV) were filled with soil mixed with sand after the soil and sand had been homogenized. The soil was collected from a TCE contaminated site known to have natural attenuation and for which the microbes had been proven to be acclimated to remediate TCE in microcosm tests. Three separate tubes were used, one for each reducing agent. Flow was established in the system using metering pumps set to deliver a flow of 0.3 feet per day in the ASV. When the flow was established through the ASV the sample reducing agents were injected on Day 0.

The ASV module used for the polylactate ester was charged with 23.1 grams of the material on Day 0. The module used for vegetable oil was charged with 24.1 grams of oil for one test and 24.6 grams for another. The module used for molasses was charged with 24.3 grams of molasses in one test and 24.6 grams in another test. Test samples of various ports along the ASV located at 6 inch intervals was performed periodically. Analysis of TCE, DCE, and VC is performed by gas chromatography using both PID and FID detectors.

For the HRC™ test each sample was also analyzed for the presence of lactic acid, pyruvic acid and acetic acid to confirm the presence of the desired lactate system. These tests are performed in a liquid chromatograph with a UV detector. The molasses system was tested for the presence of sugars and sugar degradation products (organic acids, ketones and alcohols) using liquid chromatography with a refractive index detector. The system charged with oil was also analyzed for fatty acids, glycerol and other degradation products such as alcohols, organic acids and ketones via liquid chromatography.

Experience with this type of system shows that not all ports can be sampled at the same time since the sampling at a port disturbs the flow in the system. This is not dissimilar to the case in monitoring wells in the field where pumping three well bore volumes can disturb the aquifer flow. Only about 4 ports can be sampled in the ASV without disturbing the flow, analogous to low volume field sampling. The sample size is about 25 ml for each sample at each port compared to the total liquid volume of about 10 liters. Experience has shown that mass balances based on the daughter products of bioremediation for these systems are very unreliable just as they are in filed sites. Although the soil is equilibrated with the contaminant before injection of the substrates, the daughter products can be removed by absorption. Extraction of the soil for the daughter products could only be accomplished with great difficulty and cost at the end of the test with the attendant problem of recovery of many of these volatile components. Therefore, the data are used to indicate the general trends of bioremediation without a claim that the data can be used for a meaningful mass balance.

RESULTS AND DISCUSSION

Figure 1 shows the results for the HRC™ test at the port that is 6 inches downstream from the HRC™ injection location.

Values 6 Inches Downstream

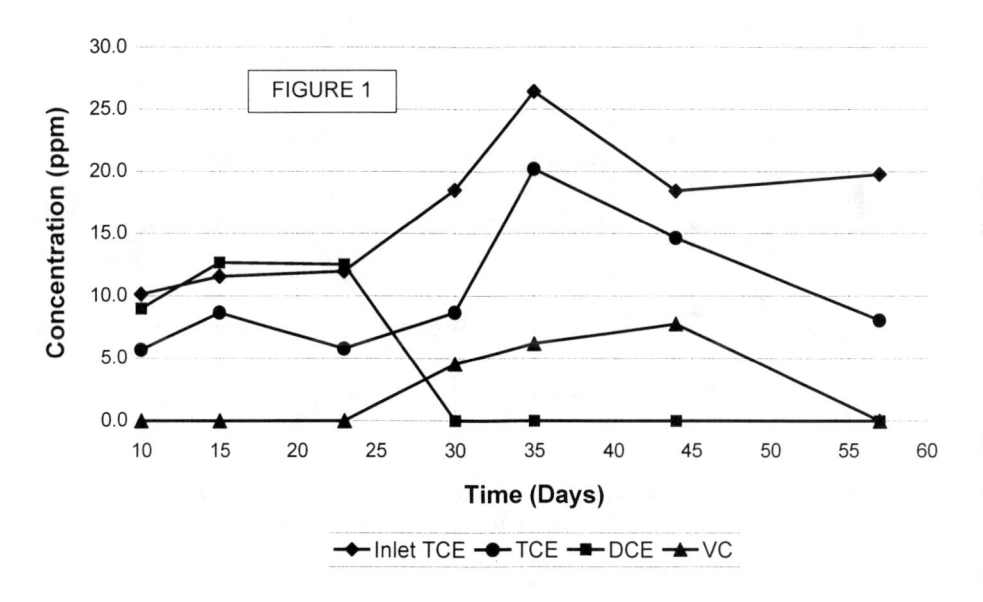

FIGURE 1

[Graph: Concentration (ppm) vs Time (Days); series: Inlet TCE, TCE, DCE, VC]

Note that the inlet values are measured upstream of the test material injection point and are influenced by a variety of flow factors as well as some remediation taking place in the upstream direct. These values are analogous to background levels entering the treatment zone at a field site and are always measured on the upstream side (about 6 inches) from the injection site.

The inlet flow eventually stabilizes at 20 ppm, the target input value of TCE. The remediation of TCE tracks the amount of TCE in the inlet region.

DCE builds up and then decreases with time. Subsequently VC builds up and decreases. By 60 days it appears that the DCE and VC are being used as fast as the TCE is degraded at this location.

Figure 2 is a similar graph for the location 30 inches downstream from the injection point. Note that VC builds up and decays and that there is a spike of DCE at about 44 days. At this flow rate (4 inches per day) the material at the inlet should reach this location in about 9 days or about 6 days after the readings at the 6 inch port. The spike in DCE occurs after the spike in TCE noted at the 6 inch port on Day 35. The complete data set for the HRC™ is given in Table 1. Transport of components can be distinguished to some extent by the analysis of adjacent ports at adjacent time periods.

The data for the vegetable oil and the molasses was not so easily obtained. Both of these materials move with the flow of the water in the ASV even though oil is insoluble. The measurement times had to be changed to accommodate these

Values 30 Inches Downstream

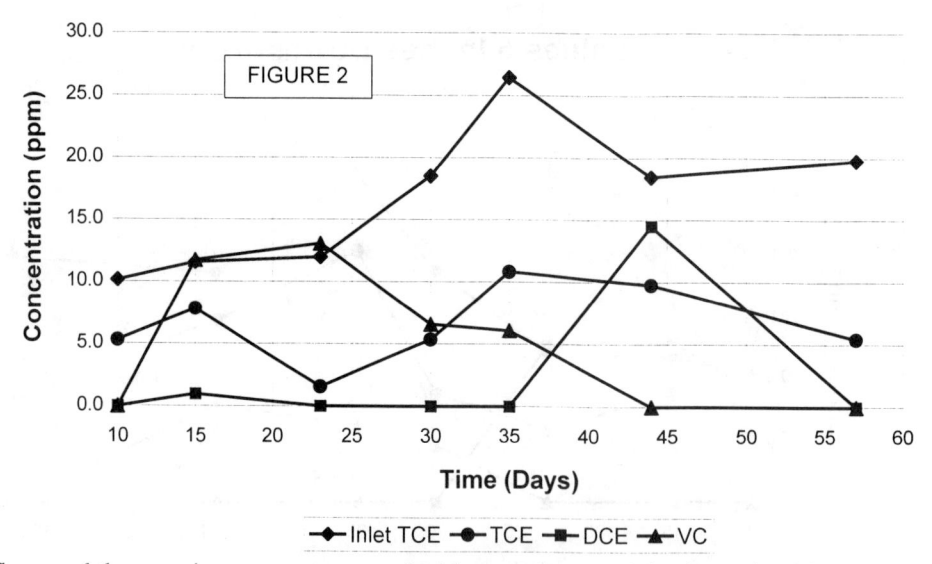

FIGURE 2

flows and the experiments were repeated with more than one injection to be able to determine any effect at all. The molasses results are given in Table 2. Within one day of injection there is an immediate drop of TCE in the vicinity of the injection site.

The build up of DCE is evidence for some bioremediation, i.e. this is not a displacement effect. However, within 3 days the effect is gone and there appears to be no further remediation. The drop in Day 7 at 18 inches may be due to the sugar migrating down steam with more TCE or it may be a flow effect. No effects were seen further down the aquifer and there was no evidence of sugar products at subsequent ports.

Laboratory Studies of Reductive Processes

The oil results are given in Table 3. Again there seems to be an effect early after the injection. Unlike sugar there are also effects down the column. The sample ports are on the bottom of the tubes and we found no evidence of oil degradation products although oil does exit the final port.

Table 1
Results for HRC™ - Concentrations (mg/l)

	Distance	Inlet	6 in.	12 in.	18 in.	36 in.	42 in.	48 in.	54 in.	60 in.	66 in.	72 in.
Day 10	TCE	10.2	5.7		7.6	5.3				0.0		
	DCE	0.6	9.0							10.3		
	VC	0.0	0.0							0.0		
Day 15	TCE	11.6	8.7		4.1	7.8		0.0				
	DCE	4.6	12.7		12.7	1.0		10.2				
	VC	0.0	0.0		0.0	11.7		0.0				
Day 23	TCE	12.0	5.8			1.6					0.0	
	DCE	1.1	12.5			0.0					1.5	
	VC	0.0	0.0			13.1					0.0	
Day 30	TCE	18.5	8.7			5.4						
	DCE	0.0	0.0			0.0						
	VC	2.1	4.6			6.6						
Day 35	TCE	26.4	20.2		15.0	10.8		7.3			3.8	
	DCE	0.0	0.0		0.0	0.0		0.4			0.5	
	VC	1.3	6.2		21.7	6.1		9.1			3.6	
Day 44	TCE	18.4	14.7		9.7	9.8		11.5		12.2		
	DCE	0.0	0.0		9.2	14.5		14.2		6.6		
	VC	1.1	7.8		0.0	0.0		0.0		0.0		
Day 51	TCE	22.4		9.7	9.1		0.0					9.0
	DCE	0.0		0.0	0.0		0.0					0.0
	VC	0.0		0.0	0.0		0.0					0.0
Day 57	TCE	19.8	8.1			5.4			4.5			0.0
	DCE	0.0	0.0			0.0			0.0			0.0
	VC	0.0	0.0			0.0			0.0			0.0

Table 2
Results for Molasses – Concentrations (mg/l)

	Distance	Inlet	6 in.	18 in.	48 in.
Day 1	TCE	18.2	8.2	20.3	20.4
	DCE	0.0	7.6	0.0	0.8
	VC	0.0	0.0	0.0	0.0
Day 3	TCE	19.8	17.0		19.5
	DCE	0.0	0.7		0.6
	VC	0.0	0.0		0.0
Day 7	TCE	20.0	20.0	14.1	20.0
	DCE	0.0	0.0	0.0	0.0
	VC	0.0	0.0	0.0	0.0

Table 3
Results for Vegetable Oil – Concentration (mg/l)

	Distance	Inlet	6 in.	18 in.	24 in.	36 in	42 in.	60 in.
Day 3	TCE	18.5	8.7		4.7			
	DCE	0.0	0.0		0.0			
	VC	2.1	0.8		1.1			
Day 8	TCE	26.4	22.2		9.0	17.3	7.7	7.6
	DCE	0.0	0.0		0.0	0.0	0.0	0.0
	VC	1.3	1.2		2.0	3.3	1.3	1.5
Day 17	TCE	18.43	14.99	21.1		21.8		24.4
	DCE	0	0	0.0		0.0		0.0
	VC	1.11	1.78	0.9		1.4		1.5

It appears that the oil moves rapidly through the tube along the upper portion of the soil providing some remediation effects that quickly are diminished. The movement of oil in field sites has not received sufficient attention in a wide variety of sites to determine the movement patterns. From petroleum oil studies we know that oil will move in plumes in aquifers.

CONCLUSIONS

Although it is difficult to get the same flow patterns in a small system as one does in the field the volumes of the ASV should be large enough to be representative of flow in a homogeneous system. AS seen in field situations the HRC™ provides a source of lactic acid and thus hydrogen for long periods of time and does not seem to migrate from the point of injection. Although vegetable oil and sugar solutions appear to be useful materials to stimulate biodegradation on a biochemical basis, they seem to require a continuous injection if they are to be useful in the field. With continuous injection of large quantities of reducing agent, the ancillary issues of "flushing" the contaminant and maintaining the appropriate conditions for dehalogenation downstream become a concern.

While many reducing agents appears to work in a confined system conditions in flowing systems should be studied to determine effectiveness. The ASV system is a scale up from test tubes and appears to present an opportunity to study effects due to water flow during remediation of materials using microbially competent soil.

REFERENCES

Farone, W.A., S.S. Koenigsberg, and J. Hughes. 1999. "A Chemical Dynamics Model for CAH Remediation with Polylactate Esters." Engineered Approaches of In Situ Bioremediation of Chlorinated Solvent Contamination, Andrea Leeson and Bruce Alleman, Editors, Batelle Press, 1999, 287-292.

Farone, W.A., S.S. Koenigsberg, T. Palmer and D. Brooker. 2000 "Site Classification for Bioremediation of Chlorinated Compounds Using Microcosm Studies." Bioremediation and Phytoremediation of Chlorinated and Recalcitrant Compounds, Godgae B. Wickramamayake, Arun R. Gavaskar, Bruce C. Alleman and Victor S. Magar, Editors, Battelle Press, 2000, 101-106.

POTENTIAL FOR BIOREMEDIATION OF GROUNDWATER CONTAMINATED WITH LANDFILL LEACHATE

David L. Freedman, Jennifer Cox, Laurin Baiden, Cristina Carvalho, Claudia K. Gunsch, and Jonathan Hunt (Clemson University, Clemson, SC, USA)
Robin L. Brigmon (Westinghouse Savannah River Corp. LLC, Aiken, SC, USA)

ABSTRACT: Solid waste was disposed of in an unlined sanitary landfill (Non-Radioactive Waste Disposal Facility) at the Department of Energy's Savannah River Site from 1974 to 1994. Some of the waste consisted of rags and wipes soaked with solvents, including tetrachloroethene (PCE) and trichloroethene (TCE). A groundwater plume containing low levels of PCE, TCE and dechlorination products has migrated beyond the landfill boundaries towards a wetland connected to the Savannah River. The objective of this study was to evaluate the potential for monitored natural attenuation (MNA) and enhanced bioremediation of the chlorinated ethenes based on historical geochemical data, using phase I of the RABITT (Reductive Anaerobic Biological In Situ Treatment Technology) assessment protocol. A high hydraulic conductivity (>0.002 cm/s) in the region downgradient of the landfill and the appearance of daughter products (dichloroethenes and vinyl chloride) resulted in a score considered satisfactory for evidence of hydrogeological and geochemical conditions favorable for dechlorination. However, the groundwater appears to be exhausted of dissolved organic carbon (<10 mg/L), and field measurements of ethene formation have not been made. Microcosms were set-up with phosphate-buffered groundwater to evaluate the potential to stimulate dechlorination activity by addition of lactate, as well as MNA. Even after 165 days of incubation, no significant reductive dechlorination was observed above autoclaved controls. There was also very little stimulation of methanogenesis. The lack of enhanced dechlorination by adding lactate may be related to a deficiency of nutrients other than phosphate.

INTRODUCTION

Solid waste was disposed of in an unlined, 70-acre trench-and-fill sanitary landfill (Non-Radioactive Waste Disposal Facility) at the Department of Energy's Savannah River Site (SRS) (Fig. 1a). The original 32-acre unlined site received sanitary waste from cafeterias, shops, construction areas, and offices. It reached capacity in 1987, at which time the 16-acre northern and 22-acre southern expansions were opened (Fig. 1b). The southern expansion reached capacity and was closed in 1993 (WSRC, 1996).

From 1974 to 1994, some of the waste sent to the original landfill consisted of rags and wipes soaked with solvents, including tetrachloroethene (PCE) and trichloroethene (TCE). Contamination of the underlying groundwater (GW) with landfill leachate containing these solvents resulted in a settlement agreement with the South Carolina Department of Health and Environmental Control (91-51-SW). A plume containing low levels of PCE, TCE, and dechlorination products has migrated beyond the landfill boundaries towards a wetland. Upper Three Runs Creek is approximately 2310 feet from the landfill and flows to the Savannah River. The flood plain and the creek are believed to be the point of discharge for GW downgradient of the landfill.

Most of the landfill has been capped with an impermeable barrier to minimize infiltration of surface water. Monitored natural attenuation (MNA) is being considered for GW remediation. Considerable amounts of geochemical data have been collected at various locations in and around the landfill site. However, field measurements of *in situ*

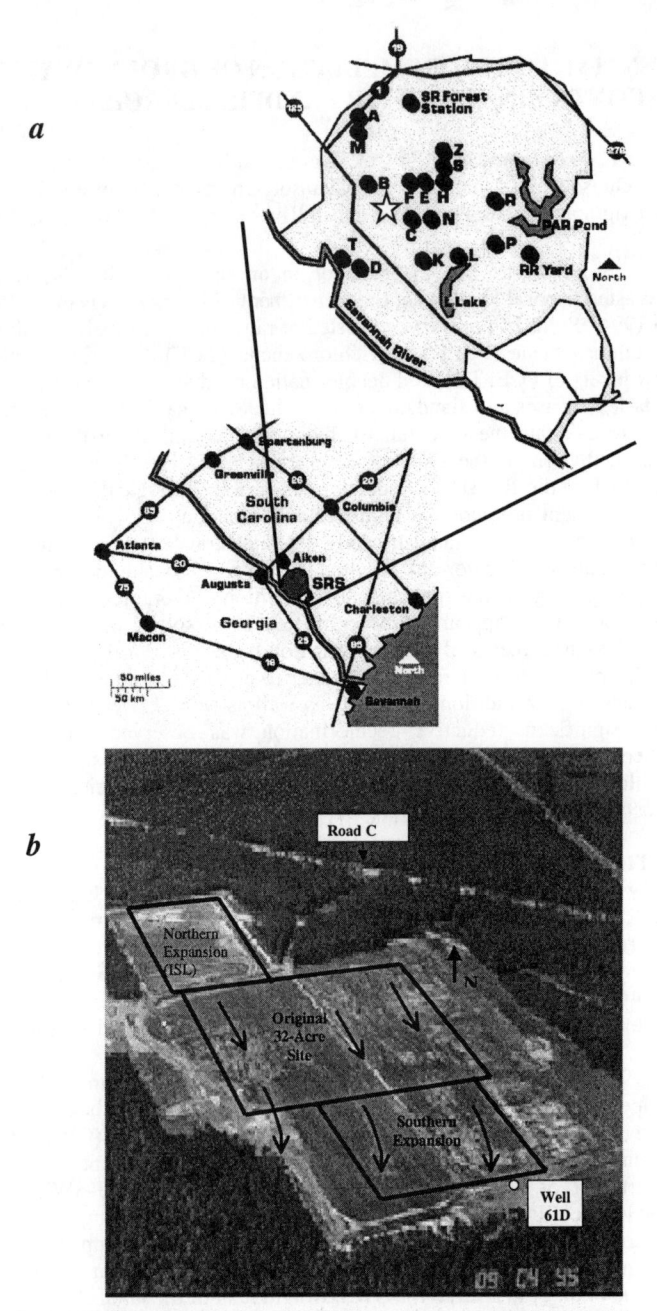

FIGURE 1. Location of the SRS sanitary landfill. (a) The star on the expanded view marks the landfill site. (b) Aerial view of the landfill showing the original unit and two expansions, plus the location of well 61D used for the microcosm study. Arrows indicate the direction of GW flow.

ethene formation have not been made. Since the landfill has aged considerably and the presence of available organic carbon has declined, enhancing dechlorination by addition of an electron donor is another remediation approach worthy of consideration.

Objective. The objective of this study was to evaluate the potential for MNA of the remaining chlorinated ethenes in SRS landfill GW, along with enhancing anaerobic reductive dechlorination through addition of an electron donor. Historical GW monitoring data were used to evaluate the site based on phase I of the RABITT protocol. Microcosms were used to assess the potential for enhancing dechlorination through addition of lactate and/or sulfate.

MATERIALS AND METHODS

RABITT Protocol. Morse et al. (1998) developed a phased approach to evaluating contaminated sites for their likely response to treatment by RABITT. The first phase involves a quantitative assessment of site potential based on changes in contaminant concentrations over time, the hydrogeological profile, and various geochemical parameters. In phase two, more intensive field monitoring is conducted in a test plot within the plume. Microcosms are used in phase three to evaluate the potential to enhance dechlorination by addition of nutrients and an electron donor. The final phase consists of a field test based on the microcosm results.

GW beneath the SRS sanitary landfill has been intensively monitored on a quarterly basis since 1988, through a network of 56 wells located within and around the site. These data were used to conduct the phase I analysis for RABITT potential.

Microcosms. The procedure outlined by Morse et al. (1998) was used to set up the microcosms, with modifications. Microcosms were prepared by adding 100-ml of GW from SRS well 61D to 160-ml serum bottles, in an anaerobic glove box, and capping them with Teflon™-faced rubber septa. It would have been preferable to also add soil from the same location, but a core sample was not available. First flush GW was used in an attempt to collect as much biomass as possible. GW from well 61D showed the highest level of chlorinated aliphatics for the most recently available data in 1998 (14.4 µg/L TCE, 59.1 µg/L 1,1-dichloroethane, 25 µg/L 1,1-dichloroethene (DCE), and 50 µg/L vinyl chloride (VC)). The pH was adjusted from 5.2 to 6.7 by adding dibasic potassium phosphate (400 mg/L), rather than using bicarbonate. Resazurin (1 mg/L) was added as a redox indicator.

Eight treatments were set-up (Table 1), based on addition of PCE, lactate, and/or sulfate. Field measurements from available monitoring wells indicate very little PCE remaining in the GW. The intent of adding PCE in some of the treatments was to provide the best opportunity to observe dechlorination activity in general, and ethene formation in particular. Dechlorination of *cis*-DCE and VC has been correlated with cultures that are actively reducing PCE (DiStefano, 1999). PCE was added from a water-saturated solution to provide an initial concentration of approximately 700 µg/L. Lactate was added as a supplemental source of organic carbon to a concentration of 0.33 mM, providing 100 times more reducing power than the stoichiometric requirement for complete dechlorination of the PCE to ethene. A higher lactate dose is specified in the RABITT protocol but did not appear to be necessary for this GW. No other electron donors were added in this round of testing, nor was addition of yeast extract and vitamin B_{12}. Addition of sulfate was examined to determine if it might accelerate the onset of obligate anaerobic conditions. Enough was added to permit complete lactate oxidation, ignoring cell synthesis (0.5 mM). Since the glove-box atmosphere contained approximately 4% H_2, all of the bottles had some readily available electron donor to support reductive dechlorination.

Two types of controls were used: autoclaved GW and deionized water, both with PCE added. All serum bottles were incubated in an inverted position in the anaerobic glove box at room temperature (22-24°C).

The amount of volatile compounds (PCE, TCE, DCEs, VC, ethene, and CH_4) present was determined by gas chromatographic analysis of headspace samples, as previously described (Freedman and Gossett, 1989). Hydrogen was monitored in a similar manner, using a thermal conductivity detector. Organic acids were measured by high performance liquid chromatography using a BIORAD Aminex HPX-87H ion exclusion column and 5 mM H_2SO_4 mobile phase.

TABLE 1. Experimental design for the microcosm study.

Treatment # (description)[a]	GW	PCE	Lactate	Sulfate
1 (as is)	√			
2 (add supplemental e.d.)	√		√	
3 (add supplemental e.a.)	√			√
4 (add supplemental e.d. and e.a)	√		√	√
5 (add PCE)	√	√		
6 (add PCE + supplemental e.d.)	√	√	√	
7 (add PCE + supplemental e.a.)	√	√		√
8 (add PCE + supplemental e.d. + e.a.)	√	√	√	√

[a] e.d. = electron donor (lactate), e.a. = electron acceptor (sulfate).

RESULTS AND DISCUSSION

Phase I RABITT Assessment. Extensive GW monitoring data were used to determine if conditions at the SRS landfill site are conducive to RABITT. The results are summarized in Table 2. The more positive the score, the more suitable the site is considered for RABITT.

Daughter products (*cis*-DCE, *trans*-DCE and VC) have been detected in the GW, although none were disposed of in the landfill. This provides direct evidence for past reductive dechlorination. PCE and TCE are the most likely parent compounds for *cis*- and *trans*-DCE. VC likely formed from *cis*- and *trans*-DCE as well as some 1,1-DCE, resulting from dehydrohalogenation of 1,1,1-trichloroethane. The appearance of VC allows for a daughter product score of +15. A higher score is assigned when ethene is detected above background, but it appears that no measurements of ethene have ever been attempted with landfill GW.

The hydrogeological profile score is based on hydraulic conductivity (K). In the saturated zone downgradient (south) of the landfill, K is 0.023 cm/s. Values above 0.001 cm/s receive the highest RABITT rating possible (+25). It should be noted, however, that conductivities in the vicinity of the landfill are highly variable. For example, west of the landfill, the average conductivity is 10^{-4} cm/s, which would drop the hydrogeological profile score to zero.

Information was available for some but not all of the geochemical parameters. Dissolved oxygen levels in excess of 3.0 mg/L have been reported for a number of samples, resulting in a RABITT score of −3. Nitrate concentrations below 1 mg/L yield the highest possible score for this parameter, since an absence of nitrate is required to establish reductive dechlorination. Low amounts of hydrogen sulfide have been detected and therefore contributed positively. Sulfate levels have consistently been below 20 mg/L, also resulting in a positive score. No data were available for redox levels or temperature, although the latter was conservatively estimated to be between 10 and 15°C. Dissolved organic carbon levels below 10 mg/L are the norm, hence the low rating for this surrogate of electron donor availability. Bicarbonate alkalinity data were not

Laboratory Studies of Reductive Processes 113

available, although it is almost always below 1000 mg/L in this part of the country, hence the –1 score. The pH values vary considerably, but generally fall in the slightly acid region, resulting is a score of –1. Only limited measurements of methane have been made with landfill GW.

The total score of +42 falls in a region considered "satisfactory," with a suggestion to proceed to microcosm testing and evaluation of the potential for complete dechlorination to ethene. However, there is reason for caution in interpreting this score. The biggest contributor is the high hydraulic conductivity. Because of the variability of K in this region, a case could be made to assign a lower score.

Furthermore, in spite of the large number of wells and high frequency of sampling, there is a paucity of data for several key parameters, especially ethene. The low level of VC found in downgradient wells at the time of this analysis (in relation to historical maximum levels of the parent compounds) suggests that dechlorination to ethene has occurred to some extent, or that other processes are responsible for mitigating accumulation of VC (e.g., biooxidation of VC under aerobic or anaerobic electron acceptor conditions). Verification of VC transformation, either by *in situ* measurement of ethene and/or microcosm data, is critical to evaluating this site for selection of MNA or RABITT.

TABLE 2. RABITT Rating System Score for the SRS Landfill.

Rating Parameter	Potential Score	Score Assigned
Evidence of daughter product formation	(-6) to (+25)	+15
Hydrogeological profile	(-50) to (+25)	+25
Geochemical profile		
Dissolved Oxygen	(-3) to (+3)	-3
Nitrate	(-3) to (+3)	+3
Hydrogen sulfide	(0) to (+3)	+3
Sulfate	(0) to (+2)	+2
Redox potential	(-1) to (+1)	-1
Temperature	(-3) to (+3)	0
Dissolved Organic Carbon	(0) to (+3)	0
Bicarbonate alkalinity	(-1) to (+1)	-1
pH	(-5) to (+3)	-1
Methane	(0) to (+3)	0
Total Point Value		+42

Microcosms. The initial concentration of PCE in treatments 1-4 (without PCE addition) was approximately 39 µg/L. This was higher than levels reported for field data, perhaps since a first-flush sample was taken rather than allowing the well to be purged. The average initial TCE concentration was 13 µg/L. Only trace levels of other volatile compounds were detected. PCE disappeared from treatments 1-4 within the first two weeks. The redox level in all of the treatments decreased below –110 mV, as indicated by a change in the resazurin color from pink to clear. This occurred within 8 days in bottles that received lactate, and within 13-20 days in the others. TCE levels decreased slightly over the 164-day study period, but there was not a commensurate increase in *cis*-DCE, *trans*-DCE, VC, or ethene. The level of dechlorination activity was no higher in bottles supplemented with lactate and/or sulfate. Hydrogen (present from the glove box atmosphere during set-up) consumption was evident only in treatments amended with sulfate. In treatments 2 and 4, most of the added lactate was consumed by day 137, with a corresponding increase in acetate. Nevertheless, no methane production occurred in any of the treatment 1-4 bottles. Although methanogens do not typically play a direct role in anaerobic reductive dechlorination, methanogenesis is often correlated with dechlorination activity (Freedman and Gossett, 1989). The lack of methanogenic activity was consistent with the absence of significant dechlorination.

Similar results were obtained for treatments 5-8, to which PCE was added. Representative trends for the bottles supplemented with lactate (treatment 6) are shown in Figure 2. The magnitude of decrease in PCE in the live bottles was no greater than in the autoclaved controls, and there was no accumulation of daughter products during 164 days of incubation. Likewise, there was no production of methane, and hydrogen remained essentially constant except for a significant decrease in the bottles supplemented with sulfate (treatments 7 and 8). All of the treatments developed redox conditions below –110 mV within the first 7-20 days. Between 55 and 100% of the lactate added (0.33 mM) was consumed by day 137, and acetate accumulated. A second addition of lactate was made on day 137, but very little additional consumption occurred over the next 47 days, indicative of very sluggish microbial activity in general.

The reason for little or no methanogenic and reductive dechlorination activity in the microcosms is not yet known. A low inoculum level may have been a factor, since soil was not available to add with the GW. Preliminary microcosms set-up with GW taken after purging well 61D (and hence having even less biomass) showed even more sluggish behavior, judging from the additional time it took for the redox level in them to drop below –110 mV, and a similar lack of methanogenesis and reductive dechlorination. This indicated the need for a higher inoculum level, which may not have been adequately met by using only a first-flush GW sample. Furthermore, field DOC data suggest that the landfill GW has been low in substrate for over 10 years, due in part to minimizing additional leachate by preventing infiltration with an impermeable cap. The absence of organic carbon for this long a time, combined with the slow growth rate of methanogens, might result in a lag in microbial activity exceeding six months (the length of the microcosm study), following addition of a substrate such as lactate. Metal toxicity is another potential concern, since metals are often present in leachate at high enough levels to inhibit methanogenesis. However, it appears that heavy metals have not been detected in significant concentrations in SRS GW samples. Sulfide toxicity is another possibility, although most of the historical GW data suggests sulfate levels were probably low (i.e., <20 mg/L), eliminating the potential for significant sulfide formation in the no-sulfate-added microcosms.

Although microcosm results often corroborate field data (and vice versa), this is not always the case. For example, Edwards and Cox (1997) reported on a lack of chlorinated ethene dechlorination in microcosms from GW beneath a waste lagoon, in spite of field evidence for conversion of *cis*-DCE to VC. They suggested that the presence of chloroform in samples used to set-up the microcosms inhibited methanogenesis and dechlorination. In microcosms reported by Odom et al. (1995), one set out of six showed no dechlorination activity, although activity was observed at all field locations. Shanke et al. (1997) also reported a lack of reductive dechlorination in a subset of microcosms used to evaluate a contaminated manufacturing site, in spite of field evidence of dechlorination activity at all locations.

Summary. This SRS site illustrates some of the complexities involved in predicting reductive dechlorination at aged landfills. Extensive field data collected over more than 10 years provides convincing evidence that reductive dechlorination of chlorinated ethenes has occurred at the SRS sanitary landfill. However, downgradient migration of VC raises some concern about the applicability of adopting MNA. There is a critical need to analyze GW samples at multiple locations and depths for ethene, to determine if the anaerobic microbial potential exists in this area for reductive dechlorination of VC to ethene. Results from the microcosms demonstrated the ease and quickness with which low redox conditions can be established in the GW. Nevertheless, methanogenesis was not observed, nor was the onset of PCE dechlorination. This suggested a nutrient

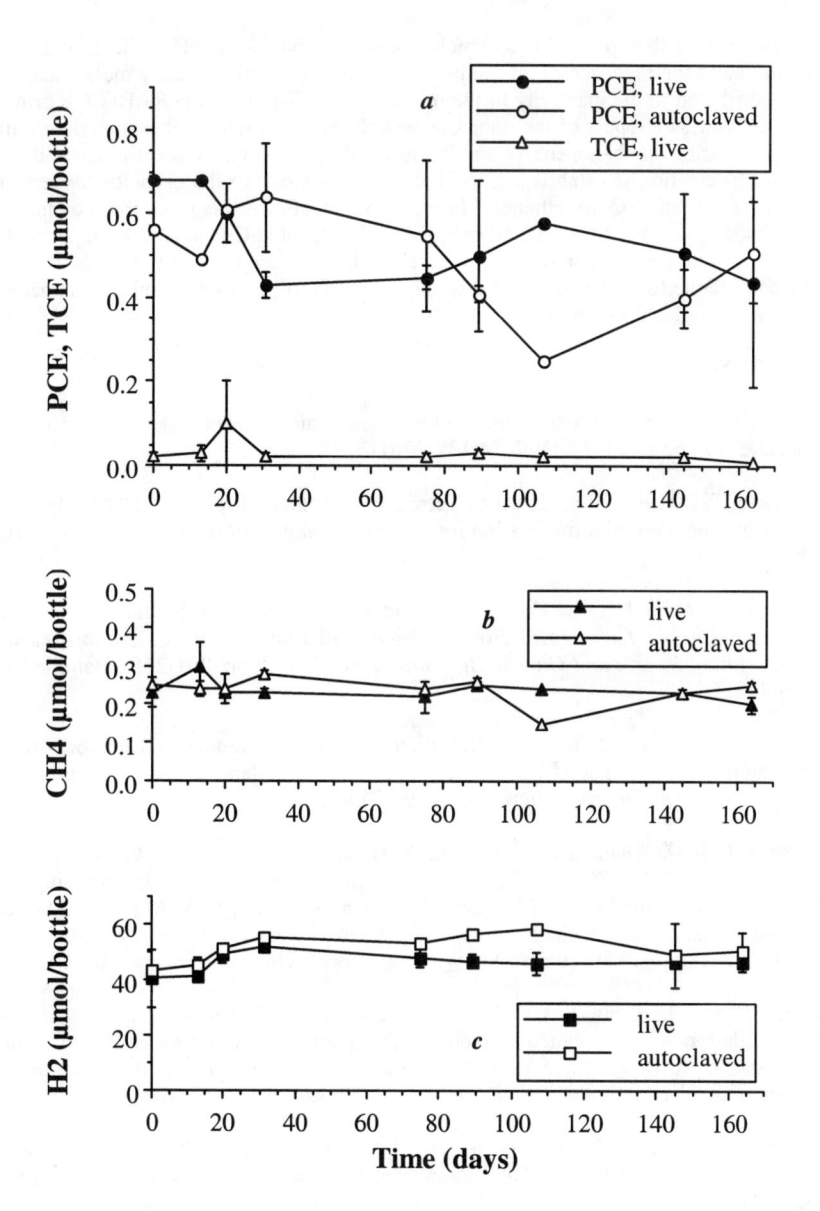

FIGURE 2. Representative GW microcosm results for addition of PCE and lactate, live bottles (treatment 6) in comparison to autoclaved controls, showing (a) PCE and TCE; (b) methane; and (c) hydrogen. Error bars are one standard deviation of duplicate bottles.

deficiency other than phosphorus, which was used to buffer the pH. A lack of adequate inoculum and the presence of an inhibitory contaminant such as heavy metals may have also contributed to the inactivity of the microcosms. The favorable RABITT scoring for this site justifies a repeat of the microcosm study, with greater emphasis on determining what is limiting methanogenesis and dechlorination activity. Once the capability for PCE dechlorination is established, the focus should move to the capacity for reductive dechlorination of VC to ethene. Information is also needed on the potential for biodegradation of VC by oxidative processes (Bradley and Chapelle, 2000), particularly in the downgradient portion of the aquifer preceding the seep line in the wetland. It may then be possible to further develop a strategy in support of MNA or enhanced anaerobic and/or aerobic dechlorination.

REFERENCES

Bradley, P. M. and F. H. Chapelle. 2000. "Acetogenic Microbial Degradation of Vinyl Chloride." *Environ. Sci. Technol. 34*(13): 2761-2763.

Distefano, T. D. 1999. "The Effect of Tetrachloroethene on Biological Dechlorination of Vinyl Chloride: Potential Implication for Natural Bioattenuation." *Wat. Res. 33*(7): 1688-1694.

Edwards, E. A. and E. E. Cox. 1997. "Field and Laboratory Studies of Sequential Anaerobic-Aerobic Chlorinated Solvent Biodegradation." in B. C. Alleman and A. Leeson (Eds.), *In Situ and On-Site Bioremediation, Vol. 3*, pp. 261-265. Battelle Press, Columbus, OH.

Freedman, D. L. and J. M. Gossett. 1989. "Biological Reductive Dechlorination of Tetrachloroethylene and Trichloroethylene to Ethylene under Methanogenic Conditions." *Appl. Environ. Microbiol. 55*(9): 2144-2151.

Morse, J. J., B. C. Alleman, J. M. Gossett, S. H. Zinder, D. E. Fennell, G. W. Sewell and C. M. Vogel. 1998. *A Treatability Test for Evaluating the Potential Applicability of the Reductive Anaerobic Biological In Situ Treatment Technology (RABITT) to Remediate Chloroethenes.* Report of the Environmental Security Technology Certification Program (ESTCP), Arlington, VA; available at www.estcp.org/technical-documents.htm.

Odom, J. M., J. Tabinowski, M. D. Lee and B. Z. Fathepure. 1995. Anaerobic Biodegradation of Chlorinated Solvents: Comparative Laboratory Study of Aquifer Microcosms," in R. E. Hinchee, A. Leeson and L. Semprini (Eds.), *Bioremediation of Chlorinated Solvents, Vol. 4*, pp. 17-24. Battelle Press, Columbus, OH.

Schanke, C. A., A. D. Bettermann and L. L. Graham. 1997. "Biological Treatability Studies for Remediation of TCE-Contaminated Groundwater," in B. C. Alleman and A. Leeson (Eds.), *In Situ and On-Site Bioremediation, Vol. 3*, pp. 267-272. Battelle Press, Columbus, OH.

WSRC. 1996. *Sanitary Landfill In Situ Bioremediation Optimization Test Final Report, Savannah River Site.* Report No. WSRC-TR-96-0065, Westinghouse Savannah River Company, Aiken, SC.

VARYING SUBSTRATE CONCENTRATION TO ENHANCE TCE DEGRADATION IN DUAL-SPECIES BIOREACTORS

John Komlos, Al Cunningham and Anne Camper (Montana State University, Bozeman, Montana, USA)
Robert Sharp (Manhattan College, Riverdale, New York, USA)

ABSTRACT: Laboratory experiments were performed to enhance biodegradation of trichloroethylene (TCE) in a porous media reactor by combining *Burkholderia cepacia* PR1-pTOM31c, a TCE-degrading organism unable to form a stable biofilm, with *Klebsiella oxytoca*, a thick biofilm-forming organism. In addition, growth rate and substrate concentration were explored to determine which variables affected the population dynamics of these two organisms in a dual-species environment. Results indicate that *B. cepacia* and *K. oxytoca* can co-exist in a dual-species biofilm and growth rate was not an adequate predictor of which organism would out-compete the other. Rather, substrate concentration was a dominant variable in controlling the population distribution of the two organisms in a biofilm. At high substrate concentrations, *K. oxytoca* was the dominant organism in the dual-species biofilm, while at lower substrate concentrations, *B. cepacia* became the dominant organism. Results from porous media column reactors showed a similar shift in population with change in substrate concentration. *B. cepacia* achieved greater population density, and higher TCE-degrading potential, at the lower substrate level. Varying substrate concentration enabled the manipulation of the population distribution in dual-species biofilm and porous media reactors.

INTRODUCTION

The utilization of bacteria for the remediation of hazardous wastes is widespread and has been implemented in such technologies as vapor phase bioreactors, pump and treat reactors and biobarrier technologies. In order to optimize these processes, it is important to establish and maintain a reactive bacterial population in the porous media matrix of these engineered systems. It is necessary for the bacterial population to produce enough biofilm material to attach to a surface in addition to carrying out the desired reaction. If the reactive bacterial population is unable to form a stable biofilm, it will not be able to colonize the porous medium in sufficient density to provide adequate biotransformation (Sharp, 1995). Therefore, the concept of using two species to develop a reactive biofilm, with a stable biofilm-forming organism providing a matrix for the reactive organism, would be a significant advance in bioreactor design. This concept could be expanded further for use in reactive/reduced permeability biobarriers. In this case, the thick biofilm-forming organism would create a matrix capable of reducing porous media permeability to prevent or slow contaminant migration, while the second organism would perform the desired reaction(s). This dual-species reactive biobarrier would offer the opportunity to separately select for microorganisms with specific characteristics (biofilm formation or contaminant degradation) which are combined into a single biofilm capable of performing multiple functions.

Since recalcitrant compounds such as trichloroethylene (TCE) show little natural attenuation, the development of a TCE reactive biofilm could lead to a significant advancement in bioremediation technology. Few organisms are known to aerobically degrade TCE. One such organism, *Burkholderia cepacia* PR1-pTOM31c, can constitutively degrade TCE (Shields and Reagin, 1992), but is unable to form a stable biofilm capable of reducing porous media permeability (Sharp, 1995). It would be desirable to combine *B. cepacia* with *Klebsiella oxytoca*, a thick biofilm-forming organism, in an attempt to create a TCE-degrading biofilm.

For the application and implementation of this technology, as well as other multi-species applications, it is essential to have an understanding of how the bacterial species interact in a biofilm and porous media environment. Though multi-species interactions have been extensively examined in both batch and chemostat (continuous culture) environments, very little research has been performed on multi-species interactions in a biofilm. In addition, the research that has been performed to date (Banks and Bryers, 1992; James et al., 1995; McEldowney and Fletcher, 1987; Siebel and Characklis, 1991; Skillman et al., 1998; Sturman et al., 1994) has resulted in conflicting conclusions about how bacteria interact with one another in a biofilm. For this reason it was important to perform preliminary experiments examining the feasibility of combining a TCE-degrading microorganism and a thick biofilm-forming microorganism in a dual-species environment in which both species can coexist in significant numbers. In addition, the effects of growth rate and substrate concentration on species composition in a dual-species biofilm were examined.

MATERIALS AND METHODS

Bacterial Strains. The bacterial isolates used for these experiments were *Burkholderia cepacia* PR1-TOM31c and *Klebsiella oxytoca*. *B. cepacia* is an aerobic bacterium that can constitutively degrade TCE in cometabolic processes using the toluene ortho-monooxygenase (TOM) pathway (Shields and Reagin, 1992). The genetic information for the degradative pathway is located on the plasmid, pTOM31c. The plasmid also encodes for the resistance to the antibiotic, kanamycin, which was used for selection of *B. cepacia*. *K. oxytoca* is a facultative bacterium and the strain used in this work was isolated from an oil field (MacLeod et al., 1988). Its ability to form thick biofilms made it an ideal candidate for use as the mucoid organism in the dual-species experiments. *K. oxytoca* was selected utilizing its resistance to streptomycin.

Bacterial Isolation and Characterization. Selective and non-selective nutrient agar plates were used to characterize the dual-species population. *B. cepacia* was selected on either modified Luria-Bertani (LBG) agar plates (Sharp, 1995) amended with 0.05 g/L kanamycin or phenol agar plates amended with kanamycin (15 g Bacto agar per liter of hydrocarbon minimal medium [HCMM2] [Sharp, 1995] with 0.0941 g/L phenol and 0.05 g/L kanamycin added 45 minutes after autoclaving). *K. oxytoca* was selected on Brain Heart Infusion (BHI) agar plates amended with streptomycin (4 g BHI media [Difco] and 15 g Bacto agar per liter of distilled water with 0.1 g/L filter sterilized streptomycin sulfate added 45 minutes after autoclaving). R2A (Difco) was used as the non-selective nutrient agar to determine total cell numbers and provide a total cell balance.

Inoculum Preparation. To prepare a viable, TCE-degrading culture of *B. cepacia* for inoculation into the biofilm or porous media reactor, a loop-full of *B. cepacia* was transferred from a frozen culture (-70°C in 2% peptone, 20% glycerol) to a phenol/kanamycin agar plate and incubated at 30°C for 48 hours. A colony was then transferred to a LBG/kanamycin agar plate and incubated at 30°C for 24 hours. A colony from the LBG/kanamycin plate was then transferred to 100 mL LBG broth (without kanamycin and agar), and incubated for 18 hours at 36°C on a horizontal shaker (150 rpm). One mL of this culture was transferred to 100 mL fresh LBG broth and incubated for 18 hours at 36°C on a horizontal shaker (150 rpm).

K. oxytoca was transferred from a frozen culture and incubated at 30°C for 24 hours on a BHI/streptomycin agar plate. A colony was transferred to a 100 mL LBG broth and incubated at 36°C on a horizontal shaker (150 rpm). After 18 hours, 1 mL was transferred to 100 mL fresh LBG broth and incubated for 18 hours at 36°C on a horizontal shaker (150 rpm).

Biofilm Experiments. A rotating disk reactor (RDR) system was used to determine the biofilm growth rates and population densities of the two organisms together for varying substrate concentrations. A RDR is a one liter glass beaker containing a magnetically driven rotor with six 1.27 cm diameter biofilm test-surface coupons. A drain spout on the side of the RDR provided a volume of 180 mL and wetted surface area (including stir plate) of 253 cm^2. A complete description of a RDR can be found in Zelver et al. (1999).

Continuous nutrient addition of 1:10, 1:100, or 1:500 diluted LBG broth (700 mg/L, 70 mg/L or 14 mg/L dissolved organic carbon, respectively) was supplied to the reactors using a peristaltic pump at a rate between 5.4 and 5.7 mL/min resulting in a detention time of 32-34 minutes. Flow rates were established using a graduated cylinder and stopwatch. This particular detention time was selected to minimize planktonic cell doubling so that any cell suspended in the reactor was the result of biofilm detachment. A cell balance on the biofilm and suspended region allows for the biofilm growth rate to be determined. Flow breaks were situated upstream of the reactor to prevent bacterial contamination of the nutrient reservoir.

Two mL of concentrated growth culture was added to each rotating disk reactor through syringe injection and left in batch mode for one hour. Nutrient media was then pumped through the reactor and three hours were allowed to wash out the high concentration of inoculated cells in suspension before sampling began. Effluent samples from each reactor were taken over time until the predetermined end of the experiment (either 48 hours or 5 days) and were used to determine when steady-state conditions were obtained. The reactors were then destructively sampled and the biofilm scraped off of the disks as described by Zelver et al. (1999). The cells were then homogenized at 13,500 rpm for 30 seconds and grown on selective agar plates. Each scenario consisted of dual species inoculation at 1:10 or 1:100 diluted LBG concentrations. The average of two destructively sampled columns (48 hour and 5 day) was taken for each experiment, which was then repeated to ensure reproducibility. An additional experiment using 1:500 diluted LBG was performed.

Porous Media Column Experiments. The porous media bioreactors used in this experiment were stainless steel columns 25.4 cm in length, 2.54 cm in diameter, and packed with 1 mm-diameter glass beads. Flow was delivered to the system using the constant head setup described in Figure 1. Stainless steel screens were inserted at both ends of the column to secure the beads, and rubber stoppers fitted with glass tubes were used to seal the columns. Silicone tubing was connected to the glass tubes at the column influent and effluent. Nutrient media (1:10 and 1:100 diluted LBG broth) was pumped into the influent constant head tank, through the silicone tubing, and up through the columns.

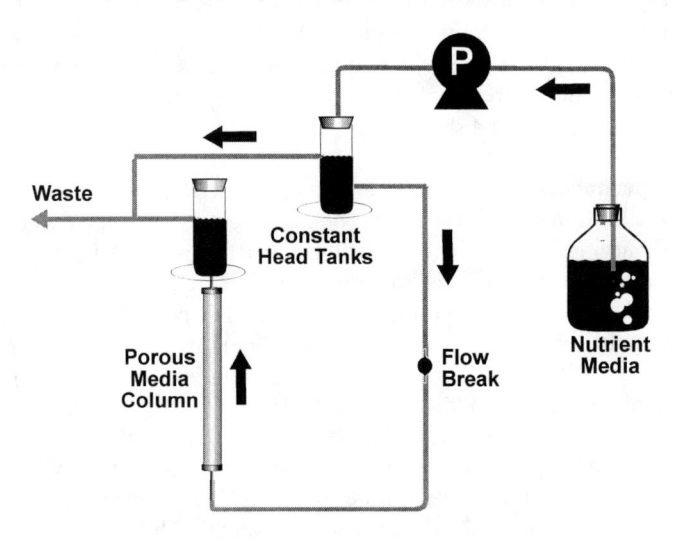

FIGURE 1. Constant head porous media bioreactor.

The columns were inoculated with *B. cepacia* and *K. oxytoca* by syringe injection in a septum just upstream of each column. With the pump turned off and the influent tubing clamped to prevent back flow, 25 mL of each growth culture was injected. The pump remained off for 12 hours, allowing the bacteria to establish within the columns. Following this assimilation period, nutrient media was pumped into the system. Periodic effluent flow rates were determined using a stopwatch and graduated cylinder. Effluent plate counts over time provided an indirect indication of population dynamics within the reactor.

Four to six days after inoculation, the columns were destructively sampled. After measuring final flow rates and plating final effluent samples, nutrient flow was suspended and the columns were drained. Approximately 10 grams of beads were removed from the beginning, middle, and end sections of each column and placed in separate test tubes. Fifteen mL of phosphate buffered saline solution (8.7 g NaCl, 0.4 g KH_2PO_4, 1.23 g K_2HPO_4 per liter distilled water) was added incrementally to each tube (5 mL, 5 mL, 3 mL, 2 mL). After each addition, the beads were vortexed for one minute and the supernatant was poured off. Microscopic analysis revealed that this "bead bashing" procedure removed virtually all of the biofilm from the beads. The resulting supernatants were homogenized and plated to determine population dynamics throughout the column. The biofilm numbers obtained from the beginning, middle and end of each column were averaged and the experiment repeated.

24-Hour TCE Disappearance Assay. To determine the TCE degrading potential of the bacterial population throughout each column, a 24-hour TCE disappearance assay was performed on bacterial samples desorbed from the glass beads. Two mL cell suspensions and 50 µL of 75 mg/L TCE were placed in 20-mm crimp-seal bottles (17 mL total capacity) yielding an aqueous phase TCE concentration of 0.64 mg/L. Controls with 2 mL sterile LBG media were used to measure abiotic TCE loss in the bottles. The bottles were capped with Teflon™-lined rubber stoppers (The West Company) and crimped before incubation at 36°C (inverted and shaken at 150 rpm). Headspace samples (200 µL) from each vial were analyzed using gas chromatography (Hewlett Packard 5890 Series 2) equipped with a 2.44 m by 2 mm I.D. glass column packed with 1% SP1000 on 60/80 Carbopak B (Supelco) and a Tracor 700A Hall Electrolytic Conductivity Detector. The injector and detector temperatures on the gas chromatograph were 200°C and 250°C, respectively. The Hall reactor temperature was 900°C and Hall electrolyte was 1-propanol. The oven temperature was initially 150°C for one minute and increased at a rate of 10°C/min to a final temperature of 200°C for two minutes. The carrier gas was helium (29 mL/min) and the reactor gas was H_2 (20 mL/min). Samples were analyzed initially and after 24 hours to determine the disappearance of TCE over a 24-hour period. This loss was compared to controls to quantify the extent of TCE degradation for each sample.

RESULTS AND DISCUSSION

Biofilm Experiments. Dual-species biofilm experiments were performed to determine the effects of substrate concentration on biofilm growth rates and biofilm population densities of *B. cepacia* and *K. oxytoca* using substrate concentrations of either 1:10 or 1:100 diluted LBG broth (700 mg/L or 70 mg/L dissolved organic carbon, respectively). The results indicate that the dual-species biofilm growth rate of *K. oxytoca* (0.82/hr ± 0.25/hr) is approximately twice that of *B. cepacia's* (0.33/hr ± 0.11/hr) for the low substrate concentration and *K. oxytoca's* growth rate (0.49/hr ± 0.16/hr) is slightly higher than *B. cepacia's* growth rate (0.28/hr ± 0.04/hr) for the high substrate concentration. Varying the substrate concentration did not statistically vary each organism's growth rate. In the high substrate experiment, *K. oxytoca's* population density (1.5e8 CFU/cm^2) was an order-of-magnitude higher than *B.* cepacia's (8.9e6 CFU/cm^2). The low substrate concentration showed the opposite of what would be expected. Now *B. cepacia's* population density (7.6e7 CFU/cm^2) was an order-of-magnitude higher than *K. oxytoca's* (7.8e6 CFU/cm^2), even though *K. oxytoca* has a growth rate higher than *B. cepacia*. This same trend was also observed by Camper et al. (1996) where slower growing organisms were able to persist at higher cell concentrations in low nutrient (oligotrophic) environments. Therefore, growth rate can predict planktonic dual-species population distribution (data not shown), but cannot be used to predict the fraction of each organism in a dual-species biofilm.

Another dual-species biofilm experiment was performed using a lower substrate concentration (14 mg/L carbon) and the results were combined with the above biofilm population density data to further show the effects of substrate concentration on biofilm population density (Figure 2). As the substrate concentration increases, the *K. oxytoca* population density increases, but the *B. cepacia* population density actually decreases. Varying the substrate concentration provided a mechanism to control the fraction of each organism in the dual-species biofilm.

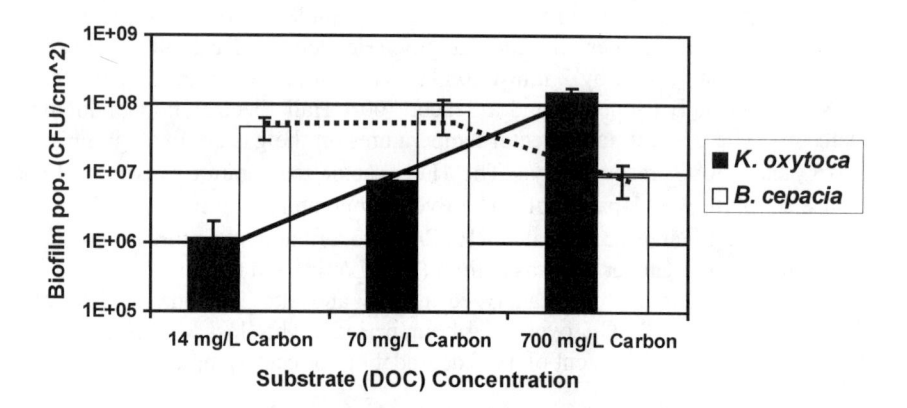

FIGURE 2. Effect of substrate concentration on dual-species biofilm population densities.

Bioreactor Experiments. The effect of substrate concentration on dual-species population dynamics was then examined using the porous media column reactor described in Figure 1. Using the high substrate concentration (700 mg/L carbon), *K. oxytoca* had a higher population density than *B. cepacia* (Table 1). But at the lower substrate concentration (70 mg/L carbon), there was the population shift observed in the biofilm experiments, with *B. cepacia* now having a higher population density than *K. oxytoca* (Table 1). A possible reason for the observed population shifts could be that at higher substrate concentrations more carbon is utilized, directly resulting in an increased loss of oxygen. Since *B. cepacia* is an aerobic organism while *K. oxytoca* is a facultative organism, it follows that *B. cepacia* would be adversely affected by the oxygen loss, while *K. oxytoca* would continue to persist in low oxygen environments.

TABLE 1. Porous media population densities (CFU/g glass beads) for low (70 mg/L carbon) and high (700 mg/L carbon) substrate concentrations.

	Low Substrate	High Substrate
Klebsiella oxytoca	1.4e8 ± 5e7	2.1e8 ± 4.7e7
Burkholderia cepacia	1.9e8 ± 1e8	3.1e7 ± 2.4e7

After the columns were destructively sampled, a 24-hour TCE disappearance assay was performed on the desorbed bacteria (Table 2). The *B. cepacia* population from the low substrate (70 mg/L carbon) experiment was able to remove approximately 60% more TCE than the high substrate (700 mg/L carbon) experiment, which coincides with the higher TCE-degrading population at lower substrate concentrations.

TABLE 2. Results from ex-situ 24-hour TCE disappearance assay.

	TCE Removal (%)
Low Substrate	84.6 ± 4.6
High Substrate	16.3 ± 5.7
Control	-1.1 ± 1.7

CONCLUSION

In a biofilm culture, the slower growing *B. cepacia* was not always outcompeted by the faster growing *K. oxytoca* indicating that biofilm growth rates could not be used to predict the dual-species population densities in a biofilm. However, substrate concentration could be used to predict the dual-species population density. At high substrate concentrations, *K. oxytoca* was the dominant organism in the dual-species biofilm. At low substrate concentrations, there was a shift in the population distribution and now *B. cepacia* was the dominant organism in the dual-species biofilm. This same shift in population distribution with change in substrate concentration was observed in porous media column reactors (bioreactors). An ex-situ 24-hour TCE disappearance assay demonstrated that the bioreactors with the lower substrate concentration had a higher TCE degradation potential.

Varying the substrate concentration can be used to regulate the fraction of each organism in these dual-species environments. The ability to manipulate the dual-species population distribution in biofilms and porous media could be useful in numerous applications including in situ remediation, reactor based remediation and potential industrial applications (enzyme/protein production, antibiotics).

ACKNOWLEDGMENT

This research has been funded in part by the Great Plains/Rocky Mountain Hazardous Substance Research Center, the National Science Foundation and MSE

Technology Applications Inc. (Butte, MT). The Biofilm Systems Training Lab (Montana State University) is acknowledged for loan of the Rotating Disk Reactors. Pamela McLeod, Laura Jennings and Allison Rhoads are acknowledged for providing technical assistance.

REFERENCES

Banks, M. K., and J. D. Bryers. 1992. "Deposition of Bacterial Cells onto Glass and Biofilm Surfaces." *Biofouling. 6*: 81-86.

Camper, A. K., W. J. Jones, and J. T. Hayes. 1996. "Effect of Growth Conditions and Substratum Composition on the Persistence of Coliforms in Mixed-Population Biofilms." *Appl. Environ. Microbiol. 62*: 4014-4018.

James, G. A., L. Beaudette, and J.W. Costerton. 1995. "Interspecies Bacterial Interactions in Biofilms." *J. Ind. Microbiol. 15*: 257-262.

MacLeod F. A., H. M Lappin-Scott, and J. W. Costerton. 1988. "Plugging of a Model Rock System by Using Starved Bacteria." *Appl. Environ. Microbiol. 54*: 1365-1372.

McEldowney S., and M. Fletcher. 1987. "Adhesion of Bacteria from Mixed Cell Suspension to Solid Surfaces." *Arch. Microbiol. 148*: 57-62.

Sharp, R. R. 1995. "Stability and Expression of a Plasmid-Borne TCE Degradative Pathway in Suspended and Biofilm Cultures." Ph.D. Thesis, Montana State University, Bozeman, MT.

Shields, M. S., and M. J. Reagin. 1992. "Selection of a *Pseudomonas cepacia* Strain Constitutive for the Degradation of Trichloroethylene." *Appl. Environ. Microbiol. 58*: 3977-3983.

Siebel, M. A., and W. G. Characklis. 1991. "Observations of Binary Population Biofilms." *Biotechnol. Bioeng. 37*: 778-789.

Skillman, L. C., I. W. Sutherland, M. V. Jones, and A. Goulsbra. 1998. "Green Fluorescent Protein as a Novel Species-Specific Marker in Enteric Dual-Species Biofilms." *Microbiology 144*: 2095-2101.

Sturman, P., W. L. Jones, and W. G. Characklis. 1994. "Interspecies Competition in Colonized Porous Pellets." *Water Res. 28*: 831-839.

Zelver, N., M. Hamilton, B. Pitts, D. Goeres, D. Walker, P. Sturman, and J. Heersink. 1999. "Measuring Antimicrobial Effects on Biofilm Bacteria: From Laboratory to Field." In R. J. Doyle (Ed.), *Methods in Enzymology: Volume 310 - Biofilms*, pp. 608-628. Academic Press.

THE FEDERAL INTEGRATED BIOTREATMENT RESEARCH CONSORTIUM (FLASK TO FIELD)

Jeffrey W. Talley (University of Notre Dame, South Bend, Indiana), Deborah R. Felt, (Applied Research Associates, Inc., Vicksburg, MS), Lance D. Hansen, (ERDC), Jim C. Spain, (Armstrong Laboratory, Panama City, FL), Hap Pritchard, (Naval Research Laboratory, Washington, D.C.), Guy W. Sewell, (USEPA, Ada, OK), James M. Tiedje (Michigan State University, East Lansing, MI).

Abstract: The Federal Integrated Biotreatment Research Consortium (Flask to Field) represented a 7-year concerted effort by several research laboratories to develop bioremediation technologies for contaminated DoD sites. The consortium structure consisted of a director and four thrust areas. A technical advisory committee (TAC), consisting of the leading biotechnology specialists assisted and advised the director. The primary project objective was to develop a field–ready biotechnology in each of four thrust areas: explosives, chlorinated solvents, polycyclic aromatic hydrocarbons (PAHs), and polychlorinated biphenyls (PCBs). The technology developed through the explosives research yielded two successful field demonstrations and is becoming the industry standard. Co-solvent extraction of chlorinated solvents was validated at a field site and a guidance document explaining the technology has been completed. Bioaugmentation enhanced PAH degradation when compared to traditional landfarming methods. Genetically engineering microorganisms (GEMs) that enhanced PCB degradation were developed, field tested, and a GEMs guidance document was written.

INTRODUCTION

The DoD has thousands of sites that have been contaminated with organic compounds that pose a serious threat to the environment. The remediation of these sites using existing technologies was problematic from an economic, technical, and political point of view. U. S. Environmental Protection Agency (USEPA) and DoD funded development and application of innovative remediation technologies to solve these problems. Of all the innovative technologies, bioremediation was considered the most promising.

Biotreatment processes had been successfully demonstrated for treatment of a wide variety of easily degraded compounds, such as low molecular weight fuels and phenols. A strong potential existed for development of biotreatment processes directed toward contaminant groups traditionally more difficult to degrade such as explosives, chlorinated solvents, polycyclic aromatic hydrocarbons (PAHs), and polychlorinated biphenyls (PCBs). All of these compounds represented a major contaminant problem to the DoD.

This project's objectives enhanced existing funded efforts in DoD programs, complimented both the USEPA and DoE research strategies, and addressed problems experienced by environmental engineers involved in Superfund and RCRA activities and international technology exchange programs.

Dr. Jeffrey Talley was the project director and was assisted by Deborah R. Felt.
Explosives. The overall objective of the explosive remediation research conducted by Dr. Jim Spain at the Air Force Research Laboratory (AFRL/MLQE) was to discover and

understand biological systems able to destroy nitroaromatic contaminants. 2,4,6-Trinitrotoluene (TNT) is the most widespread explosive contaminant of interest to the DoD. There has been a considerable amount of work in a number of laboratories worldwide that has led to the development of several strategies for the cometabolism of TNT. Dinitrotoluenes, 2,4-dinitrotoluene (2,4-DNT) and 2,6-dinitrotoluene (2,6-DNT), are the second major contaminants used not only in the production of TNT but also in the commercial production of polyurethane.

Previous work at AFRL/MLQL had shown that 2,4-DNT was biodegradable (Spanggord, et al., 1991). More recently, they have shown that 2,6-DNT is also biodegradable (Nishino, et al., 2000; Nishino, et al., 1999). Understanding the degradative process was key to designing a reliable remediation technology, therefore an effort was focused on elucidating the 2,6-DNT degradation mechanisms and pathway (Nishino, et al., 2000). A critical component of the research was to examine the bottlenecks that inhibit efficient DNT degradation in *ex situ* reactor systems (Zhang, et al., 2000) as well as *in situ* field applications. Work was later directed towards the development of *in situ* remediation strategies for DNT.

Chlorinated Solvents. Chlorinated solvents entered the environment in massive amounts during the 1950's, 60's and 70's. These contaminants have migrated through the subsurface and impacted ground water at over 1000 DoD sites. Contaminated aquifers can be remediated by removing the solvents in the porous media of the subsurface. Laboratory and pilot-scale experiments have demonstrated the potential of cosolvent-enhanced *in situ* extraction to remove Dense Non-aqueous Phase Liquids (DNAPL) in porous media. While this method is effective for mass removal, residual amounts of cosolvents and contaminants are expected to remain at levels that could preclude meeting regulatory requirements. However, with the bulk of the DNAPL extracted, *in situ* biotreatment becomes a viable polishing procedure. This was the emphasis of the work that was conducted by Dr. Guy Sewell, USEPA.

In situ biotreatment may transform the remaining contaminants to non-hazardous compounds at a rate in excess of the rate of dissolution or displacement. The efficacy of *in situ* bioremediation of chlorinated solvents is usually limited by transport and mixing considerations, i.e., supplying excess electron donors in conjunction to the chlorinated solvents at appropriate concentrations. The delivery and extraction process facilitated the co-solvency effect and supplied electron donors (cosolvent-ethanol) and electron acceptors (chlorinated solvent-PCE) to the inherent bacteria. The synergism between these abiotic and biotic processes could minimize problems associated with the individual approaches and lead to the development of a treatment train approach which could attenuate or eliminate the risks posed to human health and the environment by DNAPL sites.

PAHs. PAHs include industrial wastes such as petroleum and fuel residues, tars, and creosote that contaminate soils and sediments. Landfarming is a common treatment option for PAH-contaminated soils, but the removal of the high molecular weight (HMW) PAHs by this method is often problematic. The goal of this project coordinated by Dr. Hap Pritchard (NRL) was to modify landfarming by using bioaugmentation to improve degradation of PAHs. Bioaugmentation involved the addition of a biosurfactant-producing bacterium (strain Pa 64), a bulking agent (rice hulls), and a carbon/nitrogen source (dried blood fertilizer) (Pritchard, et al., 1999). Microcosm studies conducted at NRL validated the method and determined the degradation kinetics.

Lance Hansen conducted a pilot-scale study at ERDC implementing bioaugmentation technology for PAH remediation. The study consisted of three metal pans (10' long x 3' wide x 2' deep) each filled with approximately 1 cubic yard of PAH-contaminated soil. One pan was untreated; one received bulking agent and dried blood, and the third received bulking agent, dried blood, and the bacteria. The pan study was designed to define the sampling strategy required to measure the effectiveness of bioaugmentation and to provide a realistic cost estimate for the bioaugmentation treatment (U.S. Army Corps of Engineers, 1996). Methods were developed and refined to monitor the progress and effectiveness of bioremediation. These included molecular biological techniques to monitor for the presence of the inoculated organisms and their *in situ* activity (White and Ringelberg, 1998; Balkwill, et al., 1988), respirometric techniques that monitor relative microbial activity based on CO_2 production, and genetic techniques that monitor for the presence or absence of enzymes involved in nitrogen use and PAH degradation (Perkins, et al., 2000). These techniques were correlated to standard contaminant analytical chemistry methods and were applied to the cost-optimization of landfarming PAHs.

PCBs. Research on microbial degradation of PCBs has been ongoing for over 25 years and has shown that bioremediation requires a more sophisticated technology than the simplistic attempts that have been tried so far. This project conducted by Dr. Jim Tiedje at Michigan State University addressed key barriers to bioremediating PCBs, which are: (1) developing microorganisms that will grow on the major congeners produced by anaerobic dechlorination of PCBs, (2) improving bioavailability of PCBs through use of surfactants, and (3) optimizing field delivery of anaerobic/aerobic PCB bioremediation technologies. Genetically engineered microorganisms (GEMs) were developed capable of using PCB congeners as growth substrate under aerobic conditions. GEMS were modified to exhibit dechlorination genes that enabled the removal of chlorine before chlorocatechols are formed, avoiding toxicity (Tsoi, et al., 1999). This approach avoids the need to manage co-metabolism, which can be difficult *in situ*. These organisms can be used to remove products of anaerobic reductive PCB dechlorination, predominantly *ortho-*, and *ortho+para-* chlorinated congeners (Hrywna, et al., 1999). Vermiculite, as a carrier for the bacterial inoculum, improved survival of the GEMs in Picatinny soil.

TECHNICAL APPROACH.

Figure 1 illustrates the technical approach used to develop new biotreatment technologies during the Flask to Field project. The technical approach within the Consortium was to develop the most promising biotreatment processes at the bench scale and then validate the technology at the pilot and field scale. Engineering groups worked closely with scientists in evaluating the potential of the resulting technologies and in the transfer of technologies from bench-scale to field. The TAC periodically reviewed projects for technical merit. The recommendations of these biotechnology experts to the thrust area coordinators served to further enhance the projects. This approach ensured effective remediation technologies were developed within a reasonable time frame.

Biotreatment Process Development

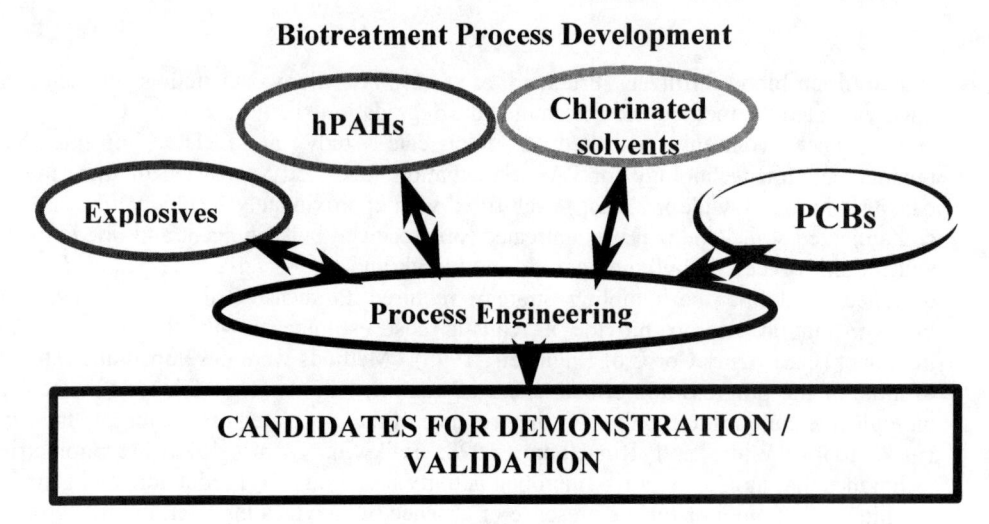

Diagram: ovals labeled "Explosives", "hPAHs", "Chlorinated solvents", "PCBs" with arrows to/from "Process Engineering" oval, pointing down to box labeled "CANDIDATES FOR DEMONSTRATION / VALIDATION"

Figure 1: Technical Approach for Flask to Field Project

THRUST AREA PROJECT RESULTS

A brief summary describing a major project from each thrust area is given below.

Explosives. The researchers at AFRL/MLQE have been very successful in their work on enzymatic reactions and the determination of biodegradation mechanisms and pathways. Pathways for the bacterial degradation of 2,4-DNT, 2-6-DNT and nitrobenzene have been characterized; the enzymes involved in the pathway have been purified and characterized, and the genes involved in the pathways have been identified and cloned. Bench scale groundwater studies established parameters necessary to design a fluidized bed reactor for a 5-month field test at Volunteer Army Ammunition Plant. Removal efficiency for TNT varied between 33 and 73% depending on hydraulic retention time. 2,6-DNT removal was low, but improved after 4 months of operation. Cost analysis showed that this technology was more cost effective than UV/ozone or granular activated carbon treatment when the total nitrotoluene removal rate exceeded 120lb/day.

Another emphasis of the explosives research was on development of pilot and field scale systems at the Badger Army Ammunition Plant (BAAP), Baraboo, WI, for degradation of mixtures of dinitrotoluenes in soil. Treatability studies using soil from two BAAP sites indicated that the soil contained indigenous populations of DNT- degrading bacteria. Figure 2 illustrates some of the results from these bench-scale studies. The left portion of the figure shows the possible microbial degradation pathway of 2,4-DNT. Microbial enzymes that eventually lead to ring cleavage mediate sequential oxidation of the aromatic ring. The right hand side of Figure 2 shows data from microcosm studies using cultures inoculated with isolated DNT degraders from BAAP compared to control cultures. 2,4-DNT concentrations fell to background levels faster in the inoculated culture compared to the control cultures (288 hr versus 576 hr) with nitrite concentrations increasing over time in both cultures. This indicates that the isolated DNT degraders enhanced 2,4-DNT transformation compared to the control culture and nitro groups are cleaved from the aromatic ring during degradation. 2,6-DNT concentrations remained relatively constant over time in both graphs, indicating that neither culture was able to appreciably degrade 2,6-DNT.

The 2,4-DNT concentrations dropped from 16,000 to 130 µg/L in 3 months during the BAAP field test. Cost savings for successful *in situ* bioremediation is expected to be $60-65 Million with a timesaving of 10 years at that site. A book outlines this technology and has been widely referenced (Spain, et al., 2000).

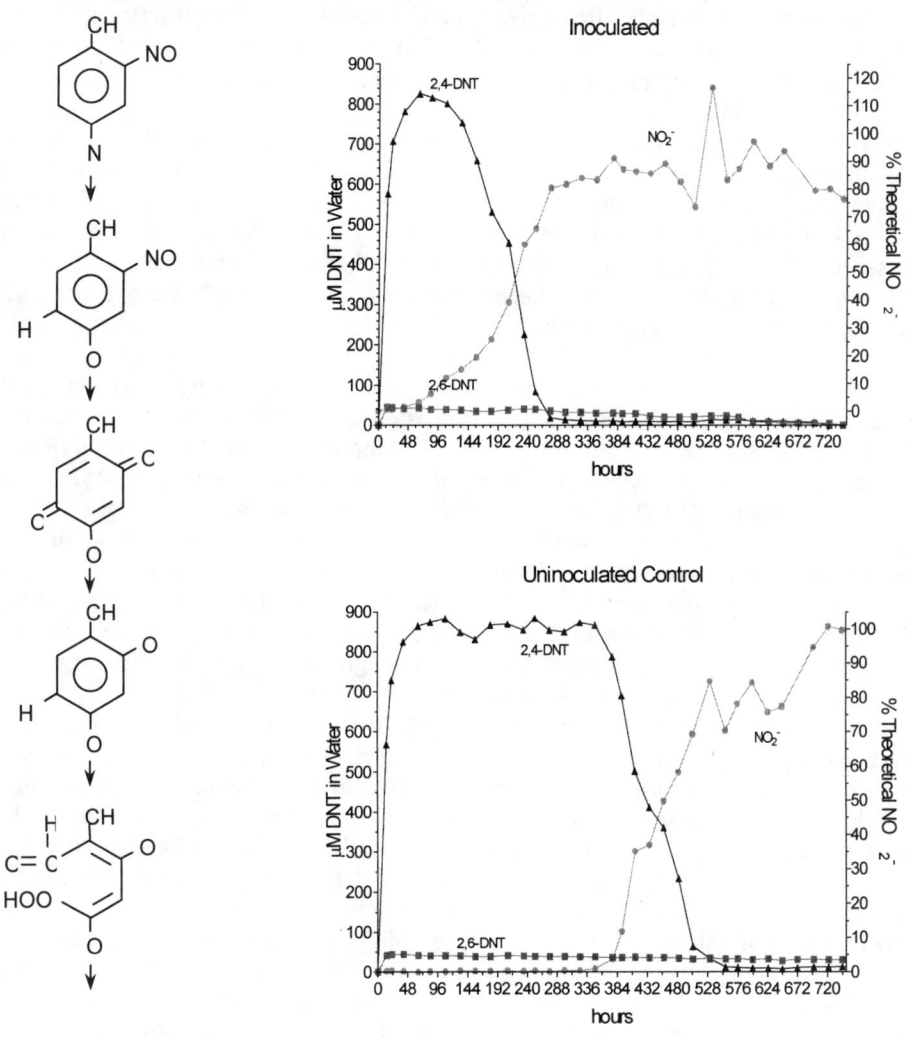

Figure 2. Illustration of DNT Degradation at BAAP Site

Chlorinated Solvents. The chlorinated solvents project, Solvent Extraction Residual Biotreatment (SERB), concentrated on the remediation of Tetrachloroethylene (PCE). SERB technology was validated in a field-scale study at Sage's Dry Cleaner site, Jacksonville, FL (Mravik, et al., 1999; Sewell, et al., 2000). PCE concentration was reduced by 70% in the aquifer using this technology. Significant levels (4 mg/L) of the dechlorination product, cis-DCE, were detected in ground water samples in the area exposed to residual ethanol after 4 months and increased to 16 mg/L after 10 months. Maximum and minimum observed rates of dechlorination (based on cis-DCE production) were 43.6 and 4.2 µg/L/day, respectively. These results indicated that over time, PCE biotransformation has been enhanced.

Microbial ecology studies using site materials indicated the site remains biologically active. Microcosm studies that indicated anaerobic microbial populations generated a reducing equivalents balance by oxidation of the co-solvent (ethanol) that was linked to the reductive dechlorination of PCE. Molecular methods indicated the presence of known groups of dechlorinators. Over all, the project was successful. SERB research is still on going at the Jacksonville site and represents an attractive alternative for chlorinated solvents remediation.

PAHs. Results from the PAH pilot-scale study implementing bioaugmentation technology indicated that bioaugmentation does enhance PAH degradation (Hansen, et al., 2000). Degradation of HMW PAHs into 4 ring compounds (including BaP toxic equivalent compounds) had been achieved. Low molecular weight PAHs were extensively degraded in the first 2-3 months and degradation of the HMW PAHs commenced in the 4th month in microcosms bioaugmented and treated with dried blood fertilizer. A reduction of total PAHs (86-87%) was realized after 16 months in the pans that had been bulked, amended with dried blood fertilizer, and bioaugmented. Comparison of the two methods indicated that the time required to achieve 50% degradation of PAHs was decreased by half through bioaugmented landfarming over traditional landfarming methods. The soil used for this research was heavily contaminated with PAHs (7,200 ppm) and would traditionally not be considered a candidate for bioremediation. This research indicates that landfarming that incorporated bioaugmentation technology may be an alternative to incineration for remediation of heavily PAH contaminated sites. Bioaugmentation would be more cost-effective than incineration because of minimal soil excavation, material handling, and energy costs.

PCBs. The PCB project has been very successful in cloning genes in bacteria that combine co-metabolism of PCBs to chlorobenzoates, and dechlorination and mineralization of chlorobenzoates as a growth substrate. Microbiological, biochemical, and physiological characterization of the selected biphenyl degraders was completed. Based on growth on PCB mixtures, toxicity testing, and survival in soil microcosms experiments, a combination of two GEMs, *Rhodococcus* RHA1 (*fcb*) and *Burkholderia* LB400 (*ohb*), were the most effective for achieving PCB degradation. The use of surfactants increased the solubility and remediation rates of the contaminant. Molecular probes were developed and used to track the bacteria in Picatinny Arsenal soils and River Raisin sediments, using both genetic and PCR-based techniques. The recombinant organisms survived in non-sterile sediment from Red Cedar River contaminated with

Aroclor 1242 and maintained degradative activity, evidenced by reducing PCB levels by 78%.

A pilot-scale study involving three different soil loadings (low, medium, and high solids reactors) is on going at ERDC. This study will evaluate the effects of different moisture content on PCB bioremediation and application of GEMS and determine the maximum soil loading rate optimum for GEMS activity to avoid/offset the subsequent costs of disposing of the stabilized soil.

Bioremediation of PCBs is effective and offers lower energy and operations costs compared to other technologies, but may take longer to remediate the soil and desorption kinetics may limit degradation rates.

ACKNOWLEDGEMENTS

We acknowledge funding and technical support for this research from the U.S. Army Engineer Research Development Center (ERDC), and the DoD through SERDP. The authors wish to acknowledge Dr. Joanne Jones-Meehan (NRL), Ms. Shirley Nishino (Tyndall AFB), Dr. Kurt Pennell (Georgia Institute of Technology), Catherine Nestler and Altaf Wani, (Applied Research Associates, Inc), and Roy Wade (ERDC) for providing support and technical collaboration, which made this research possible.

REFERENCES

Balkwill, D.L., Leach, F.R., Wilson, J.T., McNabb, J.F., White, D.C. 1988. "Equivalence of Microbial Biomass Measures Based on Membrane Lipid and Cell Wall Components, Adenosine Triphosphate, and Direct Counts in Subsurface Aquifer Sediments". *Microbial. Ecol.* 16: 73-84.

Hansen, L.D., Nestler, C., and Ringelberg, D. 2000. "Bioremediation of PAH/PCP Contaminated Soils from POPILE Wood Treatment Facility". In G.B. Wickramanayake, A.R. Gavaskar, J.T. Gibbs and J.L. Means (Eds.), *Proceedings of the Second International Conference on Remediation of Chlorinated and Recalcitrant Compounds*, pp.145-152, Battelle Press, Columbus, OH.

Hrywna Y., T. V. Tsoi, Maltseva, J. F. Quensen III, J. M. Tiedje. 1999. "Construction and Characterization of Two Recombinant Bacteria that Grow on ortho- and para-Substituted Chlorobiphenyls". *Appl. Environ. Microbiol.* 65(5): 2163-2169.

Mravik, S.C., G.W. Sewell, and A. L. Wood. 1999. *Field evaluation of the solvent extraction Residual Biotreatment Technology*. TU2A, in Abstracts of the 4th International Symposium on Subsurface Microbiology, ISSM, Vail, CO.

Nishino, S.F., G. Paoli, and J.C. Spain. 2000. "Aerobic Degradation of Dinitrotoluenes and Pathway for Bacterial Degradation of 2,6-Dinitrotoluene". *Appl. Environ. Microbiol.* 66:2139-2147

Nishino, S.F., J.C. Spain, H. Lenke, and H.J. Knackmuss. 1999. "Mineralization of 2,4- and 2,6-Dintirotoluene in Soil Slurries". *Environ. Sci. and Technol.* 33: 1060-1064

Perkins, E., Hansen, L.D., Nestler, C.C., Byrnes, J. 2001. *Changes in Abundance of In-situ Aromatic Degrading Bacteria During a Pilot Scale Landfarming of a Polycyclic Aromatic Hydrocarbon Contaminated Soil*, In, Proceedings of the Ninth International Symposium on Microbial Ecology, Amsterdam.

Pritchard, P.H., J. Jones-Meehan, J.G. Mueller and W. Straube. 1999. "Bioremediation of high molecular PAHs; Application of techniques in Bioaugmentation and bioavailability enhancement". In *"Novel Approaches for Bioremediation of Organic Pollution"*, R. Fass, Y. Flashner, and S. Reuveny, (eds.), Kluwer Academic/Plenum Publishers, New York, pp. 157-169.

Sewell, G. W., S. C. Mravik, A. L. Wood. 2000. *Field Evaluation of Solvent Extraction Residual Biotreatment (SERB)*. 7th International FZK/TNO Conference on Contaminated Soil (ConSoil 2000), Sept. 18-22, Leipzig, Germany.

Spain, J.C., J.B. Hughes, and H.J. Knackmuss (ed.). 2000. *Biodegradation of Nitroaromatic Compounds and Explosives*. Lewis Publishers, Boca Raton.

Spanggord, R. C., J. C. Spain, S.F. Nishino, and K.E. Mortelmans. 1991. "Biodegradation of 2,4-dintrotoluene by a *Pseudomonas* species". *Appl. Environ. Microbiol.* 57:3200-3205

Tsoi, T. V., E. G. Plotnikova, J. R.. Cole, W. F. Guerin, M. Bagdasarian, and J. M. Tiedje. 1999. "Cloning, Expression, and Nucleotide Sequence of the *Pseudomonas aeurginosa* strain 142 ohb Genes Coding for Oxygenolytic ortho-Dehalogenation of Halobenzoates". *Appl. Environ. Microbiol.* 65(5):2151-2162.

U.S. Army Corps of Engineers. 1996. *Bioremediation Using Landfarming Systems. Engineering and Design*. ETL 1110-1-176. Washington, D.C.

White, D.C., Ringelberg, D.B. 1998. "Signature Lipid Biomarker Analysis". In: Burlage, R.S., Atlas, R., Stahl, D., Geesey, G., Sayler, G. (Eds.) *Techniques in Microbial Ecology*. Oxford University Press, Inc., New York, pp. 255-272.

Zhang, C., S.F. Nishino, J.C. Spain, and J. B. Hughes. 2000. "Slurry-phase Biological Treatment of 2,4- and 2,6-Ditnitrotoluene: Role of Bioaugmentation and Effects of High Dinitrotoluene Concentrations". *Environ. Sci. Technol.* 34: 2810-2816

DESIGN OF AN ENHANCED ANAEROBIC BIOREMEDIATION SYSTEM FOR A LOW PERMEABILITY AQUIFER

Sean Dean[1], Lee Wiseman[1], Al W. Borquin[2], John Accashian[2], and Michael Edgar[1] ([1]Camp Dresser & McKee Inc, Orlando, Florida and [2]Denver, Colorado)
James Callender (Rockwell Automation, Cleveland, Ohio)

ABSTRACT: The surficial aquifer beneath a computer circuit board printer company is contaminated with chlorinated aliphatic hydrocarbons (CAH). The surficial aquifer extends approximately 40 feet below land surface (bls) at the site and is composed of three silty-sand units separated by layers of sandy clay. Maximum total chlorinated solvent concentrations of approximately 7,000 ug/L are present in the intermediate sand unit approximately 25 feet bls. Field characterization results indicated the groundwater is methanogenic with limited pockets of competing electron acceptors (sulfate ~300 mg/L). Reductive dechlorination is occurring with production of ethene, ethane, and native organic carbon as the primary substrate. However, the rate of attenuation is limited by low concentrations of available electron donor and the contaminant plume is nearing a downgradient surface water. Therefore, enhanced anaerobic treatment (EAB) was selected to increase contaminant removal. *In-situ* treatment cells were designed to circulate groundwater amended with lactate. Each treatment cell is a radial design with four to five extraction wells on the cell perimeter and a central injection well. Multiple cells allow independent operation of each cell based on local characteristics. The design will expedite the remediation by allowing targeted substrate loading and injection rates. One of the five treatment cells has been constructed and will be used to conducted a pilot demonstration in 2001.

INTRODUCTION

Site background. The former PEC Industries (PEC) site occupies approximately 9 acres in Orlando, Florida and has been used for the manufacturing of printed electronic circuit boards (Figure 1). In September 1987, a preliminary environmental audit was conducted at the site as part of a transactional audit. This audit revealed the presence of solvent related groundwater impacts in the shallow surficial aquifer wells in the vicinity of the drum storage area, but not in the deeper wells. Subsequent investigations defined the extent and magnitude of site groundwater contamination and set the stage for implementation of remedial action.

In late 1999, a number of technologies-including pump-and-treat, chemical oxidation, and EAB-were assessed to determine their ability to provide a technically feasible solution that could meet an aggressive remediation schedule (CDM, 2000). A review of historical data, combined with a limited groundwater sampling effort, revealed highly reduced conditions in the subsurface which would be conducive to the use of EAB at this site.

Site Hydrogeology. The site is relatively flat and is located on a gently rolling ridge which trends in a north-south direction through Orange County, Florida. Figure 2 is a hydrogeologic cross-section of the surficial aquifer.

The entire thickness of the surficial aquifer is approximately 40 feet. Three water-bearing zones have been identified within the surficial aquifer beneath the site. The uppermost zone-the shallow sand unit-extends from land surface to between 10 and 15 feet below land surface (bls) and consists of fine sand and clayey sand. The intermediate water-bearing zone occurs in the southern part of the site and is comprised of sand and clayey sand. The top of this sand unit is generally between 25 feet bls near its northern extent (MW-19D) and 18 feet bls to the south. Thickness of this unit increases from north to south and is approximately 10 feet thick in the southern part of the site. The intermediate sand unit is separated from the shallow sand unit by less than 5 feet of sandy clay. The top of the deep water-bearing zone occurs between 30 and 35 feet bls. It is comprised of approximately 5 feet of sand and clayey sand and is underlain by the regional Hawthorn aquitard, which represents the base of the surficial aquifer (Lichtler et. al, 1968). A layer of sandy clay between 5 and 10 feet thick separates the deep sand unit from the intermediate sand unit beneath the southern part of the site. It is underlain by dense clay that is interpreted to be the lower confining or semi-confining unit for the surficial aquifer.

PRE-REMEDIAL STUDIES

Pre-remedial studies including aquifer performance tests, geotechnical analyses, and extensive groundwater sampling to assess groundwater biogeochemical conditions and groundwater contamination were performed to assess the feasibility of implementing EAB at this site. These studies supported EAB as an appropriate remediation technology and the subsequent design of an EAB treatment system. This discussion focuses on the analytical results of the groundwater sampling

Biogeochemistry. Biogeochemical parameters were measured to better define environmental conditions that can affect the presence and activity of microbial populations across the site. These parameters are summarized in Table 1.

Electron acceptor and metabolic byproduct data were also collected during January 2000. All electron acceptor/metabolic byproduct data are summarized in Table 2.

As indicated in Tables 1 and 2, the hydrogeochemistry and reducing conditions are generally conducive to chlorinated hydrocarbon biotransformation. However, these data also suggest that the rate and extent of reductive dechlorination reactions may be limited by the availability of readily metabolized electron donors and the presence of competing electron acceptors (e.g., sulfate). Groundwater is generally under sulfate reducing or methanogenic conditions within the shallow and deep sand units; while methanogenic conditions prevail in the intermediate sand unit. Sulfate reducing conditions support the reductive dechlorination of the heavier molecular weight chlorinated hydrocarbons (e.g., PCE, TCE, TCA) observed at the site. Additionally, sulfate reducing conditions may support the reductive dechlorination of some of the lower molecular weight chlorinated hydrocarbons (e.g., 1,1-DCE and cis-1,2-DCE), although it is CDM's experience that the presence of sulfate at concentrations approaching 50 mg/L and higher may interfere with such reactions. In contrast, methanogenic conditions

Table 1. Surficial aquifer geochemistry

Parameter	Units	Shallow unit Range	Avg.	Surficial aquifer zone Intermediate unit Range	Avg.	Deep unit Range	Avg.
pH	Std	3.6 - 7.0	4.5	5.1 - 6.0	5.5	5.7 - 6.9	6.0
Temperature	°C	21.4 - 24.4	22.5	23.6 - 24.9	24.3	23.6 - 24.7	24.2
Conductivity	mS/cm	0.113 - 6.27	2.0	0.131 - 0.618	0.36	0.193 - 0.595	0.34
Turbidity	NTU	9 - 343	90	3 - 44	14	0 - 11	3
Chloride	mg/l	11.0 - 1,610	417	25 - 99	66	24 - 44	35
Ttl. Alkalinity	mg/l	11 - 700	216	14 - 79	46	69 - 279	137
DOC	mg/l	1.12 - 205	95.4	18.22 - 46.02	30.8	24.55 - 88.81	41.05
Ammonia N	mg/l	0.08 - 21.6	5.3	0.12 - 0.62	0.40	0.04 - 0.25	0.15
Ortho-PO_4	mg/l	0.004 - 1.07	0.16	0.022 - 1.38	0.44	0.31 - 4.02	2.07

Table 2. Surficial aquifer electron acceptors and metabolic byproducts

Parameter	Units	Shallow unit Range	Avg.	Surficial aquifer zone Intermediate unit Range	Avg.	Deep unit Range	Avg.
Nitrate N	mg/l	<0.02 - 0.24	0.08	<0.02 - 0.08	0.05	0.06 - 0.30	0.15
Nitrite N	mg/l	<0.005 - 1.00	0.15	<0.005 - 0.37	0.01	<0.005 - <0.005	<0.005
Fe(II)	mg/l	0.0 - 27.8	5.2	0.5 - 13.7	5.3	0.1 - 5.3	2.2
Sulfate	mg/l	0 - 1,060	187	0 - 78	16.3	0 - 19	3
HS	mg/l	0.0 - 4.0	0.4	0.0 - 0.5	0.1	0.0 - 0.0	0.0
Methane	ug/l	0.9 - 4,389	1,500	3.1 - 3,325	1,261	4.0 - 2,178	666
CO_2	mg/l	160 - 860	454	420 - 780	551	200 - 500	357
H_2	nM/l	NA	28.5	21.6 - 45.6	33.4	21.2 - 76.8	41.5
Redox Pot.	mV	(-62) - (-265)	(-149)	(-190) - 120	(-117)	(-169) - 72	(-80)
DO	mg/l	0.1 - 0.6	0.3	0.0 - 1.1	0.3	0.0 - 4.2	0.9

are generally required to achieve the complete reductive dechlorination of chlorinated hydrocarbons to innocuous end products such as ethene and ethane. Despite moderate to high dissolved organic carbon (DOC) concentrations and the development of methanogenic conditions within the intermediate formation, reductive dechlorination reactions appear to be incomplete within this formation. It has been hypothesized that the dissolved organic matter represented by the DOC measurements may exist as complex organic substrates that are not readily metabolized by the subsurface microorganisms. As such, reductive dechlorination reactions may be limited by electron donor availability. It is conceivable that the DOC present in the aquifer is readily biodegradable and that the aquifer is not electron donor limited. Rather, other factors such as the distribution, population, and community of dechlorinating bacteria may not be uniform, thereby resulting in variability of the extent and rate of dechlorination reactions. Additional studies will be designed to address these issues.

Volatile Organic Compounds (VOCs). Groundwater VOC concentrations recorded at all monitor wells sampled in January 2000 are summarized in Table 3.

Data presented in this table include all of the VOC concentrations that were above the analytical laboratory reporting limits using EPA Method 8260B. In general, the primary contaminants reported include numerous chloroethene and chloroethane compounds.

Table 3. Concentration range of detected VOCs in the surficial aquifer

Parameter	Units	Shallow unit Low	Shallow unit High	Surficial aquifer zone Intermediate unit Low	Intermediate unit High	Deep unit Low	Deep unit High
PCE	µg/l	ND	ND	28.5	43.4	17.6	19.1
TCE	ug/l	ND	ND	14.3	17.2	1.6	1.8
cis-1,2-DCE	ug/l	ND	3.6	11.6	86.5	ND	ND
1,1-DCE	ug/l	ND	78.5	17.3	8,010	5.0	191
VC	ug/l	3.2	371	9.0	112	ND	ND
Ethene	ug/l	0.01	36.8	0.02	3.37	0.01	0.60
1,1-DCA	ug/l	5.0	540	18.9	496	ND	2.7
1,2-DCA	ug/l	ND	3.1	2.4	2.5	ND	ND
CA	ug/l	14.6	40.9	4.5	16.6	ND	ND
Ethane	ug/l	0.01	4.77	0.01	0.09	0.01	0.05
Benzene	ug/l	ND	1.1	ND	ND	ND	ND
Toluene	ug/l	ND	ND	1.2	1.4	ND	ND
Ethylbenzene	ug/l	ND	18.4	ND	ND	ND	ND
Xylenes	ug/l	ND	4.6	1.3	1.9	ND	ND
Carbon Disulfide	ug/l	ND	4.4	ND	ND	ND	8.3

The location of site contaminants in the subsurface has changed over time, as has the speciation of the chlorinated hydrocarbons observed. These changes are partly the result of naturally-occurring intrinsic biotransformation processes, which in some instances have resulted in the complete detoxification of chlorinated hydrocarbons to innocuous end products (e.g., ethene and ethane). The distribution of chlorinated hydrocarbons, as well as the extent and rates of their biotransformation, affect the choice of corrective action appropriate for this site.

The groundwater quality data suggest that the bulk of the chlorinated hydrocarbon mass is located in the intermediate sand unit to the south and southeast of the drum storage area. These findings are significant as previous monitoring events suggested that the bulk of the contaminant mass was restricted to the shallow sand unit. Chlorinated hydrocarbon concentrations have decreased with time at all shallow formation monitor wells. The chlorinated hydrocarbons in the intermediate sand unit are generally restricted to the south and southeast of the drum storage area. Although moderate VOC concentrations were detected at the deep sand unit monitor wells, these data suggest only a limited portion of the mass in this formation relative to the intermediate sand unit.

In addition to the movement of the primary mass of contaminants from the shallow sand unit to the intermediate sand unit, the overall degree of chlorination has also been decreasing, particularly in the shallow sand unit. The overall sampling results, including the presence of both ethene and ethane, indicate that reductive dechlorination processes can mediate the complete detoxification of chlorinated hydrocarbons to innocuous end-products in some locations. Although these natural processes can mediate the complete dechlorination, current conditions in the intermediate sand unit appear to have resulted in an accumulation of 1,1-DCE and 1,1-DCA, with minimal indication of complete detoxification within this formation.

REMEDIAL DESIGN BASIS
Goals and Objectives. The objectives of the proposed remedial action are to lower groundwater contaminant concentrations to levels protective of human health and the environment and to meet regulatory requirements. Based on the data and analyses presented, the native microbial population at this site is reducing the concentration and degree of chlorination of the chlorinated hydrocarbons present in the surficial aquifer at this site. However, levels of 1,1-DCE or VC exceed natural attenuation default source concentrations (Chapter 62-777, FAC-*Contaminant Cleanup Target Levels*) in MW-5I, MW-16I, and MW-9S (downgradient monitoring wells) and it has not been demonstrated that this site meets the other criteria necessary for natural attenuation with monitoring to be an appropriate approach to remediation

The intermediate zone of the surficial aquifer at this site contains the bulk of the chlorinated hydrocarbons. In addition, the horizontal extent of the contaminant plume in this zone is clearly defined. Therefore, the proposed corrective action will focus on actively remediating the intermediate zone with the specific goal of reducing contaminant concentrations below natural attenuation default-source concentrations.

Aquifer Hydraulic Considerations. Because the primary contamination at this site is in intermediate sand unit of the surficial aquifer, remediation will target this sand unit. The transmissivity of the surficial aquifer is relatively low and will require pumping to facilitate transport of injected substrate. Although there are intervening clay layers, the shallow, intermediate, and deep sand units of the surficial aquifer are hydraulically connected. Pumping of the intermediate sand unit is anticipated to induce groundwater from the shallow and deep sand units to flow into the treatment zone.

A simple numerical groundwater model of the surficial aquifer at the site was constructed using the USGS MODFLOW (McDonald and Harbaugh, 1984) model and was used to layout the proposed extraction-injection cells. The proposed full-scale system includes five *in-situ* treatment cells with each cell consisting of a central injection well and several peripheral extraction wells. As discussed previously, each cell will have an independent chemical feed system. The preliminary modeling predicts a slight gradient (approximately an order of magnitude less than the current gradient) towards the treatment cells from all directions. Therefore, the proposed system will provide for hydraulic control of the injected fluids and the plume in the intermediate sand unit.

Design Components. The proposed full-scale groundwater recirculation system consists of five 1-inch diameter groundwater injection wells and fourteen 1-inch diameter extraction wells, each of which is planned to withdraw 0.36 gpm. A schematic of the groundwater recirculation system is presented on Figure 3. The projected annual pumping (extraction) rate is 2.65 million gallons (5.0 gpm). The proposed groundwater treatment system incorporates groundwater withdrawal and injection such that the withdrawal is balanced with injection. An alternative means of disposing extracted groundwater was incorporated into the system design in case use of the injection wells is not possible. Discharge to a nearby sanitary sewer following treatment was selected for alternate disposal to maximize operational flexibility of the system. Consequently, an air stripper was incorporated into the system to pretreat the extracted groundwater prior to discharge to the sewer. The option to discharge excess groundwater to the sewer provides for operational flexibility if it is necessary to extract at a higher rate to maintain hydraulic control, or if wet season site conditions reduce the feasible rate of injection.

Chemical Augmentation. Intrinsic biodegradation is occurring and has occurred on the site. Sodium lactate will be injected into the aquifer as an electron donor to enhance the existing biological processes. Low redox conditions must be maintained in the water used to inject the sodium lactate to avoid altering the conditions currently favoring reductive dechlorination in the subsurface.

Ammonia and ortho-phosphate, which are essential macronutrients for biological growth, do not appear to be limiting the existing intrinsic biotransformation processes. However, the non-availability of these nutrients may become a limiting factor under conditions of increased microbial activity associated with the addition of sodium lactate. Therefore, implementation of the EAB system will include amending the injected groundwater with these and other essential nutrients.

Sodium lactate is provided as a 60 percent by weight solution in 55-gallon plastic drums and will be pumped neat from the supply drum to provide an in-line concentration of 500 milligrams per liter (mg/L). The nutrient material will be fed as a liquid from a 100-gallon day tank. The nutrient supply will be manually mixed on a periodic basis. The nutrient solution will be a mix of phosphate and ammonia fed at a concentration of approximately 430 mg/L to provide an in-line concentration of 0.5 mg/L. The concentration of each component in the nutrient solution will be tested and further defined during the field pilot test. An in-line static mixer will be provided downstream of the chemical injection points. The groundwater with the substrate and nutrients will then be routed to the injection wells via 1-inch HDPE piping.

In February 2001, construction of a field pilot-scale demonstration project was initiated by constructing one of the five treatment cells and the air stripper. The treatment cell consists of one injection well and four extraction wells. The field pilot-scale demonstration will be conducted for nine months.

CONCLUSIONS

Using *in-situ* groundwater contamination and hydrogeochemical data and aquifer characteristics, CDM designed an EAB system for the PEC site. Because of the inherent complexities involved in the processes, a field-pilot scale project will be implemented to evaluate technology design considerations before implementing full-scale design. The advantages of EAB can be summarized as follows:

- Less expensive form of treatment.
- Less complex design (" the enhancement of a natural process").
- Allows for flexibility in remediating a specific and/or multiple. portions of a plume.
- Avoids extra cost associated with treatment of created byproducts.
- Enhances capability of more conventional groundwater treatment technologies.

REFERENCES

Camp Dresser & McKee Inc., 2000. Rockwell Automation Interim Groundwater Remedial Action Plan, Former PEC Industries Site. March 2000

Lichtler,W.F., W. Anderson, and B.F. Joyner, 1968. Water Resources of Orange County, Florida. Florida Geological Survey Report of Investigations No. 50.

McDonald, M.G. and A.W. Harbaugh, 1984. *A Modular Three-Dimensional Finite-Difference Groundwater Flow Model*. U.S. Geological Survey Techniques of Water Resources Investigations Book A3.

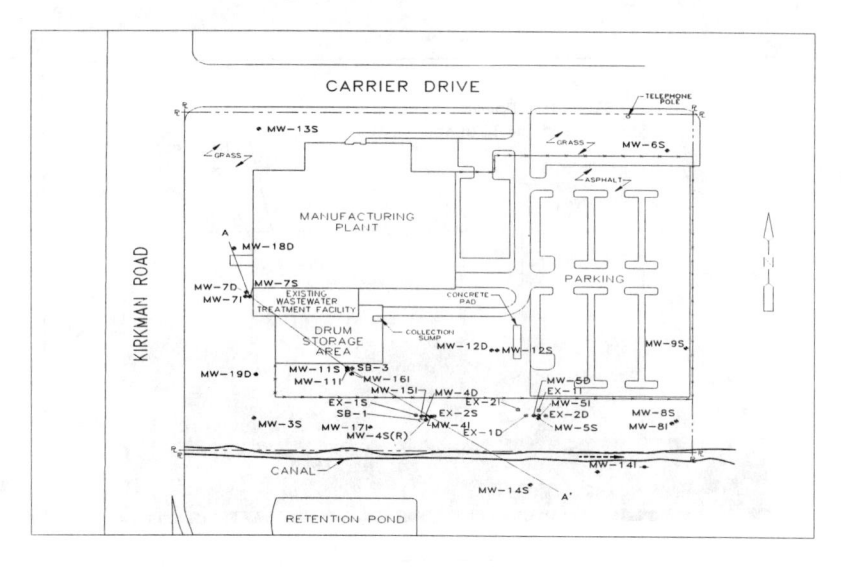

FIGURE 1. PEC Industries Site Map and Well Location Plan

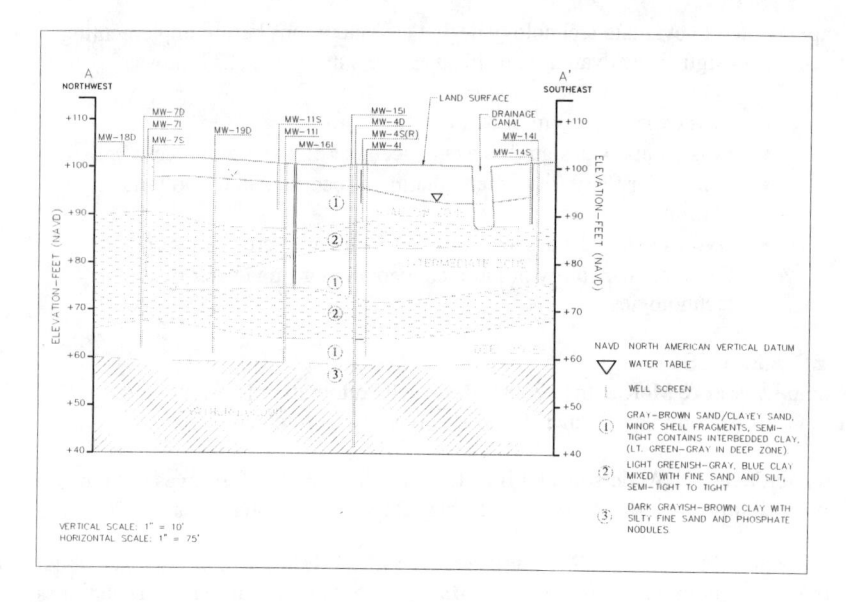

140   *Anaerobic Degradation of Chlorinated Solvents*

FIGURE 2. Site Hydrogeologic Cross-Section

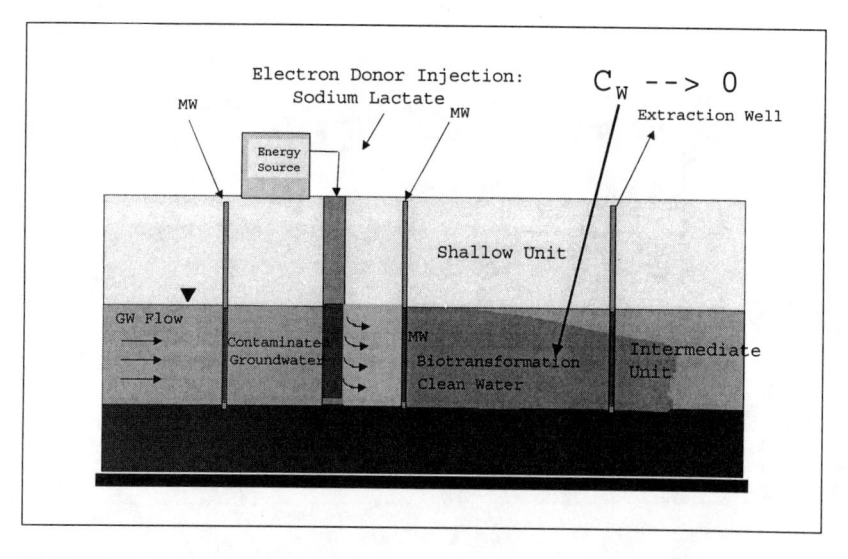

FIGURE 3. Schematic of EAB Treatment Cell

MONITORING STIMULATED REDUCTIVE DECHLORINATION AT THE RADEMARKT IN GRONINGEN, THE NETHERLANDS

A.A.M. Langenhoff (TNO-MEP, Apeldoorn, The Netherlands)
A.A.M. Nipshagen and C. Bakker (IWACO, Groningen, The Netherlands)
J. Krooneman (Bioclear, Groningen, The Netherlands)
G. Visscher (Province of Groningen, The Netherlands)

ABSTRACT: Mixed redox conditions control the intrinsic biodegradation processes at the Rademarkt site (in the centre of the city Groningen, The Netherlands), which is contaminated with perchloroethylene (PCE) and trichloroethylene (TCE). Stimulation of the degradation process was previously demonstrated in laboratory experiments. Based on these laboratory experiments a full scale experiment is designed for the Rademarkt site. The remediation strategy focuses on reductive dechlorination with both methanol and compost-extract as electron donor. An infiltration and recirculation system was designed and implemented at the site to add the electron donor to the system with the recirculated water.

The successful bioremediation process has been monitored over time by measuring changes through a unique combination of parameters. This concerns redox chemistry, dechlorination products, hydrogen measurements and MPN-PCR measurements in groundwater and in the soil.

INTRODUCTION
Mixed redox conditions control intrinsic biodegradation processes at the Rademarkt site in the centre of the city Groningen, The Netherlands, which is contaminated with perchloroethylene (PCE) and trichloroethylene (TCE). Intrinsic degradation has taken place at this site, as evidenced by the detection of degradation products (*cis*-dichloroethylene (*cis*-DCE) and vinyl chloride (VC)). This intrinsic degradation might have been enhanced by an old -temporary- leaking sewage system providing electron donors to the soil that are needed for the dechlorination processes (de Bruin et al., 1992). However, the natural transformation rates of *cis*-DCE and VC at the site are too slow to prevent migration of these hazardous compounds to areas that must be protected. The low levels of DOC in the groundwater (< 10 mg/l) indicate a lack of naturally present electron donor, and addition of electron donor is needed for complete degradation of the hazardous compounds at the site (Middeldorp et al., 1999).

The stimulation of the degradation processes was previously demonstrated in batch- and column experiments (Van Aalst-Van Leeuwen et al., 1997; Nipshagen et al., 1999). Based on these laboratory experiments, a full-scale system was designed on reductive dechlorination with both methanol and compost-extract as electron donors. The system included an infiltration and recirculation system and was implemented at the site to add the electron donor to the groundwater.

The goal of the project was to establish and/or optimize the complete reductive dechlorination of PCE to the non-chlorinated products ethene and ethane. The process was monitored using several instruments, each of which were evaluated for their relative importance for accurate monitoring of the stimulated biodegradative processes.

MATERIALS AND METHODS

Field experiment. An in situ pilot test with an anaerobic activated zone for complete reductive dechlorination was designed. The pilot system was installed at the Rademarkt site, and consisted of 10 infiltration wells on one side and 5 extraction wells at the other side of the source zone at a depth of 5 to 8 m below surface level, see Figure 1. An infiltration drain was present as well to reach deeper parts of groundwater if needed. However, this drain was never used in this project. A part of the anaerobically extracted groundwater was reinfiltrated at the site via the infiltration wells. By recirculating part of the extracted groundwater without purification, the soil was used as a bioreactor. A mixture of electron donor and nutrients was added to the unpurified reinfiltrated groundwater. Final concentrations in the infiltrated groundwater were 0.36 % methanol, 1.44 % compost leachate and 0.9 g/l ammonium chloride. The pilot test lasted for 35 weeks.

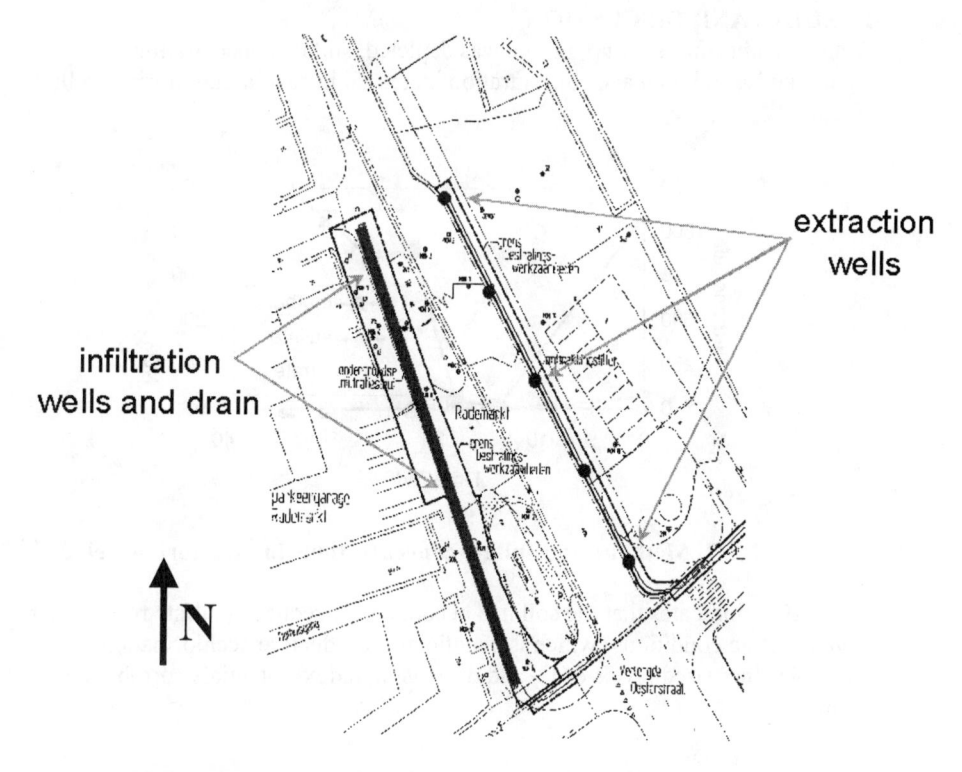

FIGURE 1 Scheme of the bioremediation system at the site

Monitoring. Monitoring included determination of the redox chemistry, groundwater levels, contaminants and their corresponding dechlorination products, methanol, and bromide concentrations. H_2-measurements were carried out directly at the site, as hydrogen is the actual electron donor involved in the dechlorination reaction and is an indicator for the prevailing redox conditions at the site (Chapelle et al., 1997). Finally, molecular detection techniques were performed in soil and groundwater samples to give additional information regarding the presence of dechlorinating organisms. A number of organisms are known to dechlorinate compounds like PCE, and most of these bacteria are related to *Desulfitobacterium spp.* Nested MPN-PCR was used to identify and enumerate bacteria belonging to the genus *Desulfitobacterium* (Nobis, 2000).

The interpretation of the degradation process at Rademarkt was based on:
1. Concentrations of contaminant and degradation products;
2. Changes in redox chemistry;
3. Changes in number of bacteria;
4. Model simulation (data not shown);
5. Combining all of the above.

RESULTS AND DISCUSSION

A few months after start up, nitrate was depleted, sulfate concentration decreased significantly, and methane concentration increased in several monitoring wells (Figure 2).

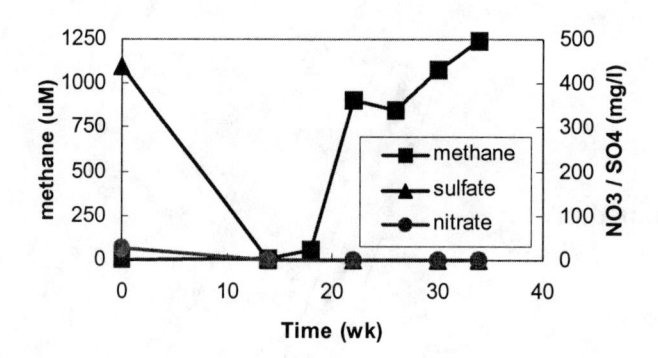

FIGURE 2. Methane and sulfate concentrations in monitoring well 2.

This indicates that the soil had become more reduced (sulfate reducing or methanogenic conditions) which is beneficial for reductive dechlorination processes. Increased levels of H_2 and decreasing redox-potentials corroborate these findings (Figure 3).

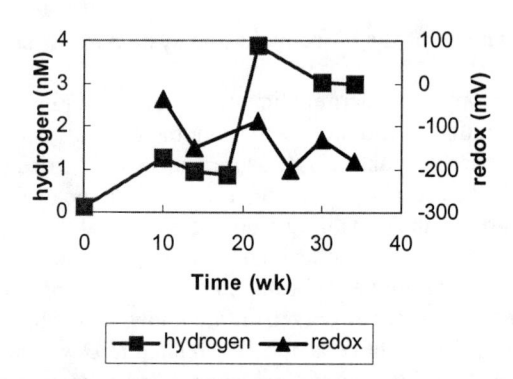

FIGURE 3. Hydrogen concentration and redox potential in monitoring well 2.

Dechlorination was established in most of the stimulated area. A high degree of dechlorination (> 75%) was found primarily in the northern part of the stimulated area and in close proximity to the infiltration wells in the southern part. Figure 4 shows an example of one of the monitoring wells where the sequence of dechlorination products was evident, and the complete dechlorinated end products ethane and ethane were produced.

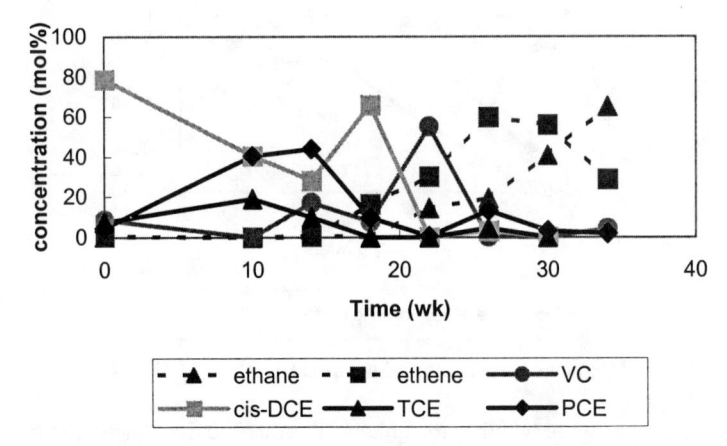

FIGURE 4. (Chloro)ethylene concentrations in monitoring well 2

Average concentrations at the start of the experiment were 1280, 1030, 2600, and 94 ug/l for PCE, TCE, *cis*-DCE and VC respectively, and 602, 10, 291, and 2263 ug/l at the end of the pilot of 35 weeks.

The percentage of dechlorination was calculated from the monthly monitoring of chloroethylene concentrations, and describes the amount of chlorine that is omitted from the source compound PCE [1].

$$\text{Dechlorination} = \frac{(1/4[\text{TCE}] + 2/4[\Sigma\text{DCEs}] + 3/4[\text{VC}] + [\text{Ethene}] + [\text{Ethane}])}{([\text{PCE}] + [\text{TCE}] + [\Sigma\text{DCEs}] + [\text{VC}] + [\text{Ethene}] + [\text{Ethane}])} * 100\% \quad [1]$$

Starting with an average dechlorination of 35 %, the dechlorination raised to 50 - 95% in the various monitoring wells.

In contrast, the percentage of dechlorination hardly increased in the middle part of the activation zone, near monitoring well 12. The redox chemistry did not show large changes, as if this area was less influenced by the infiltration extraction system. Groundwater level measurements showed that there was a groundwater flow, however, the TOC and methanol values showed that the electron donor concentrations were not sufficient to support dechlorination.

During the field experiment, groundwater samples were analyzed for the presence and increase of bacteria (Figure 5), and bacterial counts on the groundwater samples showed that in the stimulated area the number of *Desulfitobacterium* spp. increased well. This suggests that these type of bacteria were involved in the dechlorination process. Other PCE dechlorinating bacteria (e.g. *Dehalobacter* spp. and *Desulfuromonas* spp.) were not counted with this method.

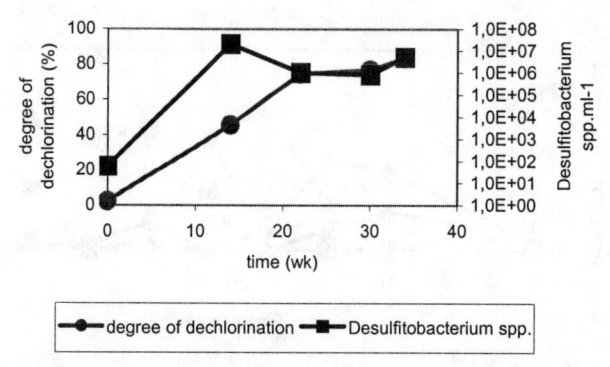

FIGURE 5. Degree of dechlorination and number of *Desulfitobacterium* spp.

To investigate whether the enhanced reductive dechlorination of PCE was indeed due to growth of *Desulfitobacterium* spp, anaerobic microcosms with groundwater from monitoring well 15 were carried out. A fast initial dechlorination from PCE to *cis*-DCE was followed by a slower dechlorination from *cis*-DCE to VC and ethane (Figure 6). The initial dechlorination of PCE to *cis*-DCE was accompanied by an increase in the number of *Desulfitobacterium* spp.

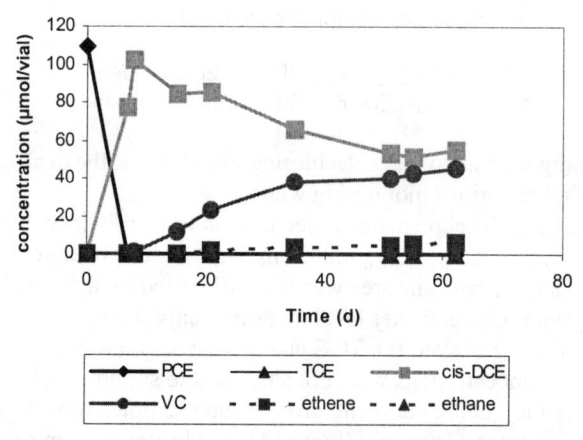

FIGURE 6. (Chloro)ethylene concentrations in the anaerobic water batches

CONCLUSIONS

In the initial phase of bioremediation at the Rademarkt site, intensive monitoring was carried out to gain more insight into the efficiency of the dechlorination process. Additionally, the importance of the selected monitoring parameters was determined.

Initial monitoring of the redox chemistry showed how conditions at the site changed, and became more reduced during infiltration. Classical redox chemistry analyses (nitrate, sulfate, methane, rodox potential etc), and hydrogen measurements showed comparable results, indicating that one of these methods is required.

Measuring the contaminants showed that the concentrations of PCE and TCE decreased over time, and that completely dechlorinated end products like ethene and ethane were formed.

The nested MPN-PCR method showed that the number of *Desulfitobacterium* spp. increased during the pilot test.

Initial monitoring of the redox chemistry, combined with measuring the biological degradation would be appropriate to obtain a reliable overview of the progress of the bioremediation process. At moments when changes in the remediation set up are performed, a more detailed monitoring strategy can be followed (e.g. enumeration of dechlorinating organisms, in-situ detection of hydrogen levels and determining contaminant degradation rates in microcosms studies).

ACKNOWLEDGEMENTS

This work has been financed by NOBIS (the Dutch Research Programme Biotechnological In-Situ Remediation), and the Province of Groningen.

REFERENCES

Chapelle, F. H., D. A. Vroblesky, J. C. Woodward, and D. R. Lovley. 1997. "Practical considerations for measuring hydrogen concentrations in groundwater." *Environmental Science and Technology* 31: 2873-2877.

de Bruin, W. P., M. J. J. Kotterman, M. A. Posthumus, G. Schraa, and A. J. B. Zehnder. 1992. "Complete biological reductive transformation of tetrachloroethene to ethane." *Applied and Environmental Microbiology* 58(6): 1996-2000.

Middeldorp, P. J. M., M. L. G. C. Luijten, B. van de Pas, M. H. A. van Eekert, S. W. M. Kengen, G. Schraa, and A. J. M. Stams. 1999.. "Anaerobic microbial reductive dehalogenation of chlorinated ethenes." *Bioremediation Journal* 3(3): 151-170.

Nipshagen, A. A. M., J. Krooneman, A. Tuinstra, and A. A. M. Langenhoff. 1999. Afbraak van per- en trichlooretheen onder sequentiële redoxomstandigheden - Fase 2: Aanvullende kolomexperimenten en aërobe veldproef. Gouda, CUR/SKB.

Nobis. 2000. Molecular detection of dechlorinating microorganisms. Report 97-4-02, Gouda.

Van Aalst-Van Leeuwen, M. A., J. Brinkman, S. Keuning, A. A. M. Nipshagen, and H. H. M. Rijnaarts. 1997. Afbraak van per- en trichlooretheen onder sequentiële redoxomstandigheden - Fase 1: Deelresultaat 2.6; Veldkarakterisatie en laboratoriumexperimenten. Gouda, Nobis.

COMPLETE PCE DEGRADATION AND SITE CLOSURE USING ENHANCED REDUCTIVE DECHLORINATION

Michael S. Maierle, *Jennine L. Cota* (ARCADIS G&M, Inc., Milwaukee, Wisconsin, USA)

ABSTRACT: A patented in-situ enhanced reductive dechlorination (ERD) process was implemented for treatment of tetrachloroethylene (PCE) and its daughter products in groundwater utilizing a dilute molasses solution as an electron donor (Lenzo, 2000; Suthersan, 2000). This case study clearly demonstrates that ERD can be used on a full-scale basis to achieve complete PCE degradation and conversion to innocuous end products in less than a two-year time frame.

Prior to implementing the ERD process, the source of PCE that caused site groundwater contamination at the former dry cleaning facility was effectively removed through soil excavation and off-site disposal. Twenty months after implementing the ERD process, PCE concentrations within the plume decreased from pre-remediation levels of approximately 1,500 to 4,000 micrograms per liter (μg/L) to non-detectable levels. As expected, there was a corresponding increase in cis-1,2-dichlorethylene (DCE) and vinyl chloride (VC) concentrations, which occurred in conjunction with the decrease in PCE concentrations. The corresponding build-up of DCE and vinyl chloride peaked at approximately 6 and 14 months, respectively, after initiating the ERD process. The DCE and vinyl chloride levels then dropped sharply over the next 6 months. Ethene and ethane levels increased over two orders of magnitude (exceeding 400 μg/L) in conjunction with the decreasing concentrations of DCE and vinyl chloride. Based on stoichiometric relationships, it is estimated that more than 90% of the PCE was degraded to ethene and ethane within the 20-month period. Regulatory approval for site closure was received in January 2001, less than 2 ½ years after initiating the ERD process. This expedited remediation time frame demonstrates the importance of source removal and proper implementation of the ERD process.

INTRODUCTION

Prior to 1998, the Washington Square Mall in Germantown, Wisconsin was a dilapidated retail center. Contamination at the site resulted from historic releases of PCE, a common dry cleaning solvent, from a dry cleaning facility that operated within the former shopping mall. The location of the former dry cleaners is presented on Figure 1.

A soil remediation program was completed in August 1998 and involved the excavation and off-site disposal of approximately 3,125 tons (2.8×10^6 kilograms) of PCE-impacted soils. The excavation extended down to the water table, which was even with the top of a saturated sand seam at a depth of approximately 14 feet (4.27 meters) below initial grade. In order to maintain suitable conditions for backfilling and to achieve additional contaminant mass removal, provisions were included for the temporary recovery of groundwater from the base of the excavation. Approximately 88,375 gallons (334,499 liters) of water were pumped from the excavation and discharged to the sanitary sewer in August 1998. It is estimated that approximately 25,000 gallons (94,625 liters) of this volume was attributable to precipitation or surface water run-in that accumulated in the excavation, and the remainder was groundwater recovered from the sand seam that was penetrated by the deep excavation.

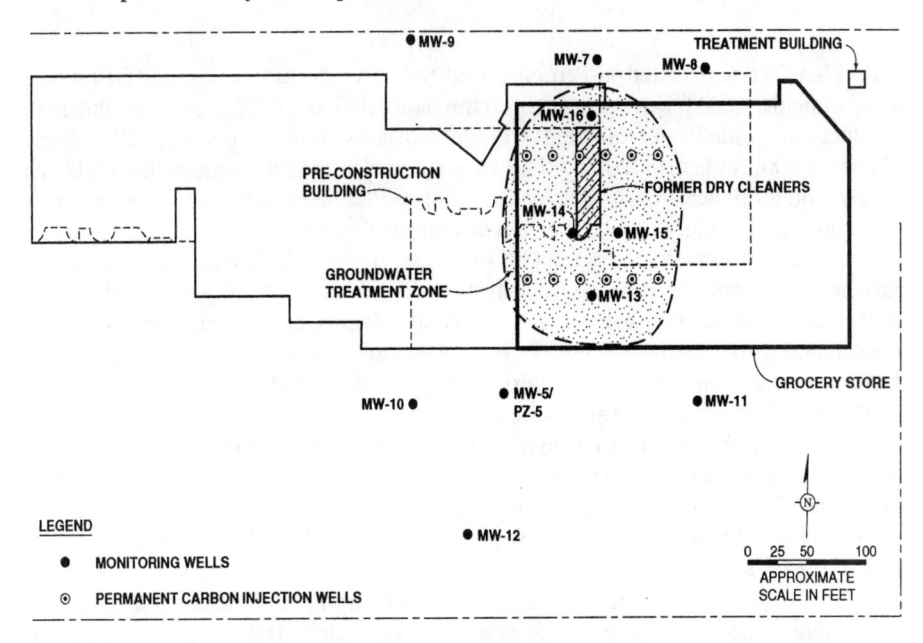

FIGURE 1. Site Layout

The lateral and vertical extent of affected groundwater was defined, and was approximately 30,000 square feet (2,787 square meters) in plan size, extending to a depth of approximately 20 feet (6.10 meters) below grade. The investigation results suggested that the affected groundwater had spread laterally from the source area primarily through a 2 to 5 feet (0.61 to 1.52 meters) thick silt and sand seam that is approximately 13 to 18 feet (3.96 to 5.49 meters) below grade. Within this seam, the extent of impacted groundwater was estimated to be 150 feet (45.72 meters) in width by 200 feet (60.96 meters) in length. The groundwater flow velocity through the seam was estimated at 0.06 feet per day (1.83 centimeters per day). A representative geologic cross section is presented in Figure 2.

The patented groundwater remediation process involved the periodic injection of an organic carbon (molasses) solution to enhance the reductive dechlorination of the groundwater contaminants (i.e., an in-situ bioremediation process). By injecting an organic carbon source, anaerobic and strong reducing conditions were created within the in-situ reaction zone. These conditions created a more suitable environment for the degrading microorganisms to promote both the desorption of the PCE from the aquifer matrix and the ERD (i.e., biodegradation) of the PCE (Payne et al, 2001). The dilute molasses solution was injected to create a reactive zone throughout the area of impacted groundwater.

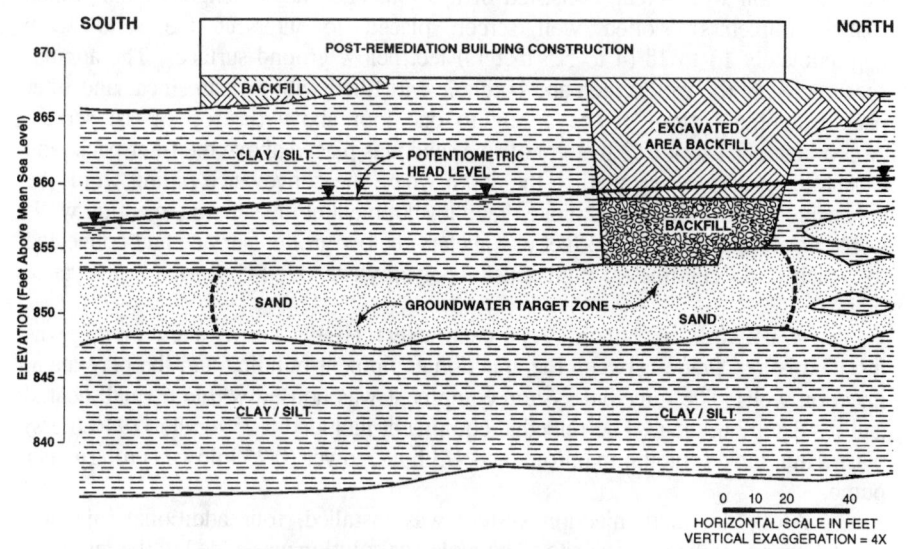

FIGURE 2. Geologic cross section

MATERIALS AND METHODS

Following completion of the soil excavation activities, the groundwater remediation program was implemented at the site. An initial injection event was conducted in August and September 1998, using 182 temporary injection wells installed by boring a hole with a Geoprobe® system. The temporary injection wells were constructed of 1-inch (2.54 centimeter) diameter polyvinyl chloride (PVC) pipe for the well screen and riser. Bentonite pellets were used to seal the temporary wells. The injection wells were advanced in a grid-like pattern across the groundwater target area. The spacing between each injection well was approximately 10 feet (3.05 meters). The injection wells were advanced to intersect the sand seam ranging from 13 to 18 feet (3.96 to 5.49 meters) below ground surface.

Edible blackstrap molasses was used for the initial injection. The edible blackstrap molasses is approximately 47% carbohydrates by weight. The molasses solution was mixed in a plastic tank on site using potable water. Approximately 15 to 25 gallons (56.78 to 94.63 liters) of the dilute molasses solution [the dilution ratio was 25 gallons (94.63 liters) of water to each gallon (3.79 liters) of molasses] were injected into each temporary well using a grout pump. Approximately 3,190 gallons (12,074 liters) of the dilute molasses solution were injected into the temporary injection points over 11 days.

A permanent injection system was installed concurrently with the initial injection event. Twelve fixed injection wells were installed at the site using conventional hollow-stem auger drilling techniques [4¼-inch (0.11 meter) inside diameter augers]. The locations of these wells are shown on Figure 1. The fixed injection wells consisted of a 2-inch (5.08-centimeter) diameter Schedule 40 PVC riser with a 2-inch (5.08-centimeter) diameter Schedule 40 PVC well screen. Each injection well screen consisted of a 5-foot (1.5 meter) length of 0.010-inch (0.25 millimeters) slotted well screen placed to intersect the sand seam approximately 13 to 18 (4 to 5.5 meters) feet below ground surface. The annular space between the well screen and borehole was filled with a clean silica sand filter pack from the bottom of the boring to one foot above the top of the screen. Approximately 1 foot (0.3 meters) of fine sand was placed above the filter pack, and a bentonite seal was installed to the depth where the conveyance piping would be connected to the well. To facilitate the redevelopment at the site, the injection wells were cut off approximately 6 feet (1.8 meters) below ground surface and connected to 1-inch (2.54 centimeter) high-density polyethylene (HDPE) buried conveyance piping.

A network of 1-inch (2.54 centimeter) HDPE conveyance piping was installed below grade between the injection equipment building and the permanent injection wells. The remedial system equipment was housed within a small heated and insulated building. The remedial equipment included a 250-gallon (946 liters) plastic mix tank, a piping manifold, and 1/3 horsepower (0.25 kilowatt) rotary gear pump.

After the fixed injection system was installed, four additional injection events were completed at the site. The molasses solution was added to the mix tank and pumped through the manifold to the injection wells at a dilution ratio of 25 gallons (94.63 liters) of water to each gallon (3.75 liters) of molasses. The molasses used for the permanent injection wells was a low-sulfur, cane juice molasses that contained approximately 66% carbohydrates by weight. A total of 2,985 gallons (11,298 liters) of the molasses solution was injected into the aquifer through the permanent injection wells during the four injection events completed over a six-month period from March 1999 to September 1999.

The quantity of the dilute molasses solution injected into the aquifer and the timing of each event was determined based on changes in biodegradation indicator parameters and the rate of reductive dechlorination determined from the groundwater monitoring data collected over time from the site monitoring well network. The optimum values for groundwater indicator parameters for the ERD process included an oxidation-reduction potential of less than –200 millivolts, total organic carbon in the range of 25-100 milligrams per liter (mg/L), and a pH above 5. Due to site redevelopment activities occurring concurrently with the groundwater remediation, post-injection groundwater monitoring did not begin until 6 months following the initial injection event.

The site monitoring well network consisted of four monitoring wells within the limits of the plume and eight monitoring wells located outside of the plume. Figure 1 shows the locations of the monitoring wells.

RESULTS

Six rounds of groundwater sampling were completed from February 1999 to April 2000 following implementation of groundwater remediation at the site. Over the 20-month period following completion of soil remediation activities and the initial carbon injection event (August 1998), PCE concentrations within the plume decreased to non-detectable levels (April 2000). As expected, a temporary increase in DCE and vinyl chloride concentrations occurred in conjunction with the decrease in PCE concentrations. The corresponding build-up of DCE and vinyl chloride peaked at approximately 6 and 14 months, respectively, after initiating the ERD process. The DCE and vinyl chloride levels then dropped sharply over the next 6 months. Figures 3 and 4 illustrate the contaminant concentration changes over time for two of the monitoring wells located within the plume.

A buildup of the non-toxic, innocuous end products of the reductive dechlorination process (e.g., ethene, ethane, carbon dioxide) indicated that the source PCE was being completely transformed. The monitoring data collected indicated significant production of ethene and ethane within the groundwater plume. Ethene and ethane concentrations in the four monitoring wells within the plume were detected approximately one to two orders of magnitude higher than the ethene and ethane levels measured in the monitoring wells located along the fringe of the plume. This was clear evidence that the reductive dechlorination process was going to completion.

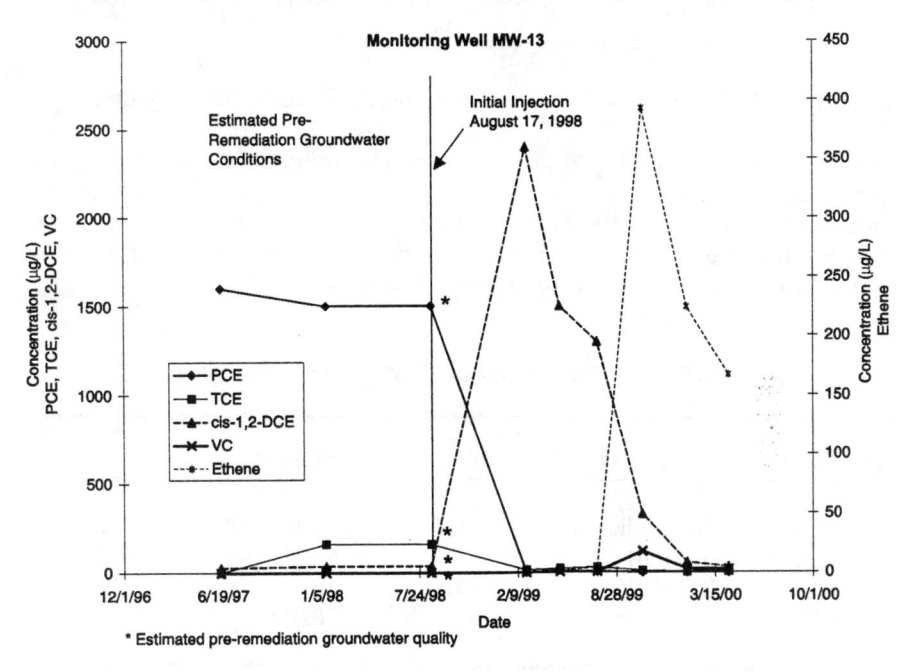

FIGURE 3. Groundwater remediation performance monitoring data

Based on the use of first-order degradation kinetics, the biodegradation rates for the chlorinated constituents at the site can be determined (U.S. EPA, 1998). Table 1 lists the average site-specific biodegradation rates determined from the collected data for each of the monitoring wells within the groundwater plume. The

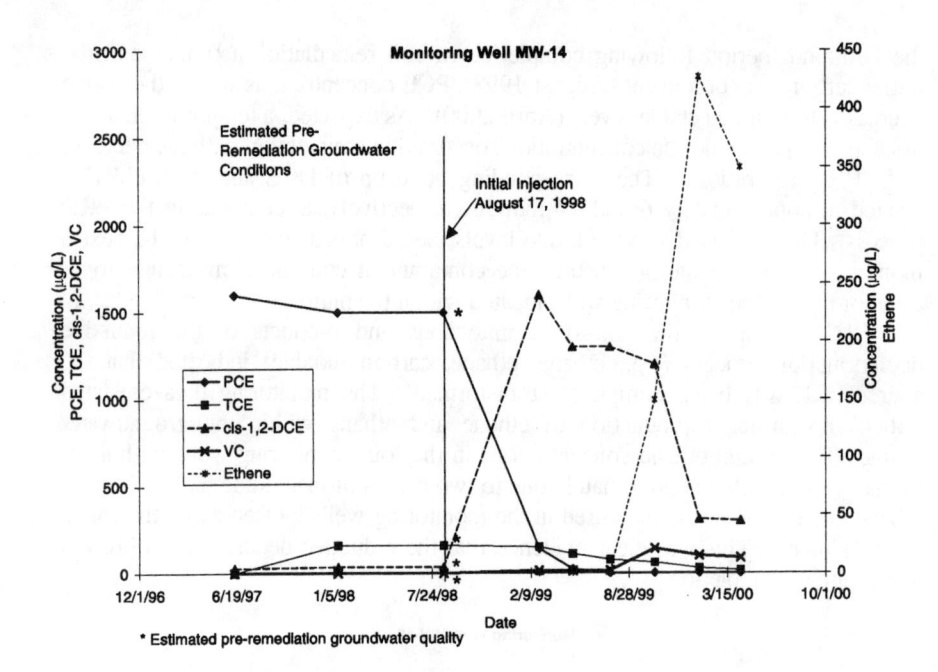

FIGURE 4. Groundwater remediation performance monitoring data

site-specific biodegradation rates are approximately two to eight times higher than average published biodegradation rates under natural conditions (U.S. EPA, 1998). This demonstrates that the ERD process can greatly accelerate biodegradation rates. Note that the total molasses solution injected was only approximately 2 percent of the total volume of groundwater in the target area, indicating that dilution effects on the observed rates were minimal.

TABLE 1. Calculated Site Biodegradation Rates (day^{-1})

Compound	MW-13	MW-14	MW-15	MW-16
PCE	Not applicable	0.027	Not applicable	0.021
TCE	0.011	0.005	Not applicable	0.023
DCE	0.010	0.004	0.011	0.017
VC	0.015	0.003	0.011	0.018

Changes in the molar concentration over time of the parent compound (PCE) and its daughter products (TCE, DCE, VC, ethene, ethane) are presented on Figures 5 and 6 for two of the monitoring wells located within the plume. These data illustrate that within 6 months of implementing the ERD process, over 90% of the PCE was degraded to DCE. In addition, within 20 months of initiating the ERD process, over 90% of the PCE in the groundwater plume was degraded to ethene and ethane, or completely mineralized to carbon dioxide, water, and chloride.

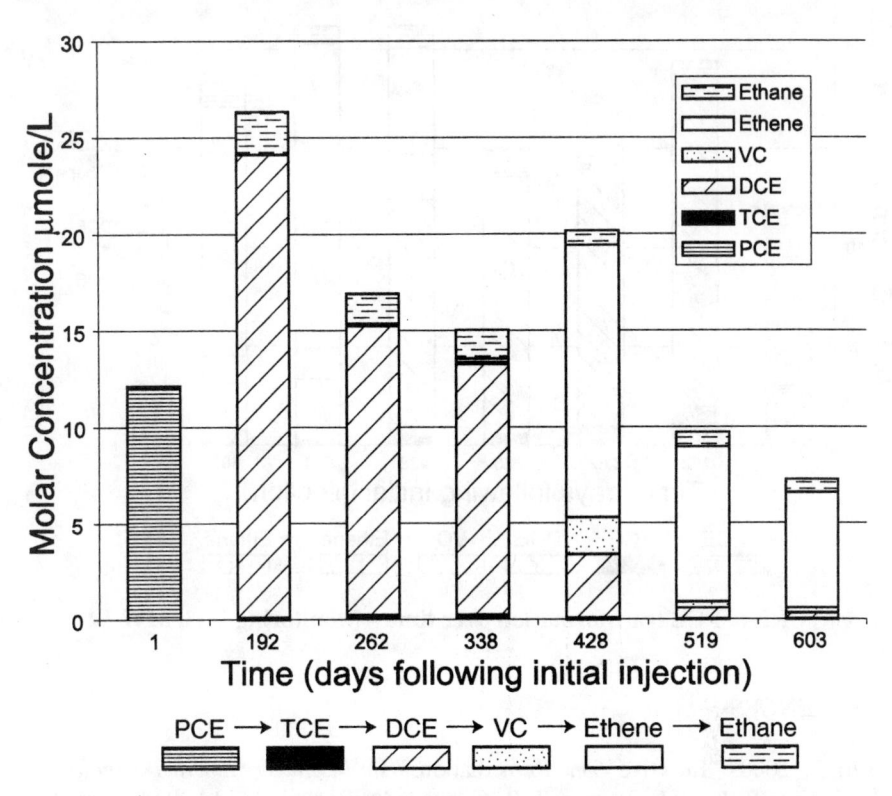

FIGURE 5. PCE transformation over time – Monitoring Well MW-13

CONCLUSION

This project demonstrates that source removal and proper implementation of the ERD process can greatly expedite the remediation time frame for PCE contaminated groundwater. Twenty months after implementing the ERD process, PCE concentrations within the plume decreased from pre-remediation levels of approximately 1,500 to 4,000 micrograms per liter (µg/L) to non-detectable levels. Based on stoichiometric relationships, it is estimated that more than 90% of the PCE was degraded to ethene and ethane within the 20-month period. Regulatory approval for site closure was received in January 2001, less than 2 ½ years after initiating the ERD process.

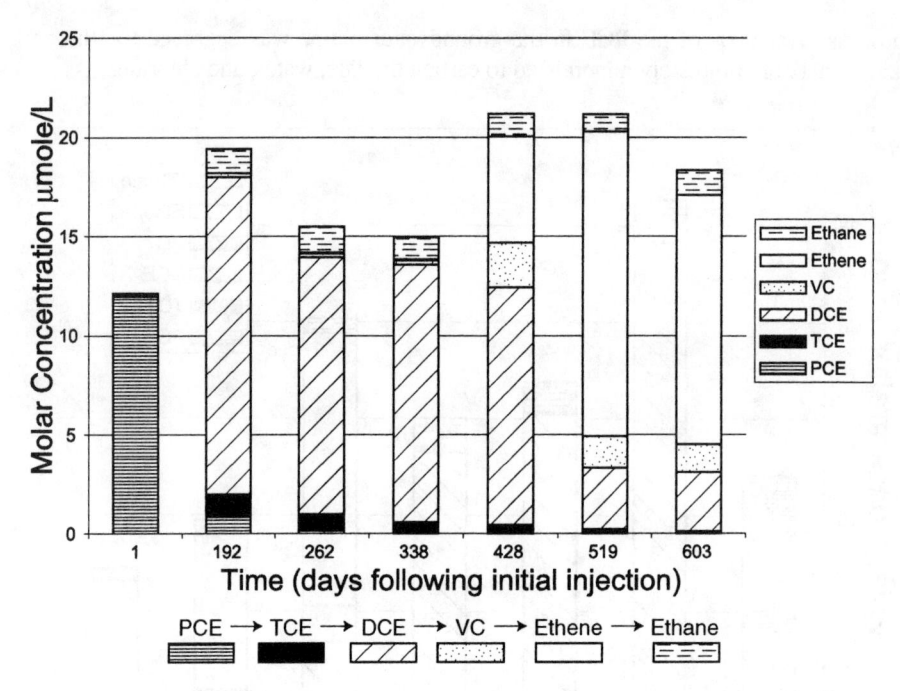

FIGURE 6. PCE transformation over time – Monitoring Well MW-14

REFERENCES

Lenzo, F. 2000. "Reactive Zone Remediation." In: *In Situ Treatment Technology, 2nd ed.* Nyer, E.K., P.L. Palmer, E.P. Carman, G. Boettcher, J.M. Bedessem, F. Lenzo, T.L. Crossman, G.J. Rorech, and D.F. Kidd. Lewis Publishers, New York.

Payne, Frederick C., S.S. Suthersan, F.C. Lenzo, and J.S. Burdick. 2001. "Mobilization of Sorbed-phase Chlorinated Alkenes in Enhanced Reductive Dechlorination." Sixth International Symposium on In Situ and On-Site Bioremediation, San Diego, California, June 4-7.

Suthersan, S.S. November 2000. *Engineered In Situ Anaerobic Reactive Zones.* U.S. Patent 6,143,177.

U.S. Environmental Protection Agency, September 1998. Technical Protocol for Evaluating Natural Attenuation of Chlorinated Solvents in Groundwater. U.S. EPA Rep. EPA/600/R-98/128. U.S. Gov. Print. Office, Washington, DC.

ANAEROBIC BIOREMEDIATION OF TRICHLOROETHENE NEAR DULUTH INTERNATIONAL AIRPORT

Robin Semer (Harza Engineering Company, Inc., Chicago, Illinois)
Pinaki Banerjee (Harza Engineering Company, Inc., Chicago, Illinois)

ABSTRACT: A pilot-study for treatment of trichloroethene (TCE) and dechlorination products, dichloroethene (DCE) and vinyl chloride (VC), is currently being conducted at a site near the Duluth International Airport. The study began in April 2000 with direct push injection of a polylactate ester (HRC®) into the aquifer immediately upgradient of a former source area monitoring well. The goal of the pilot test is to indicate if TCE, DCE, and VC concentrations can be adequately attenuated to merit full-scale application of the polylactate ester. After six monitoring events over a nine-month period, TCE decreased from 400 µg/L to <2 µg/L, cis-1,2-DCE increased from 50 µg/L to 750 µg/L, and VC increased from 10 µg/L to 20 µg/L.

The objective in using the polylactate ester was to provide a slow release hydrogen source as an electron donor to bioattenuate chlorinated compounds at a site where environmental conditions were not sufficient for natural attenuation. This study serves to quantify this objective. With respect to TCE, the rate of transformation increased from a virtually imperceptible rate of transformation to a half-life of 26 days. Alternatively, daughter products DCE and VC have accumulated. Relatively slow transformation rates for DCE and VC may be responsible for continued increases in DCE and VC concentrations. The concentration of DCE increased to a greater extent than can be accounted for by transformation of the parent compound TCE, which may indicate that additional TCE desorption from soil has occurred. It is not known whether DCE and VC are dechlorinating further downgradient due to the absence of a downgradient monitoring well in close proximity to the test well. If complete dechlorination is limited under the conditions as created, then alternative remedial methods for the transformation of DCE and VC need to be evaluated.

INTRODUCTION

Investigations at a former landfill site, Site 7 (LF-08) near the Duluth International Airport, identified the presence of volatile organic compounds (VOCs) in groundwater, primarily trichloroethene (TCE) and daughter products. The nature and extent of contamination at the site were characterized. Natural attenuation monitoring was conducted to determine if intrinsic biotransformation of TCE and daughter products is occurring and if it is an adequate remedy for contaminated groundwater. Results indicated that biotransformation of TCE in the groundwater is occurring as evidenced by consistent reports of the daughter product cis-1,2-dichloroethene (DCE), and transformation of DCE is occurring as evidenced by its dechlorination product, vinyl chloride (VC). The transformation of TCE and daughter products is occurring at a very slow rate in an anaerobic, generally iron-reducing, environment. The calculated rate of TCE transformation is so slow as to be virtually imperceptible. Modeling results predict that TCE will reach the receptor, one mile from the site, at a concentration above the federal maximum contaminant level (MCL) and Minnesota Pollution Control Agency (MPCA) cleanup goals of 5 µg/L. Therefore, natural attenuation is not a viable option for TCE and daughter product plume management at the site.

Various treatment technologies were examined for potential long-term effectiveness, implementability, short-term risks, and cost, including chemical, biological and physical treatment. Processes evaluated included a permeable iron reactive wall, chemical oxidation with potassium permanganate or hydrogen peroxide, pump and treat with air stripping, and bioremediation with various enhancements. Biological treatment was chosen with Hydrogen Release Compound (HRC®) as a potentially low cost time-release method of providing the aquifer with electron donors to enhance bioremediation.

Objective. The objective of this study is to determine whether adding polylactate ester to the groundwater at this site would enhance TCE-, DCE-, and VC-dechlorination. If successful treatment results were achieved, polylactate ester would probably be applied to other areas. The pilot study area will not be completely remediated because of continuing contaminant migration from the upgradient source area. Following the pilot study, the objective of full-scale remediation would be to achieve concentrations at or near MPCA cleanup goals. Clean-up goals for TCE, cis-1,2-DCE, and VC are 5, 70, and 0.2 µg/L, respectively.

Site Description. The former disposal area was used by the U.S. Air Force Air Combat Command (ACC) during the 1950s through the 1970s as a site for discarding general rubbish, hardfill, aircraft parts, empty drums, and drums containing unburnable and unrecoverable chemicals. The type and quantity of substances disposed were not documented. The landfill operations ceased by the end of the 1970s, after which the landfill was capped with three to four feet of local soil cover. Recharge of groundwater is primarily by direct infiltration from the ground surface and discharge is primarily to local streams, bogs and other surface waters. Potentiometric levels are within 5 to 6 feet of the ground surface. Groundwater velocity in the upper permeable zone averages about 18 ft/yr. Hydraulic conductivity in the bedrock at the site is low, supporting the conclusion that this unit forms the lower hydraulic limit of the uppermost aquifer.

Various investigations have been performed to delineate the nature and extent of contamination beginning in the late 1980s. Figure 1 shows the plume configuration at the site. The contaminant plume is relatively narrow, (300 to 400 feet at its widest point), and has migrated from its source approximately 800 feet to the northwest. The vertical extent of the contaminated zone is generally about 5 to 18 feet below the surface. The highest TCE concentrations are consistently found below wetlands about 600 feet northwest of the former source area.

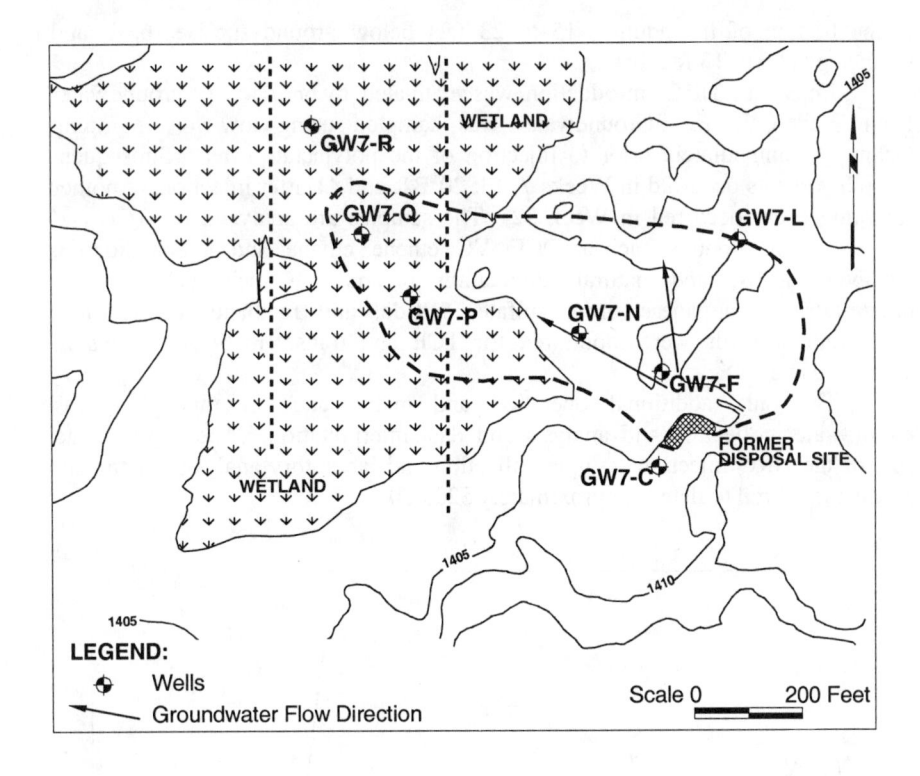

FIGURE 1. Groundwater Plume

MATERIALS AND METHODS

HRC® is a polylactate ester that releases lactic acid. The lactic acid and its metabolites (acetic, butyric, propionic and pyruvic acids) ferment over time, resulting in time-release of hydrogen. The hydrogen then supports the microbially mediated reductive dechlorination of parent and daughter products. The only operation and maintenance required after application is periodic monitoring of the results.

The amount of material to be injected was based on contaminant concentrations, electron acceptor concentrations and hydrogeologic data obtained during natural attenuation monitoring. A safety factor was also used to account for additional electron acceptors. Injection of 180 pounds of HRC into the contaminated aquifer occurred on April 2000. The material was injected immediately upgradient of the former source area and was monitored at one well (GW7-F). Polylactate ester application was via nine direct push applications placed about 5, 10 and 15 feet upgradient of the monitoring well, which was located downgradient of the former source area (Figure 2). Rods (1.25-inch outer diameter) were pushed to the bottom of the contaminated saturated zone and the material was injected as the rods were withdrawn. Injection of the material began at the bottom of the aquifer, 15 to 23 feet below ground surface (bgs), and terminated at 5 to 13 feet bgs.

Impact of HRC introduction was evaluated by monitoring groundwater from Well GW7-F. Groundwater was sampled using low-flow pumping techniques immediately prior to injection of the polylactate ester. Subsequent sampling events occurred in Weeks 7, 13, 20, 31, and 41 after injection. Another sampling event occurred in Week 52. The samples are analyzed for TCE and transformation products such as DCE, VC, ethene, ethane, and carbon dioxide; metabolic acids; and natural attenuation parameters, such as alkalinity, nitrate/nitrite, manganese, iron, sulfate, sulfide, and methane. In addition, downgradient wells were monitored for TCE and transformation products in Weeks 20 and 31.

The only additional operation and maintenance of this pilot was groundwater collection and analysis and data interpretation. The approximate cost of the HRC injection, sample collection and laboratory analysis for the six events monitored to date is approximately $20,000.

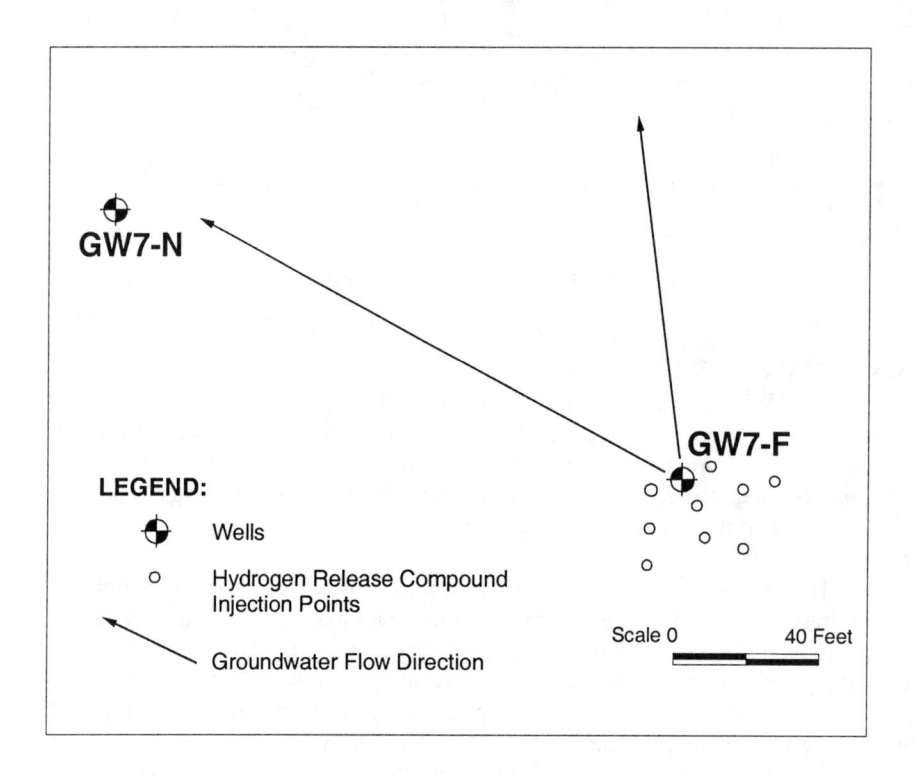

FIGURE 2. Configuration of Monitoring Well and Injection Points

RESULTS AND DISCUSSION

TCE and Transformation Products. The effects of the addition of the polylactate ester to the groundwater in the vicinity of GW7-F are shown in Figure 3. By the first sampling event (Week 7), the TCE concentration started to decrease and the concentration of its transformation product, cis-1,2-DCE, started to increase. TCE decreased from 354 µg/L to 9 µg/L in 20 weeks, a half-life of 26 days. The concentration of cis-1,2-DCE continued to rise and peaked at Week 31. There was a slight decline in DCE concentration by Week 41. VC concentration started to increase by Week 20 and peaked by Week 31.

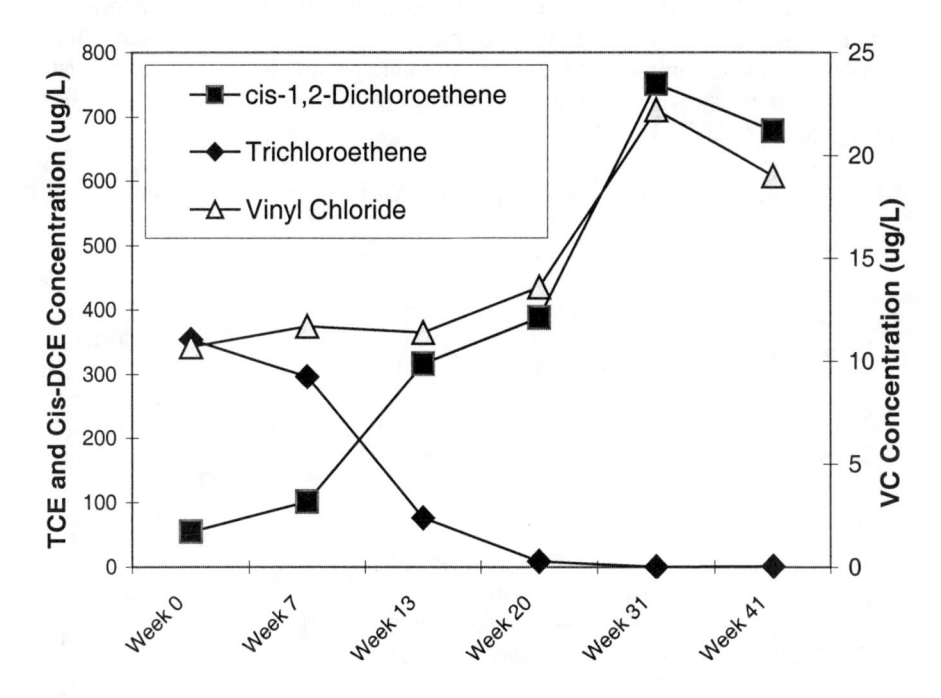

FIGURE 3. TCE, Cis-1,2-DCE, and VC Concentrations During Pilot Study

DCE concentrations in Weeks 31 and 41 were higher than expected based on the amount of TCE transformed suggesting dechlorination of previously sorbed TCE. Further increases in DCE concentration are not anticipated. Ethene and ethane were not detected in any samples. Carbon dioxide levels increased from 4.2 mg/L prior to material application to 29.5 mg/L at Week 7 and then fluctuated between those values.

Monitoring results from wells downgradient from GW7-F show no impact. Due to slow movement of groundwater (ranging from 5-33 feet per year), the injected material is not expected to reach downgradient wells during the lifetime of the pilot study.

Transformation of TCE and byproducts appear to naturally segregate into three groups (Farone et al., 2000). In the first group, the primary contaminant (such as TCE) and daughter products (such as DCE and VC) biotransform at similar rates. In the second group, the primary contaminant is reduced and the transformation products buildup prior to attenuating because of differential rates of attenuation. In the third group, the primary contaminant is reduced and there is no clear indication that the daughter products will attenuate in a reasonable time frame. It is unclear after Week 41 of the pilot study whether DCE and VC will soon attenuate, as characteristic of the second group, or if attenuation will not occur in a reasonable time frame, as characteristic of the third group.

Alternative treatment options to further transform DCE and VC will have to be evaluated if they do not decrease to acceptable levels. Future monitoring data is expected to clarify DCE and VC transformation patterns under conditions induced by HRC addition to the aquifer. Possible parallel approaches to fill data gaps include constructing another monitoring well approximately 15 to 20 feet downgradient of GW7-F.

Metabolic Acids. The metabolic acid profile over time is shown in Figure 4.

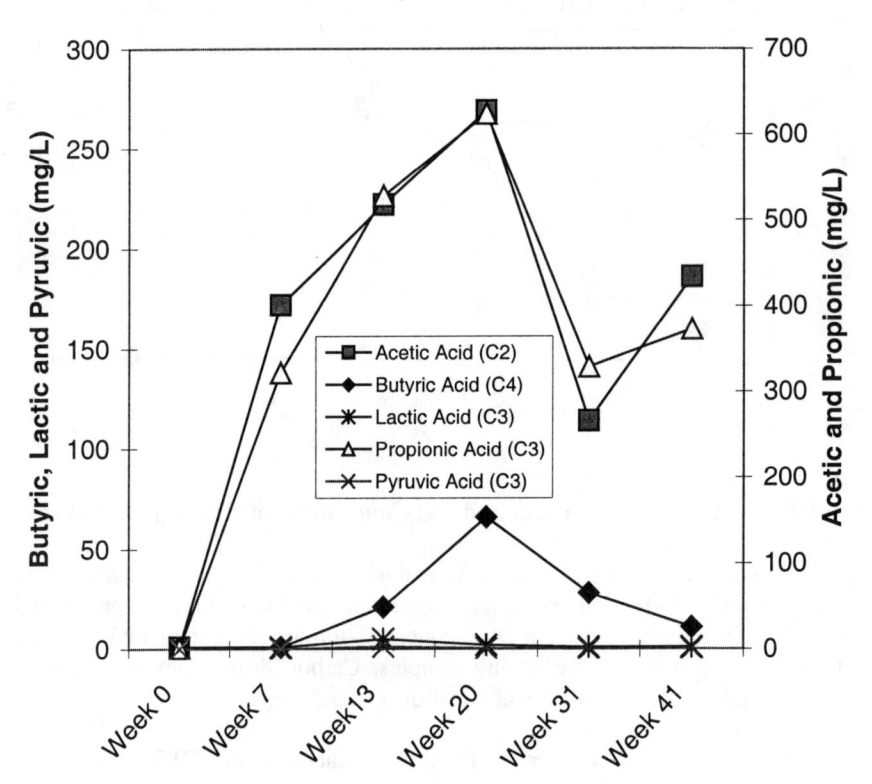

FIGURE 4. Metabolic Acids Concentrations

The primary metabolic acid released from HRC is lactic acid. Once it is released into the aquifer, it can produce a number of byproducts and derivatives including propionic, pyruvic, butyric, and acetic acids. Lactic acid is metabolized quickly at this site as indicated in Figure 4 by the lack of lactic acid. Acetic and propionic acids are apparent by Week 7. Butyric is first evident in Week 13. Acetic, propionic, and butyric acids peak in Week 20. Acetic and propionic increase slightly from Week 31 to Week 41, while butyric decreases slightly during the same period. The graph of the acid concentrations shows that the metabolic acids continue to be active 41 weeks after injection of the material.

Natural Attenuation Parameters. Total organic carbon (TOC), alkalinity, iron and manganese concentrations increased over time, peaking at Week 20, and started to show a downward trend (See Figure 5). Methane was detected at trace levels (0.06 mg/L) up to Week 41 and then increased by over an order of magnitude. Over the same period of time, oxidation-reduction potential (Eh) decreased from 88 mv to –40 mv. The data appears to indicate that an iron-reducing anaerobic environment is present at the site and that the reducing capacity of the aquifer may have peaked but continues to support active dechlorination.

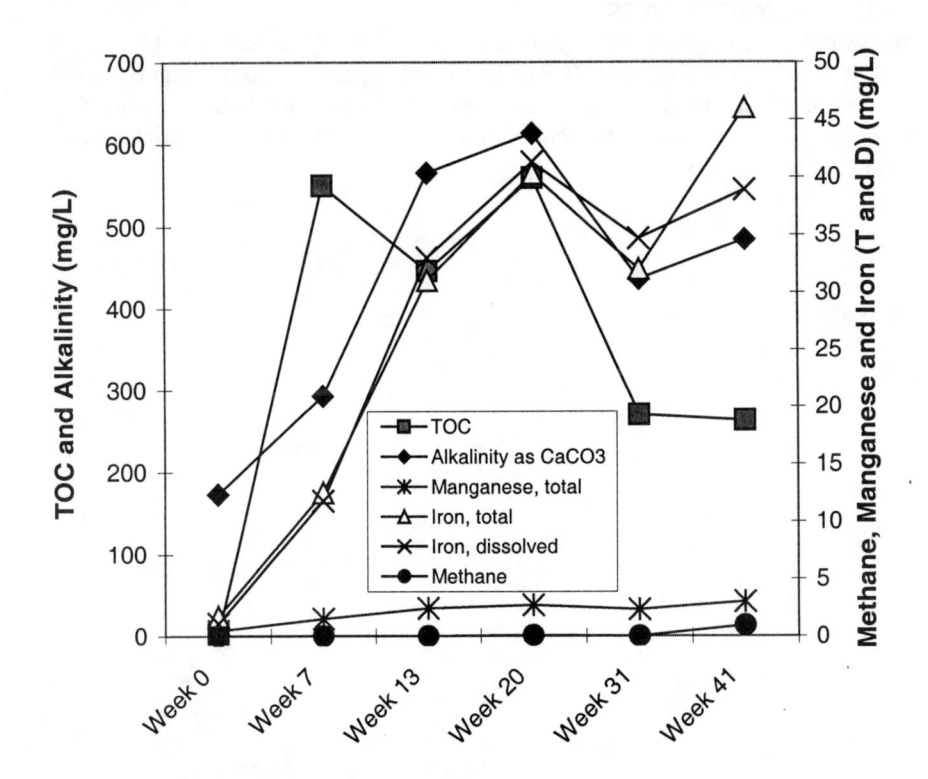

FIGURE 5. TOC, Alkalinity, Iron, and Manganese

Conclusion. The effectiveness of HRC in treating the TCE in groundwater at this site is clearly evident. It is probable that additional TCE is desorbing from the soil and being dechlorinated to DCE. DCE and VC accumulated, but DCE began to decrease towards the end of the monitoring period. This decrease may be attributable to dechlorination and to migration. If there are no signs of attenuation of DCE and VC, alternative remediation methods may have to be explored. Other indicators, such as the continued presence of metabolic acids, including propionic acid at about 400 mg/L, and an anaerobic environment show that the aquifer continues to be conducive to reductive dechlorination.

REFERENCE

Farone, W.A., S. S. Koenigsberg, T. Palmer, D. Brooker. 2000. "Site Classification For Bioremediation Of Chlorinated Compounds Using Microcosm Studies." Proceedings of The 2nd International Conference on Remediation of Chlorinated and Recalcitrant Compounds, Monterey, CA, May 22-25, 2000, Battelle Press, Columbus, OH.

ACKNOWLEDGEMENTS
We would like to thank the U.S. Air Force Air Combat Command (ACC) and the U.S. Army Corps of Engineers (USACE) for the opportunity to perform this pilot test and to share the results. The ACC and USACE are working in partnership with the Minnesota Pollution Control Agency (MPCA) to achieve groundwater cleanup goals at this site.

ENHANCED CAH REDUCTIVE DEHALOGENATION AT A FORMER WASTEWATER TREATMENT FACILITY

I. Richard Schaffner, Jr.; Amy T. Doherty; James M. Wieck; and Steven R. Lamb (GZA GeoEnvironmental, Inc., Manchester, New Hampshire, USA)

ABSTRACT: Full-scale biostimulant injection enhances chlorinated aliphatic hydrocarbon (CAH) reductive dehalogenation in overburden groundwater at a former wastewater treatment facility (WWTF). Biostimulant is injected in 125,000-gallon (GAL, 4.7E5 L) batches utilizing former WWTF infrastructure. An existing monitoring/extraction well network was used to inject biostimulant. Groundwater monitoring for CAHs and transformation indicator parameters was performed to evaluate performance. The remedial strategy includes biostimulant injection to establish an anaerobic treatment zone for stimulating parent CAH dehalogenation. Monitoring parameters include CAHs, chemical oxygen demand (COD), dissolved oxygen (DO), and oxidation-reduction potential (ORP). Biostimulation resulted in reductions in mean total CAH mass by $\leq$85%, elevated COD concentrations by $\leq$3 orders of magnitude, and depressed DO and ORP values by $\leq$96% (4.7 to 0.2 milligrams per liter, mg/L) and $\leq$156% (+168 to –94 millivolts, mV), respectively.

INTRODUCTION

CAHs are present in groundwater at concentrations of $\leq$10 mg/L at a former municipal WWTF located in New Hampshire, USA. CAHs include the reactants (parents) tetrachloroethene (PCE), trichloroethene (TCE), and 1,1,1-trichloroethane (TCA) and the products (daughters) 1,1-dichloroethene (1,1-DCE), *cis/trans*-1,2-dichloroethenes (1,2-DCEs), 1,1/1,2-dichloroethanes (DCAs), and vinyl chloride (VC). The source was a pit where CAHs were co-disposed with sludge. Substantial quantities of non-aqueous phase liquid are not present based on groundwater monitoring results. Two hydrogeologic units separated by a discontinuous clayey silt aquitard are present. An upper fine sand and silt unit extends from a depth of approximately 5 to 20 feet (ft) (1.5 to 6 m). A lower sandy and silty till unit extends from a depth of about 30 to 50 ft (9 to 15 m). The discontinuous aquitard generally extends from a depth of 20 to 30 ft (6 to 9 m). Groundwater flows westward and discharges to a river along the property boundary (Figure 1).

Two distinct redox zones with differing transformation processes are present within the plume. Anaerobic/chemically reducing conditions occur within the core, with decreasing parent and increasing daughter product concentrations. Aerobic/ chemically oxidizing conditions occur within the periphery, with decreasing parent and daughter product concentrations. Parents attenuate more slowly along the periphery than within the core. Previous site work indicated sludge disposal stimulated obligate aerobic/facultative aerobic bacteria to scavenge terminal electron acceptor (TEA) and drive conditions methanogenic. Methanogenic conditions supported parent dehalogenation within the core and evolved methane stimulated parent/daughter co-oxidation within the aerobic-

FIGURE 1- SITE PLAN

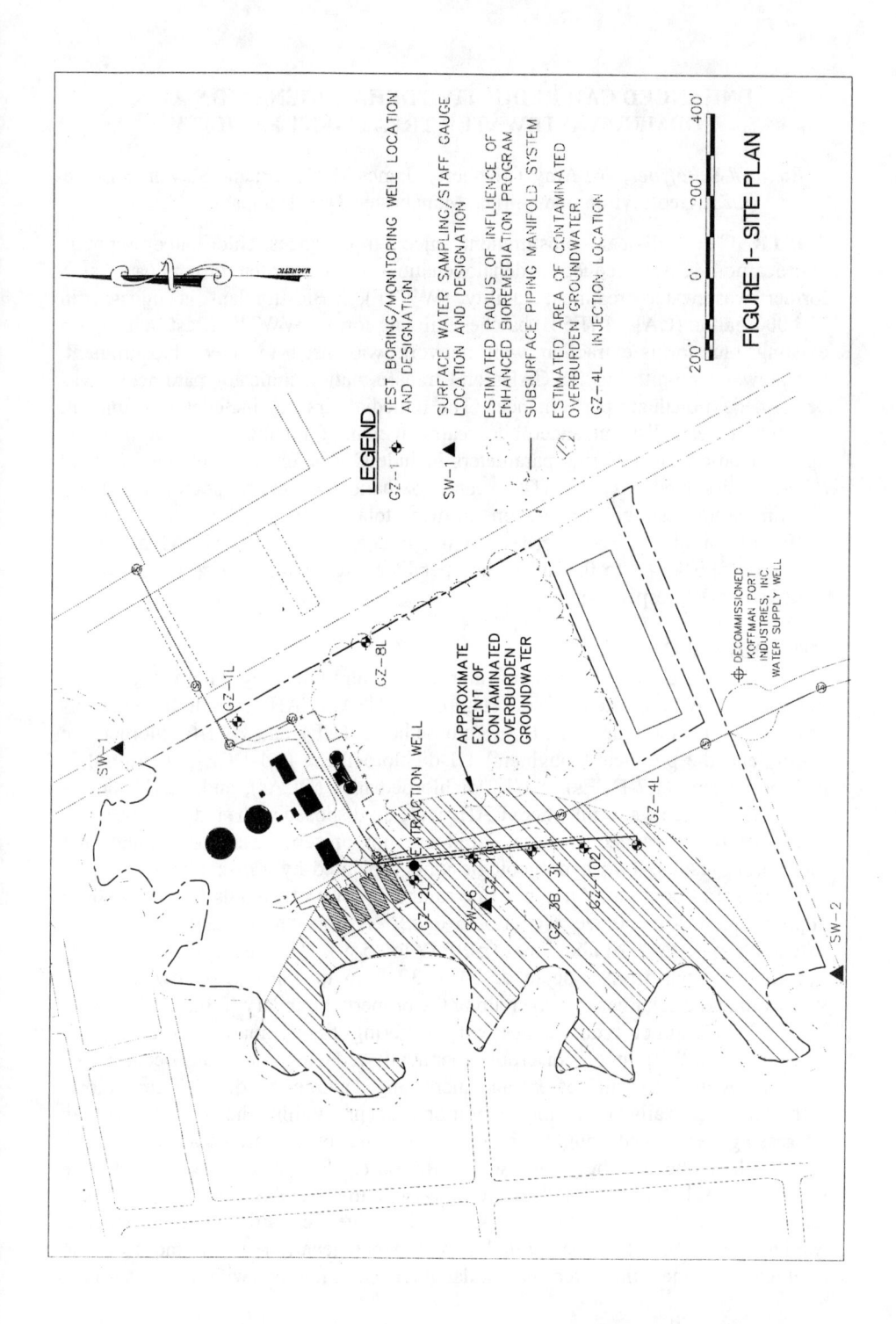

anaerobic mixing zone along the periphery (Schaffner, *et al.*, 1996). Daughter product mineralization likely occurs within that zone. Initially, residual sludge fueled CAH transformation. Recent low dissolved organic carbon (DOC) concentrations (<10 mg/L) and TEA breakthrough into the core suggest transformation was becoming DOC limited. Sludge disposal ceased about 15 years ago. Natural attenuation reduces CAH concentrations below regulatory levels upon discharging to the river based upon surface water quality data collected at downgradient sampling station SW-2 over the last 10 years.

The remedial strategy includes biostimulation to create an anaerobic treatment zone within the core of the plume and assumes daughter products will naturally attenuate in aerobic (downgradient) locations. Results of a 54-day microcosm study indicated biostimulation enhanced parent CAH dehalogenation (Schaffner, *et al.*, 1998). A Bioremediation Pilot Test (BioTest) was initiated in 1997 to develop design/operating criteria for full-scale bioremediation. The BioTest included biostimulant injection into well couplet GZ-2/2L, and evaluation of groundwater system response. BioTest results indicated that biostimulant provided DOC, which stimulated indigenous bacteria to scavenge TEAs, re-establish anaerobic, chemically reducing conditions, and provide electron donor for parent dehalogenation (Schaffner, *et al.*, 1999).

The full-scale bioremediation program involved initiating groundwater remediation by enhancing reductive dehalogenation of parent CAHs, and some intermediates, to daughter products within an anaerobic treatment zone. Study objectives included biostimulant injections at monitoring/extraction wells, and performance monitoring to evaluate groundwater system response. Wells selected for biostimulant injections were generally screened in the lower hydrogeologic unit.

METHODS
Biostimulant Injections. Biostimulation loads are summarized in Table 1. Approximately 125,000 gallons (gal, 4.7E5 L) of formation make-up water from each well was pumped at approximately 10 gallons per minute (gpm, 38 L/m) to a former WWTF sedimentation basin. Formation make-up water was used to minimize CAH dilution. An approximate 300-gal (1.1E3 L) capacity tub was used to mix biostimulant with formation water. Biostimulant generally consisted of a proprietary (patent pending) blend of lactose ($\geq$70% mass) and Yeast Extract ($\leq$30% mass). The blend was selected based upon literature search and microcosm study/BioTest results. Two countercirculating pumps connecting the tub and basin agitate the solution and cause biostimulant dissolution. A pump was used to inject solution into each well at a flow rate of approximately 10 gpm (38 L/m).

TABLE 1. Summary of Biostimulant Injection Loads.

Injection Location	Date Started	Biostimulant Load (pounds)	Approximate Volume (gal)
BioTest Injections			
Monitoring Well GZ-2L and Extraction Well	7/12/97	1,250	1,450
Monitoring Well GZ-3L	11/6/97	900	2,000
Monitoring Well GZ-4L	11/3/98	2,900	62,500
Full-Scale Injections			
Extraction Well	6/29/00	10,000	125,000
Monitoring Well GZ-3L	7/2/00	700	125,000
Monitoring Well GZ-4L	7/28/00	11,400	125,000

Performance Monitoring. Four pre-injection sampling rounds were conducted prior to the 1997 BioTest injection to establish baseline conditions. Thirteen post-injection sampling rounds were performed between July 1997 and January 2001. Table 2 summarizes the analytical program. During the BioTest, certain indicator parameters (*e.g.,* methane, ethene, ethane, nitrate, sulfate) remained conducive for CAH dehalogenation when COD concentrations exceeded 25 mg/L and DO/ORP values were depressed (*i.e.,* <0.5 mg/L DO and <+50 mV ORP). This relationship generally agrees with Wiedemeier *et al.* (1996). Therefore, transformation indicator parameters for full-scale remediation were limited to COD, DO, and ORP.

TABLE 2. Remedial Study Analytical Program.

Parameter/Method	Selection Rationale
CAHs, United States Environmental Protection Agency (EPA) Method 8021 (8010 List)	Contaminants of concern
COD, EPA Method 410.4	DOC surrogate
DO and ORP, parameter-specific electrodes	Microbial respiratory pathway

RESULTS AND DISCUSSION

Groundwater monitoring data indicate that CAH/transformation indicator parameter results vary with time. Variability is primarily attributed to surficial hydrologic events (*e.g.,* precipitation events, fluctuating river stage) impacting groundwater quality. In consideration of this variability, comparisons of pre-injection (baseline)/post-injection data were based on arithmetic/geometric mean (mean) data. Geometric means were used when data variability was an order of magnitude or greater. CAH data are summarized as total parents, total daughter products and total CAHs in Tables 3, 4 and 5 for wells GZ-2L, GZ-3L and GZ-4L, respectively. These tables also include ratios of total parents to total CAHs (parent ratios). Parent ratios were used to normalize the data based on the assumption that though concentration magnitude may vary widely, parent ratios would not.

CAHs. Post-injection mean total CAH, total parent, and total daughter concentrations are lower than baseline values, suggesting biostimulant injection stimulated CAH transformation. Reduction in mean total CAH concentrations ranged between 9% and 85%, reduction in mean total parents ranged between 44% and 93%, and reduction in mean total daughters ranged between 9% and 84%. A slight increase in VC concentration was generally noted for samples collected from most wells, suggesting parent CAH transformation to daughter products.

Calculated pre-/post-injection parent ratios generally indicate reduction in parent mass resulting from biostimulation (37%, well GZ-3L; 30%, well GZ-4L), with the exception of samples collected from well GZ-2L. The parent ratio for samples collected from that well increased 30% (3.0% to 3.9%), however, parent concentrations were generally below practical quantification limits (PQLs). (A value of one half the PQL was used to conservatively estimate total parents if the concentration was non detect.) At such low values, the total parent concentration is sensitive to changes in PQLs and may render the use of parent ratios an ineffective tool for evaluating bioremediation performance at low parent CAH concentrations.

Monitoring data for full-scale injections suggest biostimulation established anaerobic, chemically reducing conditions that stimulated parent reductive dehalogenation. Concentrations of parent and daughter products continued to meet surface water quality standards at down-gradient surface water sampling station SW-2. This suggests that natural attenuation limits impacts to surface water quality.

It is likely that volatilization results in loss of some CAH mass during injection operations. Results of long-term post-injection monitoring show continuing downward trends in parent concentrations, indicating that volatilization is not responsible for the long-term improvements in groundwater quality. This is most evident when comparing post-injection parent and daughter product data. Daughter products are more volatile then parents and would be removed before parents if mass loss was primarily due to volatilization. This further indicates that long-term loss of CAH mass is attributed to biostimulation and not volatilization.

COD. Baseline COD concentrations were relatively low, with concentrations ranging from below the <5.0 mg/L PQL to 35 mg/L. Following the first injection in 1997, COD concentrations increased up to three orders of magnitude. Post-injection concentrations ranged from 110 to 9,200 mg/L. Elevated post-injection COD concentrations reflect the presence of biostimulant, which is organic carbon enriched and exerts strong TEA demand. Post-injection concentrations fluctuated as a function of microbial utilization and hydrodynamic dispersion, as well surficial hydrologic events. Post-injection COD concentrations remained elevated above baseline values, reflecting the continued presence of biostimulant. According to Wiedemeier et al. (1996), DOC concentrations exceeding about 20 mg/L may stimulate CAH dehalogenation. Mean pre-injection COD concentrations ranged from 3 to 31 mg/L; mean post-injection concentrations ranged from 427 to 2,094 mg/L. Some individual year 2000 COD concentrations were significantly higher.

DO/ORP. Baseline overburden groundwater conditions were generally aerobic and chemically oxidizing. DO concentrations ranged between 0.2 to 7.4 mg/L, ORP values between −55 to +310 mV. There was a highly significant DO/ORP decrease in response to biostimulation. Resulting DO concentrations ranged between 0.0 to 1.8 mg/L and ORP values between +10 to -290 mV. According to Wiedemeier et al. (1996), DO concentrations exceeding about 0.5 mg/L inhibit dehalogenation while ORP values less than about +50 mV may be conducive. For example, the mean pre-injection DO concentration for well GZ-4L was 4.7 mg/L, whereas the mean post-injection concentration was <0.2 mg/L (96% reduction). The mean pre-injection ORP value for samples collected from that well was +168 mV, whereas the mean post-injection value was −94 mV (156% reduction). These reductions strongly indicate that biostimulation yielded anaerobic, chemically reducing conditions.

TABLE 3. Summary of Well GZ-2L Results.

Sampling Date	Total Parents (mg/L)	Total Daughter products (mg/L)	Total CAHs (mg/L)	Parent Ratio (%)
\multicolumn{5}{c}{Summary of Pre-Injection (Baseline) Mean Results}
5/96-4/97	0.35	14.23	14.68	3.0
\multicolumn{5}{c}{Post-Injection Results}
07/97	0.85	68.88	69.73	1.2
12/97	0.94	33.96	34.90	2.7
04/98	0.60	11.48	12.08	5.0
07/98	0.18	11.81	11.985	1.5
11/98	0.075	7.18	7.252	1.0
3/99 - 5/99	0.091	0.57	0.658	13.8
07/99	0.006	0.11	0.114	5.0
12/99	0.014	0.10	0.118	11.8
06/00	0.308	5.29	5.593	5.5
08/00	0.465	10.40	10.87	4.3
12/00	0.018	0.23	0.251	7.1
Mean Results:	0.13	3.01	3.19	3.9

TABLE 4. Summary of Well GZ-3L Results.

Sampling Date	Total Parents (mg/L)	Total Daughter products (mg/L)	Total CAHs (mg/L)	Parent Ratio (%)
\multicolumn{5}{c}{Summary of Pre-Injection (Baseline) Mean Results}
5/96-4/97	0.27	13.41	13.68	1.9
\multicolumn{5}{c}{Post-Injection Results}
12/97	0.42	14.5	14.92	2.8
04/98	0.45	14.4	14.85	3.0
07/98	0.11	21.37	21.48	0.5
11/98	0.11	10.38	10.49	1.0
3/99 - 5/99	0.15	27.98	28.13	0.5
07/99	0.15	5.35	5.50	2.7
12/99	0.04	26.41	26.45	0.1
06/00	1.22	34.49	35.71	3.4
08/00	0.08	1.84	1.92	3.9
12/00	0.03	6.21	6.24	0.5
Mean Results:	0.15	12.19	12.42	1.2

TABLE 5. Summary of Well GZ-4L Results.

Sampling Date	Total Parents (mg/L)	Total Daughter products (mg/L)	Total CAHs (mg/L)	Parent Ratio (%)
Summary of Pre-Injection (Baseline) Mean Results				
5/96-7/98	0.54	5.34	5.88	9.7
Post-Injection Results				
11/98	0.10	0.59	0.69	14.6
3,5/99	0.25	2.10	2.35	10.7
07/99	0.04	3.13	3.17	1.2
12/99	0.02	0.58	0.59	2.5
06/00	0.04	0.37	0.41	10.5
08/00	0.03	0.54	0.57	5.3
12/00	0.02	0.63	0.64	2.6
Mean Results:	0.04	0.83	0.89	6.8

CONCLUSIONS

Pre-injection parent CAH natural attenuation was DOC limited in overburden groundwater. Full-scale enhanced bioremediation via biostimulant injection supplied DOC, which stimulated native microflora to scavenge TEAs, established anaerobic and chemically reducing conditions, and provide electron donor for enhancing parent CAH reductive dehalogenation. Biostimulation resulted in reductions in CAH mass by $\leq 85\%$, elevated COD concentrations by ≤ 3 orders of magnitude, and depressed DO and ORP values by $\leq 96\%$ and $\leq 156\%$, respectively.

REFERENCES CITED

Schaffner, I.R., Wright, C.F., Wieck, J.M., Lamb, S.R. 1999. *Enhanced reductive dehalogenation of CAHs: A remedial pilot study*, in proceedings, In Situ and On Site Bioremediation, Battelle Memorial Institute, p. 171-176

Schaffner, I.R., Wieck, J.M., Lamb, S.R. 1998. *Enhanced reductive dehalogenation of CAHs at a former wastewater treatment facility: A microcosm study*, in proceedings, Northeast Focus Ground Water Conference, National Ground Water Association (NGWA), p. 115-125

Schaffner, I.R., Hawkins, E.F., and Wieck, J.M. 1996. *Screening study of intrinsic bioremediation of chlorinated aliphatic hydrocarbons at a site in southern New Hampshire*, in proceedings, The Tenth National Outdoor Action Conference, NGWA, p. 339-353

Wiedemeier, T.H., Swanson, M.A., Moutoux, D.E., Wilson, J.T., Kampbell, D.H., Hansen, J.E., and Haas, P. 1996. *Overview of the technical protocol for natural attenuation of chlorinated aliphatic hydrocarbons in ground water*, U.S. Air Force Center for Environmental Excellence, in proceedings, Symposium on Natural Attenuation of Chlorinated Organics, EPA, EPA/540/R-96/509, p. 35-63

ENHANCED REDUCTIVE DECHLORINATION OF ETHENES LARGE-SCALE PILOT TESTING

James A. Peeples (Metcalf & Eddy, Inc., Columbus, Ohio)
Joseph M. Warburton (Metcalf & Eddy, Inc., Columbus, Ohio)
Ihsan Al-Fayyomi (Metcalf & Eddy, Inc., Columbus, Ohio)
James Haff (Meritor Automotive, Heath, Ohio)

ABSTRACT: A shallow sand and gravel aquifer at an industrial site in central Ohio is contaminated with cis-1,2-dichloroethene (cDCE) and vinyl chloride (VC). Multi-phase field-scale pilot testing of enhanced reductive dechlorination began in November 1996. Pilot testing results indicated that reduction of cDCE and VC to concentrations at or near MCLs occurred within six months. Methods to expand and control the distribution of amendments in the shallow aquifer were developed. During the most recent pilot testing, a single amendment injection location was used to provide 90% or greater dechlorination within an aquifer volume of approximately 1000 yd^3 (763 m^3) with 50% or greater dechlorination achieved in 17,400 yd^3 (13,282 m^3) of aquifer.

INTRODUCTION

Chlorinated solvents are a common constituent of groundwater contaminant plumes. In situ reductive dechlorination of PCE/TCE to ethene by iron-reducing, sulfate-reducing, and methanogenic bacteria has been shown to be an effective remediation method in laboratory and field-scale studies (Beeman et al., 1994, de Bruin *et al.*, 1996, Bradley and Chappelle 1997). The presence of relatively high concentrations of cDCE and VC in groundwater at an industrial site in Ohio indicated that partial dechlorination of PCE and/or TCE had occurred. The primary source areas for chlorinated solvents and their daughter products are four former wastewater lagoons (Figure 1). The lagoons contained wastewater generated by metal machining operations from the early 1950s through 1985. The wastewater contained PCE and TCE, cutting oils, and other compounds. The liquid level in the lagoons was higher than the potentiometric surface of the underlying sand and gravel aquifer, and downward movement of lagoon waters and dissolved constituents occurred through the base of the lagoons. The lagoons were closed and filled in 1987.

A remedial investigation completed in 1993 described the subsurface hydrogeology, defined probable source areas, and delineated an area of approximately 38 acres (15.4 hectares) that was impacted by chlorinated solvents (Figure 1). Groundwater flow within the shallow sand and gravel aquifer is generally to the east, with an average gradient of 2 X 10^{-3} in the lagoon area (Figure 1). Depth to groundwater is approximately 11 feet (3.4 m), the saturated thickness of the unit is 25 feet (7.6 m), and the hydraulic conductivity is approximately 4×10^{-2} cm/sec.

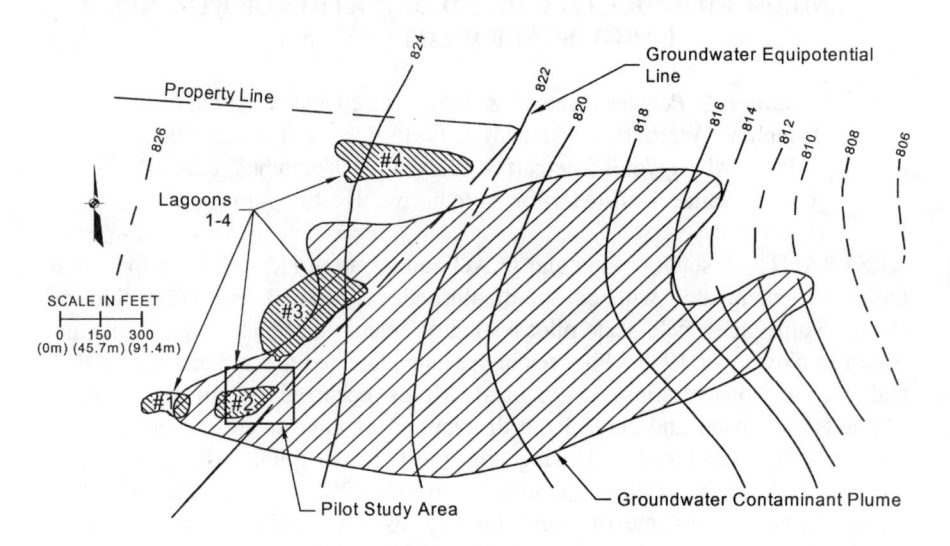

FIGURE 1. Site Area Layout

Groundwater samples were obtained to determine if reductive dechlorination of VC to ethene was occurring naturally within the aquifer. The analytical results provided evidence that reductive dechlorination to ethene was occurring at a low rate. Microcosm studies were conducted, using soil and groundwater from the site, to confirm the presence of the complete reductive dechlorination pathway and to evaluate the potential to enhance the process with amendments. Molasses was chosen as the electron donor source based on performance and cost. Field-scale pilot tests were then conducted to evaluate the potential to enhance the reductive dechlorination process in situ, to develop methods to increase the volume of treated aquifer, and to manage well biofouling and other issues associated with the injection process. The pilot testing conducted at this site is the subject of this paper.

A multi-phase approach was used to implement field-scale testing of enhanced in-situ reductive dechlorination. The Phase I test confirmed the capability of enhancing the reductive dechlorination process in situ. The Phase II test evaluated amendment distribution strategies, and the Phase III test evaluated the maximum volume of aquifer that could be treated from a single injection location.

PHASE I
Materials and Methods. The first phase of field-scale testing was conducted from November 1996 through August 1998. The purpose of this test was to demonstrate the ability to enhance reductive dechlorination in-situ, and to

Bioremediation Field Case Studies 175

determine if treatment to low concentrations could be achieved. The Phase I pilot system consisted of an extraction well (EW-1), an injection well (IW-1) and an observation well OW-1 (Figure 2). Groundwater was extracted from EW-1, amended with molasses, ammonium chloride, and monopotassium phosphate, and sodium sulfate, and reinjected into IW-1. Sodium sulfate was included with the amendments to inhibit the growth of methanogens. The amended groundwater traveled through the aquifer, distributing nutrients and a reduced carbon source to the microbial community. This groundwater was partially captured by EW-1 and recirculated. The effectiveness of the system was assessed in terms of the degradation of cDCE and VC, the formation of ethene and ethane, and the appropriate distribution of nutrients and reducing conditions.

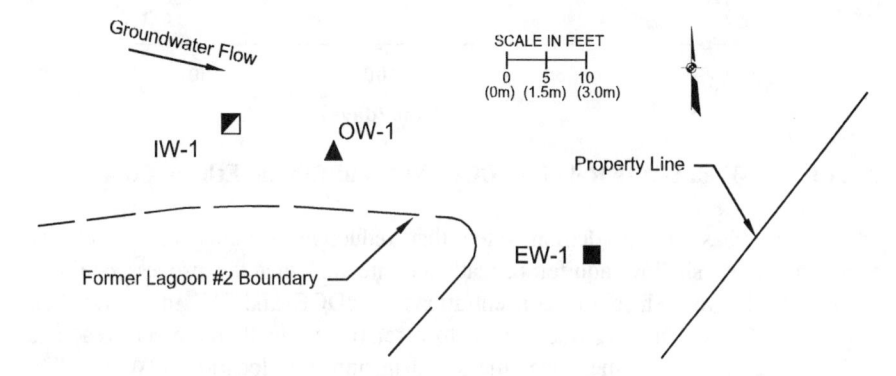

FIGURE 2. Phase I Test, System Layout

Results and Discussion. Figure 3 provides a summary of the relative concentrations of cDCE, VC, and ethene and ethane at monitoring well OW-1 during the Phase I test. The concentrations are plotted as a percentage of the total micromolar concentrations of these constituents. The concentrations of cDCE and VC in groundwater collected from OW-1 were initially 265 µg/L (2.73 µmolar) and 15 µg/L (0.24 µmolar), respectively. Prior to the start of the pilot test, 91.9 percent of the total µmolar concentration was cDCE and 8.1 percent was VC. No detectable concentrations of ethene or ethane were present at the start of the test. The relative contributions of cDCE, VC, and ethene/ethane were 6.1, 1.3, and 92.6 percent, respectively, 251 days after the start of the test, indicating a nearly complete conversion of the chlorinated ethenes to ethene/ethane. Concentrations of cDCE were reduced to less than 25 µg/L, and VC concentrations to less than 4 µg/L after 380 days.

The declining concentrations of cDCE and VC at OW-1 occurred despite the continued introduction of groundwater to IW-1 with an average cDCE concentration of 413 µg/L and an average VC concentration of 150 µg/L. VC and cDCE were largely degraded in the aquifer between the injection well (IW-1) and the monitoring well (OW-1).

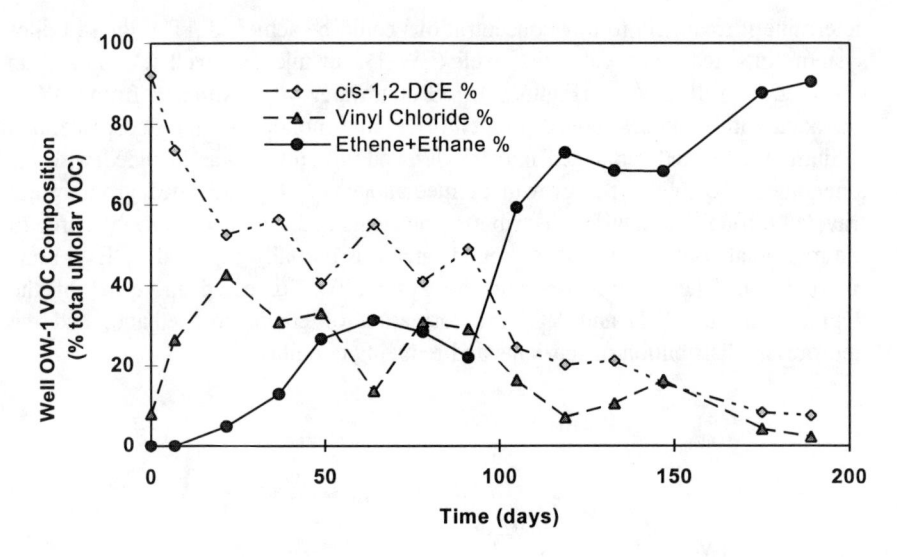

FIGURE 3. Well OW-1 Relative cDCE, VC, and Ethane/Ethene Conc.

The Phase I test demonstrated that reductive dechlorination could be enhanced in the shallow aquifer beneath the site and that the process could be applied in situ to reduce the concentrations of cDCE and VC to below their respective MCLs. The test was limited to a relatively small treatment area, and the effectiveness of treatment was measured in only one location (OW-1). The test indicated that batching nutrients slowed the process of injection well clogging, and that recirculation aggravated well clogging. Management of injection well clogging during the test was accomplished by regular cleaning using pump and surge techniques.

PHASE II
Materials and Methods. A second phase field-scale pilot test was implemented from September 1998 through April 1999. The objectives of the test were to apply the enhanced reductive dechlorination process to a larger aquifer volume, to develop methods to alleviate well clogging, and to evaluate treatment efficiency within a broader area.

Injection wells IW-3 and IW-4, and observation wells OW-3 through OW-11 were installed (Figure 4). The injection wells were located 6 ft (1.8 m) apart, along a line parallel to groundwater flow. Molasses were added to the downgradient injection well (IW-3) and groundwater was injected into the upgradient well (IW-4). Sulfate was also included in the amendments to inhibit methanogens. The injection of groundwater into IW-4 was used to move the injected nutrients away from IW-3, controlling amendment distribution and reducing biofouling in this well. Groundwater from RRW-2 (average concentration of cDCE 78 μg/L, VC 140 μg/L) and from EW-1 (average concentration of cDCE 109 µg/L, VC 160 µg/L) was injected into IW-3 at 6 gpm (23 L/min) and into IW-4 at 3 gpm (11 L/min). Nutrient injection and groundwater monitoring continued for 8 months.

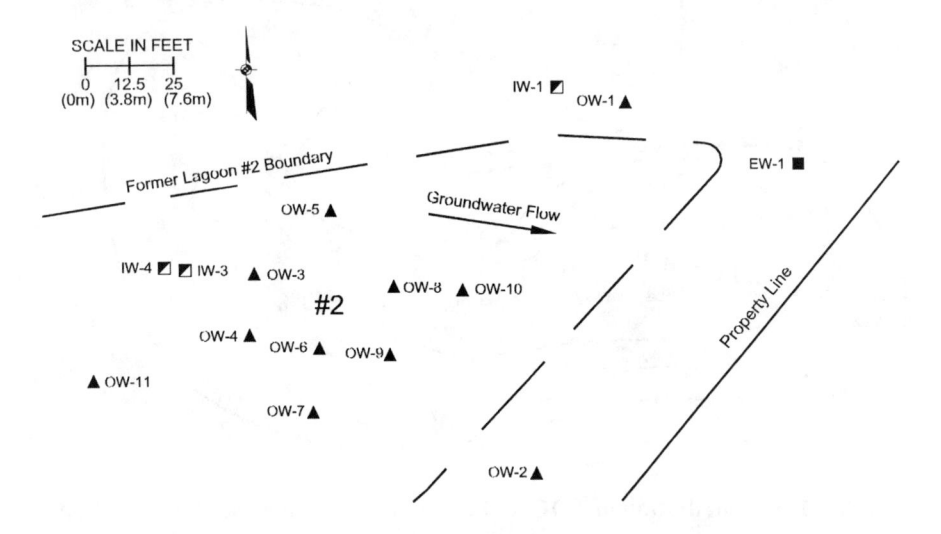

FIGURE 4. Phase II Test, System Layout

Results and Discussion. Total Organic Carbon (TOC) analysis was used to track the distribution of an electron donor supplied by the injected molasses. The area where TOC increased in groundwater during the Phase II injections is shown in Figure 5. The area was approximately 180 ft (55 m) long, and 110 ft (34 m) wide at a distance of 110 ft (34 m) downgradient from IW-3. Increases in concentrations of ferrous iron and methane, and decreases in concentrations of sulfate occurred within this area, indicating that methanogenic, iron-reducing, and sulfate-reducing bacteria were being stimulated by the injected amendments.

Reductive dechlorination of cDCE and VC occurred in most of portions of the aquifer reached by amendments (Figure 5). Treatment proceeded to near detection limits in portions of the treatment zone, while some degree of dechlorination occurred over most of the amendment distribution area. At least 90% dechlorination occurred over an area of approximately 1,200 ft^2 (111 m^2), equivalent to 1,100 yd^3 (841 m^3) of the aquifer. At least 50% dechlorination was achieved in approximately 5,300 yd^3 (4,052 m^3) of the aquifer. Clogging of IW-3 was minimized by the injection of groundwater into IW-4. The Phase II test ran for 26 weeks without shutdown, and IW-3 remained open and operable throughout the test. Amendment injection was discontinued from April 26, 1999, to June 29, 1999, while groundwater from RW-6 was injected into IW-3 and IW-4. It was assumed that the injection of this ground water, which contained cDCE (average 420 µg/L) and VC (average 220 µg/L), would recreate a VOC plume within the area downgradient of IW-3 and IW-4. This did occur in areas on the periphery of the Phase II injection zone, but within the area of the Phase II treatment zone, efficient VOC dechlorination continued without amendments.

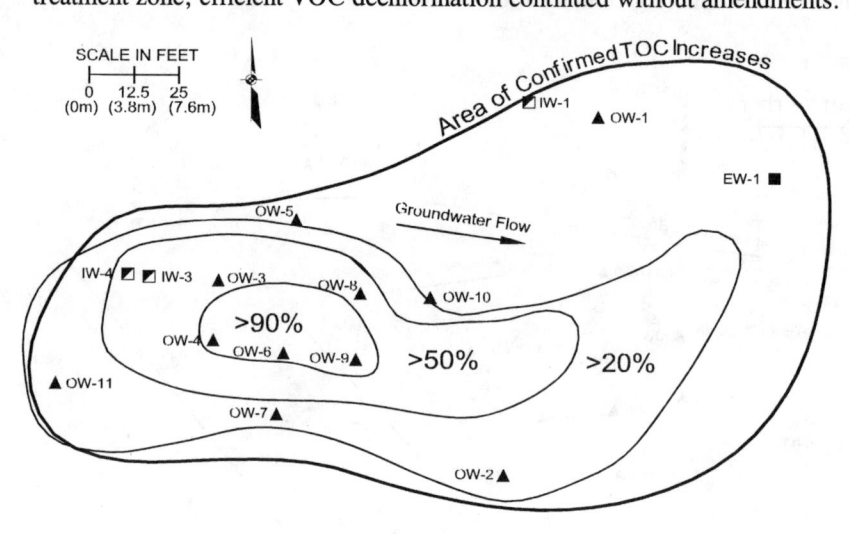

FIGURE 5. Distribution of TOC and Percent Dechlorination, Phase II Test

PHASE III
Materials and Methods. To evaluate the application of the two-well injection system to a full-scale operation, the injection method was modified for the Phase III field-scale test. During this phase, amendments were injected in varying proportions to IW-3 and IW-4, and the relative rate of injection was varied. The relative rate of groundwater injection into IW-3 and IW-4 was used to control the areal distribution and concentration of amendments within the treatment zone, and to reduce biofouling in the injection wells. The primary objective was to distribute the amendments to as large an area of the aquifer as possible from the injection locations. A combined average injection rate of 10 gpm (38 L/min) was utilized. Groundwater used for injection into IW-3 and IW-4 was pumped from RW-6, a well located approximately 560 ft (171 m) northeast of EW-1. The injected groundwater contained average cDCE and VC concentrations of 412 μg/L and 224 μg/L.

Wells OW-12 through OW-20 were installed in June 1999 (Figure 6), expanding the monitoring network used in the Phase II study to accommodate the expected increase in the size of the treatment area. Wells OW-21 through OW-31 (Figure 6) were installed in the downgradient area east of the former Lagoon 2 in September 1999 to evaluate the effects that the source area reductive dechlorination process will have on downgradient locations in the aquifer.

Amendment injections began on June 29, 1999 and continued for a total of 28 weeks. A total of 43 drums (8,950 L) of molasses and 4,100 pounds (1,860 kg) of sodium sulfate were added to the aquifer during the 28-week period.

Bioremediation Field Case Studies 179

Sodium bromide was also included with the amendments during the first week of injections to act as a conservative tracer. A total of 150 pounds (68 kg) of sodium bromide was injected.

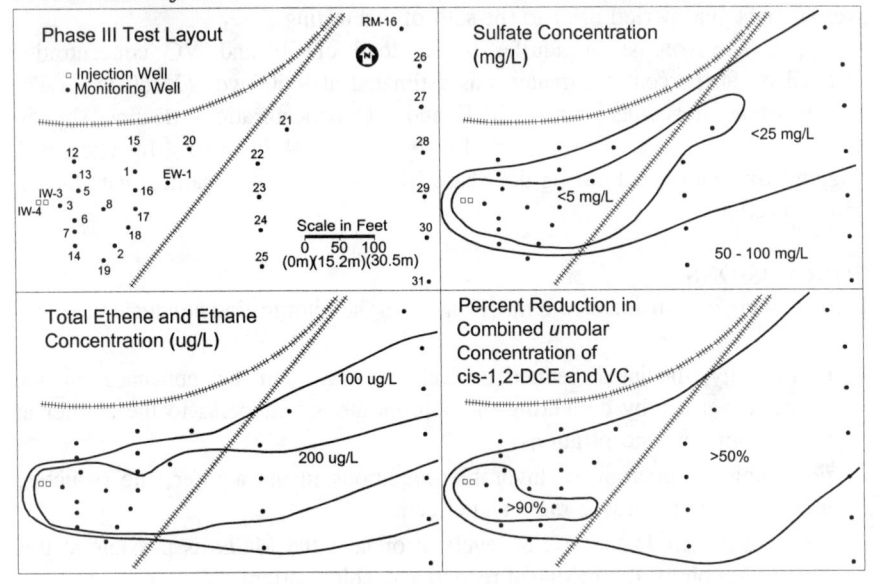

FIGURE 6. Phase III Test Set-Up and Results

Results and Discussion. The area of aquifer receiving treatment during (and following) the Phase III injections extended at least 1,000 feet (300 m) downgradient of the injection location. The area influenced by the injection system was evaluated by increases in bromide, TOC, methane, ethane, ethene, ferrous iron, and conductivity and reductions in sulfate, dissolved oxygen, ORP, cDCE and VC. Figure 6 provides distributions of zones with reduced concentrations of sulfate, increased concentrations of ethane and ethene, and reductions in cDCE and VC concentrations. Injected amendments reached a width of aquifer of at least 120 ft (37 m) at a distance of 40 ft (12 m) downgradient of the injection wells, based on concentration changes observed for a variety of constituents. This compares with a width of approximately 60 ft (18 m) achieved at this distance during the Phase II study. The width of aquifer influenced by the injected amendments increased with distance from the injection locations as shown in Figure 6. Reductive dechlorination of cDCE and VC could be verified through the source area based on trends in ethane and ethene concentrations and the presence of a sufficiently reducing environment. The presence of active reductive dechlorination in the downgradient area could not be verified. Groundwater in most downgradient areas did not reach sufficiently low ORP or low sulfate concentration to suggest that reductive dechlorination was the primary mechanism for removal of cDCE and VC. It is likely that a variety of processes contributed to the observed reductions in cDCE and VC in downgradient areas. The decrease in cDCE and VC in groundwater downgradient of the property. The decline in cDCE and VC concentration in the downgradient areas is in clear contrast to a very gradual declining trend in these parameters over an eight year period prior to the start of this testing.

The volume of aquifer where total cDCE and VC concentration declined by 90 percent or greater was estimated at 9,500 yd^3 (7,260 m^3). The volume of aquifer where total cDCE and VC concentrations declined by 50 percent or greater was estimated at 137,000 yd^3 (104,740 m^3). Injection well clogging was minimized during the Phase III test the wells remained operable for future injections.

CONCLUSIONS

The following conclusions were made based on the pilot testing conducted to date:

- Biologically mediated reductive dechlorination can be enhanced in the impacted aquifer by delivering the appropriate amendments to the aquifer at the appropriate concentrations;
- Following the creation of favorable conditions in the aquifer, the reductive dechlorination process is rapid and effective;
- Removal of cDCE and VC to levels at or near the MCLs is possible at this site using biologically mediated reductive dechlorination;
- A two-well injection system can be used to increase and control the areal distribution of the treatment zone;
- The efficiency of injection can be maximized by operating a two-well injection system, batching nutrients, and obtaining injection water from outside the treatment area;
- Full-scale implementation of enhanced reductive dechlorination at this site is feasible.

REFERENCES

Beeman, R. E., J. E. Howell, S. H. Shoemaker, E. A. Salazar, and J. R. Buttram, 1994. A field evaluation of in situ microbial reductive dehalogenation by the biotransformation of chlorinated ethenes. pp 14-175 In: Bioremediation of Chlorinated and Polycyclic Aromatic Hydrocarbon Compounds, R. Hinchee, et al. 1994 Lewis Publishers, Boca Raton, Florida.

Bradley, P.M. and F.H. Chapelle. 1997. Kinetics of cis-1,2-DCE and vinyl chloride mineralization under methanogenic and Fe(III) reducing conditions. Environ. Sci. Technol. 31:2693-2696.

deBruin, W.P., J. Michiel, J. Kotterman, M.A. Posthumus, G. Schraa, and A.J.B. Zehnder. 1992. Complete biological reductive transformation of tetrachloroethene to ethene. Appl. Environ. Microbiol. 58:1996-2000.

IN SITU ENHANCED REDUCTIVE DECHLORINATION OF PCE

Stacey A. Koch (RMT, Inc., Madison, WI, USA)
John M. Rice (RMT, Inc., Madison, WI, USA)

ABSTRACT: Tetrachloroethene (PCE) impacts were identified beneath a former dry cleaning facility. Remedial alternatives were evaluated for the site, and the approaches recommended were (1) mass reduction in the source area through excavation and enhanced bioremediation, and (2) natural attenuation of the downgradient plume. The focus of this paper is on the field application of enhanced bioremediation in the source area. A series of injections using a mixture of sodium lactate, yeast extract, and sodium sulfite was performed over a 2-year period. The results showed a 60 to 80 percent decrease in the concentration of PCE in the source area. The overall *in situ* PCE degradation rate, which accounts for both the biodegradation rate of PCE and the desorption rate of PCE from the soil matrix, was estimated from the data.

INTRODUCTION

During construction of a new city hall by a Wisconsin municipality, tetrachloroethene (PCE) impacts were identified in unsaturated soil and shallow groundwater. The shallow groundwater impacts were likely the result of PCE migrating into the underlying aquifer as a dense nonaqueous-phase liquid (DNAPL), with further migration of the dissolved phase downgradient, both horizontally and vertically. The plume was not expanding, and evidence of limited PCE degradation was observed within the primarily aerobic aquifer. The limited degradation that was observed could be attributed to reductive dechlorination in micro anaerobic zones in the source area groundwater.

After several remedial alternatives were evaluated for the site, the recommended approach included mass reduction in the source area unsaturated soil through excavation (*i.e.*, during construction of the municipal building) and enhanced bioremediation of PCE in the source area saturated soil. A natural attenuation monitoring program was implemented to address the downgradient plume. The focus of this paper is on the enhanced bioremediation portion of the overall site remedy.

The approach to bioremediation at this site was to create broader anaerobic conditions and provide the necessary substrate and nutrients to facilitate the reductive dechlorination of PCE in the source area. Laboratory research evaluating the capability of pure cultures and mixed anaerobic enrichments to degrade PCE has been ongoing for over 15 years. Researchers have used a variety of compounds as the source of the electron donor, including 3-chlorobenzoate, lactate, acetate, methanol, and hydrogen itself. To date, researchers have been successful at sustaining cultures capable of reductively dechlorinating PCE, and its breakdown products, including trichloroethene (TCE), cis-dichloroethene (cis-DCE), and vinyl chloride (VC), all the way to ethene (Vogel and McCarty, 1985; Freedman and Gossett, 1989; DiStefano et al., 1991; de Bruin et al., 1992; Tandoi et al., 1994; Ballapragada et al., 1997). This research provides the tools to understand the conditions necessary to degrade highly oxidized compounds, such as PCE. The challenge at this site was to apply those tools to provide *in situ* enhanced reductive dechlorination of PCE.

The goal of the aquifer enhancement injections was to reduce the PCE concentration in the source area to 50 percent of the concentration measured prior to the initial injection. The initial injection was performed in August 1997, followed by a year of groundwater monitoring to determine the optimal dose and frequency of injections for the full field application. Subsequent injections were performed approximately every 6 months for two additional years.

MATERIALS AND METHODS

The injection system was located in the basement of the municipal building. The system consisted of a makeup tank for the concentrated substrate and trace nutrients, a metering pump, a flow meter, and a valve controlling system plumbed into the injection well. Tap water was used to dilute the substrate and nutrients to the desired concentrations, and to provide the volume needed for the injection. The well used for the injections was an existing monitoring well, MW-4, located in the basement of the city hall (Figure 1a). The enhancement process was monitored using downgradient monitoring wells RW-1 and MW-10 (Figure 1b).

In the source area, the subsurface is composed of approximately 6 meters (20 feet) of glacially deposited sand and gravel overlying sandstone bedrock. The sandstone is weathered at its surface and interbedded with dolomite and shale with depth. The shallow groundwater flow system encompasses the unconsolidated deposits and the weathered bedrock, with a geometric mean hydraulic conductivity of 1.6×10^{-3} cm/s. The water table is located approximately 3 meters (10 feet) below the ground surface and defines the top of the shallow groundwater flow system. Shallow groundwater discharges to a perennial stream approximately 400 meters (1300 feet) to the south. The deep groundwater flow system is within the sandstone bedrock with a geometric mean hydraulic conductivity of 1.3×10^{-4} cm/s. In the source area, vertical groundwater flow is slightly downward.

The aquifer temperature ranged from 10 to 19°C over the field study period. The aquifer pH was in the neutral range, 6.5 to 7.5. Total alkalinity ranged from 280 to 430 mg/L in the shallow groundwater. Sulfate was present at 20 to 60 mg/L, and dissolved ferrous iron was present at less than 100 mg/L.

The aquifer enhancement approach was to make a dilute solution of electron donor (lactate) and trace nutrients (yeast extract) and then to pump this solution into MW-4, located upgradient from the source area[1]. For the initial injection, approximately 57,000 liters (15,000 gallons) of the dilute solution were injected at a rate of 2 liters per minute (0.5 gallons per minute) to create a cylindrical zone of influence approximately 3 meters (10 feet) in height and 12 meters (40 feet) in diameter around MW-4. This zone provided the conditions necessary to enhance the biodegradation of PCE in the treatment zone around MW-4, and at downgradient locations as the treatment chemicals were transported

---

[1] The processes described herein are disclosed in U.S. Patent Number 6,001,252.

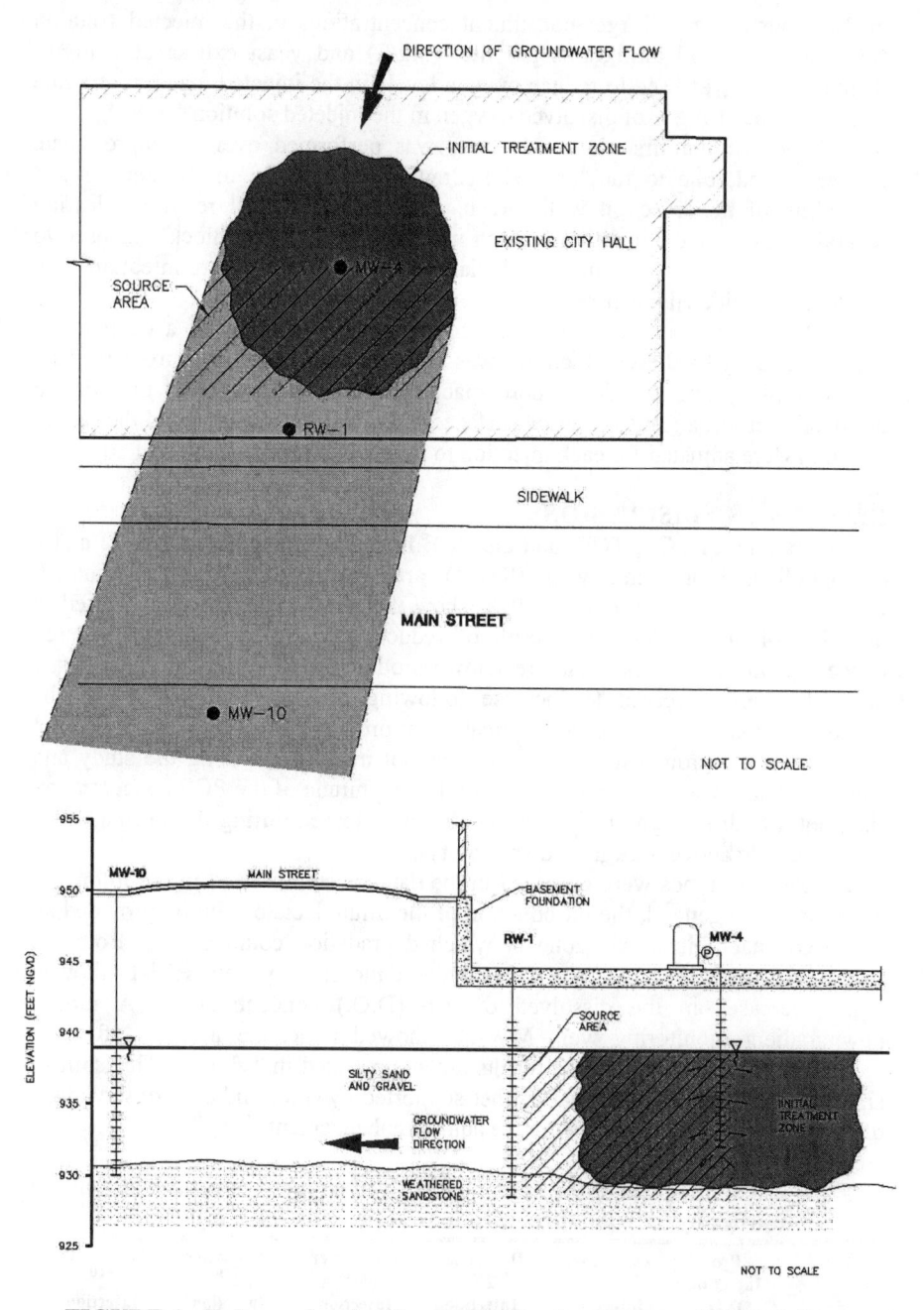

FIGURE 1. Enhanced reductive dechlorination system. (a) Plan view showing injection well (MW-4) and downgradient monitoring wells (RW-1 and MW-10). (b) Cross section showing system layout, injection well, and initial treatment zone.

in the groundwater. Target constituent concentrations in the injected solution were lactate at 200 milligrams per liter (mg/L) and yeast extract at 2 mg/L. Sodium sulfite was used to reduce oxygen levels in the injected material at a rate of 10 mg/L per 1 mg/L of dissolved oxygen in the injected solution.

Each of the first three injections was performed over an approximate 2-week period, due to the lowered hydraulic conductivity of the aquifer and biofouling of the injection well screen. The injection well required clearing several times over the duration of each injection, and a surge block was used to periodically clean the screen. To alleviate this condition in future injections, the material was injected under pressure.

Over the course of this enhanced bioremediation project, a decrease in porosity and hydraulic conductivity was inferred from field measurement data. This was likely due to reduced pore space caused by the increased growth and decay of microorganisms in the subsurface. Therefore, the volume and the rate of injection were adjusted for each injection to maintain a similar treatment zone.

RESULTS AND DISCUSSION

Concentrations of PCE, TCE, and cis-DCE in the injection well (MW-4) and a downgradient monitoring well (RW-1) are shown on Figures 2 and 3, respectively. Temporal trends in PCE show that the concentration decreased in both MW-4 and RW-1, as a result of reductive dechlorination. In general, concentrations of the anaerobic breakdown products of PCE, including TCE and cis-DCE, were observed to increase following injections, depending on the relative rates of degradation. The breakdown products VC and ethene were not observed in the groundwater. Lactate was not monitored during the study and chloride results were inconclusive due to the magnitude of the PCE concentration (i.e., not a high enough chlorine concentration released during dechlorination to be measurable above background variability).

Several trends were observed in the data, as summarized in the following paragraphs. In general, the introduction of the dilute lactate solution provided an increased anaerobic/anoxic zone in which degradation could occur. Both the injection well, MW-4, and the downgradient monitoring well, RW-1, showed rapid decreases in the dissolved oxygen (D.O.) concentration. A further downgradient monitoring well, MW-10, showed a decline in D.O. after the second lactate injection. These results are summarized in Table 1. This shift to an anaerobic/anoxic aquifer was further supported by color and odor observations of groundwater samples following the aquifer enhancement.

TABLE 1. Comparison of the average dissolved oxygen concentrations in the aquifer by location, pre-injection versus after each injection.

Location	Pre-Injection D.O. (mg/L)	D.O. After 1ST Injection (mg/L)	D.O. After 2ND Injection (mg/L)	D.O. After 3RD Injection (mg/L)	D.O. After 4TH Injection (mg/L)	D.O. After 5TH Injection (mg/L)
MW-4	5 to 6	0.88	0.73	0.35	0.7	0.33
RW-1	4.2	0.88	0.8	0.85	1.07	1.5
MW-10	5.5	5.83	2.38	0.9	2.0	2.0

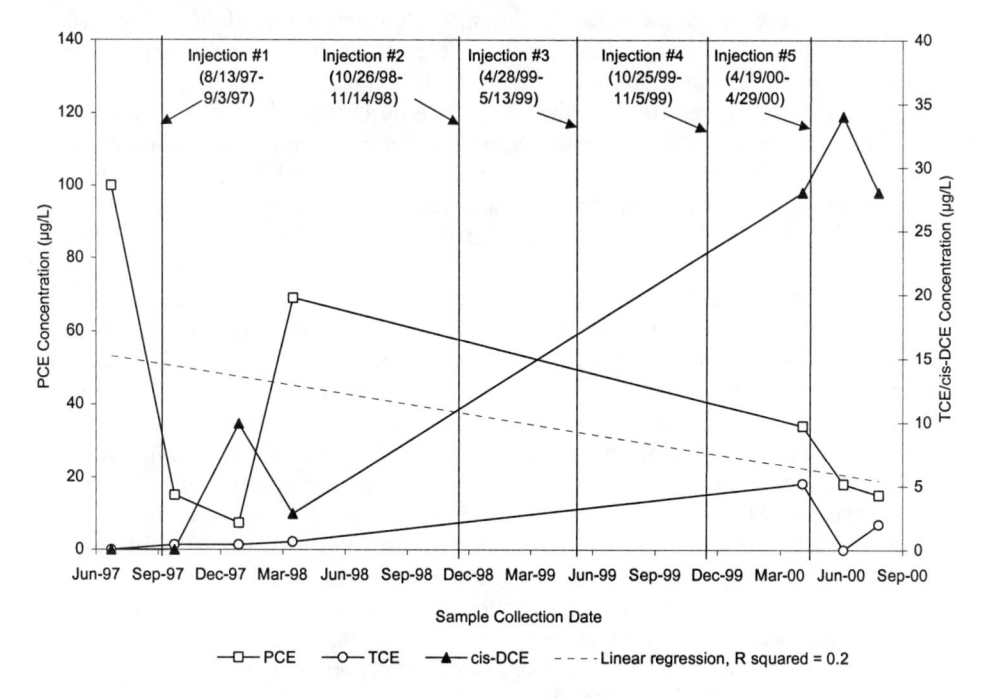

FIGURE 2. PCE, TCE, and cis-DCE concentrations in injection well MW-4.

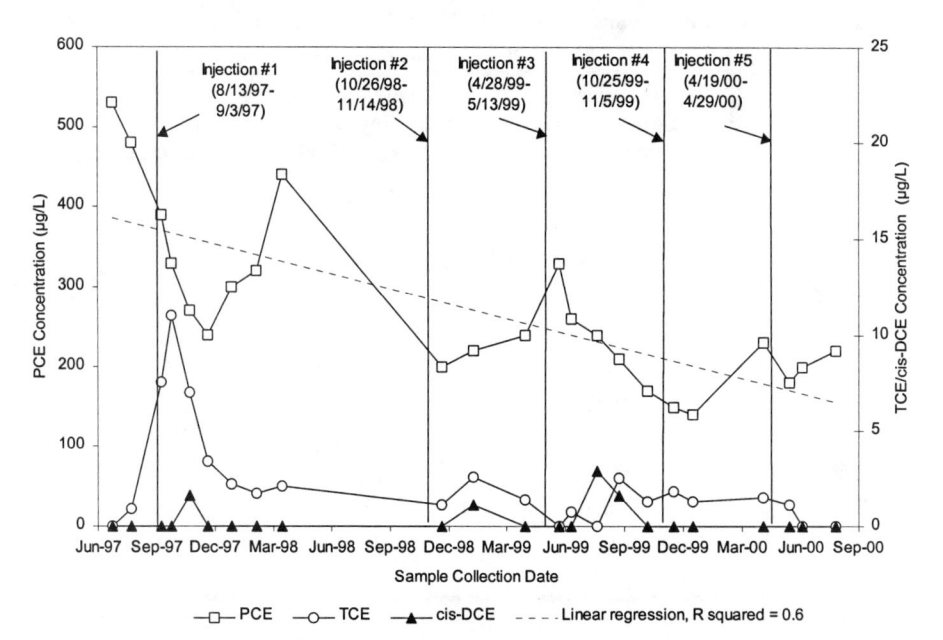

FIGURE 3. PCE, TCE, and cis-DCE concentrations in downgradient monitoring well RW-1.

Immediately following an injection, the concentration of PCE declined (Figures 2 and 3). During these periods, the electron donor concentration was relatively high, spurring dechlorination activity. This resulted in the rate of biodegradation being greater than the rate of PCE desorption, so that the dissolved PCE concentration declined. However, as the electron donor was consumed in the aquifer, the degradation rates declined, and the dissolved PCE concentration rebounded. A consistent pattern of rebounding concentrations with respect to PCE, TCE, and cis-DCE was not always observed due to the timing of the monitoring events.

A decreasing trend in PCE concentrations was observed in the shallow groundwater. This decrease likely indicates an overall depletion of the PCE sorbed to the subsurface material, as the dissolved PCE was degraded. The effect cannot be attributed solely to dilution, as the injections were performed upgradient of the source area. This effect was consistent with the primary goal of the aquifer enhancement injections, which was to reduce the PCE concentration in the source area to 50 percent of the concentration measured prior to the August-September 1997 injection. A consistent downward trend in the PCE concentrations was observed in MW-4, RW-1, and MW-10. Table 2 summarizes the pre-injection PCE concentrations, the average concentrations after the 5th injection, and the overall percent reduction in the PCE concentration. The 50 percent goal was met and exceeded at each of the three monitoring locations within a 2-year period. The concentrations of the breakdown products TCE and cis-DCE are also provided for reference.

TABLE 2. Measured average PCE, TCE, and cis-DCE concentrations and the overall percent reduction following aquifer enhancements.

Location	Pre-Injection Concentration (avg. µg/L)	Concentration After 5TH Injection (avg. µg/L)	Percent PCE Reduction
MW-4			
PCE	100	16.5	83.5 %
TCE	< 0.38	2	NA
cis-DCE	5.5	31	NA
RW-1			
PCE	505	200	60.4 %
TCE	0.91	1.1	NA
cis-DCE	< 0.47	< 0.46	NA
MW-10			
PCE	1,800	380	78.9 %
TCE	2.1	1.5	NA
cis-DCE	1.7	1.4	NA

NA – Not Applicable

The overall *in situ* PCE degradation rate was estimated from the field data, using linear regression (Figures 2 and 3). The overall degradation rate accounts for both the biodegradation rate of PCE and the desorption rate of PCE from the soil matrix. Therefore, the overall degradation rate is slower than what would be expected from biodegradation alone. The overall *in situ* PCE degradation rate was found to be linearly proportional to the original aquifer PCE concentration, as shown on Figure 4. Rates ranged from 0.7 µg/L-day, at an original PCE concentration of 1,800 µg/L, to 0.03 µg/L-day, at an original PCE concentration of 100 µg/L.

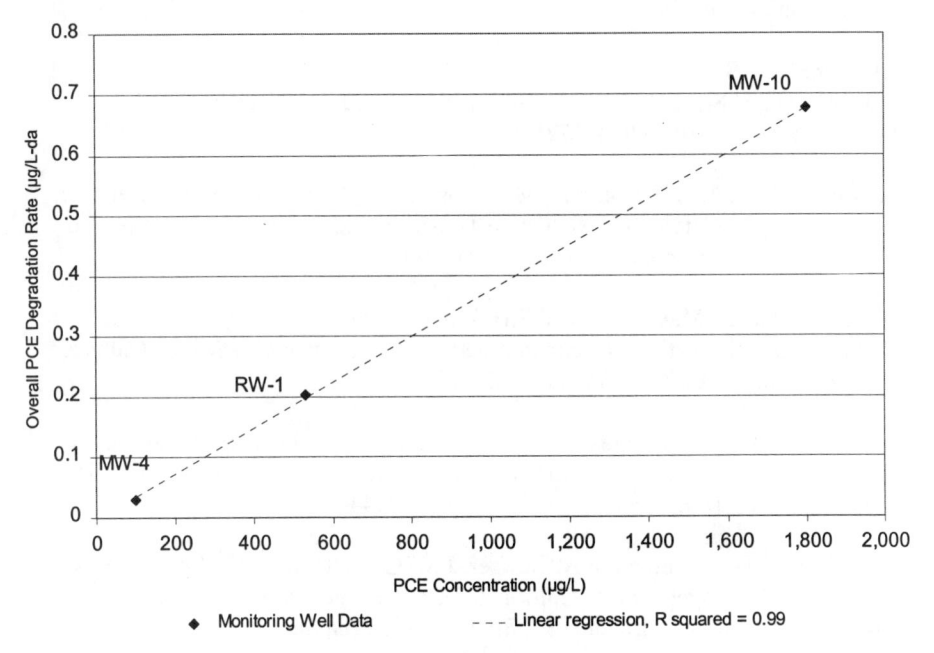

FIGURE 4. Overall PCE degradation rate versus the original concentration of PCE in the aquifer.

CONCLUSIONS

An innovative remedial solution was used to create temporary anaerobic conditions and provided the necessary substrate and nutrients to enhance reductive dechlorination of PCE in the source area. A series of injections using a mixture of sodium lactate, yeast extract, and sodium sulfite was performed over a 2-year period. The results showed a decrease of between 60 and 80 percent in the concentration of PCE in the source area. Increases in the anaerobic breakdown products of PCE, including TCE and cis-DCE, were measured. VC and ethene were not observed.

Groundwater monitoring of the downgradient wells was used to verify that reductive dechlorination was enhanced at the site. Through this monitoring, a rebound pattern was observed in the PCE data. This rebound effect is likely attributed to PCE desorbing from the soil matrix. This factor needs to be taken into account when estimating the overall time to reach remedial goals.

The overall *in situ* PCE degradation rate was estimated from the field data. The overall degradation rate accounts for both the biodegradation rate of PCE and the desorption rate of PCE from the soil matrix. Therefore, the overall degradation rate is slower than what would be expected from biodegradation alone. The overall *in situ* PCE degradation rate was found to be linearly proportional to the original aquifer PCE concentration. These rate data, although not correlated to the biomass present at this site, can be used in the planning stages as a preliminary method to estimate the duration of an enhanced bioremediation remedy.

REFERENCES

Ballapragada, B.S., H.D. Stensel, J.A. Puhakka, and J.F. Ferguson. 1997. *Environ. Sci. Technol. 31*(6): 1728-1734.

de Bruin, W.P., M.J.J. Kotterman, M.A. Posthumus, G. Schraa and J.B. Zehnder. 1992. "Complete Biological Reductive Transformation of Tetrachloroethene to Ethane." *Appl. Environ. Microbiol. 58*(6): 1996-2000.

DiStefano, T.D., J.M. Gossett, and S.H. Zinder. 1991. "Hydrogen as an Electron Donor for Dechlorination of Tetrachloroethene by an Anaerobic Mixed Culture." *Appl. Environ. Microbiol. 58*(11): 3622-3629.

Freedman, D.L. and J.M. Gossett. 1989. "Biological Reductive Dechlorination of Tetrachloroethylene and Trichloroethylene to Ethylene under Methanogenic Conditions." *Appl. Environ. Microbiol. 55*(9): 2144-2151.

Tandoi, V., T.D. DiStefano, P.A. Bowser, J.M. Gossett, and S.H. Zinder. 1994. "Reductive Dehalogenation of Chlorinated Ethenes and Halogenated Ethanes by a High-Rate Anaerobic Enrichment." *Environ. Sci. Technol. 28*: 973-979.

Vogel T.M. and P.M. McCarty. 1985. "Biotransformation of Tetrachloroethylene to Trichloroethylene, Dichloroethylene, Vinyl Chloride, and Carbon Monoxide under Methanogenic Conditions." *Appl. Environ. Microbiol. 49*(5): 1080-1083.

EFFECTIVE ENHANCEMENT OF BIOLOGICAL DEGRADATION OF TETRACHLOROETHENE (PCE) IN GROUND WATER

Robert W. North, Sharon E. Burkett, and M. Jennifer Sincock, ENVIRON International Corporation, Princeton, New Jersey, USA

ABSTRACT: Injection of a time-release source of lactic acid (HRC®) into a plume of PCE-contaminated ground water in a shallow aerobic aquifer beneath a dry cleaning facility resulted in (1) a significant increase in the rate of biologically-mediated degradation of PCE and its daughter products and (2) a greater than 95% reduction in PCE concentrations near the source area in 12 months. The injection of a lactic acid source created an anaerobic and nutrient-rich environment, thus accelerating reductive dechlorination of the chlorinated solvents. Enhanced degradation occurred most rapidly near the source area where an acclimatized bacterial population may have been present due to a historic discharge of No. 2 fuel oil creating anaerobic, reducing conditions in the past. Lower rates of degradation and a lag effect for the onset of accelerated degradation was observed in other areas of the treatment zone. Fresh water influx due to differential recharge near the margins of the plume appears to have somewhat limited the effectiveness of the lactic acid source in altering the aquifer geochemistry near the plume margins. Ongoing ground water monitoring indicates that, more than one year after injection, most of the treated area remains anaerobic and PCE concentrations continue to decrease.

INTRODUCTION

Injection of a time-release source of lactic acid (Hydrogen Releasing Compound [HRC®]) was performed at a dry cleaning facility in central New Jersey to enhance natural degradation of chlorinated hydrocarbons in ground water. HRC® is a polylactate ester (glycerol tripolylactate) designed to provide a carbon energy source and create anaerobic, reducing conditions in ground water to stimulate reductive dechlorination of chlorinated solvent compounds such as tetrachloroethene (PCE) and trichloroethene (TCE). The results indicate that successful alteration of groundwater chemistry and resulting accelerated degradation of contaminants have been achieved.

The site is located at the end of a strip mall in a residential area and is bordered on the north and east by residential properties (Figure 1). Subsurface geology consists of fine to medium silty sand and gravel with an average depth to ground water of approximately 10 feet (3 m). The saturated thickness of the aquifer ranges from 5 to 12 feet (1.5 to 3.7 m) across the plume area. The average hydraulic conductivity for the shallow aquifer is 5×10^{-3} cm/sec. The shallow sand aquifer is underlain by a 30-foot (9 m) thick dense silty clay that provides an effective barrier to downward migration of contaminants. The ground water flow direction is generally toward the north-northwest but appears to vary

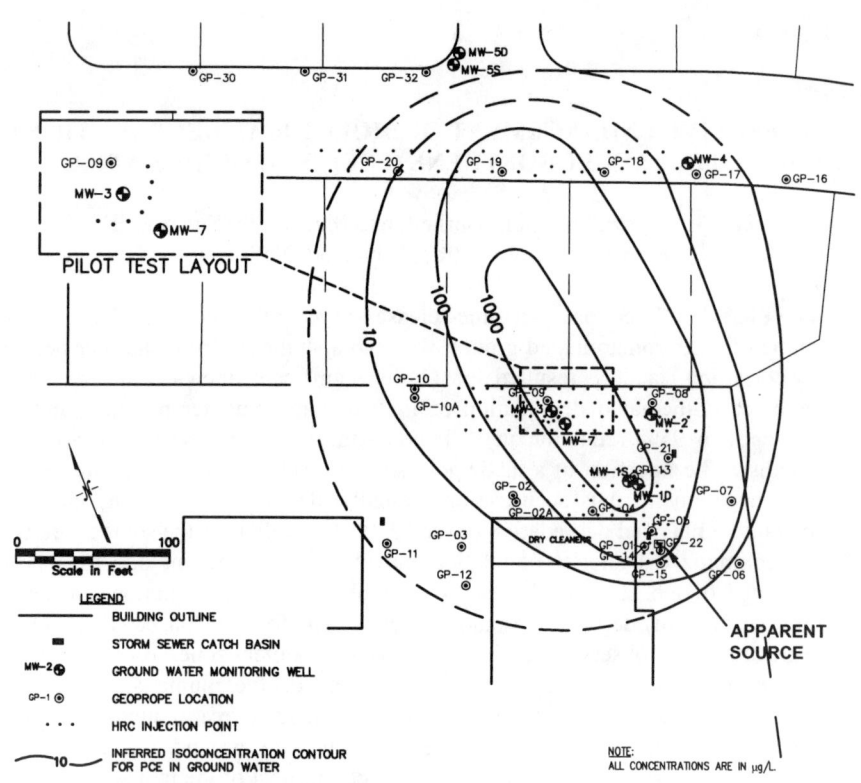

FIGURE 1. Extent of PCE In Ground Water Prior To Full-Scale HRC Application and Location of Pilot Test

from north to west-southwest. Water levels have a strong seasonal influence (± 1.5 feet [0.5 m]) and differential recharge between the fully paved shopping center and the residential yards appears to affect ground water flow.

Ground water sampling identified a plume of chlorinated hydrocarbons (PCE and degradation products TCE and cis-1,2-dichloroethene [cis-1,2-DCE]) emanating from the rear of the dry cleaners and extending offsite (see Figure 1). PCE concentrations between 1,000 and 3,000 ug/L were detected near the source area and up to 4,000 ug/L were detected near the downgradient property boundary, approximately 120 feet from the source area. Soil sampling indicated a source area near a storm sewer catch basin behind the dry cleaners where concentrations of PCE in soil exceeded the state cleanup criterion. Source area soils with elevated levels of PCE were excavated and disposed of offsite to mitigate a potential continuing source of VOCs to ground water. At the same time, an adjacent underground fuel oil tank that was no longer in use was also removed. Evidence of a minor fuel oil release was observed during the removal of the UST. This fuel oil may have provided a carbon source and created reducing conditions in ground water in the immediate vicinity of the UST.

The presence of degradation products of PCE (TCE and cis-1,2-DCE) in site ground water indicated that degradation either had occurred previously or was currently occurring onsite, but at a very slow rate. Vinyl chloride was not detected in any of the ground water samples from the study area. PCE to TCE ratios were approximately 50 to 1 and PCE to *cis*-1,2 DCE ratios were approximately 10 to 1. The concentration of the TCE and *cis*-1,2-DCE relative to the PCE concentrations increased away from the source area providing further evidence that some degree of natural degradation was occurring. However, the ground water geochemistry parameters (dissolved oxygen [DO], oxidation reduction potential [ORP], nitrate [NO_3^-], sulfate [SO_4^{2-}], and total and ferrous iron) indicated that the aquifer was slightly to moderately aerobic with DO ranging from 1 to 3 mg/L and ORP around +200 mV. Physical observations of the sandy aquifer during subsurface activities suggested that the aquifer does not naturally contain significant amounts of organic carbon to provide an energy source for microbes. Such conditions are not considered favorable for biologically mediated natural attenuation of chlorinated compounds.

Various alternatives were considered for remediation of the PCE plume, including pump-and-treat, a reactive barrier wall, monitored natural attenuation (MNA) and enhanced degradation. Cost considerations, site disruption and potential installation difficulties, and operating costs made pump-and-treat or a reactive barrier wall system the least desirable remedies. Although natural degradation appeared to be occurring despite the moderately aerobic conditions, the VOC concentrations were greater than the state would likely allow for MNA and the apparent degradation rates were very slow. For example, using the method described by Buscheck and Alcantar (1995), the calculated half-life for PCE was approximately 24 years and the half-life for *cis*-1,2 DCE was 9 years.

Enhanced degradation was considered for several reasons including the probability of significantly reducing the time for VOC concentrations to reach state standards, the relatively low cost, the lack of site disruption, the low visibility to the public, and the lack of O&M other than monitoring. In addition, the site provided ideal subsurface conditions for reductive dechlorination of VOCs using enhanced degradation including: (1) a shallow, sandy water table aquifer; (2) an effective underlying aquitard; (3) a limited saturated zone; and (4) a limited plume extent.

METHODS

Field Pilot Test. A pilot test was undertaken to determine the viability of HRC application and to gather data necessary to design the full-scale application. HRC was injected within the core of the plume upgradient of an existing well (MW3) and ground water monitoring was performed to assess the effects on VOCs and ground water geochemistry. HRC was injected in an arc around MW3 at a rate of 5 pounds per foot (~7 Kg/m) of saturated thickness. A total of 210 pounds (96 Kg) of HRC was injected. Monitoring well MW7 was installed 10 feet (3 m) upgradient of MW3 to monitor ground water chemistry upgradient of the test area (Figure 1). The effects of HRC injection on ground water geochemistry in this limited portion of the aquifer were monitored over a 10-week period. VOCs and several key geochemical indicator parameters were monitored. These indicator parameters included DO, ORP, several terminal electron acceptors including total and dissolved iron, sulfate, and nitrate, metabolic acids (lactic, acetic, butyric, propionic, and pyruvic), and total organic carbon (TOC).

The initial PCE concentration in MW3 was approximately 300 ug/L prior to the pilot test, and remained at approximately this level throughout the test period. PCE concentrations in upgradient well MW7 increased from 490 ug/L to ~ 1,500 ug/L during the test period. Similar increases in TCE (from 20 t 50 ug/L) and *cis*-1,2-DCE (from 190 to 42 ug/L) were also observed. The increased levels at MW7 indicated an increased flux of chlorinated compounds across the pilot test area. The fact that PCE and daughter product concentrations in MW3 remained relatively constant during the test period, despite significant increases upgradient suggested that these compounds were being degraded during the pilot test period.

DO measurements fluctuated widely (probably due to the inherent difficulty in obtaining consistently accurate field DO readings), but typically indicated aerobic conditions at both MW3 and MW7. At MW3, ORP decreased from 106 mV prior to the test to –106 mV at 72 days. The negative redox measurements at MW3 indicate that HRC was effective in creating moderately reducing conditions in the aquifer. Ferrous iron concentrations increased from non-detect to 5,200 ug/L, indicating the iron-reducing conditions had been achieved. Concentrations of both lactic acid and TOC increased rapidly at MW3 following injection. For example, TOC in MW3 increased from 1,000 ug/L to 3,125,000 ug/L at 16 days and then stabilized at 225,000 ug/L 72 days after the injection. The presence of lactic acid and elevated levels of TOC in MW3 indicated the effective release of these compounds under the site specific conditions and that a sufficient carbon energy source was available for dechlorinating bacteria. Overall, the results of the pilot test indicated that the HRC was capable of creating an anaerobic, nutrient-rich environment capable of supporting enhanced biodegradation of chlorinated VOCs.

Full-Scale Application. Based on the results of the pilot test, a full-scale application grid of HRC was designed to completely cover the on-site plume area and to create a downgradient barrier at the off-site leading edge of the plume. HRC was injected into borings at a total of 118 points on-site and 41 points offsite using direct-push techniques (see Figure 1). Injection points were not installed between the on-site grid and the off-site barrier array due to concerns associated with work on the residential properties. The final grid was also designed to avoid multiple underground utilities. The HRC was injected over the measured saturated thickness of aquifer (3 to 12 feet [~1 to 4 m]) plus 2 feet (0.6 m) above the water table to account for water table fluctuations. The application rate was 3 to 6 pounds per vertical foot (4.4–8.8 Kg/m). In total, approximately 5,000 pounds (2,275 Kg) of HRC was injected.

Ground water monitoring was performed prior to and following the injection to monitor the effects on aquifer geochemistry and the increase in biodegradation rates. Four wells have been regularly monitored during the remedial process: MW1S (source area); MW2 (lateral plume margin); MW3 (plume core); and MW6 (leading edge of plume). Ground water samples were collected from these wells monthly during the first six months for analysis of VOCs and geochemical indicator parameters. Ground water sampling has continued quarterly since then.

RESULTS AND DISCUSSION

The application of HRC resulted in a significant increase in the rate of degradation of chlorinated solvents and a greater than 95% reduction in PCE concentrations near the source area during the initial 6-month period and continued reduction up to one year following application. Changes in VOC and geochemical parameter concentrations during the application period are provided in Table 1. The concentration trends for chlorinated compounds in each well are shown on Figure 2. Lactic acid and ferrous iron trends are shown on Figure 3.

TABLE 1: Representative Ground Water Monitoring Data from Full Scale Treatment

	MW1S			MW2		
	Initial	6 mos.	12 mos.	Initial	6 mos.	12 mos.
PCE	1400	30	2.3	550	310	190
TCE	ND	48	1.6	11	220	93
cis-1,2-DCE	110	270	240	ND	28	49
Vinyl Chloride	ND	ND	ND	ND	ND	ND
Ethene	ND	ND	NA	ND	ND	NA
Ethane	ND	ND	NA	ND	ND	NA
DO	5.3	1.0	0.28	2.42	9.6	0.7
ORP	240	-66	-146	223	-5	37
TOC	ND	720,000	NA	ND	2,400	NA
Ferrous iron	ND	>10	5.0	ND	5.0	4.0
Sulfate	50	2	NA	61.1	67	NA
lactic acid	ND	394,000	NA	ND	ND	NA

	MW3			MW6		
	Initial	6 mos.	12 mos.	Initial	6 mos.	12 mos.
PCE	280	42	51	31	14	63
TCE	5.9	570	110	0.6	0.4	19
cis-1,2-DCE	41	140	390	3	0.96	22
Vinyl chloride	ND	ND	ND	ND	ND	ND
Ethene	ND	ND	NA	ND	ND	NA
Ethane	ND	ND	NA	ND	ND	NA
DO	NA	1.4	0.7	2.73	0.4	0.71
ORP	106	-167	-130	287	-141	-46
TOC	1,000	64,200	NA	ND	5,800	NA
Ferrous Iron	5.0	>10	3.8	2.0	10	5.6
Sulfate	47.6	26	NA	36.4	51	NA
lactic acid	ND	1,700	NA	ND	ND	NA

All concentrations in ug/L
ORP in mV   DO in mg/L
ND - Not Detected
NA - Not Analyzed

The conclusion that the rate of reductive dechlorination in the treatment areas has increased dramatically is based on the decrease in PCE concentrations, the increase and subsequent decrease of TCE concentrations, and the increase in cis-1,2-DCE concentrations within the plume area. For example, at MW1S, the PCE concentration was consistently between 1,000 and 2,000 ug/L in the two years prior to the full-scale treatment (1,400 ug/L immediately prior to treatment). The

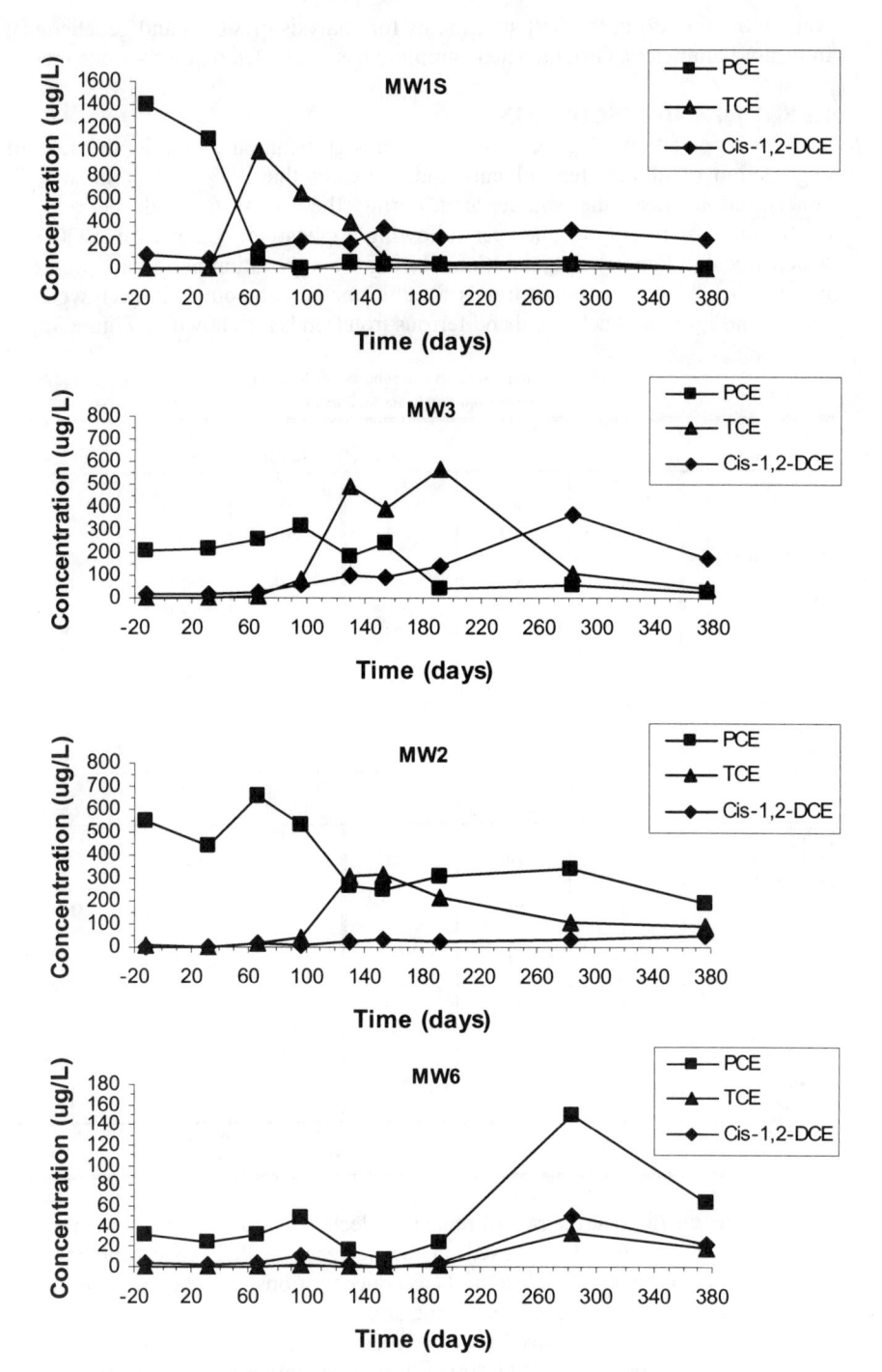

Figure 2. VOC Concentration Trends

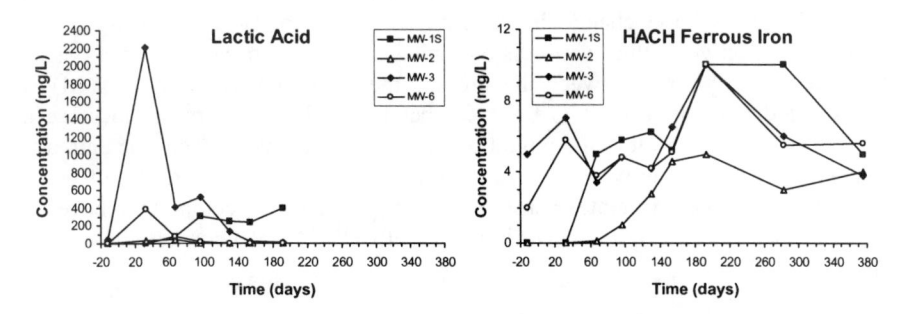

Figure 3: Lactic Acid and Ferrous Iron Trends

PCE concentration in MW1S decreased to 75 ug/L at 67 days after treatment and was last measured at 2.3 ug/L at 376 days after treatment. Concurrently, TCE concentrations increased from < 1 ug/L to 1,000 ug/L at 67 days, but have since decreased to 1.6 ug/L at 376 days. *Cis*-1,2-DCE concentrations initially increased and have remained relatively stable over the last six months, indicating that the *cis*-1,2-DCE is also degrading. However, vinyl chloride has not been detected in any wells suggesting either extremely rapid degradation or an alternate degradation pathway for *cis*-1,2-DCE. Unfortunately, the data collected as part of this remedial action are not sufficient to determine the degradation pathway for *cis*-1,2-DCE or vinyl chloride.

Geochemical parameters in MW1S all indicated the development of favorable anaerobic, reducing conditions. DO has remained below 0.5 ppm since early in the monitoring period; ORP readings have remained less than –100 mV; and there have been significant increases in ferrous iron concentrations and some sulfide formation. The TOC concentration increased dramatically after the first month following treatment and has since remained above 200,000 ug/L. Similar, though not as marked concentrations trends were observed in MW3, which is also in the core of the plume.

MW2 is located closer to the plume margin than MW1s or MW3 and may be affected by fresh water influx due to differential recharge between the unpaved residential areas and the paved shopping center. PCE concentrations at MW2 have decreased by 50% since the injection, and TCE and *cis*-1,2-DCE concentrations have steadily increased indicating reductive dechlorination is occurring, but at a slower rate. However, unlike MW1S and MW3, the PCE concentration remained higher than that of either of its daughter compounds. Changes in the aquifer geochemistry were also less at MW2 than in the core wells. DO readings have fluctuated between anaerobic and aerobic conditions and ORP has been consistently near 0 mV, which is significantly less reducing than at either MW1S or MW3. TOC concentration increases were also much less at MW2, with a maximum TOC concentration or 38,000 ug/L and average concentrations less than 10,000 ug/L.

At off-site well MW6, the concentration trends have been less clear. PCE initially decreased after treatment, but subsequently increased to concentrations higher than the initial concentration. The increased concentrations of TCE and cis-1,2-DCE and the change in the ratios of PCE to TCE, and cis-1,2-DCE indicate reductive dechlorination is occurring. However, because a large portion of the off-site plume (between the site and the offsite barrier array near MW6) was not directly treated, it is believed that the PCE concentration increase at MW6 is simply a result of the migration of contaminated water from the untreated area toward the barrier wall. Concentrations in MW-6 are expected to decrease as trated ground water migrates across this area. Chlorinated compounds have not been detected at levels of concern in a sentinel well (MW5S) downgradient of the barrier wall.

CONCLUSIONS

Injection of a polylactate ester (HRC) into plume of PCE-contaminated ground water beneath a dry cleaning facility effectively changed the chemistry of the aquifer to an anaerobic and nutrient-rich environment, thus accelerating reductive dechlorination of chlorinated solvent compounds. In the core of the plume (MW1S and MW3), PCE concentrations have decreased by up to 99% in the one year since injection of the HRC. Ongoing ground water monitoring indicates that more than one year after injection, the majority of the treated area remains anaerobic and PCE concentrations continue to decrease. Based on results to date, vinyl chloride is either not being produced or is degrading very rapidly, as it has not been detected. The results from this site provide further evidence that in-situ methods of altering ground water chemistry to create a favorable environment for biodegradation can be an extremely effective means of remediating ground water contamination, particularly at sites with a reasonably permeable aquifer. In treatment design applications, it appears that both full plume coverage and barrier wall-type systems can be effective. Where practicable, complete coverage of the source and nearby areas will likely be more effective at mitigating impacts than a barrier wall only design. In addition, the potential for the influx of untreated aerobic water near the plume margins should be evaluated to ensure adequate amounts of the treatment agent are injected in areas where fresh water influx may be significant. Given the apparent impact of a fresh water flux on the effectiveness of the HRC at creating favorable conditions, the economics of treating an area larger than the plume itself compared to the cost of potential multiple injections should be considered in designing an injection approach.

REFERENCES

Buscheck, T.E., and C.M. Alcantar. 1995. "Regression Techniques and Analytical Solutions to Demonstrate Intrinsic Bioremediation" in *Intrinsic Bioremediation,* ed.; R.E. Hinchee, J.T. Wilson and D.C. Downey. Batelle Press: Columbus, OH.

ENHANCED BIOREMEDIATION OF CHLORINATED SOLVENTS

Willard Murray (Harding ESE, Wakefield, MA)
Maureen Dooley and Stephen Koenigsberg (Regenesis, San Clemente, CA)

ABSTRACT: Enhanced in situ bioremediation of chlorinated solvents in groundwater has been successfully demonstrated at many sites by supplying lactic acid as an electron donor. The source of lactate for these successful pilot tests is Hydrogen Release Compound (HRC™), a polylactate ester specially formulated for slow release of lactic acid upon hydration. HRC has been delivered to chlorinated solvent groundwater plumes by being contained in perforated canisters hung in wells, by being injected into the contaminated aquifer through the screened section of monitoring wells, or by being directly injected into the contaminated aquifer using a direct push technology such as Geoprobe®. This paper presents the results from selected pilot tests where HRC has been used as a slow release electron donor to enhance natural biological destruction of chlorinated solvents. Completed pilot tests show that HRC can effectively enhance the natural attenuation of chlorinated solvents with very efficient degradation half-lives, an obvious requirement for economic site cleanup.

INTRODUCTION

During the period 1990 through 1993, microbiologists at Harding ESE (then ABB Environmental Services) conducted bench scale testing of enhanced biodegradation of perchloroethylene (PCE) under the EPA Superfund Innovative Technology Evaluation (SITE) – Emerging Technology Program. This research was intended to demonstrate a sequential anaerobic/aerobic biodegradation process using various electron donors. Subsequent laboratory tests showed lactic acid to be the most effective electron donor. Saturated soil column simulations of anaerobic aquifer conditions demonstrated complete degradation of 20 mg/L PCE to 60% vinyl chloride (VC) and 40% ethylene with a contact time of 2 hours. Aerobic soil column simulations demonstrated that VC was degraded by methanotrophic bacteria when oxygen was provided through the addition of hydrogen peroxide (ABB-ES, 1994).

The successful bench scale studies led to a field pilot test also sponsored by the EPA SITE Program. During the period 1994 through 1997 a field test of the sequential anaerobic/aerobic biodegradation of chlorinated solvents was designed and conducted. A site with contaminated groundwater containing 1,500 µg/L of PCE, 16,000 µg/L of trichloroethylene (TCE), 3,500 µg/L of dichloroethylene (DCE) and 100 µg/L of VC was found in Watertown, Massachusetts that could be used for the field pilot. The pilot test was conducted by isolating a small circulating groundwater cell (approximately 30 ft in diameter) in the central area of a larger solvent plume by using extraction and injection wells. Amendments consisting of lactic acid, ammonia chloride, potassium tripolyphosphate, yeast extract, and sodium hydroxide (to neutralize the acid) were continuously fed into the circulating flow. After a 4-month acclimation period, a significant reductive dechlorination period began which over the next 4 months reduced the PCE to approximately 100 µg/L and the TCE to approximately 1000 µg/L, with DCE remaining essentially unchanged at approximately 3000 µg/L and VC increased to approximately 1000 µg/L. At this time, the circulating cell was transformed to an aerobic condition by hanging ORC™ socks in the injection wells. Methane was also added periodically. After a short period of time to remove the residual oxygen demand from the anaerobic phase, biodegradation again began with significant reductions (total ethenes reduced by half) in TCE, DCE and VC over a 3 month period until a premature ending of the test due to an accidental destruction of the equipment shed. Total ethenes were reduced by approximately 70% in 11 months (HLA, 1999).

During the latter stages of the sequential pilot test, discussions with Regenesis led to the testing of their new product for enhancing reductive dechlorination during 1997-98. The site in Watertown was allowed to return to its original level of contamination, and then HRC-containing canisters were suspended in the injection wells and the circulating cell was once more initiated. After 206 days under anaerobic conditions created by the HRC, the initial concentrations of 1,000 µg/L PCE, 13,000 µg/L TCE, 3,000 µg/L DCE, and 200 µg/L VC were reduced by 97% in total mass (Dooley, et. al., 1999). Sulfate reducing and iron reducing conditions appeared to be the predominant microbiological conditions across the treatment cell. VC rose from approximately 200 µg/L to 3,000 µg/L after the first 6 months, but then rapidly decreased to less than 200 µg/L. No significant accumulation of VC occurred and complete biodegradation of PCE, TCE and daughter products was demonstrated (as evidenced by the reduction of PCE to 8.5 µg/L, TCE to 95 µg/L, DCE to 163 µg/L, and VC to 157µg/L).

After this successful demonstration of total biodegradation anaerobically, other projects using HRC were initiated. In the following, results from five enhanced biodegradation projects are described.

MICHIGAN SITE

At a site in Michigan, a plume of DCE and VC is found in a silty sand aquifer. A passive application of HRC was implemented by using Geoprobe injection points in a fence-like pattern across the width of the plume. Monitoring wells were placed upgradient and downgradient of the fence to monitor the performance. After 60 days, the oxidation-reduction potential (ORP) decreased to –150 mV at the monitoring well downgradient of the fence. It is presumed that increased biological activity increased the desorption off aquifer solids, and the concentrations of both DCE and VC increased initially during the first 60 to 90 days. After this initial period, the concentration levels of both DCE and VC continuously decreased for the next 6 to 7 months (see Figure 1). The DCE reduction during this time exhibited a half life of approximately 80 days; the VC reduction was much less pronounced, but indicated a half life of approximately 200 days if no production of VC from the reductive dechlorination of DCE was considered. Large concentrations of lactic acid fermentation products, acetic acid and propionic acid, were detected during this active destruction period. However after Day 255, no organic acids were detected and the oxidation reduction potential began to rise, indicating that the HRC supply was becoming depleted; consequently the enhanced biodegradation process was slowing and the concentrations of DCE and VC began to increase. Although it was apparent that additional HRC was needed to continue the destruction of contamination, the passive application of HRC was effective at this site in destroying both DCE and VC without accumulation of VC.

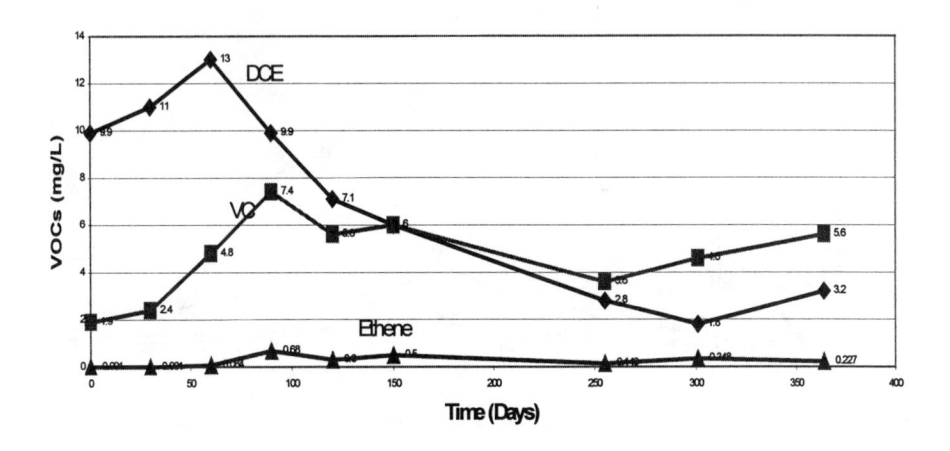

FIGURE 1. Michigan Site – VOC Concentrations

KANSAS SITE

At a site in Kansas, a plume of primarily PCE, with low concentrations of daughter products, was found in an aquifer consisting of clayey soils. Concentrations of PCE in the treatment area were in excess of 6,000 µg/L. HRC was injected as a fence across the migrating plume. Downgradient behavior of VOCs is shown in Figure 2 (monitoring well approximately 10-ft downgradient of the HRC injection points). Within 30 days, the PCE was almost totally degraded. TCE and VC remained at low levels initially, but DCE increased markedly for 120 days to concentrations that were approximately 4 times the original PCE concentration (5 times on a molar basis). This again was presumed to be due to increased rates of desorption caused by the increased biological activity. The large increases indicate that a significant amount of contamination was sorbed to aquifer solids. After 120 days the levels of DCE began to decrease, as did the total mass of contamination. By the end of one year the total mass of dissolved contamination had been reduced to approximately 25% of the maximum attained on Day 120, and the levels of all constituents, including VC, were decreasing. Although a significant amount of DCE was produced due to the increased desorption effect, a continued supply of HRC has caused biological activity to ultimately destroy the contamination without an excessive amount of VC being produced.

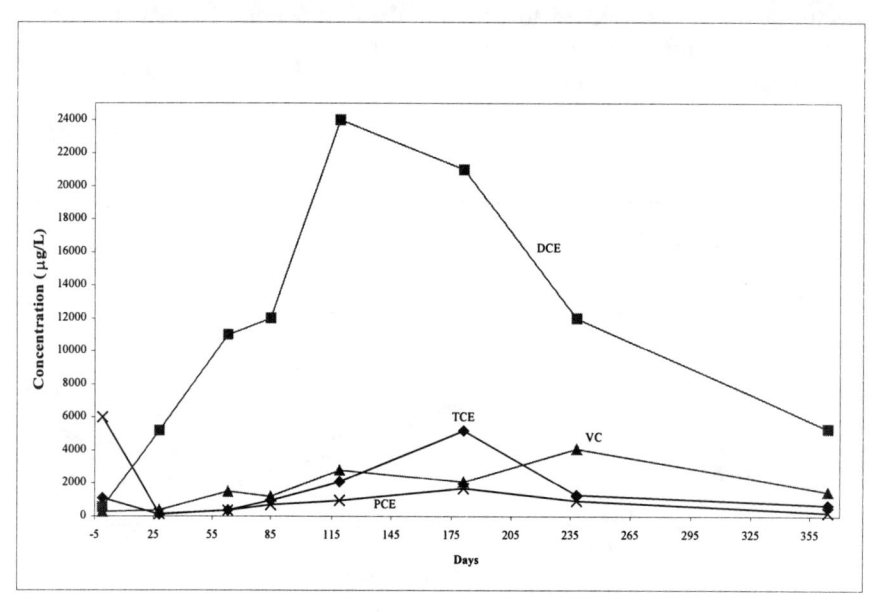

Figure 2. Kansas Site – VOC Concentrations

TENNESSEE SITE

At a site in Tennessee a plume of primarily PCE and TCE was slowly migrating in a clayey aquifer. Two areas were the subject of an enhanced biodegradation test – a source area where a grid of HRC injection points was installed, and a fence line of HRC injection points near the property boundary. Figure 3 shows the results from a monitoring well in the source area on the downgradient fringe of the injection grid. Groundwater samples from this well show complete destruction of PCE and TCE within 4 months, however DCE levels increased to 110 mg/L, which is 10 times greater than the initial total ethene concentration. Once again this indicates the increased desorption effect caused by the enhanced microbial activity, and perhaps a microbial surfactant effect acting on the residual source material. It was apparent that the residual PCE and TCE located in the source area were continually being degraded to DCE and VC, which both remained at elevated concentrations. The high concentration of DCE (greater than 100 mg/L) was equivalent to more than 130 mg/L of TCE (12% of TCE solubility) and equivalent to a PCE concentration greater than its solubility. These facts suggest that there is residual dense non-aqueous phase liquid (DNAPL) within the aquifer in the source area. Although the enhanced biodegradation had been very successful in converting the PCE and TCE into daughter products DCE and VC, the concentrations of DCE and VC can be expected to remain elevated until the residual DNAPL is dissolved.

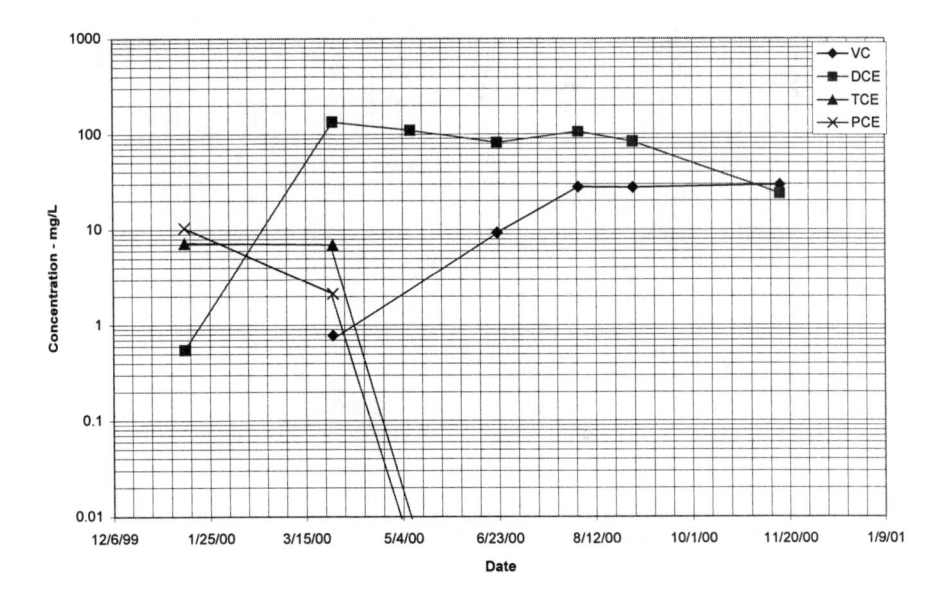

Figure 3. Tennessee Site – Source Area VOCs

Figure 4 shows results from a monitoring well located immediately downgradient of the injection fence line. At this location, initial concentrations of PCE at 12000 mg/L and TCE at 6800 mg/L were reduced to 0.075 mg/L and 0.38 mg/L respectively within 80 days. They were completely degraded after 10 months. Concentrations of DCE initially increased and remained at elevated levels for approximately 5 months after injection, and then began to decrease. The total chlorinated solvent concentrations showed an 89% reduction in total ethenes from January to November. Although there was a significant increase in DCE during this period, VC levels remained relatively low and were not detected in the November sampling event. The overall results from this project indicated that the enhanced biodegradation technology was very effective at stimulating indigenous microbes to degrade the chlorinated solvent contamination. It is apparent that although there is significant generation of DCE in the reductive dechlorination process, excessive VC accumulation does not appear to be a problem.

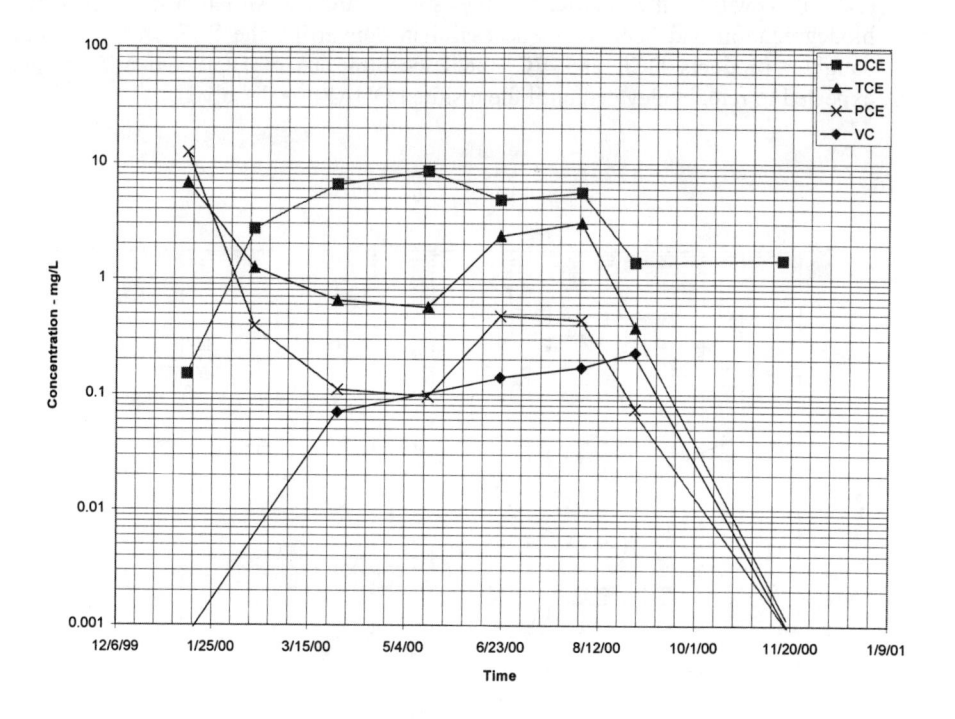

Figure 4. Tennessee Site – Downgradient Fence VOCs

FLORIDA SITE

At a dry cleaner site in Florida, a grid injection of HRC has been applied to the entire site. Results from a monitoring well within the chlorinated solvent plume are shown in Figure 5. The data show a "typical" behavior as a result of the injection of HRC. The injection in April 2000 caused an initial increase in the concentration of total ethene (as PCE). However, subsequent sampling episodes revealed a sharp decrease in the concentration of PCE and TCE with modest increases in DCE and VC before they too decreased. It can be seen that there was a sharp decrease in total ethenes over the period from June 2000 to January 2001.

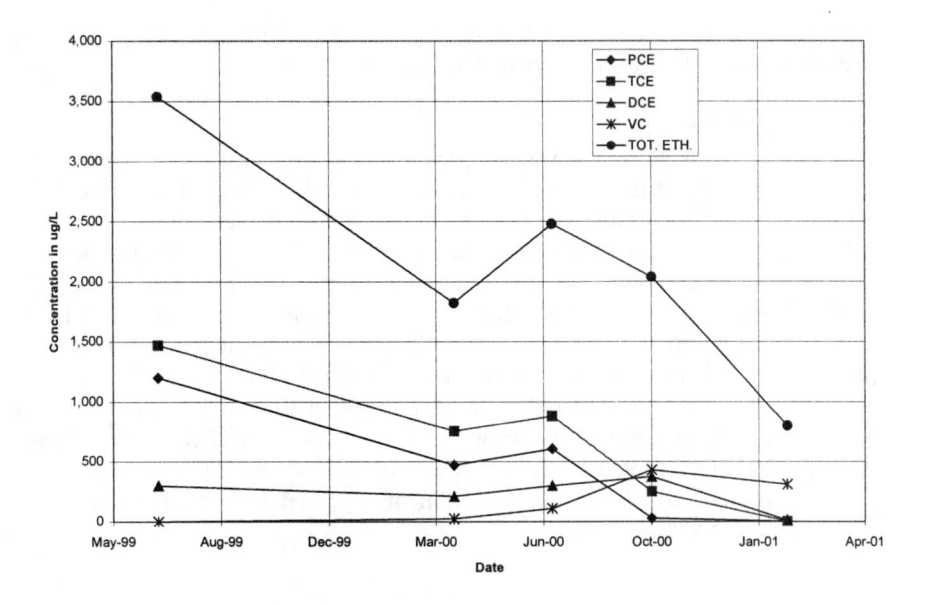

Figure 5. Florida Site – VOC Concentrations

CENTRAL MASSACHUSETTS SITE

At a Brownfields site in the Blackstone River Valley in central Massachusetts, a new high profile pilot test was recently initiated under the EPA's Superfund Innovative Technology Evaluation (SITE) program. A new technique for initiating the anaerobic process by preceding the normal, highly viscous HRC injection with several wellbore volumes of a new and experimental (more flowable) variant of HRC was demonstrated. This pre-step was designed to sweep the aquifer clean of competing electron acceptors more rapidly to promote earlier onset of reductive dechlorination. Initial results showed that after 3 months the oxidation-reduction potential had decreased markedly in all monitoring wells within 25 feet downgradient of the HRC injection wells, some to less than –400 mV. However, this initial large decrease in oxidation-reduction potential was not accompanied by early onset of reductive dechlorination. Significant reductive dechlorination began 2 months later when the oxidation-reduction potential in most of the treatment area approached the range of –50 mV to –150 mV. Sampling conducted 7 months after injection revealed large decreases in TCE (on average from approximately 1,000 µg/L to approximately 200 µg/L), large increases in DCE (on average from approximately 400 µg/L to approximately 1,200 µg/L), but no change in VC, which remained less than 5 µg/L. Total ethenes increased in three monitoring wells from 11% to 26%, and since these wells are also associated with large conversions from TCE to DCE, the increases are attributed to the desorption effect of increased biological activity. Three other monitoring wells showed decreases in total ethenes of 38% to 63%, indicating total mass destruction with no accumulation of VC.

CONCLUSIONS

The results from the various projects described herein all have similar implications. The first is that with HRC-enhanced biodegradation, it can be expected that initial results will show increases in total contamination. This is a result of increased rates of desorption of PCE and TCE off the aquifer solids due to their reduced concentrations in the groundwater. Then as the additional PCE and TCE appear in the groundwater, they are quickly biodegraded to DCE. There may also be microbial surfactants created in the aquifer to contribute to the production of increased concentrations where residual DNAPL is present. This should not be a concern but rather an assurance that the natural microbes have been stimulated into action. The second is that as long as the HRC electron donor does not become depleted, the total destruction of the chlorinated ethenes proceeds without significant accumulation of VC, and in some cases with no accumulation of VC.

REFERENCES

ABB Environmental Services (ABB-ES). 1994. "Biological Removal of Perchloroethylene from Saturated Soils." Final Report for Cooperative Agreement No. CR 816820-01-0, SITE Emerging Technologies. Risk Reduction Engineering Laboratory, USEPA, Cincinnati, Ohio.

Harding Lawson Associates (HLA). 1999. "Anaerobic/Aerobic Sequential Biodegradation of PCE." Final Report for Cooperative Agreement No. CR 822714-01, SITE Emerging Technologies. Risk Reduction Engineering Laboratory, USEPA. Cincinnati, Ohio.

Dooley, M. A., W. A. Murray, and S. Koenigsberg. 1999. "Passively Enhanced In Situ Biodegradation of Chlorinated Solvents." *Engineered Approaches for In Situ Bioremediation of Chlorinated Solvent Contamination.* Editors: A.Leeson and B. C. Alleman. Proceedings from the Fifth International In Situ and On-Site Bioremediation Symposium held April 19-22, 1999, in San Diego, California.

ACCELERATING THE REDUCTIVE DECHLORINATION PROCESS IN GROUNDWATER

Daniel South, P.G. (Harding ESE, Denver, Colorado, USA)
Joe Seracuse, C.H.G. (Harding ESE, Denver, Colorado, USA)
Kevin Garrett, Ph.D. (Harding ESE, Denver, Colorado, USA)
Dong Li, Ph.D. (California, USA)

Abstract: The objective of this pilot test program was to assess the effectiveness of a relatively low cost, easy to implement, and low maintenance remedial technology that could accelerate the in-situ reductive dechlorination process that was already occurring at the site. To accelerate this process, an environmentally safe polymer was injected into the aquifer and the volatile organic compounds (VOCs) were monitored for one year between September 1999 and September 2000. Direct-push techniques were used to inject the polymer into a portion of the aquifer where limited in-situ reductive dechlorination of the chlorinated VOCs, primarily perchloroethene (PCE), were present in groundwater. Within the first month following the polymer injection, PCE concentrations in groundwater decreased over 97% in the treated area. Based on the data collected throughout the test (including decreasing PCE concentrations and increasing ethene concentrations), the polymer provided the microorganisms with the proper conditions to accelerate the dechlorination of the chlorinated VOCs detected in groundwater. This pilot test program demonstrated that the application of this easy to implement bioenhancement remedy can accelerate the in-situ biodegradation of PCE at the site at lower costs relative to other more traditional remedies.

INTRODUCTION

Chlorinated solvents have long been one of the most difficult contaminants to remediate. Many remediation technologies including pump-and-treat systems and zero-valence barriers typically have high construction and/or maintenance costs. Recent developments in the enhanced bioremedation field have lead to cost-effective remediation alternatives.

Harding ESE (formerly Harding Lawson Associates) conducted a pilot test to assess the viability of accelerating the bioremediation of chlorinated volatile organic compounds (VOCs) at a site where some biodegradation of chlorinated VOCs was occurring. The pilot test involved placement of a commercially available polymer (Hydrogen Release Compound [HRC®]) into the impacted aquifer and monitoring the degradation for a year. Figure 1 shows the configuration of the monitoring well network and the injection array of the pilot test.

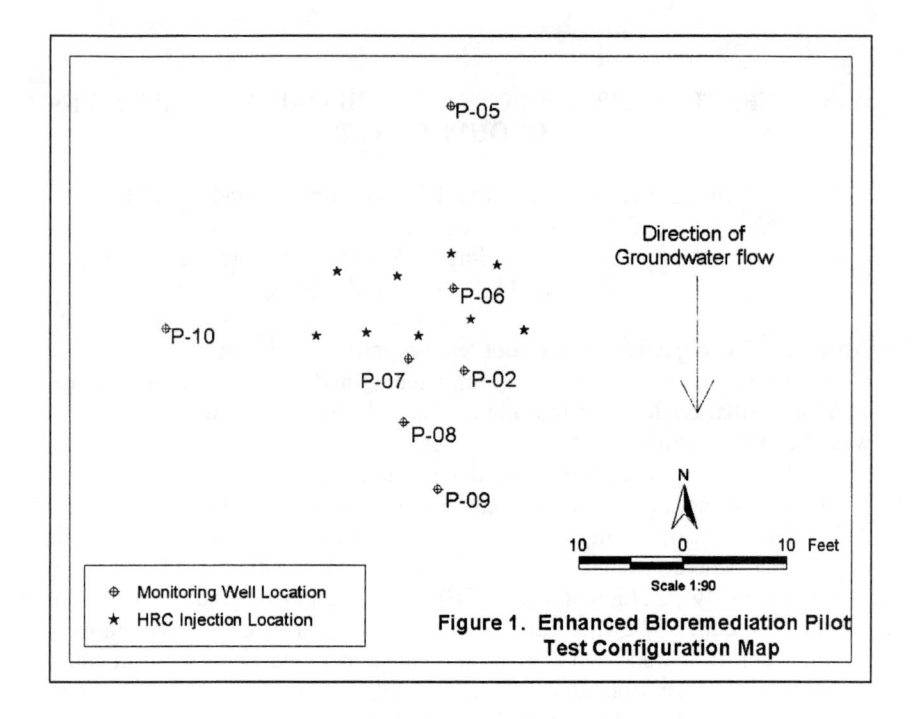

Figure 1. Enhanced Bioremediation Pilot Test Configuration Map

Site Background. The site selected for the test was a closed industrial facility that had previously handled various chlorinated solvents (chlorinated VOCs) and other petroleum products. Based on data from previous investigations, groundwater beneath the site contained chlorinated VOCs (primarily PCE and trichloroethene [TCE]), and petroleum hydrocarbons including benzene, toluene, ethylbenzene, and xylenes (BTEX). In portions of the site, the petroleum hydrocarbon plume overlays a chlorinated VOC plume. Investigations in these areas of the site have indicated that reductive dechlorination of the chlorinated VOCs is occurring in the reduced environment caused by the petroleum hydrocarbons. The indication of reductive dechlorination was supported by detections in groundwater of ethene in another area of the site away from the pilot test area and the detections of other chlorinated VOC daughter products of PCE including TCE, cis-1,2-dichloroethene (cis-1,2-DCE), and vinyl chloride in the test area. In the areas of the chlorinated VOC plume where the petroleum hydrocarbons are absent, lower concentrations of chlorinated VOC daughter products were observed suggesting a slower rate of biodegradation of the chlorinated VOCs in these areas of the site. As presented in Table 1, before the test was initiated, the concentrations of PCE in the test area were as high as 35,000 micrograms per liter (µg/l). In addition, prior to the start of the pilot test no ethene was detected in groundwater in this area of the site.

The geologic setting at the pilot test area consists of approximately twenty to twenty-five feet of unconsolidated silt and clay with occasional thin lenses of fine sand and sandy clay overlaying a severely weathered claystone.

Approximately 20 feet below the silt/clay and claystone interface, the weathering decreases to little or none. Groundwater in the pilot-test area generally flows in a south-southwest direction. The depth to groundwater fluctuates from approximately 7 to 10 feet below ground surface (bgs). The hydraulic conductivity (k) of the unconsolidated silt and clay has been calculated from pumping tests to range from approximately 2.4 to 5.0 feet per day. The gradient (i) in the unconsolidated silt and clay is estimated to be approximately 0.01. With an assumed effective porosity of 10 percent, the seepage velocity is calculated to be approximately 0.5 feet per day. Based on data from the site, groundwater primarily flows through the unconsolidated materials with some of the flow also occurring through the severely weathered claystone. With depth, the weathering decreases and the claystone has a very low hydraulic conductivity. The unweathered claystone behaves more like an aquitard, limiting groundwater flow.

MATERIALS AND METHODS

The pilot test was designed to be a one time injection of the polymer (HRC®) with regular groundwater monitoring throughout the year. Based on numerous direct-push groundwater samples, elevated concentrations of PCE, lower concentrations of the daughter products, and little or no BTEX characterized the selected pilot test location at the site. In addition, the pilot-test area is downgradient from a suspected PCE source area. The pilot test was initiated in September 1999 and was concluded in September 2000.

The polymer used for this pilot test (HRC®) is a polylactate ester that upon hydration degrades to lactic acid. Lactic acid is then metabolized into several other organic acids and eventually into hydrogen, which in turn drives the reductive dechlorination of chlorinated VOCs. In September 1999 the pilot test monitoring well network was installed as shown on Figure 1. The monitoring well network consisted of seven groundwater monitoring wells.

To assess the conditions of the groundwater prior to the injection of the polymer, the wells were sampled using a combination of a peristaltic pump with flow cell and a disposable bailer. Field personnel measured the dissolved oxygen concentrations, pH, temperature, oxidation/reduction potential (ORP), conductivity, and ferrous and total iron concentrations. In addition, groundwater samples were collected and analyzed at an off-site laboratory for VOCs, ethane, ethene, methane, sulfate, and volatile acids using standard methods.

Once the baseline samples were collected, the polymer was injected into the ground using a Strataprobe™ direct-push rig. There were nine injection locations (Figure 1). The polymer was injected in 1-foot intervals from approximately 30 to 5 feet bgs as the direct-push rod was withdrawn. Approximately 0.5 gallon of the polymer, which equates to approximately 5 pounds (lbs.) of the polymer, was injected in each 1-foot interval. A total of approximately 1,200 lbs. (approximately 200 gallons) of the polymer were injected into the injection array for this test. The groundwater in this test area contained PCE in groundwater at a concentration of between 35,000 and 3,900 µg/l (less than 1 pound of PCE in groundwater at the pilot test area).

Groundwater samples were collected and analyzed regularly following the injection. The same groundwater parameters analyzed during the baseline testing were analyzed following the injection.

RESULTS AND DISCUSSION

Prior to injection of the polymer, the groundwater in the pilot test area contained low to moderate concentrations of dissolved oxygen (0.75 to 2.68 milligrams per liter [mg/l]), a general absence of nitrate (<0.1 to 0.4 mg/l), and moderate concentrations of sulfate (56.6 to 74.5 mg/l). The ORP readings prior to the injection of the polymer were generally positive or slightly negative ranging from -25 to +269 millivolts (mV).

In general, the dissolved oxygen and nitrate concentrations in groundwater remained similar to initial concentrations throughout the pilot test. The concentrations of sulfate in groundwater decreased to less than 10 mg/l in samples collected from the downgradient monitoring wells within 180 days after injection. Within approximately one month following the injection of HRC®, the ORP in the downgradient wells (P-02 and P-06 through P-09) had become strongly negative (-429 mV) while the ORP in the upgradient well (P-05) and cross-gradient well (P-10) remained strongly positive (+335 mV).

A summary of the chlorinated VOC analytical results are presented in Table 1. Figure 2 presents the analytical data in graphic form for the representative downgradient Well P-09. As can be seen in Table 1, the concentrations for PCE decreased by approximately 97 percent in Well P-06 within the first month following the injection of the polymer. In addition, the concentrations of PCE in groundwater from Well P-09 decreased from 3900 µg/l prior to the injection to <5 µg/l by the end of the pilot test. Likewise, the TCE concentrations in groundwater from Well P-09 also decreased from a concentration of 640 µg/l to non-detectable levels at the end of the pilot test. As expected, the cis-1,2-DCE and vinyl chloride concentrations in groundwater generally increased in the downgradient wells following the injection and began to decrease at the end of the pilot test. The vinyl chloride concentrations were generally an order of magnitude less than the cis-1,2-DCE concentrations from the corresponding samples.

Ethene concentrations in groundwater generally increased in the downgradient wells (P-02 and P-06 through P-09) during the pilot test. The concentrations of ethene in groundwater were less than 10 µg/l prior to the injection of HRC® and increased to as high as 1,310 µg/l at the end of the pilot test. The data in Table 1 show generally similar trends in the other downgradient wells, with minor variation based on distance from the injection array and localized heterogeneity.

Table 2 presents the molar percentage ratios of PCE, TCE, cis-1,2-DCE, vinyl chloride, and ethene for the monitoring network. Figure 3 presents a graph of the data for Wells P-05 (upgradient) and P-09 (14 feet downgradient from the injection array). This representation of the data assists in the evaluation of the reductive dechlorination process. As observed in Table 2, the molar percentages for the control upgradient well (P-05) and crossgradient well P-10 remain similar through time for the different VOCs monitored even though the concentrations of VOCs change throughout the test. However, the molar percentages of VOCs in the downgradient wells change significantly throughout the test.

At the start of the test, PCE comprised the highest concentration and largest molar percentage of the chlorinated VOCs in groundwater in the pilot test area. For example, the molar percentage of PCE in the initial sample from P-09 was approximately 78 percent. Prior to the injection of HRC®, the majority of the other VOCs in groundwater in the pilot test area were TCE and cis-1,2-DCE with a minor amount of vinyl chloride. Following the HRC® injection, the reductive dechlorination process was accelerated and the dechlorination process continued to the generation of ethene and ethane. By the conclusion of the pilot test, the molar percentages of PCE and TCE in groundwater from Well P-09 had essentially decreased to zero, the molar percentage of cis-1,2-DCE had decreased to approximately 8 percent, and vinyl chloride had decreased to approximately 6 percent (Table 2 and Figure 3). The molar percentage of ethene in groundwater from Well P-09 increased steadily throughout the pilot test up to approximately 86 percent. The trends of increasing and decreasing molar percentages for the other downgradient wells were generally similar to those observed in P-09, though less extensive.

Volatile acids are breakdown products of the HRC®. Lactic acid released from the HRC® is the carbon source used by the microbes during the reductive dechlorination process. The following volatile acids were monitored to assess the persistence of the reducing conditions: acetic, butyric, lactic, propionic, and pyruvic acids. At the end of one year of monitoring, the concentrations of lactic and propionic acids were continuing to increase in most wells with maximum concentrations of 4,590 µg/l and 1,810 µg/l, respectively. Therefore, the initial injection of HRC® in the pilot-test area was sufficient to support the reductive dechlorination process in the pilot-test area throughout the one-year monitoring period and no additional injections were needed.

CONCLUSIONS

Based on the observed chlorinated VOC degradation trends throughout the pilot test, the application of the polymer HRC® successfully accelerated the reductive dechlorination process that was already occurring at the Site. The sulfate and ORP data also support the assessment that the reductive dechlorination process was accelerated by the injection of the HRC®. In addition, based on the lactic acid concentrations, and VOC data from the last groundwater sampling round (September 26, 2000), the reductive dechlorination process was continuing at the end of the pilot test. Therefore, the process of reductive dechlorination can continue for a period of time greater than one year. As demonstrated by the increasing concentrations and molar percentages of ethene through the pilot test, the addition of the HRC® also promoted and enhanced the dechlorination of vinyl chloride to ethene.

TABLE 1. Volatile organic compound analytical data.

	Injection Date:	9/28/99							
	Sampling Date	9/27/99	10/26/99	11/30/99	12/22/99	1/25/00	3/27/00	5/23/00	9/26/00
	Days After Injection	-1	28	63	85	119	181	238	364
Site ID	Compound	Concentration ($\mu g/l$)							
P-02	PCE	6400	4800	510	2700	1700	2600	1800	41
	TCE	1400	890	960	2300	3300	1800	8200	240
	cis-1,2-DCE	490	1800	6100	6500	10000	7800	7000	17000
	Vinyl chloride	180	510	280	390	880	3600	1300	2400
	Ethene	<10	<10	<30	<20	25	373	243	140
P-05	PCE	35000	49000	55000	45000	45000	10000	8300	47000
	TCE	2600	4000	4400	3900	3100	990	560	3200
	cis-1,2-DCE	1400	2500	3800	3700	4100	710	370	3000
	Vinyl chloride	250	240	120	110	290	170	160	130
	Ethene	<10	<20	<20	<20	<20	<7	<20	<30
P-06	PCE	6000	150	360	700	960	1700	940	230
	TCE	1100	120	370	960	2100	5200	1300	680
	cis-1,2-DCE	550	5200	11000	12000	24000	21000	12000	5300
	Vinyl chloride	290	380	1500	1200	2800	2100	4100	1500
	Ethene	<10	<30	<20	<20	20	44	203	880
P-07	PCE	6500	210	160	180	250	270	430	310
	TCE	840	68	210	200	540	1300	2200	410
	cis-1,2-DCE	560	3900	8000	10000	15000	12000	14000	8900
	Vinyl chloride	160	140	390	360	1100	540	2200	770
	Ethene	<10	<20	<20	<20	<10	89	180	240
P-08	PCE	9000	950	140	56	31	35	46	47
	TCE	1300	140	220	40	80	60	370	350
	cis-1,2-DCE	860	4000	5300	4900	9200	5600	5900	5600
	Vinyl chloride	170	250	930	920	2000	2400	2300	1000
	Ethene	<10	<20	<30	<20	130	876	1150	580
P-09	PCE	3900	1700	140	1600	170	110	49	<5.0
	TCE	640	390	64	310	220	120	210	<5.0
	cis-1,2-DCE	160	1700	1600	1000	1100	1100	1900	430
	Vinyl chloride	<100	120	440	290	530	570	1200	190
	Ethene	<10	<20	<30	87	447	580	1200	1310
P-10	PCE	5300	2600	610	2200	2000	940	1200	500
	TCE	200	120	72	110	120	42	84	67
	cis-1,2-DCE	140	98	95	100	97	24	62	47
	Vinyl chloride	<100	<10	4	<1	1.2	<1	<1	<1
	Ethene	<10	<10	<30	<20	<20	<10	<20	<20

TABLE 2. VOC molar percentages.

Site ID	Compound	mol. Wt. (g/mol)	% of Total Moles						
Injection Date:			9/28/99						
Sampling Date			09/27/99	10/26/99	11/30/99	12/22/99	1/25/00	3/27/00	5/23/00	9/26/00
Days After Sampling			-1	28	63	85	119	181	238	364
P-02	PCE	165.83	67	46	4	15	7	9	6	0
	TCE	131.39	19	11	9	16	16	8	36	1
	cis-1,2-DCE	96.94	9	30	81	63	67	45	41	79
	Vinyl chloride	62.50	5	13	6	6	9	32	12	17
	Ethene	28.05	0	0	0	0	1	7	5	2
			100	100	100	100	100	100	100	100
P-05	PCE	165.83	85	83	82	80	79	77	82	83
	TCE	131.39	8	9	8	9	7	10	7	7
	cis-1,2-DCE	96.94	6	7	10	11	12	9	6	9
	Vinyl chloride	62.50	2	1	0	1	1	3	4	1
	Ethene	28.05	0	0	0	0	0	0	0	0
			100	100	100	100	100	100	100	100
P-06	PCE	165.83	66	1	2	3	2	3	3	1
	TCE	131.39	15	1	2	5	5	13	5	4
	cis-1,2-DCE	96.94	10	87	80	80	79	72	58	47
	Vinyl chloride	62.50	8	10	17	12	14	11	31	21
	Ethene	28.05	0	0	0	0	0	1	3	27
			100	100	100	100	100	100	100	100
P-07	PCE	165.83	73	3	1	1	1	1	1	2
	TCE	131.39	12	1	2	1	2	7	8	3
	cis-1,2-DCE	96.94	11	91	90	92	87	84	70	78
	Vinyl chloride	62.50	5	5	7	5	10	6	17	10
	Ethene	28.05	0	0	0	0	0	2	3	7
			100	100	100	100	100	100	100	100
P-08	PCE	165.83	72	11	1	1	0	0	0	0
	TCE	131.39	13	2	2	0	0	0	2	3
	cis-1,2-DCE	96.94	12	79	76	77	72	45	43	59
	Vinyl chloride	62.50	4	8	21	22	24	30	26	16
	Ethene	28.05	0	0	0	0	4	24	29	21
			100	100	100	100	100	100	100	100
P-09	PCE	165.83	78	31	3	32	3	2	0	0
	TCE	131.39	16	9	2	8	4	2	2	0
	cis-1,2-DCE	96.94	5	54	66	34	30	27	23	8
	Vinyl chloride	62.50	0	6	28	15	22	21	23	6
	Ethene	28.05	0	0	0	10	41	48	51	86
			100	100	100	100	100	100	100	100
P-10	PCE	165.83	92	89	70	88	86	91	85	75
	TCE	131.39	4	5	10	6	7	5	8	13
	cis-1,2-DCE	96.94	4	6	19	7	7	4	8	12
	Vinyl chloride	62.50	0	0	1	0	0	0	0	0
	Ethene	28.05	0	0	0	0	0	0	0	0
			100	100	100	100	100	100	100	100

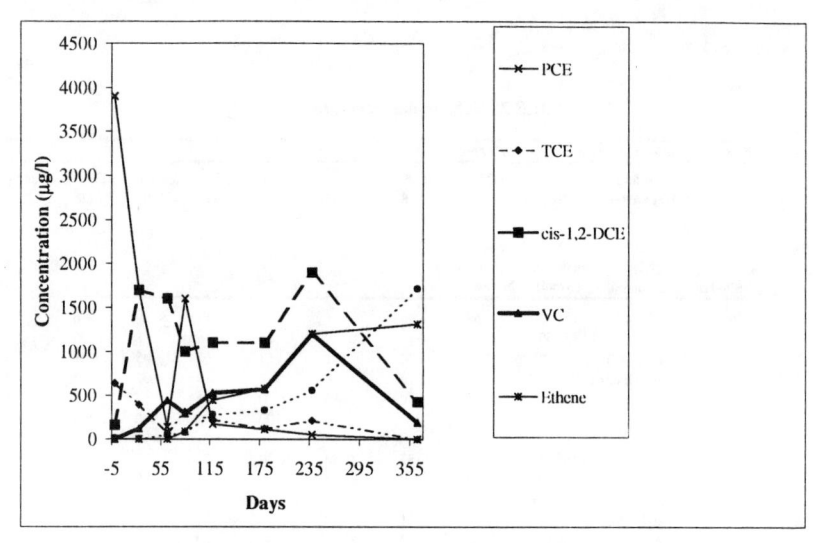

FIGURE 2. Volatile organic compound analytical data for Well P-09.

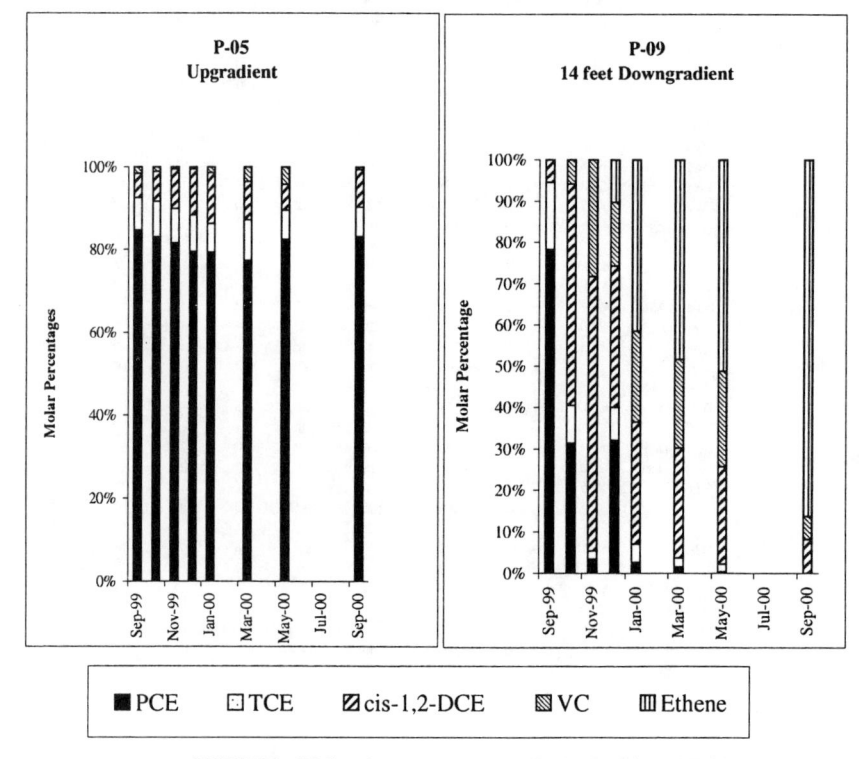

FIGURE 3. VOC molar percentage data for Wells P-05 and P-09.

BIOLOGICALLY-ENHANCED REDUCTIVE DECHLORINATION

George J. Skladany (ERM, Inc., Princeton, New Jersey)
Dick Brown (ERM, Inc., Princeton, New Jersey)
David A. Burns (ERM, Inc., Richmond, Virginia)
Mike Bell (Coats North America, Toccoa, Georgia)
Michael D. Lee (Terra Systems, Wilmington, Delaware)

ABSTRACT: Enhanced reductive dechlorination is being evaluated as a method to treat chlorinated solvent plumes at a site in Virginia containing predominantly tetrachloroethene (PCE), trichloroethene (TCE), *cis*-1,2-dichloroethene (cDCE), and vinyl chloride (VC). Site data suggests that the natural attenuation of VOCs is already occurring, but not at rates sufficient to prevent off-site migration. A microcosm study was used to confirm that site microbial populations could completely dechlorinate PCE at significantly faster rates when provided with additional substrates (lactate or soybean oil) as well as inorganic nutrients. A subsequent field pilot test of enhanced reductive dechlorination was begun in May 2000. A nutrient solution (containing soybean oil, inorganic nutrients, and a surfactant) was injected in a grid pattern in the presumptive PCE source area. Monitoring data from 3 and 7 months post-injection showed a reduction in the concentration of target VOCs, consistent with the action of reductive dechlorination processes. This paper provides site background, results of the bench-scale microcosm study, implementation details for the field pilot test, as well as data generated from the 3 and 7-month post-injection monitoring samples.

INTRODUCTION

A former zipper manufacturing facility in Newport News, Virginia is the site of four distinct ground water plumes covering a total area of approximately 3,720 m². Subsurface contamination is limited to the shallow coastal plain aquifer, the top of which is approximately 1.5 m below grade and the bottom of which is confined by a thick layer of clay at approximately 3 m below grade. The former facility is known to have used tetrachloroethene in both degreasing and dry cleaning operations.

Investigations performed during a transfer of property ownership in 1998 indicated that ground water contamination posed unacceptable risks to potential on-site receptors. The remedial action plan identified specific ground water concentration goals that would avoid or minimize land-use restrictions on the property (i.e., no prohibition of on-site ground water usage or deed restrictions which could prevent disturbance of the subsurface). The risk assessment identified the most critical exposure scenario as the migration of contaminants volatilized from ground water into enclosed building spaces. Even though contaminant vapors were not previously found within buildings, a practical goal of the remediation project was to reduce contaminant concentrations to levels that would preclude this exposure pathway. The former property owner had entered into Virginia's Voluntary Remediation Program (VRP) of the Department of

Environmental Quality (DEQ) with the goal of remediating the site and obtaining a Certificate of Satisfactory Completion from the VRP with few or no restrictions on the use of the property.

Various remediation approaches were considered in 1999. Air sparging was eliminated when geophysical data revealed that the dense and silty soil would prevent air from passing efficiently through any significant volume or depth of soil. Enhancing the existing subsurface biological activity on site was selected as the remediation process most likely to achieve the desired cleanup levels in the most efficient and cost-effective manner.

SITE CONDITIONS

Total ground water VOC concentrations exceeded 1 mg/L in most of the contaminated areas. The two larger plumes (each covering 1,765 m^2) contained total VOCs above 10 mg/L. In the center of the pilot test area where dry cleaning operations were known to occur, PCE levels up to 120 mg/L had been observed, along with total VOC levels of approximately 145 mg/L.

Groundwater data collected since site operations ceased (1990) indicated that the attenuation of various chlorinated solvents was occurring throughout contaminated areas of the site. A natural attenuation evaluation conducted in 1999 supported the belief that biodegradation was playing a dominant role in the natural attenuation of chlorinated compounds for several reasons: (1) The concentration of PCE had steadily decreased since 1995, and the concentrations of various daughter compounds (TCE, cDCE, and VC) had either remained the same or risen slightly over the same time period; (2) The concentrations of various electron acceptors in the ground water were consistent with anaerobic biodegradation taking place. Dissolved oxygen (DO) concentrations were low in areas containing contamination. The well with the highest VOC concentration exhibited a DO of only 0.66 mg/L. Additionally, the DO levels across the central portion of the site, and in the direction of ground water flow, were generally below 1 mg/L. According to Weidemier et al. (1995), reductive dechlorination will not occur if DO concentrations are significantly above 1 mg/L. The DO levels in one of the background wells and the furthest down-gradient well were approximately 2 mg/L. In addition, the concentrations of sulfate and nitrate, the next two most favorable electron acceptors, were also lower in the wells that contain high VOC levels compared to their concentrations in wells with lower VOC levels; and (3) Chloride levels in ground water from wells in the presumptive source area were 2 to 6 times higher than concentrations observed in background and downgradient well waters.

However, the observed natural attenuation rate was not sufficient to prevent the off-site migration of contaminants. Enhanced reductive dechlorination was considered to accelerate biodegradation and control plume migration.

LABORATORY BIOTREATABILITY STUDIES

The purpose of the study was to determine whether the site reductive dechlorination rate(s) could be increased through the addition of organic substrates, inorganic nutrients, or known dechlorinating cultures. The objectives of the biotreatability studies were to: (1) determine if and to what extent the native microbial population could degrade the chlorinated solvents with and without additional substrate; (2) evaluate lactate and soybean oil as potential substrates; and (3) evaluate the potential need for bioaugmentation with a dechlorinating enrichment culture. The microcosm studies were conducted by Terra Systems (Wilmington, Delaware) using aquifer soil and ground water slurries incubated in sealed serum bottles over a twelve-week period.

The microcosms amended with lactate or soybean oil showed degradation of PCE to ethene, as well as the transformation of dichloroethane (DCA) to chloroethane (CA). While the addition of the *Pinellas* dechlorinating culture enhanced the biotransformation rates of PCE and TCE to ethene, bioaugmentation did not appear to be necessary as the indigenous cultures were also able to successfully transform PCE and TCE to ethene. The study demonstrated that microcosms amended with soybean oil showed significantly increased rates of ethene production, as shown Figure 1. Natural attenuation alone seemed insufficient to produce a timely remedy for the site. Based on the positive results of the microcosm study, a field pilot test using soybean oil was approved.

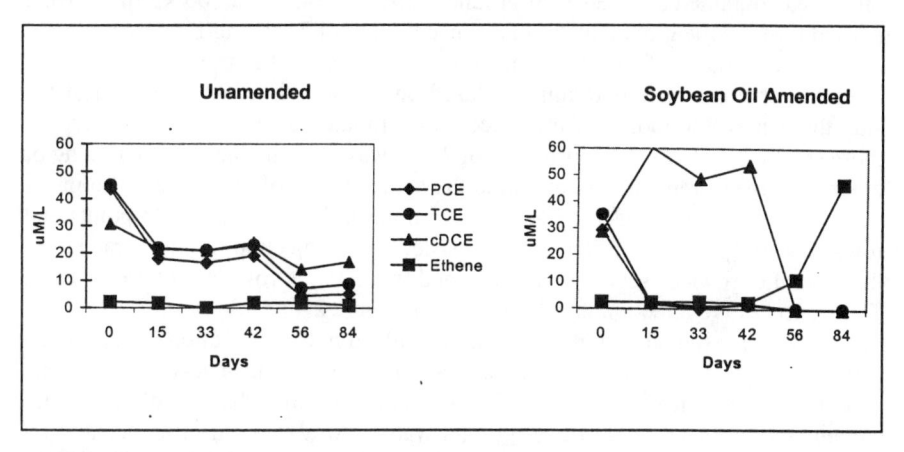

FIGURE 1. Microcosm study results showing changes in selected VOC molar concentrations over the 84-day study.

EMULSIFICATION TESTS

Prior to initiating the field pilot tests, ERM's Remediation Technology Center (West Chester, Pennsylvania) was tasked with formulating a nutrient solution suitable for subsurface injection. The desired solution was designed to create conditions in the ground water following injection similar to those in the soybean oil-amended slurry microcosm (7,000 mg/L soybean oil, 25 mg/L nitrogen, 2.5 mg/L phosphorus, 25 mg/L yeast extract [as carbon], and 500 mg/L sodium bicarbonate). The final solution required emulsification to keep the oil in suspension and had to be non-toxic to indigenous microorganisms. A final mixture consisting of 1:1 volumes of oil and water and a 1% concentration of non-antibacterial surfactant (dish washing detergent) was recommended for field use.

FIELD PILOT TEST

The pilot test was conducted at the location of the former site dry cleaning unit, immediately adjacent to an existing occupied building. This plume covers approximately 150 m^2, and contains the highest VOC groundwater levels observed at the site. Six piezometers were located within the area to be treated (PZ-6 was preexisting, and PZ-8, PZ-11, PZ-12, PZ-13, and PZ-14 were installed prior to the pilot test). Piezometer PZ-10 (preexisting) was located outside of the area to be treated, and was used to provide background concentrations. Figure 2 provides a localized site map showing the locations of the former dry cleaning unit, the existing building, and the piezometers.

Nutrient injections occurred during the last week of April 2000, following notification of state and federal environmental agencies. Permission from the Underground Injection Control (UIC) office of EPA Region III was received, and no UIC permit was required.

Determination of Injection Grid and Volume Requirements. Based on the estimated volume of soil to be treated and assuming a soil porosity of 35%, the pilot area contained 317,184 L of ground water. A total of 2,200 kg of oil were needed to spike the ground water to a concentration of 7,000 mg/L.

The number of injection points and the volume of nutrient solution to be injected per point were determined. Based on previous injections of fluid at the site, the achievable radius of influence was estimated to be 3.7 m. However, a more conservative radius of influence of 2.4 m was used in place of dye tracing or other field measurements to determine the actual radius of influence. The injection grid pattern was offset at each row with 4.3 m between injection points and rows spaced 4.3 m apart. The grid consisted of 33 injection points, arranged in three parallel rows of seven points each and two parallel rows of six points each. This conservative injection grid is depicted in Figure 2.

It was assumed that at each point the added nutrient solution would treat a soil column with a radius of 2.4 m and a height of 1.5 m. The volume of this cylinder is approximately 27 m^3, and at a soil porosity of 35%, the column would contain approximately 9,450 L of ground water. At a soybean oil concentration of 7,000 mg/L, each injection point would require approximately 66 kg of oil. Since the oil had a density of 0.92 kg/L, approximately 72 L of oil were needed per injection point. The emulsified nutrient solution is a 1:1 volume ratio of soybean oil to water, so approximately 144 L of nutrient solution were required per injection point.

Field Production of Soybean Oil Nutrient Solution. Batches of the nutrient injection solution were prepared in 208-L (55-gallon) drums. Each batch contained 76 L of soybean oil, 76 L of water, 1.9 L of detergent, 4.5 kg of sodium bicarbonate, 0.29 kg of yeast extract, and 1.5 kg of plant fertilizer. Solid plant fertilizer containing 20% N and 20% P was used to supply inorganic nutrients. Commercial-grade sodium bicarbonate and laboratory-grade yeast extract were also added as solids. Each batch was mixed using a hand mixer.

Bioremediation Field Case Studies 217

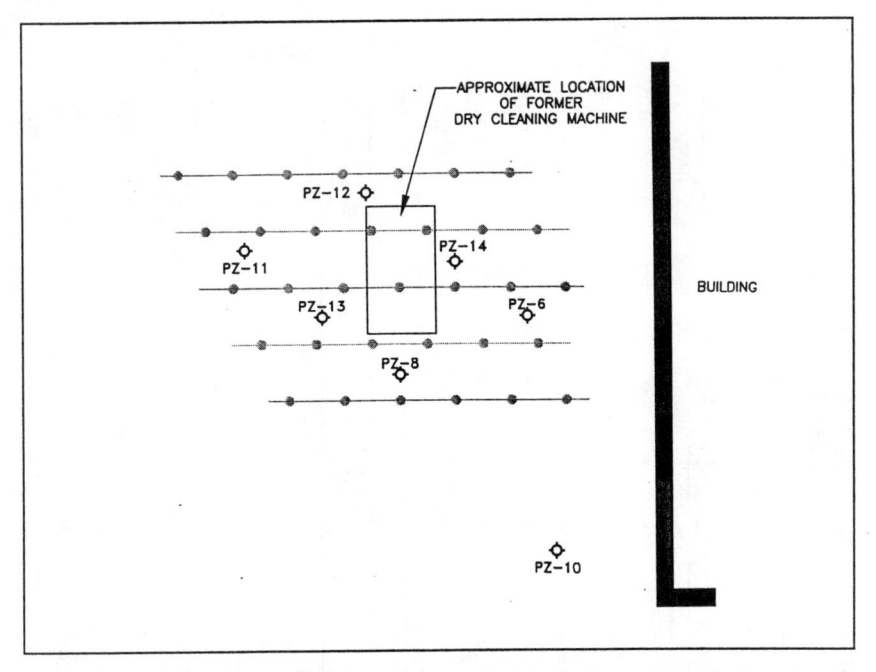

FIGURE 2. Site map showing relative locations of the former dry cleaning machine, piezometers, nutrient injection points, and occupied building.

Injection. A Geoprobe® and a high-pressure (1,000-psi) grout pump were used to inject approximately 150 L of the nutrient solution 3 m below grade at each point. The majority of injection points received the entire 151 L of solution, though at several points lesser volumes were injected due to the emergence of the solution through cracks in the surface concrete and asphalt. The soybean oil solution was observed at the ground surface at distances up to 3.7 m from the actual injection point, well beyond the conservative 2.4 m radius of influence used in designing the injection point spacing.

PILOT TEST GROUND WATER MONITORING

Ground water samples were collected from the seven piezometers one week prior to nutrient injection (T = 0) and at three and seven months post-injection. Table 1 presents the concentrations for the chlorinated VOCs, light gases, and Total Organic Carbon (TOC) detected in samples above their respective analytical detection limits. PZ-10 is a background well for comparative purposes.

Ground water samples were analyzed for chlorinated VOCs using US EPA SW-846 Method 8260B. Only samples collected prior to nutrient injection and seven months post-injection were analyzed for dissolved light hydrocarbon gases (methane, ethane and ethene) using Microseeps Method AM18. In the field, each

Table 1. Enhanced reductive dechlorination field pilot test results. Only those analyte concentrations above the method reporting limit are shown. Ethene, ethane, and methane concentrations were not determined at 3 months.

Compound	PZ-10 (Background) T = 0 (ug/L)	3 Months (ug/L)	7 Months (ug/L)	PZ-6 T = 0 (ug/L)	3 Months (ug/L)	7 Months (ug/L)	PZ-8 T = 0 ug/L	3 Months ug/L	7 Months ug/L	PZ-11 T = 0 ug/L	3 Months ug/L	7 Months ug/L
Methylene Chloride		6			47		500	82		460	320	
1,1-Dichloroethane	6	6		300	598		800	69	27	330	17	130
1,2-Dichloroethane									83		160	
1,1,1-Trichloroethane											0	
Vinyl Chloride	16	1	10	26,000	19,580	27,000	3,600	307	470		9	
1,1-Dichloroethene				87	110	110						
cis-1,2-Dichloroethene	19	10		44,000	46,290	63,000	9,500	242	470			
trans-1,2-Dichloroethene	0			280	492							
Trichloroethene		1			110	43	450	21	40			
Tetrachloroethene				67			1,600	13				
Ethene	3		43	1,328		1,400	1,771		1,900	19		23
Ethane	12		11	52		48	112		51	47		14
Methane	170		38	1,640		920	9,270		5,400	6,610		2,300
Total Organic Carbon (mg/L)	11	7	17	10	10	870	330	245	270	110	813	180

Compound	PZ-10 (Background) T = 0 (ug/L)	3 Months (ug/L)	7 Months (ug/L)	PZ-12 T = 0 (ug/L)	3 Months (ug/L)	7 Months (ug/L)	PZ-13 T = 0 (ug/L)	3 Months (ug/L)	7 Months (ug/L)	PZ-14 T = 0 (ug/L)	3 Months (ug/L)	7 Months (ug/L)
Methylene Chloride		6			282	250	120	18		4,000	2,350	
1,1-Dichloroethane	6	6		120	8	140	460	39	8.9	300	159	140
1,2-Dichloroethane					83				62	2,200	1,050	960
1,1,1-Trichloroethane										100		
Vinyl Chloride	16	1	10	40			140	23.5	21	1,300	3,300	4,600
1,1-Dichloroethene										500	486	300
cis-1,2-Dichloroethene	19	10					340	22	28	14,000	17,510	22,000
trans-1,2-Dichloroethene	0										114	
Trichloroethene		1					160			8,000	17,240	18,000
Tetrachloroethene				150			620			120,000	100,000	51,000
Ethene	3		43	13		50	399		440	623		870
Ethane	12		11	122		21	176		45	113		120
Methane	170		38	19930		7100	9.33		4400	10		3,600
Total Organic Carbon (mg/L)	11	7	17	160	283	560	380	149	160	910	591	470

well was monitored for dissolved oxygen (DO), Oxidation/Reduction Potential (ORP), temperature, specific conductivity, and pH.

INDOOR AIR QUALITY (IAQ) MONITORING

Because of concern that VOCs (especially vinyl chloride) may volatilize from ground water into enclosed building spaces, the pilot test monitoring program included initial IAQ monitoring of the building adjacent to the test plot. Three indoor "background" samples were collected in April prior to the start of the pilot project. Each sample was collected over a 480-minute period and sent for off-site analysis using the OSHA 7 Method (charcoal trap and subsequent analysis using gas chromatography with flame ionization detection). Samples were analyzed for 1,1-DCE, 1,2-DCE, PCE, TCE, and VC. None of the target compounds was detected above its respective detection limit in any of the samples. Similar testing was performed in July at 3 months post-injection, and produced the same non-detectable results for each of the target compounds.

In addition to the stringent monitoring efforts described above, on-site monitoring of vinyl chloride and methane concentrations within the building were conducted weekly over the first three months of the pilot test. Neither vinyl chloride nor methane was ever detected in these tests. Based on the consistent negative findings, the monitoring frequency for these tests was first reduced to bi-weekly and eventually to monthly.

MONITORING OF SUBSURFACE SOIL GASES

Changes in the concentration of oxygen, carbon dioxide, and methane in the seven piezometer casings were determined in the field weekly from 2 May through 3 August, and biweekly through 15 September. The gases from three soil gas survey points installed between the injection grid and the building were also monitored at these times. As the concentration of oxygen within wells decreased, a corresponding increase in carbon dioxide and methane levels was observed. These changes became pronounced in June and July, approximately 2 to 3 months post-injection.

RESULTS

Table 1 presents the Time = 0 (pre-injection) as well as the 3- and 7-month analytical results using standard mass per volume units. While such data are frequently expressed in molar concentrations to aid in determining whether dechlorinated daughter products are produced in stiochiometric amounts, this transformation is not necessary to discern the overall trends in the project data.

The data provide strong support that reductive dechlorination is occurring. PZ-14, the most heavily contaminated ground water, showed a 58% reduction in PCE concentration (from 120,000 to 51,000 µg/L) over the 7 month test, while the concentrations of TCE, cDCE, and VC showed significant increases. Ethene and ethane levels increased 40% and 6%, respectively, during this period.

In PZ-8, PZ-12, and PZ-13, PCE was reduced to non-detectable levels. In PZ-8, the concentrations of TCE, cDCE, and VC were significantly reduced, and ethene levels rose from 1,771 to 1,900 µg/L. In PZ-12, TCE, cDCE, and VC were non-detectable after 7 months, and ethene increased from 13 to 50 µg/L. In PZ-13, TCE was non-detectable after 7 months, cDCE and VC concentrations decreased significantly, and the ethene level rose from 399 to 440 µg/L.

PZ-10 (considered the background well) and PZ-11 contained very low initial levels of chlorinated VOCs. With one exception, the concentration of these VOCs continued to decline during the field pilot test.

While PZ-6 had non-detectable levels of PCE and only low levels of TCE at T = 0, high concentrations of cDCE (44,000 µg/L) and VC (26,000 µg/L) were present. While the cDCE level increased 43% over the test, the VC level was stable. Ethene and ethane levels were also essentially unchanged. The concentrations of oxygen, carbon dioxide, and methane in this casing from May to September showed little change, in contrast to other piezometers where oxygen concentrations decreased and carbon dioxide and methane levels increased. The TOC concentration remained at only 10 mg/L over the first three months of the test, before increasing to 870 mg/L. It is possible that the nutrient solution may not have reached the vicinity of this well immediately after injection, and that reductive dechlorination in this area is delayed.

Except for PZ-6, the piezometer casing gas monitoring data indicated that stable anoxic conditions had been established in the test area, with correspondingly elevated methane and carbon dioxide levels. Methane generation is attributed to the anaerobic biodegradation of the added soybean oil.

Drinking water standards have been met at some of the monitoring points. However, the test has been extended to more fully evaluate the most heavily contaminated parts of the plume. Data from samples to be collected in June 2001 (14 months post-injection) will be used to determine whether: (1) full-scale implementation of enhanced reductive dechlorination can be expected to achieve the drinking water standards; (2) additional soybean oil injections may be required to successfully treat the most heavily contaminated areas of the site; and (3) some restrictions on ground water or land-use should be considered as the entire site is remediated.

CONCLUSIONS

Bench- and pilot-scale tests were used to show that indigenous microorganisms, when sufficiently stimulated by the addition of soybean oil and inorganic nutrients, could successfully dechlorinate PCE and its daughter products to ethene. While drinking water standards have been attained at 7-months post-injection at some monitoring points, additional sampling in June 2001 (14 months post-injection) will be evaluated before a final decision is made to use enhanced reductive dechlorination as a full-scale treatment technology at this site.

REFERENCES

Wiedemier, T., M. Swanson, D. Moutoux, E. Gordon, J. Wilson, B. Wilson, D. Kapbell, P. Hass, R. Miller, J. Hansen, and F. Chapelle. 1995. *Technical Protocol for Evaluating Natural Attenuation of Chlorinated Solvents in Ground Water.* National Risk Management Research Laboratory, US EPA, Cincinnati, Ohio.

ENHANCED BIOREMEDIATION IN CLAY SOILS

Ms. Zahra M. Zahiraleslamzadeh (FMC Corporation, San Jose, CA)
Jeffrey C. Bensch (GeoTrans, Rancho Cordova, CA)

ABSTRACT: This case study evaluates the full-scale in-situ application of an electron donor to enhance biodegradation of chlorinated solvents in clay soils and groundwater beneath an active light industrial property in a congested urban setting. A pilot test was successful by varying degrees in three test areas as shown by the production of biodegradation daughter products and microbial end products. The results of the pilot test were used to design and implement the full-scale remediation with minimal disturbances to ongoing business operations. After twelve months of remediation monitoring, the trichloroethylene (TCE) is degrading to vinyl chloride (VC) and ethylene.

Organic acid, oxidation-reduction potential, and hydrogen monitoring results indicate that the microbial environment for reductive dechlorination was improved throughout the remediation area. TCE concentrations were significantly reduced, while cis-1,2 dichloroethylene concentrations increased then declined during the twelve months following the full-scale application. Elevated sulfate concentrations also declined coincident with the reductions in TCE concentrations. VC and ethylene concentrations, however, have increased significantly. It is expected that VC will continue to degrade to ethylene, based on the observed increases in ethylene and other microbial end products. Additional applications of the electron donor are anticipated to complete the remedy. If necessary, a sequential in-situ aerobic remediation process may be required to fully degrade the VC.

INTRODUCTION

The biological reductive dehalogenation process of chlorinated solvents, such as tetrachloroethylene (PCE) and trichloroethylene (TCE), is an accepted viable groundwater remediation process. Various enhancements are available to stimulate biological activity and accelerate the dehalogenation process. Applying these enhancements to the subsurface for effective remediation can be difficult and uncertain. This paper presents the results of a full-scale remediation involving the injection of hydrogen release compound (HRC®) into a silty clay soil to stimulate biodechlorination of TCE in groundwater.

The site is a 4.1-acre, relatively flat, property with a 76,000 square-foot light-industrial retail building and paved areas in a congested commercial/light industrial neighborhood. The site was used as agricultural land before the 1960s, then for various heavy manufacturing purposes through 1988. The building is currently leased to various light-industrial tenants.

Site soils are homogenous silty clays from ground surface to a depth of approximately 45 to 50 feet. A gravelly sand unit, approximately 30 to 35 feet thick, underlies the silty clay. The depth to groundwater is approximately 7 to 10 feet below ground surface in wells screened in the silty clay, and the groundwater flow velocity is approximately 10 feet per year.

The site is impacted with volatile organic compounds (VOCs) in soil and shallow groundwater in the northeastern corner of the property (Figure 1).

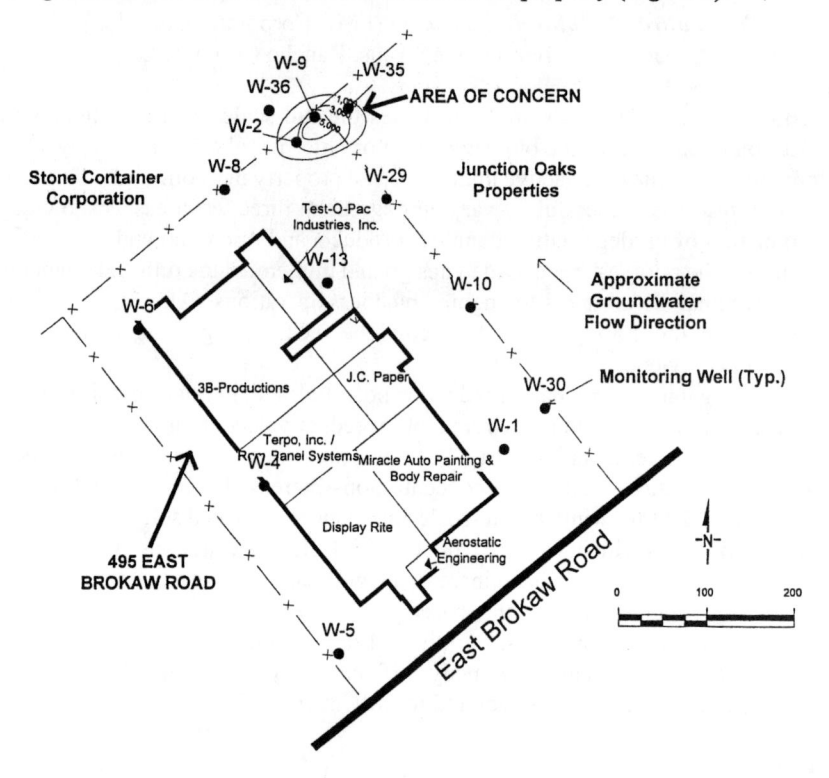

FIGURE 1. Site Features

Anaerobic biodegradation of VOCs occurs in environments free of oxygen, where the parent chlorinated compound is progressively dechlorinated into daughter products. The process at this site is the dechlorination of TCE into subsequent daughter compounds: cis-1,2 dichloroethylene (*cis*-1,2-DCE), vinyl chloride (VC), and ethylene. Microorganisms mediate this process using chlorinated compounds as electron acceptors and a source of hydrogen as the electron donor.

In the anaerobic biodechlorination process, the *cis*-1,2-DCE isomer is produced preferentially to the *trans*-1,2-DCE and 1,1-DCE isomers (Wiedemeier, et.al., 1996). As such, evidence for reductive dehalogenation of TCE can be obtained by observing the formation of *cis*-1,2-DCE in excess of *trans*-1,2-DCE and 1,1-DCE, e.g., a *cis/trans* ratio greater than one. Increasing concentrations of VC and the microbial end product ethylene suggest the process is continuing toward completion.

TCE is among the most susceptible VOCs to reductive dechlorination because it is well oxidized (i.e., contains more chlorine atoms than hydrogen atoms). VC is the least susceptible to reductive dechlorination because it is the least oxidized of the daughter product compounds (Wiedemeier, et.al.,1996). Anaerobic destruction of VC has been observed in study cases (Cornuet, et.al., 2000), although it is more easily degraded in aerobic environments (Morse, et.al., 1998). The slower dechlorination of VC than TCE can result in an accumulation of VC.

Intrinsic biodegradation testing and evaluations were conducted at the site in 1998 using methods similar to the Air Force Center for Environmental Excellence (AFCEE) protocols (Wiedemeier, et.al.,1996). Field parameters, organic and inorganic parameters, microbial end products, microbial community structure, and dechlorinated daughter products were evaluated. The dissolved oxygen, pH, and oxidation reduction potential (ORP) indicated that the site was anaerobic, slightly reducing, and suggested that iron reduction may be the dominant redox process. The total organic carbon (TOC) concentrations and biomass as measured by phospholipid fatty acids (PLFA) analyses indicated the site could support a microbial population suitable for anaerobic reductive dechlorination. The nitrate and sulfate concentrations were elevated with respect to desirable levels for reductive dechlorination. Sulfate concentrations ranging from 130 to 320 milligrams per liter (mg/L) indicated that sulfate reducing bacteria may compete for available hydrogen and hinder the reductive dechlorination processes. The dechlorination daughter products indicated formation of cis-1,2-DCE in groundwater at the site. The ratio of cis/trans-isomers indicated that this formation was likely due to a reductive dechlorination process. Microbial end-products were not present during the initial investigations. This indicated that reductive dechlorination was not continuing through completion. In summary, the 1998 investigation indicated that biodegradation of VOCs occurred in the past, but it was very slow or in a dormant stage at the time of the evaluation.

A pilot study was implemented to evaluate the reductive dechlorination process in an isolated area at the site (Zahiraleslamzadeh, 2000). HRC was applied in three adjacent areas with varying degrees of TCE impacted groundwater. Following HRC application, significant reductions of TCE concentrations were observed with the accompanied increases in concentrations of daughter products and microbial end-products. The results of the pilot study indicated that reductive dechlorination could be stimulated at the site and a full-scale HRC application was warranted.

MATERIALS AND METHODS

The scope of the full-scale remediation included injecting approximately 12,000 pounds of HRC through 103 direct-push points in the northeast corner of the property and along the northern property boundary to remediate an apparent TCE source area and to provide a barrier to downgradient TCE migration, respectively. As shown in Figure 2, the injection points were located on a 5-foot by 10-foot grid in most areas and a 5-foot by 5-foot grid in the apparent TCE source area.

The HRC application was conducted during 10 days of field work between May 16 and May 31, 2000. HRC injections were completed using a top-down approach starting approximately eight feet below ground surface and ending at 28 feet below ground surface. Top-down injection was accomplished using a steel rod perforated near the bottom four feet of pipe. After the perforated rod was driven to

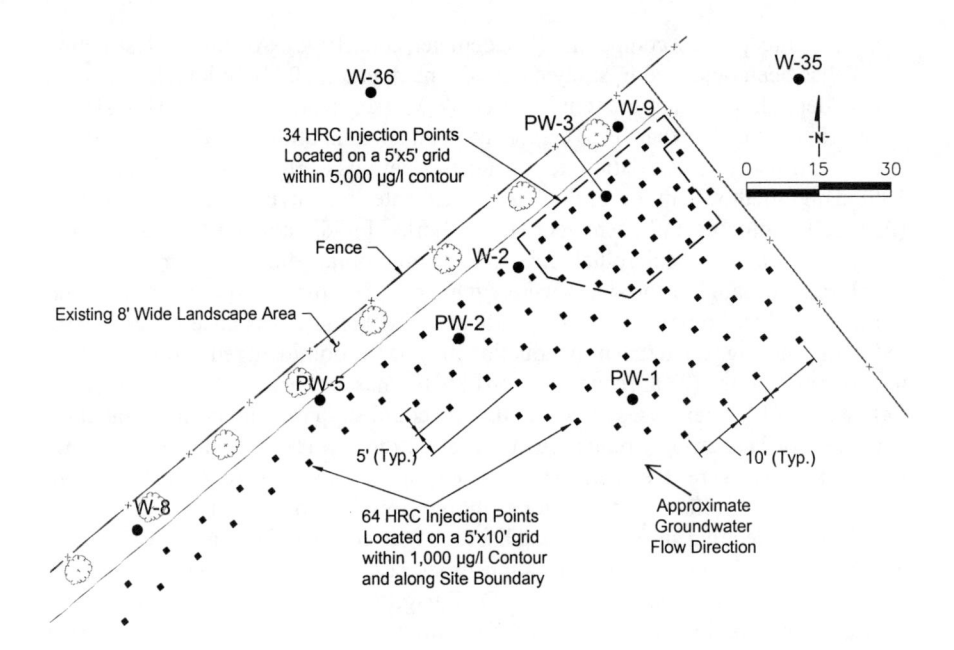

FIGURE 2. HRC Injection Locations and Monitoring Wells

the desired depth, it was coupled to a high pressure hose and pump. A specific amount of HRC was injected and the rod was then driven to the next depth interval. Approximately 12 gallons of HRC were applied to each injection point.

Groundwater samples for baseline testing were collected prior to HRC injection from monitoring wells. As shown on Figure 2, wells PW-3 and W-2 are located in the area where TCE concentrations are approximatley 5,000 micrgrams per liter (ug/L), and wells PW-2 and W-9 are close by where TCE concentrations are approximately 3,000 ug/L. Well PW-1 is in the upgradient area of impacts and wells PW-5 and W-8 are along the property boundary where TCE concentrations are less than 1,000 ug/L. Wells W-35 and W-36 have TCE concentrations in the 2,000 ug/L and 200 ug/L ranges, respectively. These wells are located outside the HRC application area.

Monitoring was conducted every other month for six months from June through November 2000, then at three month intervals in February and May 2001 (5 events after HRC injection).

RESULTS AND DISCUSSION

Time series trend evaluations for TCE, cis-1,2-DCE, VC, and ethylene provide an indication of the remediation effectiveness at each monitoring well. A time series graph for well PW-1, and selected monitoring data for monitoring wells PW-1 and W-9 are presented to illustrate the remediation performance in areas of high and low biodechlorination activity.

High Biodechlorination Activity Wells. Significant TCE biodechlorination was observed at wells PW-1, PW-2, PW-3, and W-2. The TCE and daughter product concentrations in well PW-1 changed dramatically by July 2000, when the level of TCE dropped an order of magnitude. A corresponding rise in the concentration of daughter products *cis*-1,2-DCE and VC further indicated enhanced TCE breakdown. By September 2000, the concentration of daughter product *cis*-1,2 DCE also began to decrease. The increase of ethylene concentrations in July, September, and November 2000 indicated that the TCE dechlorination process was continuing through VC to completion. Ethylene concentrations declined, however, in February and May 2001.

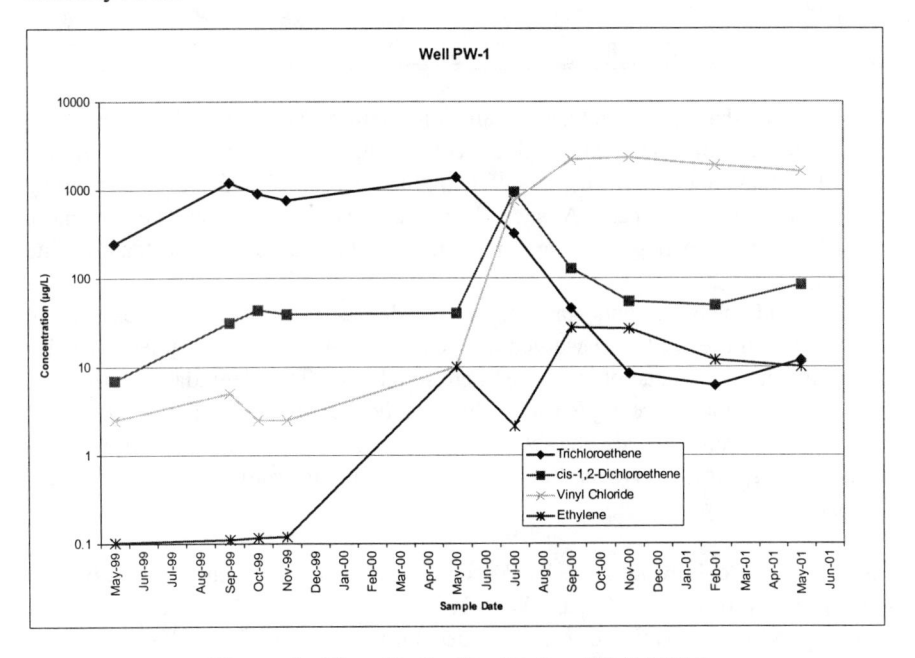

Figure 3. Time Series Evaluation Well PW-1

The hydrogen concentrations in some wells were extremely high during the six months following HRC injection. Well W-2 contained 153,248 nano-moles (nM) and PW-3 contained 1,161 nM in July 2000; PW-2 contained 6,600 nM in November 2000. These elevated hydrogen concentrations together with the initially low ORP measurements, indicate that methanogenic conditions existed through May 2001 in many of the high activity wells. This is further supported by increasing methane concentrations in these wells. Hydrogen concentrations have declined, and the corresponding dechlorination performance versus methane production is being monitored.

Sulfate concentrations dropped to less than 10 mg/L in some wells coincident with high hydrogen concentrations and the significant declines in TCE concentrations. This observation is consistent with Yang and McCarty (1998), where

Table 1. Selected Data for Well PW-1

Parameter	Units	May 2000	July 2000	Sep 2000	Nov 2000	Feb 2001	May 2001
TCE	ug/L	1400	320	46	8	6	12
cis-1,2 DCE	ug/L	40	950	130	55	49	84
Vinyl Chloride	ug/L	<20	760	2200	2300	1900	1600
Ethylene	ug/L	<20	2.15	27.4	27	12	10
Sulfate	mg/L	273	127	23	30	18	91
Methane	ug/L	34	93.2	117	4100	7000	6000
Hydrogen	nM	1.75	12.3	4.05	5.6	3.2	1.8
ORP	mV	-89	-253	-30	135	-165	-258

competing methanogenic and dehalogenation reactions were studied with respect to hydrogen concentrations. Although the competing reactions may be an inefficient use of the available hydrogen, sulfate reduction has eliminated a competing biological reduction process. As such, the remaining hydrogen should become more available for methanogenises and reductive dechlorination rather than sulfate reduction.

Field measurements gave further evidence of enhanced biodegradation occurring at these wells. Dissolved oxygen readings were typically less than 1.0 mg/L, indicating an anaerobic environment suitable for TCE degradation, while the pH measurements were typically within the desirable range of 5 to 9. ORP measurements remained less than 50 millivolts (mV) through May 2001 in these high activity wells, except in November 2000 where a faulty instrument reported elevated values.

Low Biodechlorination Activity Wells. Low biodechlorination activity was observed in wells PW-5, W-8, and W-9. Results from well W-9 show concentration trends with little indication of reductive dechlorination of TCE or sulfate reduction until November 2000; although cis-1,2-DCE, VC, ethylene, and methane concentrations increased slightly from May through November 2000. Substantial reductions in TCE and sulfate concentrations were observed in February 2001; followed by an apparent rebound effect observed in May 2001. This may be indicating enhanced biodegradation of TCE at well W-9, while the TCE concentrations are maintained through desorption of TCE from the soil matrix, or this rebound effect could be migration of dechlorination by products into the vicinity of well W-9.

Hydrogen concentrations at well W-9, and other low biodechlorination activity wells, were elevated above 2 nM, but did not achieve the levels or sustain the elevated concentrations for as long of duration as in wells PW-2, PW-3, and W-2. The TCE reduction in W-9 occurred from November 2000 through February 2001 when hydrogen concentrations were significantly lower than their peak concentration of 123 nM in September 2000. This is also consistent with the findings of Yang and

McCarty (1998) where the dechlorination bacteria were shown to compete best against methanogens within a hydrogen concentration range of 2 to 11 nM. The TCE reduction also correlates closely with the increase in methane concentration in February 2001, indicating that the reductive dechlorination in the vicinity of W-9 may be a cometabolic reaction under methanogenic conditions.

Table 2. Selected Data for Well W-9

Parameter	Units	May 2000	July 2000	Sep 2000	Nov 2000	Feb 2001	May 2001
TCE	ug/L	3,620	4,400	4,500	4,500	410	600
cis-1,2 DCE	ug/L	81	230	280	1,000	1,300	2,800
Vinyl Chloride	ug/L	17	81	100	350	1,100	2,000
Ethylene	ug/L	<20	0.21	1.4	9.1	16	60
Sulfate	mg/L	161	195	200	189	24	72
Methane	ug/L	108	175	210	310	1,900	1,600
Hydrogen	nM	1.71	2.13	123	4.5	4.7	1.2
ORP	mV	145	-28	50	197	-88	-242

Field measurements in these low activity wells indicate that the environment was capable of supporting enhanced biodegradation. DO measurements were close to 0.5 mg/L, and ORP measurements were below 50 mV. The ORP at well W-9 dropped from 145 during the baseline monitoring to -28 mV in July 2000. The ORP remained below 50 mV throughout the year except for the anomalous reading in November 2000.

Given historical groundwater flow rates less than 10 feet per year, wells W-35 and W-36 lie well outside of the area of influence from the full-scale HRC implementation. As expected, wells W-35 and W-36 did not show evidence of enhanced bioremediation from the addition of HRC. These wells have been used successfully as control wells to evaluate the remedy effectiveness.

CONCLUSIONS

The HRC application enhanced biological reductive dechlorination of TCE in all of the monitoring wells within the HRC application area. The enhanced biodechlorination effectiveness is shown to be a function of the amount and proximity of HRC addition to a monitoring well. A high volume of HRC injected close to a well resulted in greater evidence of the dechlorination of TCE. Wells PW-1, PW-3, and W-2 are located in the highest density of HRC injection locations and these wells exhibit the greatest evidence of biodechlorination. The limited biodechlorination at well W-9 appears to be due to a delayed development of the reductive dechlorination processes.

The accumulation of VC in the site groundwater is a recognized concern. It is expected that VC will accumulate through the dechlorination of cis-1,2 DCE, which is faster than the dechlorination of VC to ethylene. The elevated concentrations of VC should continue to decrease with time, as illustrated in well PW-1. The VC concentrations at other wells also appear to have peaked, while the ethylene concentrations are increasing in most wells. If VC does not biodechlorinate to ethylene as anticipated, an aerobic environment may be needed to complete the destruction of this daughter product. Further monitoring efforts are ongoing to determine the long term effectiveness of the HRC application.

REFERENCES

Cornuet, Thomas S., C. Sandefur, W. Eliason, S. Johnson, C. Serna, 2000. *Accelerated Bioremediation of Chlorinated Compounds in Groundwater - Aerobic and Anaerobic Bioremediation of cis-1,2-Dichloroethene and VC*, International Conference on Remediation of Chlorinated and Recalcitrant Compounds, 2nd:2000, C2-4, Battelle Press, 2000.

Morse, Jeff J., Alleman, B.C., Gossett, J.M., Zindler, S.H., Fennell, D.E., Sewell, G., Vogel, C.M., 1998. *Draft Technical Protocol - A Treatability Test for Evaluating the Potential Applicability of the Reductive Anaerobic Biological In Situ Treatment Technology (RABITT) to Remediate Chloroethenes*, Department of Defense Environmental Security Technology Certification Program, 1998.

Wiedemeier, Todd H., M. Swanson, and D. Moutoux, E. K. Gordon; Drs. J. Wilson, B. Wilson, and D. Kampbell; J. Hansen and P. Haas; Dr. F. Chapelle, 1996. *Technical Protocol for Evaluating Natural Attenuation of Chlorinated Solvents in Groundwater*. Air Force Center for Environmental Excellence, November 1996.

Yang, Yanru, and P. McCarty, 1998. *Competition for Hydrogen within a Chlorinated Solvent Dehalogenating Anaerobic Mixed Culture*, Environmental Science & Technology, 1998.

Zahiraleslamzadeh, Zahra M., J. Bensch, 2000. *Enhanced Bioremediation Using Hydrogen Release Compound (HRC^{TM}) in Clay Soils*, International Conference on Remediation of Chlorinated and Recalcitrant Compounds, 2nd:2000, C2-4, Battelle Press, 2000.

AEROBIC AND ANAEROBIC BIOREMEDIATION OF 1,1-DCE AND VINYL CHLORIDE IN GROUNDWATER

John R. Larson, P.G., M.P.H. and *Vincent J. Voegeli, P.G.*
(TranSystems Corporation, Kansas City Missouri)

ABSTRACT: A three-tiered remediation approach is being used for successful field treatment of a 1,1-dichloroethene (1,1-DCE) groundwater plume. The plume is characterized by a chemical source area subject to anaerobic conditions (<1.0 mg/L dissolved oxygen [DO]) with high levels of dissolved phase 1,1-DCE (>8.0 mg/L); an interior plume area of moderate aerobic (<4.0 mg/L DO) and chemical concentrations (<1.0 mg/L DCE); and a 300-foot perimeter plume area of high aerobic (>4.0 mg/L DO) and low chemical concentrations (<0.01 mg/L DCE). Previous studies from 1998 and early 1999 indicated that downgradient of the source area, elevated DO levels and intrinsic aerobically-degrading microbes appeared to promote natural attenuation processes of chemical degradation of 1,1-DCE to non-toxic endpoints such as carbon dioxide in the absence of vinyl chloride. Low, but persistent, concentrations of 1,1-DCE along the plume perimeter indicated the rate of plume movement exceeded the rate of natural attenuation. In late 1999 and mid 2000, two groundwater circulation wells were installed near the source area, in part, to facilitate aerobic transformation of 1,1-DCE to its primary daughter product carbon dioxide. Also in late 1999, perimeter areas of the plume with high levels of DO and low 1,1-DCE were stimulated using a nutrient mixture of Oxygen Release Compound (ORC©) and simple table sugar. Pilot studies conducted in 1998 indicated the nutrient mixture, when injected at perimeter wells, provided a carbon substrate (sugar) for microbial activity to degrade the 1,1-DCE while the ORC offset the reductive dechlorination transformation process of 1,1-DCE to vinyl chloride. Since late 1999 the in-situ bioremediation injection technique at the plume perimeter has stimulated biological activity without the creation of vinyl chloride. While this was an effective technique, excess sugar applications over time apparently inhibited aerobic 1,1-DCE degradation by driving the aquifer more anaerobic. Groundwater concentrations at the plume source have diminished with a corresponding increase of the daughter product vinyl chloride. Monitoring of plume geochemistry and geometry indicate vinyl chloride is naturally degraded by the groundwater system aerobic microbes downgradient from the source area. Closure of the site subject to state risk assessment requirements is expected in 2002.

INTRODUCTION

Historical releases of chlorinated solvents at a manufacturing facility in Houston, Texas impacted shallow groundwater, creating a plume that extends approximately 500 feet in length and 300 feet in width (Figure 1). Groundwater sampling since 1994 indicated that chlorinated hydrocarbons, predominantly 1,1-dichloroethene (DCE), tetrachloroethene (PCE), trichloroethene (TCE), and vinyl chloride (VC) exceeded Texas-regulated groundwater cleanup criteria. Remedial implementation of groundwater restoration to prescribed Texas Natural Resource Conservation Commission (TNRCC) Voluntary Cleanup Program (VCP) cleanup criteria commenced in 1999 (Phase I remediation) after a pilot study of 1998 indicated enhanced biodegradation at the site was feasible by stimulating indigenous microbes in the groundwater.

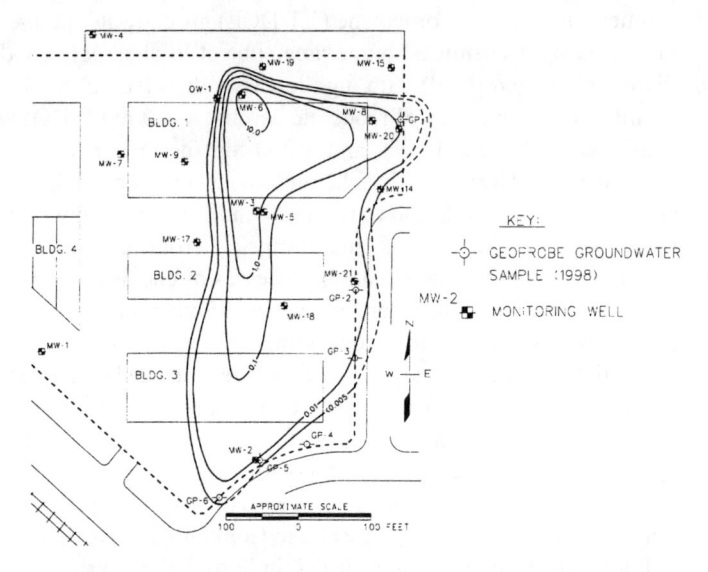

FIGURE 1. Pre-remedial action 1, 1-DCE plume (mg/L).

A three-tiered remediation approach for meeting the TNRCC cleanup objectives included:
- Sparging near the source to remediate source area chemical concentrations;
- Bioremediation at the property boundary via oxygen and nutrient injection to ensure no off-site migration of chemicals; and,
- Periodic monitoring and testing of wells to ensure compliance of no plume growth.

Site Description. The site is located in a commercial-industrial area of Houston, Texas. A groundwater plume exists beneath the site (see Figure 1). The primary chemical of concern (COC) is 1,1-DCE. The chemical 1,1-DCE or DCE is located at several monitoring wells and is present at relatively high concentrations in the groundwater near the suspected source area, MW-6, and at very low concentrations and/or non-detect at property boundary locations. Other volatile organic compounds (VOCs) have been detected in groundwater but at relatively low concentrations. By 1997, the lateral and vertical extent of subsurface impacts were delineated. In 1998, a pilot testing program was conducted to assess proposed methods for treating the groundwater plume.

Site hydrogeologic features include the presence of an extensive transmissive sand beginning at approximately 25 feet below ground surface (bgs) extending to approximately 50 feet bgs in several downgradient, exterior portion plume wells. The sand thins upgradient toward the source well, MW-6. Based on correlation of well data, including gamma log results of site stratigraphy from Phase I injection points, it appears the transmissive sand is part of a fluvial system that regionally thins toward the northwest across the site. Hydraulic conductivities range from 1.01E-03 to 2.69E-04 cm/sec based on slug and tracer tests.

Plume geometry and biodegradation processes on the site are strongly associated with the hydrogeologic regime for this site. Characteristics of the transmissive system include a more robust flow regime along the eastern portion of the site. These physical characteristics influence the chemical processes as evidenced by the elevated dissolved oxygen levels and aerobic nature of the plume along the eastern plume boundary.

The groundwater plume is distributed in an elongated trend along the groundwater flow direction from north (upgradient) to south (downgradient), as depicted in Figure 1. The toe of the plume, or furthest measured downgradient extent, is approximately 500 feet south of MW-6 at MW-2. The concentration of DCE at MW-2 was 0.084 mg/L in February, 2001. Other property boundary well concentrations of DCE in February 2001 were 0.066, 0.091, ND, and ND (mg/L) at OW-2, OW-3, OW-4, and OW-5, respectively.

The area of highest COC concentrations was located at MW-6, the suspected chemical source area. DCE maximum concentration was 15.5 mg/L at MW-6 in March, 2000. In January 2001 the DCE concentration at MW-6 was 5.70 mg/L. Other COCs detected at MW-6 included TCE, PCE, 1,1-DCA, total-1,2-DCE (cis, trans), 1,1,2-TCA, 1,2-DCA, and VC. Vinyl chloride concentrations were not detected at property boundary locations, and have been historically absent outside of the source region of the plume. The lack of VC across the site, particularly at downgradient well locations is due to the aerobic conditions that exist in the plume outside of the source region.

REMEDIAL SYSTEM DESCRIPTION

The site was screened for the COCs and their degradation products in 1997. At that time it became apparent that the site exhibited unique characteristics for decay of DCE under aerobic conditions in the exterior portions of the plume and anaerobic conditions at the plume interior. Laboratory and in-situ field chemical and geochemical analyses indicated the presence of biodegradation processes occurring in the groundwater. Table 1 illustrates the correlative biodegradation indicators at this site noted prior to the 1998 pilot test study. These indicators were evaluated further in subsequent geochemical testing during the remediation phases of 1999 to 2001 (USEPA-ORD, 1998).

Table 1 reflects the footprint of the plume as anaerobic to aerobic indicators transition from the plume source region to exterior portions of the plume. The table is a conceptual model of plume geochemistry that is based on data collected from wells at various portions within and outside of the plume.

TABLE 1. Observed biodegradation indicators.

Plume Area	1,1-DCE Conc. (ppb)	Depleted NO_3	Elevated Fe^{2+}	Depleted SO_4	Elevated Methane	Elevated Alkalinity ($CaCO_3$)	Redox Status
Upgradient (Outside)	ND	NO	NO	NO	NO	NO	Aerobic
Upgradient (Outside)	ND	NO	NO	NO	NO	NO	Aerobic
Cross-gradient (Outside)	ND	**YES**	NO	NO	NO	NO	Mildly anaerobic
Down-cross gradient (fringe)	ND	NO	NO	NO	**YES**	NO	Mixed
Source (Inside)	6,100	**YES**	**YES**	**YES**	**YES**	**YES**	Anaerobic
Downgradient closest to source (Inside)	375	NO	NO	**YES**	**YES**	NO	Anaerobic
Downgradient (Inside)	27	NO	NO	**YES**	**YES**	**YES**	Anaerobic
Downgradient (Inside)	19	NO	NO	NO	**YES**	NO	Mixed

Pilot Study – 1998. A pilot test was conducted to determine the ratio of nutrients (or carbon loading) to oxygen that was needed for plume containment and ultimate plume shrinkage. A mixture of simple table sugar and ORC was injected and monitoring of the groundwater was performed to assess the ratio of bacteria to oxygen required in the aquifer to enhance bioremediation. The injection and monitoring process covered a 17-day study period in two wells, MW-2 and MW-8, in Fall, 1998. The sugar plus oxygen mixture was injected in MW-2 and MW-8 every odd day beginning Day 1. Groundwater samples from MW-2 and MW-8 were collected every even day beginning Day 2 and submitted for laboratory analyses for biological oxygen demand (BOD), chemical oxygen demand (COD), 1,1-DCE, VC, and field measured for dissolved oxygen (DO) and redox potential (Eh).

The pilot study hypothesis was that DCE degraded more readily in an oxygen-rich environment; however, it occurred to us that attention should be paid to the amount of carbon in the system. The hypothesis was that a sufficient amount of carbon is required to sustain microorganisms such that oxygen can be utilized. Reliance on the contaminants alone to provide carbon in an oxygen-rich environment was seen as limiting. Consequently, sugar additions were designed to rectify this limitation. As a result, the technique was to inject the appropriate amount of sugar and ORC such that the microbial population could flourish and bulk up on the available sugar, then consume the DCE, thereby reducing DCE concentrations. ORC is a formulation of magnesium peroxide that slowly releases molecular oxygen when hydrated. Ultimately the objective is to keep the sugar readily available to stimulate the microbes in presence of sufficient oxygen to degrade DCE.

The pilot test results indicated that DO increased significantly via ORC injection for the first five days. Carbon doses balanced oxygenation and DO levels stabilized accordingly over the study period. DCE substantially diminished from 0.014 mg/L to non-detect in MW-2 during the test period. 1,1-DCE concentration in MW-8 declined by 50% in week 1, then stabilized. The test indicated that exterior plume regions undergo some attenuation via aerobic decay processes. High DO and colony forming units (CFUs) at these locations indicated that bacteria is stimulated with addition of an appropriate mixture of oxygen and carbon. Most importantly, the DCE concentration declines were not accompanied by detections of VC. Therefore, the pilot test indicated that given the appropriate injection mixture, plume containment can be achieved at the property boundary by decreasing DCE without the formation of stable VC.

Bacterial plate counts measured as CFUs were submitted from three groundwater samples MW-2, MW-8, and MW-18; and two soil samples from T-1; one at the capillary fringe and one in the saturated zone to identify heterotrophic and DCE degrading bacteria. Results indicated populations of facultative anaerobes in the source region transitioning to aerobes at the plume exterior.

Phase I – 1999-2000. A groundwater circulation well (GCW), S-1, was chosen for installation near the source (MW-6) location based on the physical and geologic properties of the site, relative to successful applications of this technology reported elsewhere (Mueller et al, 1999; USEPA, 1998). The coaxial system uses pressure from a compressor to circulate groundwater within the wellbore that allows upward air flow. The air bubbles rise within the well via an air-lift pump to an uphole carbon filter where treated air is vented. The system is an effective technology for removal of dissolved phase groundwater VOCs which are transferred from the liquid to the gas phase via the circulation process. The process also serves to stimulate aerobic bioremediation via oxygenating groundwater in the radius of influence of the GCW. Soil air VOCs in the capillary fringe are also treated.

The bioremediation/injection procedure was conducted using direct push technology (i.e., Geoprobe truck and equipment). Three distinct injection arrays were installed immediately upgradient to existing wells MW-2, MW-20 and MW-21 (see Figure 2). The injection arrays were approximately four feet upgradient from each monitor well. Each array consisted of several boring injection locations spaced approximately 12-feet apart. For each array, the vertical depth injection interval was measured prior to the injection procedure using a slim-hole gamma logging tool to determine the maximum permeability interval for injection. A top-down injector rod was used to introduce the oxygen and nutrient slurry into the groundwater. After pumping a proportionate amount of chemical into the formation, the probe rod was driven another two feet and the process repeated. Based on pilot test results, it was determined that approximately 30 pounds of ORC and 1 pound of sugar should be injected per boring over the saturated interval.

Phase II – 2000-2001. Implementation of Phase II bioremediation included the following: 1) Installation of additional monitoring wells to provide permanent injection and observation well points; 2) Periodic bio-slurry injections at designated injection wells to provide site-specific calculated slurry mixtures to facilitate biodegradation of DCE, and, 3) installation of a second (sparging) GCW, S-2, was installed south of the Building 1 source area (see Figure 2).

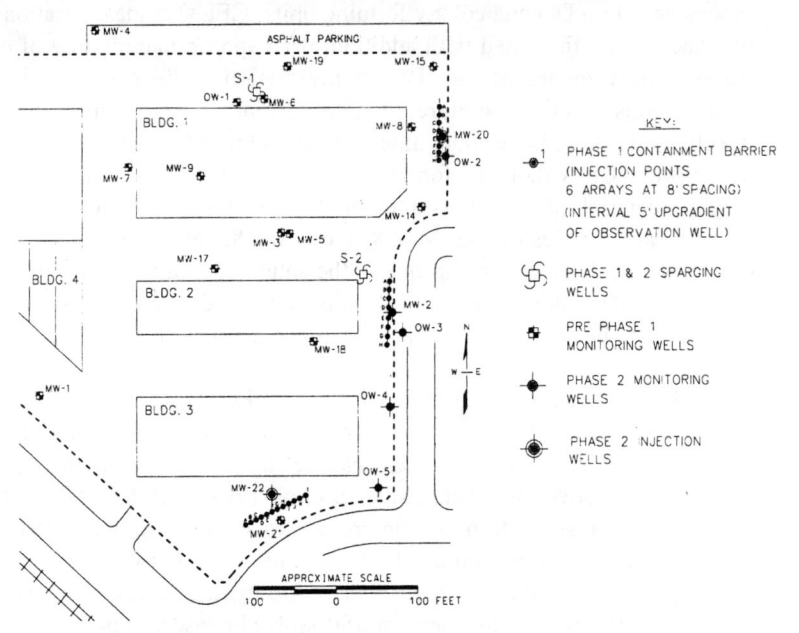

FIGURE 2. Phase I and Phase II remedial activities.

Based on the pilot test, during Phase I a 30:1 ratio of ORC to sugar was introduced into the groundwater via direct push probes at injection arrays parallel to the property boundary. Observations of 1,1-DCE decay occurred in the first 30 to 60 days, then diminished, presumably as oxygen was depleted. The microbes were stimulated by the carbon substrate, which provided a food source for the microbes to feed upon thereby increasing DCE degradation while the food source was present. Once the sugar was exhausted, DCE degradation diminished markedly. Because the ORC has a six-month time released duration of effectiveness, the aerobic plume regions became potentially saturated with oxygen, thus, DO levels were significantly elevated post-injection.

In order to optimize the 1,1-DCE degradation process at property boundary locations, periodic injections in monitoring wells were conducted during Phase II. Also, the injection ratio of ORC:carbon was much more aggressive than the Phase I injection ratio. ORC and sugar slurry was injected every 60 days at MW-20 and MW-22. The ratio of ORC:carbon slurry mix was dependant upon the measured DO level response in observation wells, MW-2 and OW-2.

Calculations were made to derive the appropriate amounts of ORC and sugar to be injected. First, using basic stoichiometry, a ratio of 1.7 units oxygen to 1 unit carbon (table sugar) was calculated as required for aerobic respiration. Thus, for every 1 unit of carbon in the substrata, 1.7 units of oxygen are required by the microorganism to consume that carbon. This ratio (1.7:1) is the ratio required under ideal conditions. A second calculation based on benzene (a six carbon ring structure) was performed to obtain a more conservative 3:1 ratio. The calculated ratio varied during Phase II in order to compensate for field conditions such as the presence of oxygen sinks.

During the Phase II injection period, the ratios changed based on responses in the observation wells that indicated non-background oxygenated or reductive conditions that required inducement of additional sugar or diminished oxygen to guard against the production of VC (reductive conditions) or the increase of DCE concentrations (oxygenated conditions). Since field conditions vary (i.e., DCE concentration, porosity, etc.), decisions were made in the field as to the exact amounts of ORC and sugar to be injected over each two-foot interval of saturated zone. For example, one injection may require, based on porosity, DCE concentration, depth to confining clay, that 120 pounds of ORC and 1.5 pounds of sugar, and another injection may only require 40 pounds of ORC and 0.5 pounds of sugar. A spreadsheet was developed to aide in calculating the amount of ORC and sugar. Other adjustments to the injection criteria were made based on conditions encountered in the field.

Observation wells were analyzed for field indicator parameters (e.g. dissolved oxygen, redox potential) on a semi-monthly to quarterly basis. The observation wells were also chemically analyzed for VOCs on a frequent monitoring schedule to assess chemical-specific trends. In addition, key geochemical parameters (e.g. nitrate, sulfate, carbon dioxide, and alkalinity) were measured bi-annually for six wells and the entire suite of geochemical parameters were measured annually in eight wells to assess in-situ aquifer conditions. In addition, to assess the sparge well effectiveness, pressure transducer tests were conducted and off-gas VOCs were measured periodically.

RESULTS AND DISCUSSION

In order for the compound 1,1-dichloroethene ($C_2H_2Cl_2$, DCE) to be degraded efficiently at this site (especially at perimeter plume areas), the microbes in the subsurface plume need to be stimulated in their natural environment. The objective of the bioremediation approach was to balance the amount of carbon and oxygen in the system so microbes utilize the naturally elevated oxygen. It has been observed that distinct regions of the plume are subject to different degradation processes, largely due to the microbial population and geochemical variances that are present throughout the groundwater plume.

Along the perimeter regions of the plume, natural elevated levels of DO are present, which have prevented the creation of VC. The high DO levels provided for aerobic degradation of DCE favoring the formation of CO_2 without creation of VC. However, the rate of plume movement is greater than the rate of DCE degradation. As a result, it was necessary to provide carbon substrate to facilitate microbial activity in perimeter portions of the plume, while using oxygen to offset VC production and potentially increase DCE decay. In the source region of plume, sparging wells were used to remediate DCE via oxygenating the aquifer and stripping the COCs.

Perimeter and Southern Portion of Plume. Measured responses of VOC and geochemical groundwater concentrations in perimeter wells indicated that carbon source was the limiting factor. Therefore, the amount of sugar was increased in injection wells at the beginning of Phase II, until parameters indicated high biological activity. An example of this effect was redox potential decline observed in the toe of plume area (see Figure 3).

In general, steadily declining DCE concentrations should occur in conjunction with a robust aerobic microbial degrading environment. To accomplish this, sugar and oxygen must be balanced in the aquifer to facilitate indigenous microbial activity. Background DO levels in all perimeter wells are normal to very high for groundwater (3 ppm for MW-20 to >10 ppm for MW-2). At MW-2, initial (i.e., Phase I) excellent DCE degradation response at the 30:1 ratio in 30 days occurred due to sufficient carbon source. DCE concentration declined to non-detect at the beginning of Phase I. Later, the influx of ORC caused potentially high oxygen saturation at this ratio.

At MW-2, additional sugar was added with no oxygen during Phase II because DO levels were observed to be elevated to support DCE degradation during Phase I. The DO meter maximum reading was 35 ppm that suggested sufficient oxygen levels were present. Phase II sugar injections were aggressive beginning at a 0:1 ORC:Carbon ratio. DO remained elevated, however, redox levels dropped significantly indicating biologic activity due to additional carbon source. The Phase II results indicated the sugar injection-only technique inhibited decay of DCE as a result of creating anaerobic (low redox) instead of preferred aerobic conditions. The lag effect of measured responses such as DCE concentration and redox, is a function of the rate of microbial activity acting upon the injected carbon food source.

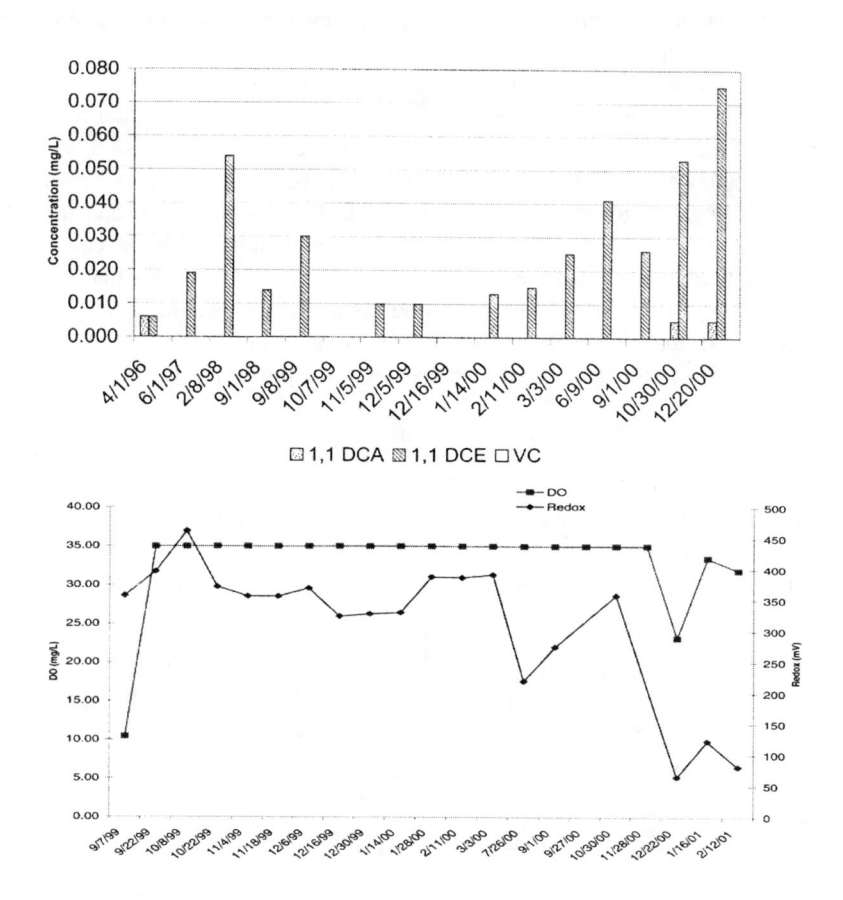

FIGURE 3. MW-2 laboratory (top) and field observations (bottom).

Geochemistry varied across the plume with the highest microbial activity observed generally closer to the source region. CO_2 produced by the metabolism of microorganisms in MW-20 increased significantly from non-detect to 104 mg/L in six months; CO_2 levels in MW-21 increased from non-detect to 75 mg/L in three months. Denitrification processes occurred in MW-20 and MW-21 as nitrate (NO_3) and nitrite (NO_2) via reduction of NO_3 levels in MW-20 in six months and MW-21 in three months.

Source and Northern Portion of Plume. The effectiveness of the sparge well is demonstrated by the Graph Data for MW-6 and MW-8 (see Figure 4). In approximately 12 months, the DCE concentration in MW-6 near the source, declined from 15.5 mg/L to 5.0 mg/L. The modeled radius of influence (ROI), the horizontal distance from the center of the well to the limit of the cone of depression, or the approximate maximum distance from the well where vacuum of pressure occurs appears to be up to 100 feet.

The historical persistent detected concentrations of VC at the source is attributed to the mass of DCE and its associated decay by-product to VC in the presence of elevated anaerobic conditions. VC concentrations at the source have remained constant while DCE concentrations have declined due to the sparge well oxygenating the groundwater. Increasing DO levels in MW-6 and MW-8 also suggest that mechanically induced oxygenated conditions have contributed to DCE degradation. Evidence from Phase I and Phase II indicates that DCE at this site is degraded under aerobic conditions by oxygenating the aquifer by mechanical sparging wells at the plume source and by stimulating the aquifer microbes with injections of ORC and sugar at the plume perimeter.

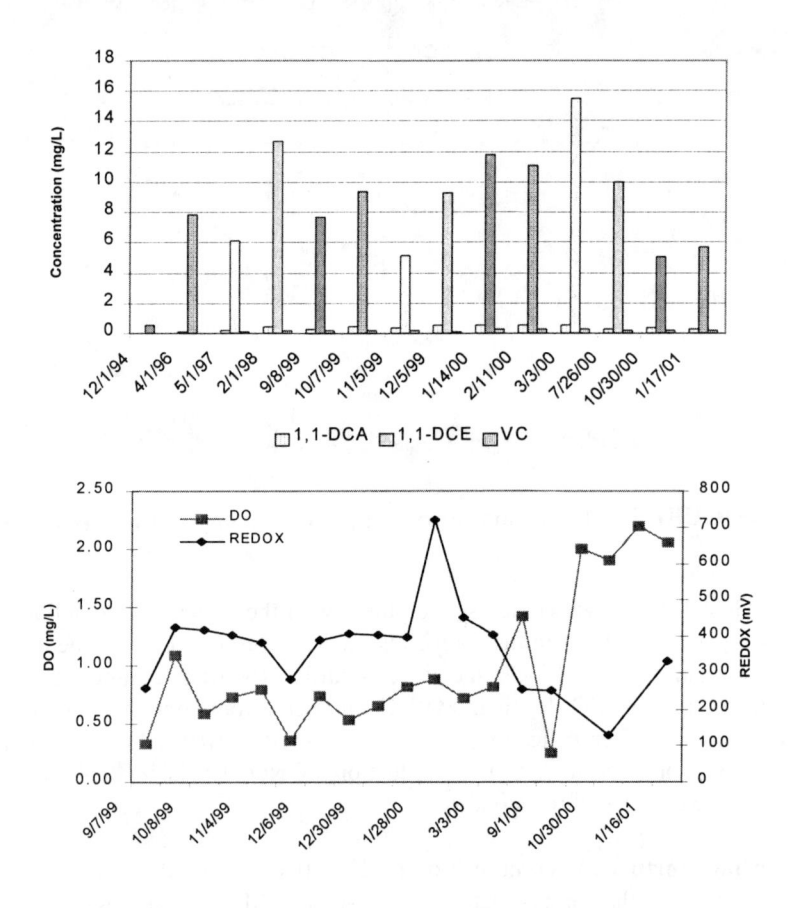

FIGURE 4. MW-6 laboratory (top) and field observations (bottom).

REFERENCES

Miller, R., and Diane S. Roote. 1997. "In Well Vapor Striping." *Technology Overview Report* TO-97-01. GWRTAC Groundwater Remediation Technologies Analysis Center.

Mueller, J., M., Ohr, D. Wardell, and F. Lakhwala. 1999. "mGCW Technology for Remediation of Chlorinated Hydrocarbons." *Soil & Groundwater*. Oct/Nov 1999: 8-14.

USEPA, Office of Research and Development. 1998. "Technical Protocol for Evaluating Natural Attenuation of Chlorinated Solvents in Ground Water." *EPA Status Report* 600-R-98-128.

USEPA, Solid Waste and Emergency Response. 1998. "Field Applications of In-Situ Remediation Technologies: Ground-Water Circulation Wells." *EPA Status Report* 542-R-98-009.

TECHNICAL PROTOCOL FOR ENHANCED REDUCTIVE DECHLORINATION VIA VEGETABLE OIL INJECTION

Todd H. Wiedemeier (Parsons Engineering Science, Inc., Denver, Colorado)
Bruce M. Henry (Parsons Engineering Science, Inc., Denver, Colorado)
Patrick E. Haas (Mitretek Systems, Inc., San Antonio, Texas)

ABSTRACT: Chlorinated solvents dissolved in groundwater are one of the United States' biggest environmental problems. It is likely that the number of sites contaminated with chlorinated solvents is second only to those contaminated with petroleum products. Because of the relative recalcitrance of chlorinated solvents in the subsurface, it is likely that the cost of remediating these sites will significantly exceed the cost of remediating sites contaminated with petroleum products. Reductive dechlorination is known to degrade the common chlorinated solvents including tetrachloroethene (PCE), trichloroethene (TCE), carbon tetrachloride (CT), and trichloroethane (TCA). This paper describes a novel technique to enhance reductive dechlorination by injecting food-grade vegetable oil into the subsurface. This approach has been termed the VegOil process and, at a minimum, has been implemented at six sites across the county. At the sites where historical data are available, contaminant concentrations and mass declined after vegetable oil injection.

The Air Force Center for Environmental Excellence (AFCEE), in conjunction with Parsons Engineering Science, Inc. (Parsons ES) is developing a technical protocol to aid in the implementation of the VegOil process (Wiedemeier et al., 2001). This paper provides a brief overview of the protocol. Included is a description of the VegOil process, design considerations for the injection and monitoring system, and data evaluation techniques that can be used to determine the effectiveness of the VegOil process on a site-specific basis.

INTRODUCTION

Extensive research into remediating soil and groundwater at sites contaminated with chlorinated solvents has been conducted. Over time, two bioremediation strategies have emerged: reductive dechlorination and oxidative cometabolism. Experience has proven that an obstacle to the development of competitive bioremediation processes is often the cost of nutrient addition. For example, although oxidative cometabolism has been shown to be an effective means of bioremediation, the cost of nutrient and (or especially) oxygen addition has severely limited its commercial acceptance. Reductive dechlorination usually requires less nutrient mass and does not require oxygen. For these reasons, reductive dechlorination has the potential to be more cost effective than oxidative cometabolism. The most common approach utilized to date to stimulate reductive dechlorination has been addition of a carbon source dissolved in groundwater. Materials that have been added to the subsurface to stimulate dechlorination include, but are not limited to, propionate, lactate, butyrate, molasses, and hydrogen. However, to date no widely accepted process for *in situ* bioremediation of chlorinated solvents has emerged. One reason for this is that, because of their physical properties, all of these materials require continuous injection, or at a minimum, multiple injections and thus may have difficulty competing with groundwater extraction and treatment.

This paper describes the use of vegetable oil (which costs $0.20 to $0.50 per pound) as a source of organic carbon to stimulate reductive dechlorination of chlorinated solvents including the chlorinated ethenes, ethanes, methanes, and benzenes. This approach has been termed the VegOil process and is designed to create the oxidation-reduction and electron donor conditions necessary to promote microbial reductive dechlorination of chlorinated solvents. A secondary benefit of the VegOil process is the partitioning of the aqueous-phase chlorinated solvents into the vegetable oil nonaqueous phase liquid (NAPL). Thus, the vegetable oil acts as a "sponge" to quickly remove solvents from the groundwater. This is beneficial because aqueous-phase chlorinated solvent concentrations will be rapidly lowered until steady state conditions are reached. After the solvent has partitioned into the vegetable oil it will be released slowly into the groundwater at the same time as vegetable oil and thus will be situated right in the bioreaction core. One disadvantage of the VegOil process is that toxic intermediates such as vinyl chloride may be produced. In addition, it is not known how the vegetable oil will behave over the long-term.

Vegetable oil can be injected directly into an aquifer utilizing conventional wells or using direct push techniques. This allows significant volumes of organic carbon to be added at multiple locations, and to be well distributed in the contaminated aquifer. The separate phase nature of vegetable oil allows for slow dissolution into groundwater thus, making it a slow release carbon source. Vegetable oil is an inexpensive, innocuous, food-grade carbon source that is not regulated as a contaminant by the United States Environmental Protection Agency (USEPA). Because vegetable oil is a NAPL, the potential exists that a single, low cost, injection could provide sufficient carbon to drive reductive dechlorination for many years. This will significantly lower operation and maintenance (O&M) costs compared to aqueous phase injection, and will allow injection of a much greater quantity of carbon than solid phase carbon emplacement. This is the key to the cost effectiveness of the VegOil process; the oil can be injected once, and if properly placed it will serve as a carbon source for many years with little or no ongoing O&M cost. Although the process itself requires no O&M, process monitoring will likely be required, as discussed below.

The VegOil process depends upon successful injection of vegetable oil into an aquifer, its transport as a NAPL to the proper location, and its subsequent dissolution into groundwater. Vegetable oils are available in a wide spectrum of densities, viscosities, and solubilities. Understanding these properties and how they impact the behavior of injected vegetable oil is key to developing a viable process design. Natural vegetable oils are all less dense than water, and thus are light nonaqueous-phase liquids (LNAPLs).

DESIGN CONSIDERATIONS FOR THE INJECTION AND MONITORING SYSTEM

Vegetable oil has been successfully injected into the subsurface using conventional wells and direct push techniques including Geoprobe and cone penetrometer. The most common method being used to inject the oil into the subsurface is an air driven diaphragm pump. Vegetable oil can be injected as either pure vegetable oil or as an emulsion of oil and one of the following: water, lecithin, and various emulsifiers/surfactants. Vegetable oil present in the subsurface is acted upon by three forces including gravity, capillary pressure, and hydrodynamic pressure. The migration and ultimate distribution of vegetable oil in the subsurface is determined by the interaction between these forces and the physical and chemical properties of the oil and the properties of the aquifer matrix. As in most problems of fluid flow in naturally occurring porous media, the heterogeneity of the aquifer will exert an important and, perhaps, the dominant influence on the distribution of the injected vegetable oil.

Figure 1 shows a hypothetical layout for a pilot-scale vegetable oil injection system. This figure shows three vegetable oil injection wells, a well upgradient from the injection wells, and numerous wells downgradient from the injection wells. The well upgradient from the pilot testing system is intended to monitor the chemistry of groundwater flowing into the area influenced by the pilot test. The wells downgradient from the injection wells are intended to monitor changes in groundwater chemistry brought about by the VegOil process. If the system is to be installed a short distance from potential receptors, then contingency monitoring wells (as described in Wiedemeier et al., 2000) should be installed and a contingency plan developed.

The scale of the final system will be dependent on many variables including the size and distribution of the solvent plume and the behavior of the vegetable oil once it is injected into the aquifer. Perhaps the biggest consideration when designing the system is the radius of influence of the vegetable oil. As discussed above, aquifer heterogeneity which likely be the dominant variable influencing the final distribution of the oil. It is also desirable to emplace the vegetable oil as an immobile, residual phase to create a

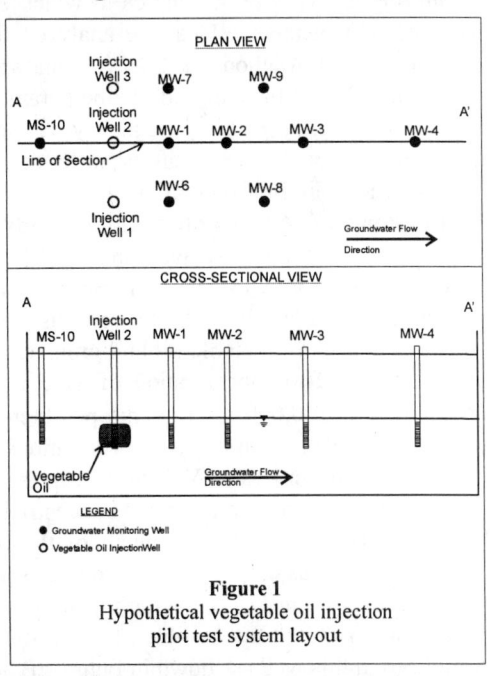

Figure 1
Hypothetical vegetable oil injection pilot test system layout

controlled reaction zone. This can be accomplished using straight oil injection followed by a water flush, or by the use of an oil in water emulsion (preferably with native groundwater). In many cases the results of a pilot test can be useful for full-scale system design.

EVALUATING THE EFFECTIVENESS OF VEGETABLE OIL INJECTION

Process monitoring should be conducted to understand how well the VegOil process is working. The following sections describe sampling protocols and various data evaluation techniques.

Sampling Protocol and Schedule: Groundwater samples should be collected prior to vegetable oil injection to provide baseline conditions. These groundwater samples should be analyzed for chlorinated solvents and their degradation products, dissolved oxygen, nitrate, Fe(II), Mn(II), sulfate, methane, ethane, ethane, oxidation-reduction potential (ORP), alkalinity, pH, temperature, dissolved organic carbon, volatile fatty acids, hydrogen, and chloride. This list of analytes in similar to those for evaluating biodegradation found in the USEPA protocol for evaluating natural attenuation of chlorinated solvents (USEPA, 1998). After injection of the vegetable oil, samples of groundwater and vegetable oil (if present in a well) should be collected on multiple occasions for process monitoring to evaluate the degree to which biodegradation has been stimulated. Samples of vegetable oil can be analyzed for chlorinated compounds using a modification of Method 8260B. Groundwater samples collected for process monitoring should be analyzed for the parameters listed earlier in this paragraph. Process monitoring sampling frequency will depend on many things including, but not limited to: well spacing, groundwater seepage velocity, aquifer heterogeneity, and the efficacy of biodegradation.

Partitioning of Chlorinated Compounds into Vegetable Oil: Because chlorinated compounds are hydrophobic, partitioning of these compounds into the vegetable oil is likely to occur. Table 1 shows data from a site in Florida where vegetable oil was injected. Shown on this table are concentrations of chlorinated compounds in the vegetable oil and groundwater beneath the oil before and after oil injection. The concentration of TCE in groundwater in the injection well dropped from 100,000 micrograms per liter (µg/L) to 84 µg/L within about 2 months of oil injection. The concentrations of *cis*-1,2-dichloroethene (*cis*-1,2-DCE) and vinyl chloride (VC) also decreased rapidly. It is important that this rapid reduction in contaminant concentration not be attributed entirely to biodegradation. In this case much of the observed reduction in contaminant concentrations was caused by partitioning of the chlorinated compounds into the vegetable oil (Table 1). Although much of the observed decrease in contaminant concentration was attributable to partitioning, some biodegradation was stimulated in this well as shown in Figure 2B and discussed in the next section.

Table 1
Concentrations of chlorinated compounds in vegetable oil and groundwater in a vegetable oil injection well (concentrations in micrograms per liter).

Compound	6/8/99[a]		8/24/99		10/19/99		11/17/99		12/14/99	
	Oil	Water	Oil	Water	Oil	Water	Oil	Water	Oil	Water
TCE	ND	100,000	44,000	84	99,000	230	47,000	130	68,000	160
cis-1,2-DCE	ND	48,000	15,000	230	29,000	660	13,000	340	22,000	460
VC	ND	330	3,900	<10	<1,000	56	<1,000	<10	<1,000	<10

a/ Oil injected on June 15, 1999

Changes in Contaminant Concentrations and Molar Fractions: When evaluating the effectiveness of the VegOil process important considerations are reductions in contaminant concentration and changing molar fractions of dissolved constituents over time. An historical database showing a reduction in contaminant mass before and after vegetable oil injection can be used to show that the VegOil process is working to destroy contaminants. In addition, a change in the molar fractions of parent compounds to daughter products can be useful in evaluating if VegOil process is working to remove contaminant mass from the aquifer via biodegradation.

When evaluating changes in contaminant concentrations and molar fractions it is important to take into account

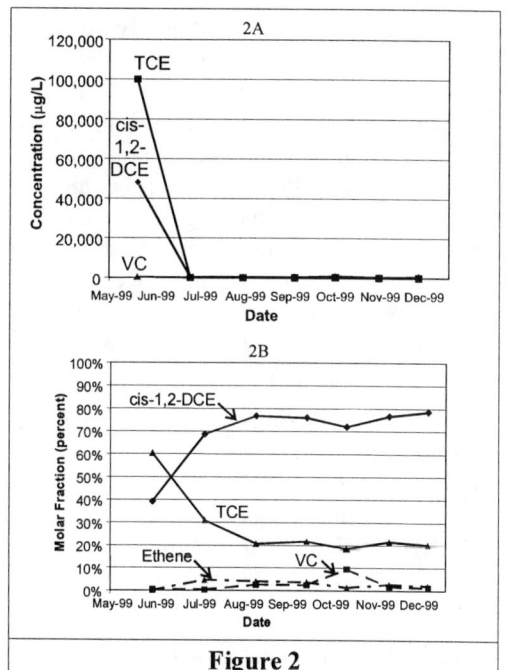

Figure 2
Changes in aqueous-phase concentrations (A) and molar fractions (B) of chlorinated compounds at an injection well. Changes are caused by a combination of partitioning into the vegetable oil and biodegradation.

the effects of partitioning. As discussed in the previous section, if vegetable oil NAPL is present then partitioning of contaminants into the oil will occur. Figure 2 illustrates the effect of the vegetable oil on contaminant concentrations. Note the rapid drop in contaminant concentrations in the groundwater caused by partitioning into the oil (Table 1). Figure 2B shows that some biodegradation has been stimulated by the change in molar ratios of the chlorinated ethenes. In particular, the molar mass of the parent compound TCE has been significantly reduced relative to the daughter product cis-1,2-DCE. The molar ratios seem to stabilize after about three months, likely because the rate of TCE release from the oil into groundwater has come into equilibrium with the rate of biodegradation. It is important to keep in mind that there will be preferential partitioning into the vegetable oil of TCE over DCE by a factor of about two.

Figure 3 shows measured changes in concentrations and molar fractions of chlorinated compounds in a well located downgradient from the area with vegetable oil. Dissolution of the vegetable oil into the groundwater has clearly stimulated biodegradation in the aquifer around this well by significantly reducing the TCE concentration and elevating the concentration of *cis*-1,2-DCE.

Changes in Groundwater Geochemistry: Biodegradation causes measurable changes in groundwater geochemistry that can be used to evaluate the effectiveness of the VegOil process. For reductive dechlorination to be an efficient process the groundwater typically must be sulfate reducing or methanogenic. Thus, groundwater in which reductive dechlorination is occurring should have the following geochemical signature:

- depleted concentrations of dissolved oxygen, nitrate, and sulfate;
- elevated concentrations of Fe(II), Mn(II), methane, ethane, ethane, hydrogen, and chloride; and
- elevated alkalinity

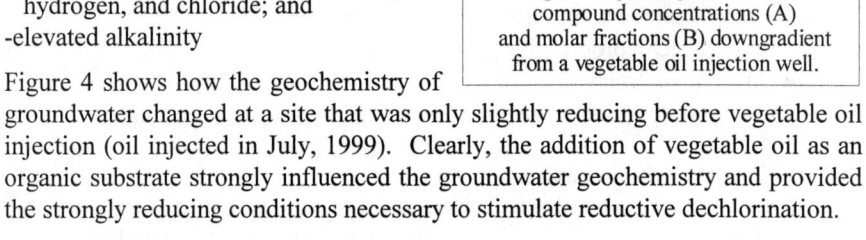

Figure 3
Changes in aqueous phase chlorinated compound concentrations (A) and molar fractions (B) downgradient from a vegetable oil injection well.

Figure 4 shows how the geochemistry of groundwater changed at a site that was only slightly reducing before vegetable oil injection (oil injected in July, 1999). Clearly, the addition of vegetable oil as an organic substrate strongly influenced the groundwater geochemistry and provided the strongly reducing conditions necessary to stimulate reductive dechlorination.

CONCLUSIONS

Using the techniques described in this paper, the injection of food-grade vegetable oil as an organic substrate has been shown to stimulate biodegradation and reduce contaminant concentrations at six sites across the USA. In addition, laboratory studies being conducted by Cornell University under subcontract to Parsons ES show conclusively that the addition of vegetable oil to microcosms stimulates biodegradation of chlorinated solvents (personal communication, 2000). Continued research is necessary to determine the longevity of the emplaced vegetable oil for stimulating reductive dechlorination and ultimately degrading chlorinated solvents to innocuous byproducts.

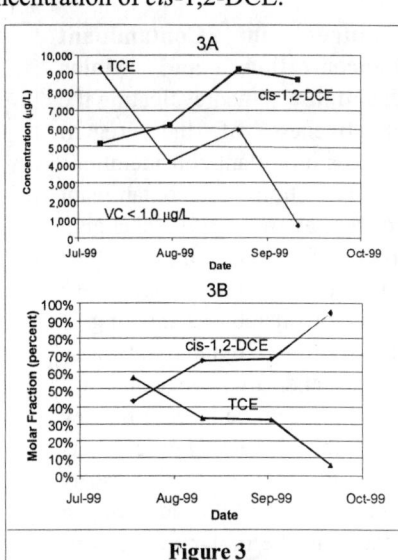

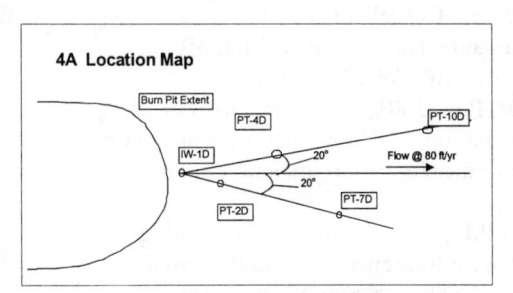

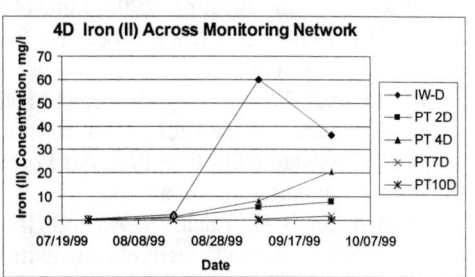

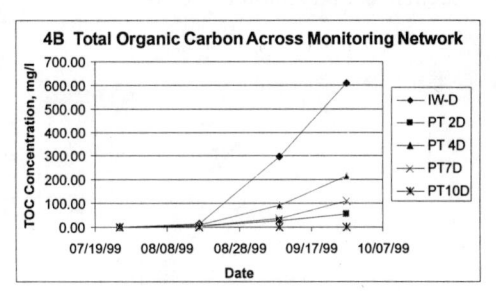

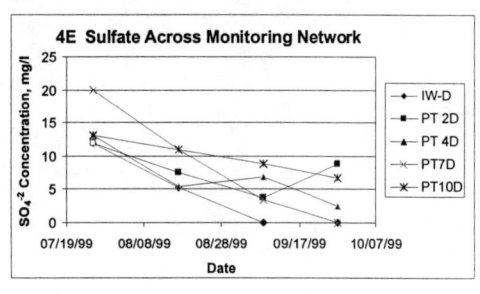

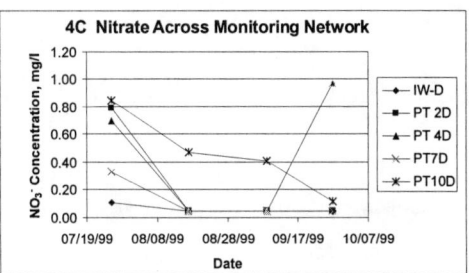

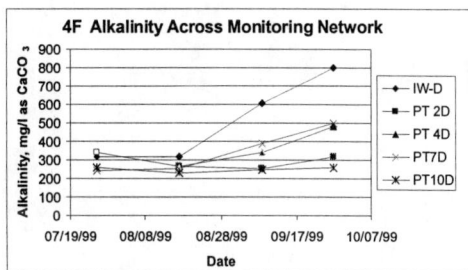

Figure 4
Changes in groundwater geochemistry
caused by vegetable oil injection.
Well IW-D is the injection well.

ACKNOWLEDGEMENTS

The authors would like to thank Rob Hinchee, Rick Johnson, Dave McWhorter, Jim Gossett, and Don Banks for their help in developing the VegOil process.

REFERENCES

Personal Communication, 2000, James Gossett, Cornell University.

USEPA, 1998, Technical Protocol for Evaluating Natural Attenuation of Chlorinated Solvents in Ground Water: EPA/600/R-98/128.

Wiedemeier, T.H., Henry, Bruce, and Haas, P.E., 2001, A Stream-Lined Field Feasibility Test for In Situ Bioremediation of Chlorinated Solvents via Vegetable Oil Injection: Air Force Center for Environmental Excellence, San Antonio, Texas.

Wiedemeier, T.H., Lucas, M.A., and Haas, P.E., 2000, Designing Monitoring Programs to Effectively Evaluate the Performance of Natural Attenuation: Air Force Center for Environmental Excellence, San Antonio, Texas.

EFFECTIVE DISTRIBUTION OF EDIBLE OILS - RESULTS FROM FIVE FIELD APPLICATIONS

Michael D. Lee (Terra Systems, Inc., Wilmington, DE, USA)
Bob Borden, M. Tony Lieberman, Walt Beckwith, and Terry Crotwell (Solutions Industrial & Environmental Services, Raleigh, NC, USA)
Patrick E. Haas (AFCEE, Brooks AFB, TX, USA)

ABSTRACT: Edible oil was injected at five sites to promote reductive dechlorination. The edible oil was injected as a non-aqueous phase liquid (NAPL) or as an emulsion of oil, an emulsifier, and water. At Dover Air Force Base, injection of the emulsion into a barrier resulted in greater distribution of the total organic carbon (TOC) and more rapid dechlorination than direct injection of the edible oil into another barrier. Direct addition of oil using a Geoprobe™ direct push rig promoted reductive dechlorination in two sites with shallow groundwater contamination. The emulsion moved at least 7.6 m at a site on Long Island, NY. An increase in the intermediate and final degradation products ethene and ethane was noted over the first four months after injection. The emulsion was distributed throughout a 7.6 m zone surrounding the four injection wells in a deep aquifer at Edwards Air Force Base. Little dechlorination has been observed at this site after two months. Direct oil injection is generally less expensive when the contamination is shallow and site conditions allow rapid installation of Geoprobe™ points. Injection of the oil-in-water emulsion is more cost effective when the contamination is deeper or where a larger area must be treated. However, injection of the emulsion requires an additional handling step and a suitable water supply, but results in a superior product that can be readily introduced into the subsurface.

INTRODUCTION

Numerous laboratory and field studies have shown that chlorinated solvents such as perchloroethene (PCE), trichloroethene (TCE), and 1,1,1-trichloroethane (1TCA) can be biodegraded by naturally occurring microorganisms when provided with an appropriate organic substrate (Ellis et al., 2000; Lee et al., 1997; Lee et al., 1998). The key to successful implementation of this technology is developing an effective low-cost method of distributing substrate throughout the treatment zone (Quinton et al., 1997). Previous studies using lactate, molasses, and other soluble substrates have successfully treated chlorinated solvent-contaminated aquifers by recirculating groundwater amended with the dissolved substrate. However capital costs are substantial because of the required tanks, pumps, mixers, injection and pumping wells, and related process controls. In addition, operation and maintenance (O&M) costs are high because of problems associated with clogging of injection wells and the labor for extensive monitoring and process control. High O&M costs are a particularly important issue in aquifers where cleanup rates are limited by slow dissolution of NAPLs or diffusion of contaminants from low permeability zones.

One potentially very cost-effective approach for enhancing anaerobic processes is using immobilized edible oils to promote anaerobic biodegradation in permeable reactive barriers (PRBs). Lee et al. (2000) demonstrated that a variety of edible oils can support reductive dechlorination in laboratory and column studies from five sites. The edible oils can be purchased for as little as $0.40 per pound. As solvents or other contaminants migrate through the barrier, the immobilized oils slowly dissolve, enhancing contaminant biodegradation, leaving uncontaminated water to emerge from the downstream side. When applied at full scale, O&M costs will be minimal since the oils are selected to last three to ten years between reinjections.

The key to successful implementation of this technology is developing an effective low-cost method of distributing and immobilizing the oils throughout the treatment zone. Two different injection and distribution approaches are being evaluated: (1) injection of the edible oil as a non-aqueous phase liquid (NAPL); and (2) injection as an oil-in-water emulsion. Performance results from five different sites are evaluated (Table 1). The cost and effectiveness of each approach is site specific and dependent on aquifer lithology, depth of contamination, drilling costs, and available water supply.

RESULTS

Dover Air Force Base. Two barriers were installed in a shallow aquifer contaminated with PCE, TCE, 1TCA, and daughter products (Zenker et al., 2000). A direct oil barrier was prepared by injecting soybean oil into ten injection wells 0.6 m apart. The oil was not injected under pressure. Seven hundred sixty liters of groundwater were used to chase the oil. Four injection wells in the second barrier received soybean oil that had been emulsified with groundwater and an emulsifying agent using a high-shear mixer. The emulsion was injected under pressure. About 7,600 L of bromide-amended groundwater were used to distribute the oil throughout the area. Six monitoring wells were installed around each barrier.

Nine months after the oil injection, the distribution of dissolved total organic carbon (TOC) was much better in the emulsion barrier than the direct oil barrier (Figure 1). The oil did not penetrate very deep in the direct oil barrier. More oil was injected at depth in the emulsion barrier. There has been limited production of *cis*-1,2-dichloroethene (cDCE), vinyl chloride (VC), and chloroethane (CA) in the emulsion barrier after nine months. Little dechlorination has been observed to date in the direct oil barrier, possibly as a result of poor distribution of the oil to depth or the need for bioaugmentation to complete the dechlorination reaction. Methane has accumulated at the groundwater table near both barriers, but has not reached the land surface.

Eastern VA. Soybean oil was injected into the area surrounding a dry cleaning machine at an industrial facility. The shallow unconsolidated aquifer was

TABLE 1. Site Descriptions.

Site	Geology	Contaminants	Depth to Water (m)	Treated Depth (m)	Project Size Length x Width m (# injection points)	Evidence for Dechlorination
Dover Air Force Base, DE	Fine to medium sand	PCE, TCE, cDCE, VC. 1TCA, 1DCA, 1DCE, CA	3.3	4.6-13.7	Direct oil 6.1 x 6.1 (10) Emulsion 6.9 x 6.1 (4)	Little dechlorination Limited production of cDCE, VC, and CA
Eastern VA	Unconsolidated clay, silt, sand, and limestone	PCE, TCE, cDCE, VC. 1TCA, 1DCA, 1DCE, CA	1.5	1.5-3.0	18 x 37 (35 direct oil)	Decreases in PCE, TCE, 1DCA, 1DCE, 2DCA, tDCE; increase in cDCE, VC, and ethene
Eastern NC	Silty-sand	TCE, 1TCA, 1DCA, 1DCE	3.0	3.0-7.3	8000 m^2 (185 direct oil)	Little change after 3 months
Long Island, NY	Fine to coarse sand, some silt and gravel	PCE, TCE, cDCE, VC. 1TCA, 1DCA, 1DCE, CA	3.0	6.7-14.6	15.2 x 6.1 (6 emulsion, 1 direct oil)	Decreases in cDCE, VC, 1TCA, 1DCA, and 1DCE; increase in CA and ethane
Edwards Air Force Base, CA	Clay, silty clay, sandy clay, silt, clayey sand, and fine to coarse sand	TCE	14.6	13.1-19.2	6.8 x 8.7 (4 emulsion)	No dechlorination after 3 months

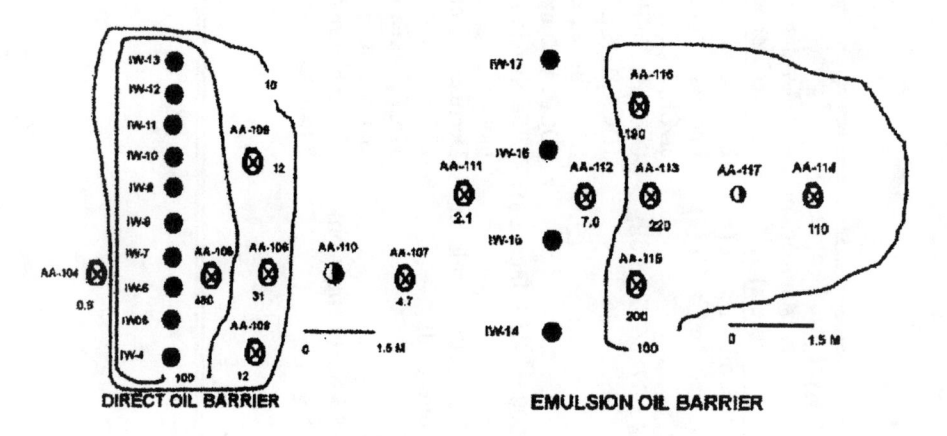

252   *Anaerobic Degradation of Chlorinated Solvents*

FIGURE 1. TOC Distribution DAFB Barriers after 9 Months.

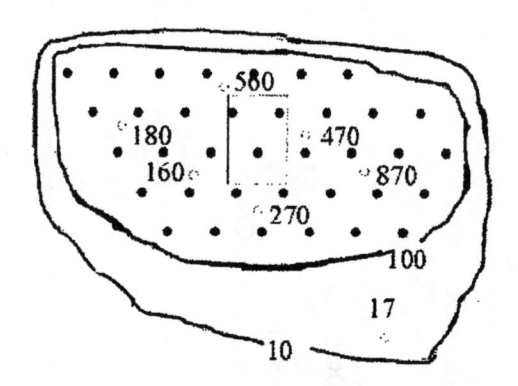

FIGURE 2. TOC Distribution Eastern VA Direct Oil barrier after 7 Months.

contaminated with PCE, 1TCA, and daughter products. A total of 36 injections of soybean oil were made using a Geoprobe™. After seven months, TOC levels in the area ranged from 160 to 870 mg/L in piezometers within the treated area (Figure 2). The TOC in a piezometer outside of the treated area was 17 mg/L. Daughter products including cDCE, VC, and ethene increased in several wells with decreases in the PCE and TCE concentrations. Concentrations of 1TCA, 1,1-dichloroethane (1DCA), 1,1-dichloroethene (1DCE), 1,2-dichloroethane (2DCA), and *trans*-1,2-dichloroethene (tDCE) decreased in several piezometers, but without a concomitant increase in chloroethane and ethane increases. The soybean oil injection has improved groundwater quality at the site.

Eastern NC. TCE, 1TCA, and daughter products were found in the groundwater beneath a former manufacturing facility. Soybean oil was injected into 185 points at the facility including beneath the floor of the facility and as barriers to treat two separate areas of the plume. Decreases in the dissolved oxygen and redox potential have been noted over the seven months since oil injection. Ferrous iron and methane levels have increased as the groundwater becomes more reducing. However, little change in the dissolved contaminants was observed in the three months of data available after oil injection.

Long Island, NY. This industrial facility was contaminated with high concentrations of 1TCA, PCE, and their daughter products as well as a number of other organic constituents. A emulsion of soybean oil and water was injected into six points between 6.7 and 15.2 m bgs. Another point received direct injection of oil. The emulsion moved at least 7.6 m from an injection point to a shallow monitoring well. Three and a half months after the injection of the emulsion, the TOC ranged from <1 to 1,990 mg/L in shallow and deep monitoring points surrounding the barrier (Figure 3). Sulfate concentrations are declining and methane levels are increasing in a number of the nearby monitoring wells. There is evidence for complete dechlorination in many of the wells. For example, in shallow monitoring point S-2, cDCE has declined from 1.5 µM to <0.002 µM, VC from 2.8 to 0.6 µM, and ethene has been constant at 7.9 µM. Concentrations of 1TCA have declined from 24 to 7.5 µM, 1DCA from 41 to 12 µM, and 1,1-DCE from 1.1 to < 0.02 µM, but CA has increased from 19 to 47 µM and ethane from <0.2 to 1.3 µM.

Edwards Air Force Base, CA. TCE was the only chlorinated contaminant found in a relatively deep aquifer (water table at 15.2 m) at Edwards Air Force Base. A fine emulsion of soybean oil and groundwater was injected into four wells screened between 15.2 and 19.8 m bgs. Five monitoring wells were installed up to 4.6 m away. The emulsion reached all of the monitoring points except the well the furthest away. TOC distribution was greater than 1,000 mg/L in the injection wells and one downgradient well after two months (Figure 4). TOC levels in the remaining wells ranged between 29 and 320 mg/L versus the pre-treatment average of 9 mg/L. No evidence of dechlorination has been observed in the two months following substrate injection.

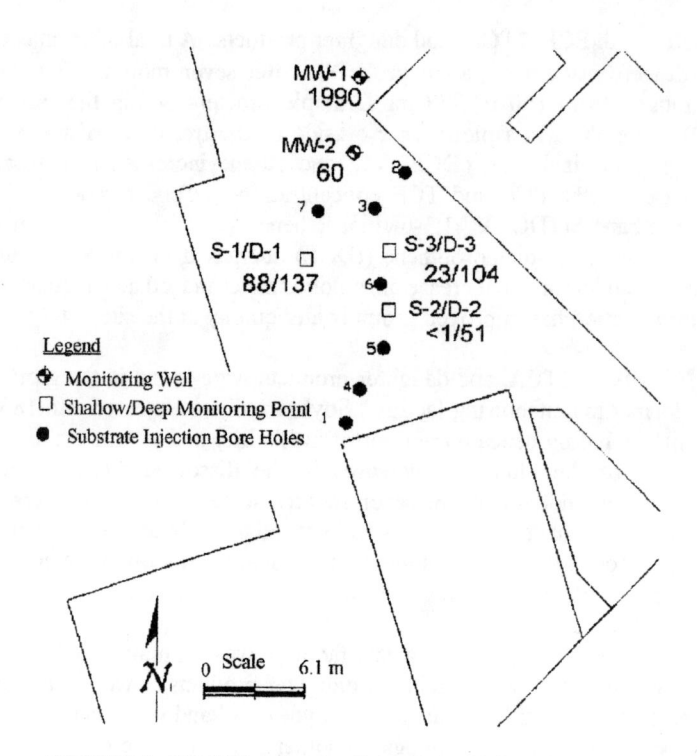

FIGURE 3. TOC Distribution Long Island Pilot after 3.5 Months.

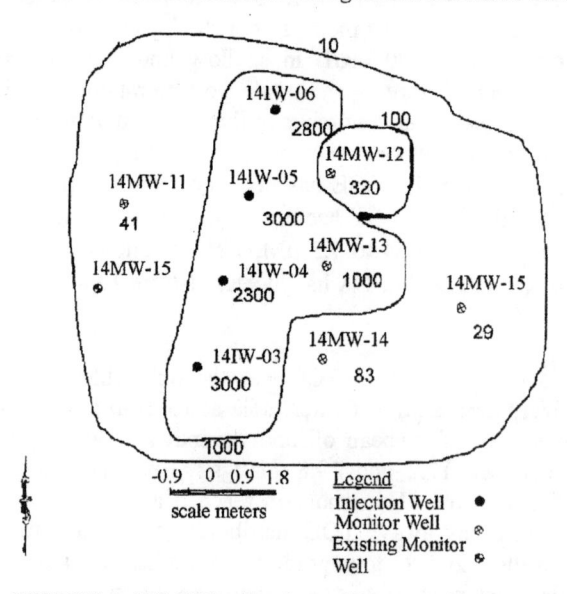

FIGURE 4. TOC Distribution at Edwards Air Force Base Emulsion Pilot after 2 Months.

DISCUSSION

Direct oil injection as a non-aqueous phase liquid (NAPL) is often less expensive when the depth of the contamination is low and site conditions allow rapid installation of Geoprobe™ points at relatively close spacing. However, it may be more difficult to effectively distribute NAPL oil over large vertical intervals because of density effects. NAPL oil in the pore spaces may also result in a greater loss of permeability and contaminant bypassing around the oil treated zones.

Injection of the oil as an oil-in-water emulsion is more cost effective when the depth of contamination is greater and drilling costs are higher. When properly prepared, oil-in-water emulsions may be distributed over substantial distances (at least 7.6 m) significantly reducing drilling costs and reducing the impact on aquifer permeability. The emulsion can provide over 1,000 mg/L TOC. However, injection as an emulsion requires additional handling and requires a suitable water supply.

Edible oils can be used to treat an entire plume, a source area, or could be applied as a barrier to intercept a plume. The oils should be useful for fractured bedrock sites where the oil can coat the surfaces of the fractures and support reductive dechlorination. Laboratory studies (Lee et al., 2000) have shown that the oils can be used in conjunction with bioaugmentation at sites where dechlorinating organisms are not present. Bioaugmentation may be necessary at the Dover AFB, Eastern NC, and Edwards AFB sites as limited dechlorination has been observed at these sites after two to nine months. Solutions Industrial & Environmental Services and Terra Systems, Inc. have a patent pending on the use of oil emulsions to support reductive dechlorination.

ACKNOWLEDGEMENTS

Funding for the Air Force projects was supplied by the Air Force Center for Environmental Excellence. The cooperation of Dover Air Force Base, Edwards Air Force Base, the Eastern Virginia, Eastern North Carolina, and Long Island facilities is greatly appreciated.

REFERENCES

Ellis, D. E., E. J. Lutz, J. M. Odom, R. J. Buchanan, Jr., C. L. Bartlett, M. D. Lee, M. R. Harkness, and K. A. DeWeerd. 2000. "Bioaugmentation for Accelerated *In Situ* Anaerobic Bioremediation." *Environ. Sci. Technol.* 34(11). 2254-2260.

Lee, M. D., J. M. Odom, and R. J. Buchanan, Jr. 1998. "New Perspectives on Microbial Dehalogenation of Chlorinated Solvents. Insights from the Field." *Ann. Rev. Microbiol.* 52: 423-452.

Lee, M. D., G. E. Quinton, R. E. Beeman, A. A. Biehle, R. L. Liddle, D. E. Ellis, and R. J. Buchanan. 1997. "Scale-up of an I*n Situ* Anaerobic PCE Bioremediation Project." *Journal of Industrial Microbiology and Biotechnology* 18(2/3):106-115.

Lee, M. D., R. J. Buchanan, and D. E. Ellis. 2000. "Laboratory Studies Using Edible Oils to Support Reductive Dechlorination." In G. B. Wickranayake, A. R. Gavaskar, B. C. Alleman, and V. S. Magar (Eds.), *Bioremediation and Phytoremediation of Chlorinated and Recalcitrant Compounds.* pp. 77-84. Battelle Press, Columbus, OH.

Quinton, G. E., R. J. Buchanan, Jr., D. E. Ellis, and S. H. Shoemaker. 1997. "A Method to Compare Groundwater Cleanup Technologies." *Remediation* 1997 (Autumn): 7-16.

Zenker, M. J., R. C. Borden, M. A. Barlaz, M. T. Lieberman, and M. D. Lee. 2000. "Insoluble Substrates for Reductive Dehalogenation in Permeable Reactive Barriers." In G. B. Wickranayake, A. R. Gavaskar, B. C. Alleman, and V. S. Magar (Eds.), *Bioremediation and Phytoremediation of Chlorinated and Recalcitrant Compounds.* pp. 47-53. Battelle Press, Columbus, OH.

TIME-RELEASE ELECTRON DONOR TECHNOLOGY: RESULTS OF FORTY-TWO FIELD APPLICATIONS

Stephen Koenigsberg (Regenesis, San Clemente, CA)
Craig Sandefur (Regenesis, San Clemente, CA)
Kevin Lapus (Regenesis, San Clemente, CA)

ABSTRACT: Hydrogen Release Compound (HRC®) is a food grade, polylactate ester that, upon being deposited into the aquifer, is slowly hydrolyzed to release lactic acid and other organic acid derivatives. The organic acids are fermented to hydrogen, which in turn donates electrons that drive reductive bioattenuation processes. HRC delivers electrons in a time-release fashion for about one year. The material is applied to the aquifer by push-point injection or backfill-auguring and is normally indicated for treatment of dissolved phase plumes and hydrophobically sorbed contaminant. It has long been known that enhancing bioremediation can facilitate desorbtion of the residual, sorbed phase; now this has been specifically established in an HRC-mediated environment. HRC has now been used on over 120 sites, which we believe makes it the most widely understood electron donor for accelerating bioattenuation. Of these applications, forty-two are in a position to be evaluated. From among these there were nine sites that displayed exceptional results demonstrating very rapid and complete dechlorination. At twenty-two sites results are very positive, displaying accelerated degradation rates with varying degrees of daughter product formation depending on the age of the data set. Finally, nine sites are showing moderately accelerated rates of dechlorination with varying degrees of daughter product formation depending on the age of the data set, and two sites were unresponsive to a single treatment.

INTRODUCTION

An emerging and highly desirable strategy for the management of groundwater contaminated with anaerobically degradable compounds is to simply provide organic substrates to the aquifer. The organic materials can donate electrons that facilitate the destruction of contaminants, such as chlorinated hydrocarbons, nitroaromatics and oxyanions, by microbially mediated chemical reduction. The electron donating processes are typically linked to the production of hydrogen by the fermentation of the organic matter. One option in implementing this technology is to use a time-release hydrogen/electron donor, which can eliminate or reduce major design, capital and operational costs, as well as allow for the engineering of a low-impact application and a subsequently invisible remediation process.

Hydrogen Release Compound (HRC) is one option currently available to deliver electrons in a time-release fashion. HRC is a food grade, polylactate ester that can be applied to the aquifer by push-point injection or backfill-auguring. It will slowly hydrolyze over a period of about a year and generate readily fermentable lactic acid and its derivatives. This technology is most applicable to the passive, long term and low cost treatment of dissolved phase plumes, associated hydrophobically sorbed contaminants, and moderate residual DNAPL.

RESULTS

Published History: The basic HRC chemistry and laboratory performance verification was published by Koenigsberg and Farone (1999). A mathematical model for HRC performance was later described by Farone et al. (1999). The first field tests were performed by Kallur and Koenigsberg (1999) and Wu (1999). In these studies HRC laden canisters were placed in a well and significant reductive dechlorination was observed. Soon after, Dooley et al. (1999) used these canisters in a recirculation well system and added to the proof-of-concept for this technology. At this point, direct injection into the aquifer was executed and this also generated favorable results. Upon the addition of 240 pounds of HRC in a 60 square foot area there was an 80% reduction in PCE after four months with a classical pattern of rise and fall in daughter products and expected changes in geochemical parameters such as a decrease in sulfate and an increase in iron (Sheldon et al., 1999).

Other early work includes a series of microcosm-to-field comparative studies (Farone et al. 2000); a single canister study with extensive monitoring (Sheldon and Armstrong 2000); three scaled-up pilot barrier applications (Anderson et al. 2000, Dooley and Murray 2000, Schuhmacher et al., 2000); a pilot source treatment (Zahiraleslamzadeh and Bensch 2000) and two full scale source treatments (Boyle et al. 2000, Lodato et al. 2000). A comprehensive compilation of these studies and subsequent field work can also be found in Koenigsberg and Sandefur (1999).

Summary of Results Related to Wide-Scale Use: At this writing, HRC has now been used on over 120 chlorinated hydrocarbon sites; which we believe make it the most widely used electron donor for bioattenuation. Of these applications, 42 were mature enough to be evaluated. From among these there were 9 sites that displayed exceptional results, defined as demonstrating very rapid and complete dechlorination (type A). At 22 sites results are very positive, displaying accelerated degradation rates with varying degrees of daughter product formation depending on the age of the data set (type B). Finally, 9 sites show moderately accelerated or mixed rates of dechlorination with varying degrees of daughter product formation depending on the age of the data set (type C), and 2 of the sites were not responding at all after a single application (type D).

Representative Cases from Results Summary: Three representative examples from the A-C data sets not being reported on at the conference are as follows:

Type A- Industrial site in New Jersey: 1,080 lbs. of HRC was injected via 23 direct-push points, arranged in a grid, covering an area of 1,100 square feet. After 124 days, PCE concentrations decreased by an average of 85%. Total cost for this effort (product and application) was $15,000 and the site is scheduled for closure. This represented a significant savings over projected ongoing operations and maintenance at the site.

Type B- Dry Cleaner in Oregon: A barrier containing 2,300 pounds of HRC was installed across the plume in two locations. After 143 days there was a 53% reduction in PCE. The total cost of the project was $31,000 versus an estimated $150,000 for multiphase extraction.

Type C- Industrial site in Illinois: Three barriers were installed that used 2,500 pounds of HRC. After three months there were positive results at two barriers and a third performed less efficiently; it was in a high sulfate zone, where sulfate reduction may have competed with reductive dechlorination. A re-application of HRC was made and sulfate was reduced dramatically after another three months; reductive dechlorination was expected to follow. Total accumulated costs are in the range of $60,000 and represent about an order of magnitude lower cost than the alternatives that were considered.

Performance Characteristics: HRC and Desorbtion of DNAPL

While it is clear that bioremediation is effective against dissolved phase contamination and normal levels of hydrophobically sorbed material, which is readily available for microbial consumption, there has been debate about how easily a residual sorbed DNAPL can be treated. It has been clear for a while that microorganisms can accelerate sorbed contaminant removal by several mechanisms. Primarily, when microbes consume the "newly born" dissolved phase that is outside an actively desorbing source - they maintain a concentration gradient; the flux from the sorbed material will increase with the steepness of this gradient. Also, microbes may actively secrete "biosurfactants" which can facilitate desorbtion. Consequently, there are at least two mechanisms to support the observed biological impact on residual sources.

Our laboratory work was modeled after Carr et al. (2000). In the studies we used TCE as a representative chlorinated solvent and studied the action of HRC on some or all of the phases as discussed. The work was performed by Applied Power Concepts (Anaheim, CA). Field results were then examined for corroborative evidence. Three types of experiments were performed:

A visible drop of TCE (about 0.5 grams) was placed in a flask. Water from a second flask containing soil and HRC was slowly recirculated through the flask containing the pure TCE. After 12 days there was disappearance of the "free product". Even projecting these results out an order of magnitude, with respect to actual field conditons, what is illustrated is that moderately large globules of DNAPL can be completely removed relatively quickly from the aquifer solely by microbial consumption

In a simple version of the Carr-Garg-Hughes experiment cited above, a column of acclimated, microbially active soil was flushed with water saturated with TCE (175 mg/L) until the soil was saturated with TCE and in equilibrium with the dissolved phase. HRC was added to the soil and the aqueous phase circulating through the system was analyzed on a regular basis. The results are shown in Figure 1 and essentially represent the action of HRC on the dissolved phase and the hydrophobically sorbed material that replenishes the dissolved phase as it diminishes.

In a modified version of the experiment replicate samples were created in which there was excess TCE. HRC was applied and the data was accumulated by complete analysis of a given set of tubes in a time series. In each test tube there was 10 grams of soil, 0.5 grams of TCE, 1.5 grams of HRC and 130 ml of distilled water. Figure 2 presents the change in total TCE mass in the system over time, noting that the actual measured Day 1 results are less than the intended delivered concentration and also subject to one day of consumption (value unknown). Once again, what is illustrated is that small globules of DNAPL can be completely removed relatively quickly from the aquifer solely by microbial consumption. Also, in contrast to the "Disappearing Drop" experiment, in which a mild flow was present, in this instance the system was completely static and still there was a rapid removal of residual source under microbial action.

All of these laboratory experiments showed that the pure phase TCE was remediated very effectively with HRC assisted bioremediation. Proving that bioremediation accelerates desorption in the field is a more daunting proposition, due to the heterogeneity of the system, the costs and uncertainly involved. Still, there is some indirect evidence in the data sets for bioremediation that can be extracted from some of our fieldwork.

Examining some of the field results, if we consider a simple sequential dechlorination of TCE to the first daughter product DCE, we can accumulate some indirect evidence for desorption events. Given that TCE tends to degrade faster than DCE, we can hypothesize that if desorption is occurring that we will see an excess of DCE in the system over time. So, as we get "turnovers" of dissolved phase TCE, where the dissolved phase is fed by desorption, the DCE levels should systematically increase. Once again, the DCE build-up is a function of the kinetic disparity - the slower rate of removal of DCE relative to the TCE.

This, in fact, is the case in a number of the forty-two data sets, indicating areas where an unknown residual was being treated. This is not uncommon due to the fact that it is often hard to precisely locate a DNAPL source. As a result, if we see a small reduction on TCE and a much larger increase in DCE in the same time period we are probably "turning over" the TCE pool by dechlorination.

A site in Kansas was contaminated with PCE at levels reaching 7,000 ug/L in a silt and clay aquifer with a groundwater velocity of approximately 0.03 ft/day. Depth to groundwater ranges from 5 to 9 ft bgs. HRC was applied to the area using 15 injection points. Figure 3, displaying the results from a single downgradient sentinel well, clearly shows that the DCE daughter product appears in excess of the PCE parent material over time. From Day 0 to Day 27, PCE concentrations decrease from 6,500 ug/L to 210 ug/L (97%). Past Day 27, PCE levels remain stable. TCE decreases from 840 ug/L at Day 0 to 540 ug/L at Day 118. DCE rises from a baseline concentration of 560 ug/L to a final concentration of 15,000 ug/L at Day 118. Translating this into moles, the removal of 38 moles/L of PCE and 2 mol/L per liter of TCE produced 149 mol/L of DCE. The 3.8X differential of DCE mass could be attributed to a desorbtive turnover of PCE and TCE.

Bioremediation of Pure TCE in Soil

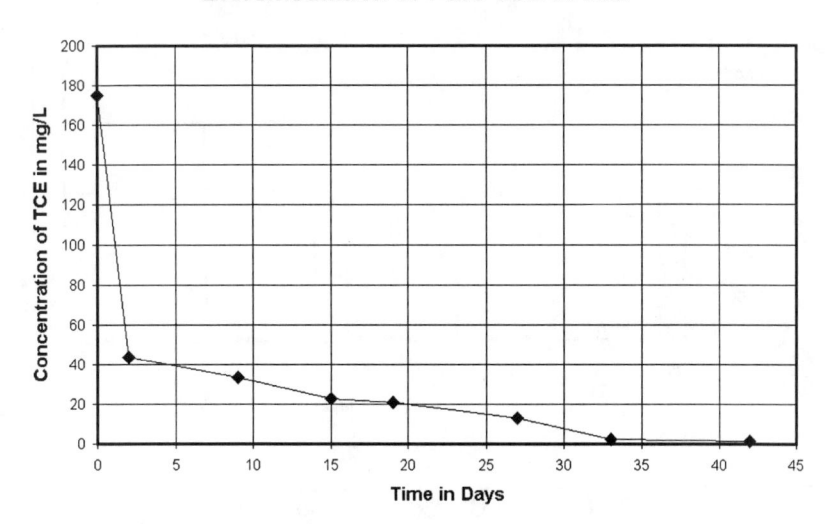

FIGURE 1. Bioremediation of Pure TCE in Soil.

Total TCE in System (in mg)

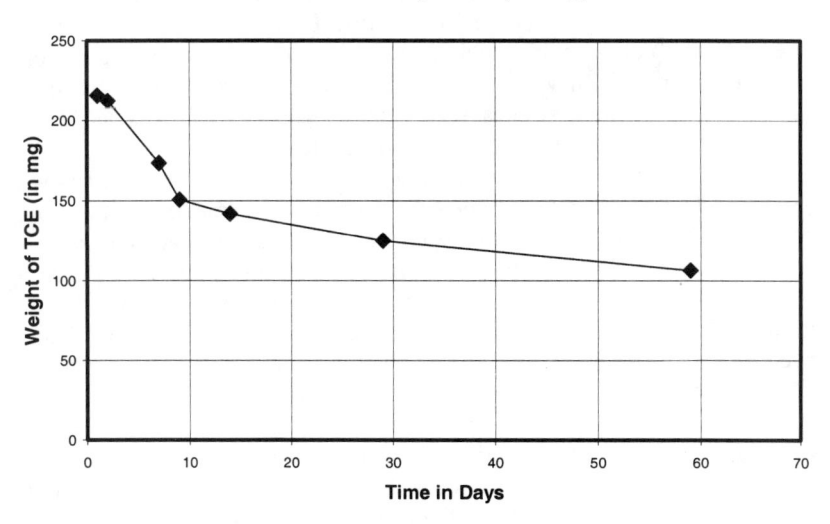

FIGURE 2. Total TCE in System (in mg).

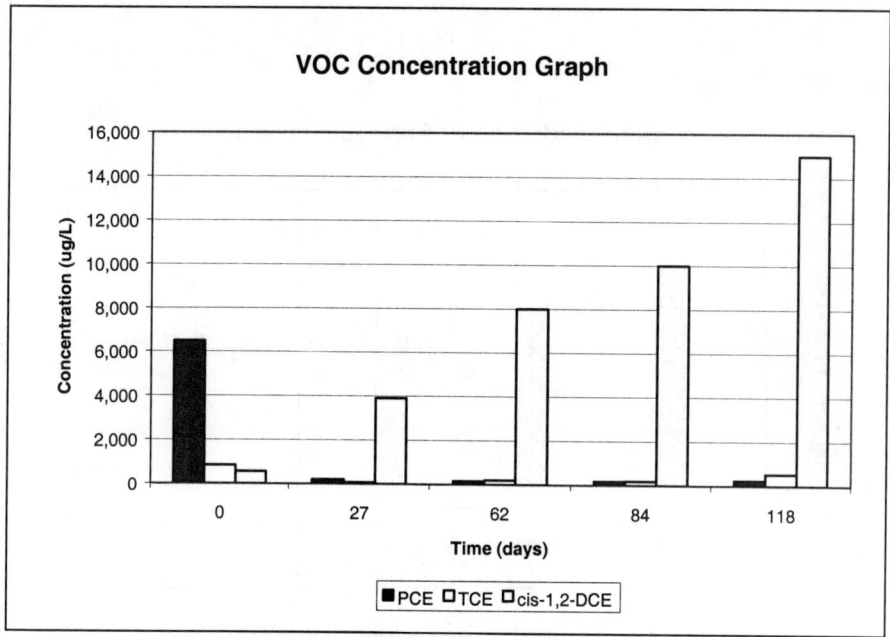

FIGURE 3. VOC Concentration Graph.

CONCLUSION

In spite of our technological sophistication, we are still humbled by the fact that aquifer remediation is a macroscopic, multi-variable problem. There are exceptions; but, in general, groundwater contamination problems are hard to characterize, understand, monitor, and solve. Brute force is not usually the answer—at least not at an affordable price and monitored natural attenuation has serious limitations. The new paradigm being advanced here is that a sensible and proactive thing to do to close sites is to engineer accelerated bioattenuation.

Through this terminology – accelerated bioattenuation - we establish that nature offers a solution while invoking the promise that there is a way to accelerate it. Unassisted natural remediation processes operate on a time scale that does not satisfy many that are responsible for environmental health and safety.

Hydrogen Release Compound has been demonstrated to be a simple, passive, low-cost and long-term option for the anaerobic bioremediation of chlorinated hydrocarbons (CHs) via reductive dehalogenation. HRC should be viewed as a tool for the enhancement of natural attenuation at sites that would require high levels of capital and operating expense.

HRC is best utilized for the remediation of dissolved phase plumes and the associated hydrophobically sorbed contaminant. While HRC has been shown to be applicable to addressing residual DNAPL source, its use would be contraindicated for free-product conditions unless the total mass to be remediated is within the scope of economic feasibility in comparison to alternative treatments. Results of on-going and future research as well as the results of commercial applications, where permission is granted, can be accessed on the web at www.regenesis.com.

REFERENCES

Anderson, D. 2000. "Remedial Action Using HRC Under a State Dry Cleaning Program". In: Wickramanayake, G.B., Gavaskar, A.R., Chen, A.S.C. (eds.), *Bioremediation and Phytoremediation of Chlorinated and Recalcitrant Compounds*, pp. 213-219. Battelle Press, Columbus, OH.

Boyle, S.L., V.B. Dick, M.N. Ramsdell and T.M. Caffoe. 2000. "Enhanced Closure of a TCE Site Using Injectable HRC^{TM}". In: Wickramanayake, G.B., Gavaskar, A.R., Chen, A.S.C. (eds.), *Bioremediation and Phytoremediation of Chlorinated and Recalcitrant Compounds*, pp. 255-262. Battelle Press, Columbus, OH.

Carr, C.S., S. Garg and J.B. Hughes. 2000. "The Effect of Dechlorinating Bacteria on the Longevity and Composition of PCE-Containing Non-Aqueous Phase Liquids under Equilibrium Conditions". *Environmental Science and Technology*. 34(6): 1088-1094.

Dooley, M., W. Murray and S. Koenigsberg. 1999. "Passively Enhanced In Situ Biodegradation of Chlorinated Solvents". In: Leeson, A., Alleman, B.C. (eds.), *Engineered Approaches for In Situ Bioremediation of Chlorinated Solvent Contamination*, pp. 121-127. Battelle Press, Columbus, OH.

Dooley, M. and W. Murray. 2000. "HRC-Enhanced Bioremediation of Chlorinated Solvents". In: Wickramanayake, G.B., Gavaskar, A.R., Chen, A.S.C. (eds.), *Bioremediation and Phytoremediation of Chlorinated and Recalcitrant Compounds*, pp. 287-294. Battelle Press, Columbus, OH.

Farone, W.A., S.S. Koenigsberg and J. Hughes. 1999. "A Chemical Dynamics Model for CAH Remediation with Polylactate Esters". In: Leeson, A., Alleman, B.C. (eds.), *Engineered Approaches for In Situ Bioremediation of Chlorinated Solvent Contamination*, pp. 287-292. Battelle Press, Columbus, OH.

Farone, W.A., S. Koenigsberg, T. Palmer and D. Brooker. 2000. "Site Classification for Bioremediation of Chlorinated Compounds Using Microcosm Studies". In: Wickramanayake, G.B., Gavaskar, A.R., Chen, A.S.C. (eds.), *Bioremediation and Phytoremediation of Chlorinated and Recalcitrant Compounds*, pp. 101-106. Battelle Press, Columbus, OH.

Kallur, S. and S. Koenigsberg. 1999. "Enhanced Bioremediation of Chlorinated Solvents- A Single Well Pilot Study". In: Leeson, A., Alleman, B.C. (eds.), *Engineered Approaches for In Situ Bioremediation of Chlorinated Solvent Contamination*, pp. 181-184. Battelle Press, Columbus, OH.

Koenigsberg, S.S. and W. Farone. 1999. "The Use of Hydrogen Release Compound (HRC™) for CAH Bioremediation". In: Leeson, A., Alleman, B.C. (eds.), *Engineered Approaches for In Situ Bioremediation of Chlorinated Solvent Contamination*, pp. 67-72. Battelle Press, Columbus, OH.

Koenigsberg, S.S. and C.A. Sandefur. 1999. "The Use of Hydrogen Release Compound for the Accelerated Bioremediation of Anaerobically Degradable Contaminants: The Advent of Time-Release Electron Donors". *Remediation Journal*. 10(1): 31-53.

Lodato, M., D. Graves, and J. Kean. 2000. "Enhanced Biological Reductive Dechlorination at a Dry-Cleaning Facility". In: Wickramanayake, G.B., Gavaskar, A.R., Chen, A.S.C. (eds.), *Bioremediation and Phytoremediation of Chlorinated and Recalcitrant Compounds*, pp. 205-211. Battelle Press, Columbus, OH.

Schuhmacher, T., W. Bow and J. Chitwood. 2000. "A Field Demonstration Showing Enhanced Reductive Dechlorination Using Polymer Injection". In: Wickramanayake, G.B., Gavaskar, A.R., Chen, A.S.C. (eds.), *Bioremediation and Phytoremediation of Chlorinated and Recalcitrant Compounds*, pp. 15-22. Battelle Press, Columbus, OH.

Sheldon, J.K., S.S. Koenigsberg, K.J. Quinn, and C.A. Sandefur. 1999. "Field Application of a Lactic Acid Ester for PCE Bioremediation". In: Leeson, A., Alleman, B.C. (eds.), *Engineered Approaches for In Situ Bioremediation of Chlorinated Solvent Contamination*, pp. 61-66. Battelle Press, Columbus, OH.

Sheldon, J.K. and K.G. Armstrong. 2000. "Barrier Implants for the Accelerated Bioattenuation of TCE". In: Wickramanayake, G.B., Gavaskar, A.R., Chen, A.S.C. (eds.), *Chemical Oxidation and Reactive Barriers: Remediation of Chlorinated and Recalcitrant Compounds*, pp. 347-352. Battelle Press, Columbus, OH.

Wu, M. (1999). "A Pilot Study Using HRC™ to Enhance Bioremediation of CAHs". In: Leeson, A., Alleman, B.C. (eds.), *Engineered Approaches for In Situ Bioremediation of Chlorinated Solvent Contamination*, pp. 177-180. Battelle Press, Columbus, OH.

Zahiraleslamzadeh, Z.M. and J.C. Bensch. 2000. "Enhanced Bioremediation Using Hydrogen Release Compound (HRC™) in Clay Soils". In: Wickramanayake, G.B., Gavaskar, A.R., Chen, A.S.C. (eds.), *Bioremediation and Phytoremediation of Chlorinated and Recalcitrant Compounds*, pp. 237-244. Battelle Press, Columbus, OH.

FAVORING EFFICIENT IN SITU TCE DECHLORINATION THROUGH AMENDMENT INJECTION STRATEGY

Jennifer P. Martin, Kent S. Sorenson, Jr., and Lance N. Peterson
Idaho National Engineering and Environmental Laboratory, Idaho Falls, ID, USA

ABSTRACT: An enhanced in situ bioremediation field test has been ongoing at the Test Area North (TAN) Facility of the Idaho National Engineering and Environmental Laboratory (INEEL) since November 1998 to determine whether anaerobic reductive dechlorination (ARD) of a trichloroethene (TCE) source area could be enhanced through the addition of an electron donor (lactate). The results collected at TAN have led to five conclusions. (1) Propionate degradation in the absence of lactate fermentation supports a microbial population capable of more efficient TCE degradation. (2) Lactate injection strategy can be manipulated to create conditions that favor this population. (3) This strategy has successfully led to the complete dechlorination of TCE to a radius of 15-30 m around the injection well. (4) The area of active ARD does not encompass the entire residual source in the downgradient direction. (5) Additional lactate injection locations are required to achieve the necessary distribution of electron donor.

INTRODUCTION

Site Background. Historical waste disposal activities have resulted in a nearly 3-km-long trichloroethene (TCE) plume in the groundwater at the TAN Facility of the INEEL. Facility process waste consisting of liquid organic, inorganic, and low-level radioactive waste along with sanitary sewage wastewater was injected directly into the Snake River Plain Aquifer (SRPA) via injection well TSF-05 from the mid-1950s to 1972.

Previous Remedial Activities. The 1995 Record of Decision (ROD) selected pump and treat as the default remediation technology with the provision to evaluate five innovative technologies, including enhanced bioremediation. Continuous pump and treat operations began in November 1996 and operated for approximately 18 months prior to preparing for the bioremediation field evaluation.

Enhanced Bioremediation Field Evaluation. A field evaluation was conducted from November 1998 to September 1999 to determine whether anaerobic reductive dechlorination (ARD) of TCE could be enhanced through the addition of an electron donor (lactate). A complete description of the operations and data generated during the field evaluation is presented in Sorenson (2000).

ARD is the process by which chloroethene compounds (in this case) are transformed to ethene through the sequential removal and replacement of chlorine atoms with hydrogen atoms via the following pathway (Freedman and Gossett, 1989): tetrachloroethene (PCE) → TCE → dichloroethene (DCE) → vinyl chloride

(VC) → ethene. During this process, the chloroethenes act as electron acceptors in the microbial respiration process. The respiration process is carried out in the presence of an appropriate electron donor, providing microorganisms with the energy for growth and cell maintenance. The electron donor selected for the evaluation was lactate (in the form of sodium lactate) based on the results of laboratory studies (Barnes et al., 1998).

Lactate injection began at Well TSF-05 in January 1999 (Figure 1). The injection strategy created electron donor concentrations in the thousands of milligrams per liter surrounding the injection well shortly after the initial injection. The effect on redox conditions and ARD reactions was also rapid in source area wells. Sulfate reduction began within 4 weeks of injection and methanogenic conditions were achieved within 5 months (Sorenson, 2000). ARD of TCE to cis-DCE began coincident with sulfate reduction in source area wells (Sorenson et al., 2000a). Complete ARD of cis-1,2-DCE to VC and ethene corresponded exactly to the onset of methanogenic conditions. After 8 months of lactate injection, complete ARD was observed wherever electron donor was sufficiently distributed to provide strongly reducing (methanogenic) conditions (greater than 40 m from the injection well in some cases).

Objectives. Following the 8 month Field Evaluation Phase, when lactate was injected weekly, concentrations of electron donor up to 4,500 mg/L were present in the aquifer treatment cell (Sorenson, 2000). At this time, lactate injections were discontinued for 5 months in order to (1) Determine the persistence of electron donor and ARD reactions within the treatment cell and (2) Evaluate the efficiency of ARD reactions in the prolonged presence of acetate, propionate, and butyrate (fermentation products) as electron donors. This period of time is referred to as Pre-Design Phase I (PDP-I). Lactate injections were renewed once the electron donor in the treatment cell was depleted (~ 5 months later). The objective of this phase, referred to as Pre-Design Phase II (PDP-II), was to recreate the conditions that favored more efficient ARD in the absence of lactate injections during PDP-I by manipulating the lactate injection strategy.

MATERIALS AND METHODS

Lactate Injection. The strategy for lactate injection during the Field Evaluation Phase consisted of weekly injections, providing the system with a steady supply of electron donor. In PDP-II the goal was to provide a large mass of lactate on an infrequent basis. In order to achieve this, a mass of lactate corresponding to twice that of a single Field Evaluation Phase injection was injected every 8 weeks. The injection location (Well TSF-05) was the same throughout the study.

Monitoring Locations, Frequency, and Analytes. The treatment cell consisted of an injection well (Well TSF-05), an extraction well located approximately 150 m downgradient (Well TAN-29), and 11 monitoring locations (Figure 1). Groundwater monitoring was conducted at 8 locations on a biweekly basis and at all 11 locations on a monthly basis. Analytes consisted of electron donors

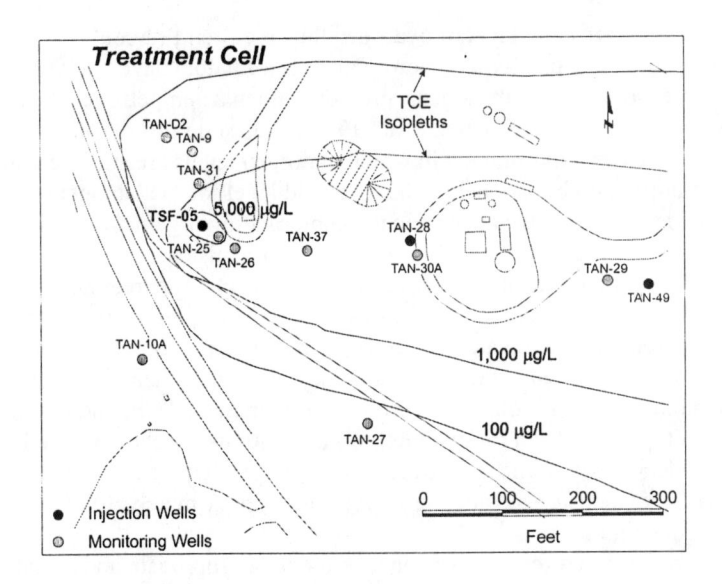

FIGURE 1. Bioremediation treatment cell.

(lactate, propionate, acetate, butyrate, and chemical oxygen demand), bioactivity indicators and nutrients (carbon dioxide, alkalinity, ammonia, and phosphate), redox indicators (dissolved oxygen, nitrate, ferrous iron, sulfate, methane, and oxidation-reduction potential), contaminants and degradation products (PCE, TCE, 1,1-DCE, cis-DCE, trans-DCE, VC, ethene, and ethane), water quality indicators (temperature, pH, and conductivity), and tritium, a co-contaminant that was used as a tracer.

RESULTS AND DISCUSSION

Electron Donor Utilization and Effect on ARD Reactions. Lactate can be fermented via two pathways (as presented in Fennell and Gossett, 1998):

$$3 \text{ Lactate}^- \rightarrow \text{Acetate}^- + 2 \text{ Propionate}^- + \text{HCO}_3^- + \text{H}^+ \quad (1)$$
$$\text{Lactate}^- + 2 \text{ H}_2\text{O} \rightarrow \text{Acetate}^- + \text{HCO}_3^- + \text{H}^+ + 2 \text{ H}_2 \quad (2)$$

Pathway 1 produces propionate and acetate at a stoichiometric ratio of 2:1. Following lactate fermentation via Pathway 1, propionate can be fermented via Pathway 3 (as presented in Fennell and Gossett, 1998):

$$\text{Propionate}^- + 3 \text{ H}_2\text{O} \rightarrow \text{Acetate}^- + \text{HCO}_3^- + \text{H}^+ + 3 \text{ H}_2 \quad (3)$$

These fermentation pathways produce the free hydrogen necessary to support ARD. Hydrogen can either be produced during fermentation of lactate via Pathway 2, or from the fermentation of propionate (Pathway 3), which is produced via Pathway 1.

An understanding of the dominant pathways being utilized is important because populations performing certain reactions may be more or less desirable than others in terms of facilitating ARD. It has been hypothesized that the level of hydrogen produced is important because hydrogenotrophic organisms each have a minimum threshold level of hydrogen they require. Previous investigations indicate that hydrogenotrophic dechlorinators have a lower hydrogen threshold than methanogens, providing them a competitive advantage in hydrogen-limited systems (Smatlak et al., 1996). Based on this, it may be desirable to create an environment in which hydrogen levels are sufficient to sustain dechlorination but are below the threshold level for methanogens. This would provide a competitive advantage to dechlorinating organisms while minimizing electron donor demands of competing organisms.

The microbial utilization pathway of lactate within the treatment cell at TAN was determined by examining the relative concentrations of the lactate degradation products propionate and acetate. The corresponding efficiency of ARD reactions was correlated to changes in the dominant utilization pathway using the molar concentrations of the primary contaminant, TCE, and the final ARD product, ethene. Data for two monitoring locations are shown. Well TAN-25 is located within the source area 8 m downgradient of the injection well. Well TAN-37A is located outside the source area, approximately 40 m downgradient (Figure 1).

Figure 2a shows the relative concentrations of propionate and acetate, expressed as the propionate to acetate ratio (P:A), and the molar concentrations of TCE and ethene at TAN-25 during the three phases of the study. During the Field Evaluation Phase, lactate was injected on a weekly basis and the P:A remained relatively stable at approximately 1.5:1. This indicates that the predominant lactate fermentation pathway being utilized was that of Pathway 1 as it produces a P:A of 2:1. TCE dechlorination began approximately 6 weeks after the initial lactate injection. Complete ARD as evidenced by significant concentrations of ethene began after approximately 6 months. In September 1998, lactate injections were discontinued (marking the beginning of PDP-I) and the P:A decreased rapidly as propionate fermentation via Pathway 3 became the dominant utilization pathway. During this time of propionate utilization, TCE decreased and significant quantities of ethene were produced. This indicates that the efficiency of ARD reactions under the propionate-fermenting conditions of PDP-I increased compared to the Field Evaluation Phase when lactate fermentation was a dominant activity.

Based on these results, the goal of PDP-II was to recreate the conditions observed during PDP-I that supported efficient ARD. The lactate injection strategy for PDP-II was designed to supply a large volume of lactate on an infrequent basis, thus minimizing the time of lactate fermentation and maximizing the period of propionate fermentation, the conditions under which ARD was most efficient. As shown in Figure 2a, lactate injection resulted in an initial increase in the P:A followed by a decrease as propionate fermentation to acetate via Pathway 3 was occurring. This response was repeated with each subsequent lactate injection. After a slight increase at the start of PDP-II, TCE concentrations remained near the detection limits. The production of ethene continued, indicating continued efficient ARD; however, concentrations decreased relative to their PDP-I values. This may indicate that the conditions created during PDP-I were not completely recreated using the PDP-II lactate injection strategy.

TAN-25

(a)

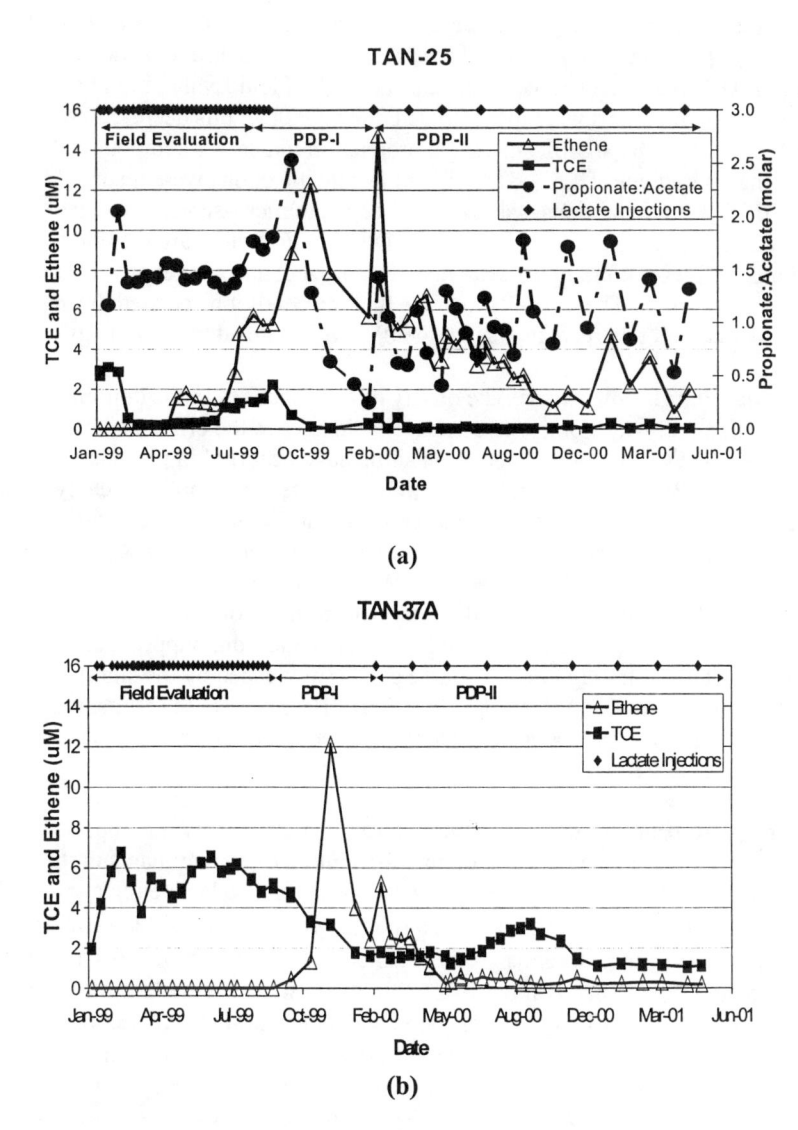

TAN-37A

(b)

FIGURE 2. Concentrations of TCE and ethene and propionate:acetate for wells TAN-25(a) and TAN-37A (b).

Alternatively, it may indicate that the flux of contaminants from the source has decreased by the activities performed to date.

Well TAN-37A was located downgradient of the source area outside the zone of active ARD (Figure 1) and provided an indication of the conditions along the flowpath downgradient of TAN-25. The travel time from Well TSF-05 to Well TAN-37A is approximately seven days. Figure 2b presents the molar concentrations of TCE and ethene in TAN-37A during each phase of the test.

Electron donor (propionate and acetate) is utilized before it reaches this monitoring location. A consistent decrease in TCE concentrations indicating the onset of ARD was not observed until the end of the Field Evaluation Phase. When lactate injections were discontinued during PDP-I, this decrease in TCE was accompanied by significant ethene production indicating complete ARD was occurring upgradient of TAN-37A. When lactate injections were resumed in PDP-II, TCE concentrations stabilized and began to increase while ethene production ceased. Results at the end of PDP-II indicate that after its slight rebound, TCE concentrations declined and stabilized, and ethene production remained minimal. These results indicate that the conditions created during PDP-I were not completely recreated using the PDP-II lactate addition strategy.

Microbial Population Shifts. The observation of changes in the dominant fermentation pathway along with concurrent changes in the efficiency of ARD reactions suggests microbial population shifts have occurred in response to a change in the available electron donor. When lactate was supplied weekly during the Field Evaluation Phase, lactate was abundant and lactate fermentation was a dominant activity. While these conditions did support enhanced ARD of TCE, TCE was still present in the source area. When lactate injection was stopped during PDP-I, propionate was the dominant electron donor and propionate fermentation became a dominant activity. These conditions supported more rapid and efficient ARD throughout the source area as evidenced by the significant quantities of ethene in TAN-25 and TAN-37A. Based on these results, it appears that the conditions under which the propionate fermentation activity was dominant resulted in more efficient ARD.

Impact of Bioremediation Remedy. Figures 3a and 3b show the impact of almost 2 years of lactate injection on the distribution of contaminants in the treatment cell. After 18 months of pump and treat prior to lactate injection, TCE was present in the source area at concentrations greater than 3200 µg/L and downgradient at concentrations greater than 800 µg/L (Figure 3a). After 2 years of lactate addition, concentrations of TCE in the source area are below 5 µg/L (Figure 3b). Downgradient, the large area with concentrations greater than 800 µg/L seen in Figure 3a is greatly reduced in size and the maximum concentration has been reduced by half to around 400 µg/L. This bioremediation effort has been extremely successful compared to the pump and treat remedy. However, a zone of relatively high TCE concentrations remains immediately downgradient of the source area. The lactate addition strategy employed during PDP-II does not appear to be completely effective in impacting the full extent of the source area as observed during PDP-I (Sorenson et al., 2000b).

CONCLUSIONS

The results collected at TAN have led to five conclusions. (1) Propionate degradation in the absence of lactate fermentation supports microbial activity capable of more efficient TCE degradation. (2) Lactate injection strategy can be manipulated to create conditions that favor this population. (3) This

Electron Donor Injection Strategies 271

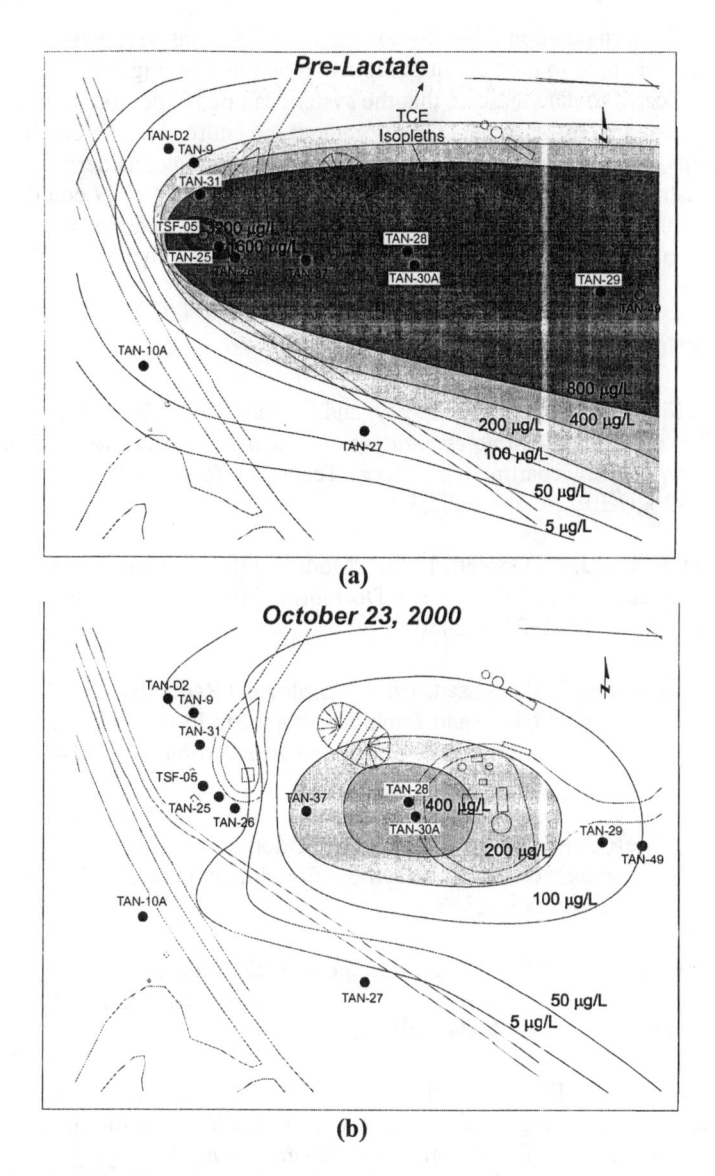

FIGURE 3. Concentrations of TCE after 18 months of pump and treat (a) and after almost 2 years of lactate injection (b).

strategy has successfully led to the complete dechlorination of TCE to a radius of 15-30 m around the injection well. (4) The area of active ARD does not encompass the entire residual source in the downgradient direction.
(5) Additional lactate injection locations may be required to achieve the necessary distribution of electron donor.

The bioremediation field test conducted at TAN has been extremely successful compared to the default pump and treat remedy (Figures 3a and 3b). Results collected to date indicate that the system can be further manipulated to achieve more efficient ARD of TCE throughout the entire source area. In order to achieve this, additional lactate injection locations are being considered to achieve a better distribution of electron donor. In particular, Well TAN-37 could be utilized as an injection location in order to distribute electron donor within the downgradient portion of the source area, the area not currently being effectively impacted.

REFERENCES

Barnes, J. M., G. E. Matthern, R. L. Ely, and C. Rae. 1998. *Microbial Studies Report Supporting Implementation of In Situ Bioremediation at Test Area North.* U. S. Department of Energy Technical Report, INEEL/EXT-99-00736, INEEL, Idaho Falls, ID.

Fennell, D. E., and J. M. Gossett. 1998. "Modeling the Production of and Competition for Hydrogen in a Dechlorinating Culture." *Environ. Sci. Technol.* 32(16): 2450-2460.

Freedman, D. L. and J. M. Gossett. 1989. "Biological Reductive Dechlorination of Tetrachloroethylene and Trichloroethylene to Ethylene Under Methanogenic Conditions." *Appl. Environ. Microbiol.* 55: 2144-2151.

Smatlak, C. R., J. M. Gossett, and S. H. Zinder. 1996. "Comparative Kinetics of Hydrogen Utilization for Reductive Dechlorination of Tetrachloroethene and Methanogenesis in an Anaerobic Enrichment Culture." *Environ. Sci. Technol.* 30(9): 2850-2858.

Sorenson, K. S. 2000. *Intrinsic and Enhanced in situ Biodegradation of Trichloroethene in a Deep, Fractured Basalt Aquifer.* Ph.D. Dissertation. University of Idaho, Idaho Falls, ID.

Sorenson, K. S., L. N. Peterson, and R. L. Ely, 2000a. "In Situ Biostimulation of Reductive Dehalogenation – Dependence on Redox Conditions and Electron Donor Distribution." *Groundwater 2000*, P. L. Bjerg, P. Engesgaard, and Th. D. Krom, eds., A. A. Balkema Publishers, Rotterdam, Netherlands, pp. 379-380.

Sorenson, K. S., L. N. Peterson, and R. L. Ely, 2000b. "Enhanced In Situ Bioremediation of a TCE Source Area in Deep, Fractured Rock." *Contaminated Site Remediation: From Source Zones to Ecosystems*, Proceedings of the 2000 Contaminated Site Remediation Conference, C. D. Johnston ed., Centre for Groundwater Studies, Wembley W. A., Australia, pp. 621-628.

DESIGN OF A NOVEL INJECTION SCHEME FOR ENHANCED ANAEROBIC BIOREMEDIATION

Michael E. Miller and John T. Drake, Camp Dresser & McKee Inc., Cambridge, Massachusetts, USA

ABSTRACT: Following contaminant source removal, low-grade concentrations of chlorinated solvents remain in the groundwater beneath a former factory basement floor. Microbially mediated natural attenuation occurred for a time, but ultimately stopped before the chlorinated contaminants were completely degraded due to depletion of the carbon source. The geochemistry of the contaminated aquifer was investigated with respect to the chlorinated solvent history and biodegradation indicator parameters. Injection of a metabolic carbon substrate/electron donor was selected to overcome the chemical limitations and promote enhanced anaerobic bioremediation of the chlorinated solvents, which themselves behaved as terminal electron acceptors. A novel substrate injection scheme is described, which employs a sequential process of substrate addition across the contaminated zone. In the process, a fraction of the contaminated groundwater was extracted, mixed with a carbon substrate, and re-injected, all with minimal contact with the outside air.

INTRODUCTION

Chlorinated solvent contamination is widespread in soil and groundwater, since these chemicals found common use in cleaning applications ranging from industrial machinery to fine apparel. Tetrachloroethene, also known as perchoroethylene (PCE), was one of the most generally used solvents and the primary site contaminant.

In situ bioremediation of PCE in groundwater relies on anaerobic processes in which the chlorinated solvent and its breakdown products serve as the terminal electron acceptor (TEA) of the electron transport chain. Under favorable aquifer conditions, including sufficient reducing potential, anaerobic bioremediation of PCE can occur naturally. However, if one or more necessary elements are missing, natural biodegradation cannot occur. Enhanced anaerobic bioremediation (EAB) overcomes such chemical limitations with the controlled addition of any limiting materials directly into the contaminated groundwater.

The chemistry of a PCE-contaminated aquifer is described below. The design of the EAB system, tailored specifically to site hydrogeological and chemical conditions, is also detailed below. The system was installed in late April/early May 2001. The remediation is an ongoing project, evaluated by periodic groundwater monitoring, which will continue for at least one year after installation. Results of the groundwater remediation will be published in the future.

SITE BACKGROUND

The site, located in New Jersey, contains a former manufacturing facility where glass containers for the food industry were produced until 1991. A well-defined area located beneath the basement floor of the compressor and boiler room of a former factory building contained chlorinated ethene-contaminated groundwater. The cause of contamination was thought to be a spill and/or leak of PCE and possibly trichloroethene (TCE) in the former compressor area.

Source removal was undertaken in January 1995, with the removal of the basement floor in the source area and the excavation of approximately 45 cubic yards (34.4 cubic meters) of solvent-contaminated soil from directly beneath the floor. Further residual groundwater contamination was removed during the subsequent pumping and activated carbon treatment of approximately 89,000 gallons (337,000 liters) of contaminated groundwater between April and May 1996. The basement floor was replaced and sealed in February 1998. The residual groundwater contaminants were PCE, TCE, and the further breakdown products cis-1,2-dichloroethene (DCE), and vinyl chloride (VC).

In the contaminated area, the average depth to the water table was 1.5 ft (0.5 m) below the basement floor. The contaminated portion of the aquifer lay in a moderately permeable silty sand and was underlain by an iron-cemented sandstone layer, 1.5 to 2 ft (0.5 to 0.6 m) thick, and approximately 8.5 ft (2.6 m) below the basement floor. The sandstone likely provided a relatively impermeable layer to downward migration of the contaminants. Groundwater hydraulic conductivity in the vicinity of the contaminated area was determined, by rising head methods, to be 1.00×10^{-4} cm/sec. Based on water levels, the local hydraulic gradient was approximately 0.0054. With an assumed effective porosity of 0.3, the groundwater velocity in the area was calculated at 1.9 ft/yr (0.58 m/yr). Due to the minimal groundwater migration, the contaminated zone remained small, with a footprint of approximately 20 ft by 40 ft (6 m by 12 m).

CONTAMINATED GROUNDWATER CHEMISTRY

Figure 1 shows chlorinated organic compound concentrations measured in groundwater from a monitoring well located in the contaminated zone from 1995 through 2000. After the 1995 soil removal, groundwater concentrations of chlorinated ethenes PCE, TCE, and DCE showed a steady decline at similar disappearance rates, probably via volatilization through the area of the removed basement floor. At the same time, VC concentrations remained below the method detection limit. Groundwater pumping in 1996 resulted in further concentration decreases of the three chlorinated ethenes, but these concentrations recovered through 1997 as the contaminants desorbed from soil particles and diffused back into the monitored groundwater.

After the basement floor was replaced in early 1998, PCE and TCE concentrations began to decrease again, DCE concentrations increased at first and then began to drop, and following a lag of several months, measurable VC concentrations were observed for the first time. This last set of changes was attributable to anaerobic biodegradation of the chlorinated ethenes. With the replacement of the basement floor, the fresh supply of oxygen to the near-surface

Electron Donor Injection Strategies

groundwater was removed. It is believed that a small quantity of fuel oil from the boilers originally located in the same room was discarded into a floor drain or leaked from piping in the contaminated area at around the time of the original solvent spill. The fuel oil co-contamination then provided a limited-quantity of hydrocarbons that served as electron donors to trigger dechlorination of the chlorinated ethenes after the aquifer became anaerobic. By early 1999, the chlorinated ethene concentrations had stabilized within the range of 10 µg/L to 40 µg/L, where they showed no further statistical change through 2000. In May 2000, concentrations of the chlorinated ethenes were as follows: 25 µg/L PCE, 42 µg/L TCE, 40 µg/L DCE, and 11 µg/L VC.

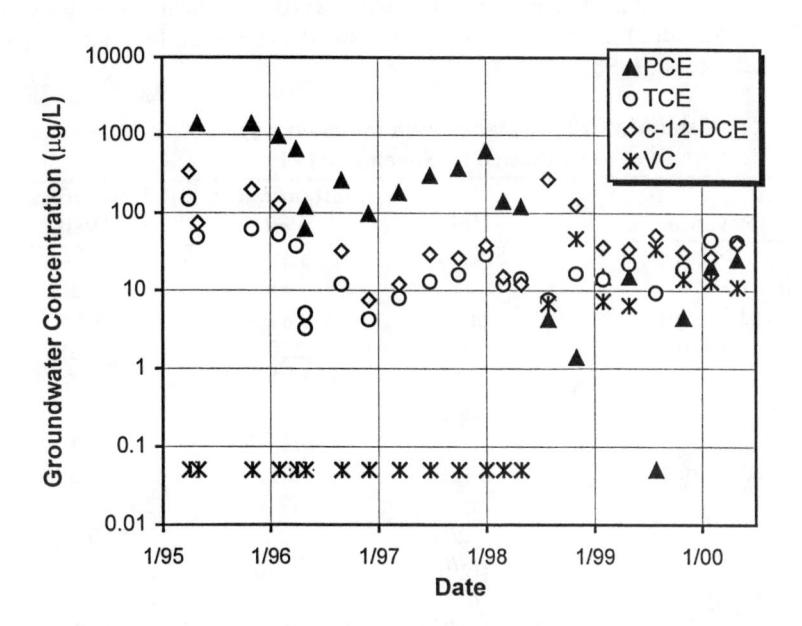

FIGURE 1. Concentrations of chlorinated ethenes (log scale) measured during quarterly groundwater monitoring within the contaminated zone.

The geochemistry of the contaminated groundwater was well suited for anaerobic biodegradation of the chlorinated ethene contaminants. In November 1999, biodegradation indicator parameters were measured in groundwater samples from the contaminated zone well, and a background well located approximately 100 ft (30.5 m) up-gradient of the contaminated zone (Table 1). The differences in geochemistry between water from the contaminated and background wells demonstrated the effect of the contaminants on groundwater conditions.

The contaminated groundwater was strongly reducing, with a redox potential of –220 mV, compared to the background value of +261 mV. The fuel oil co-contamination most likely served as the electron donor that depleted various electron acceptor concentrations and created reducing conditions in the contaminated zone. The contaminated well yielded low dissolved oxygen at 0.8 mg/L, compared to the background of 3.7 mg/L. Also in the contaminated well water, the TEA nitrate was largely exhausted (< 0.2 mg/L vs. the background value of 14 mg/L), while elevated concentrations of ferrous iron (76 mg/L), dissolved manganese (0.8 mg/L), and methane (12 mg/L) demonstrated that TEAs ferric iron, manganese dioxide, and carbon dioxide have been used. Only sulfate remained at the same level (57 mg/L) in the two test wells. Thus, the contaminated groundwater was anaerobic, reducing, and sulfate reducing/methanogenic, all conditions that are favorable for reductive dechlorination of chlorinated ethenes. The concentration of dissolved hydrogen measured in the contaminated groundwater, 2.8×10^{-9} moles/liter (vs. 0.8×10^{-9} mol/L background), was also conducive to reductive dechlorination (see Wiedemeier et al., 1998).

TABLE 1. Groundwater biodegradation indicator parameters from samples collected in November 1999.

Parameters	Units	Contaminated Well	Up-gradient Well
pH	Standard	6.56	5.04
Alkalinity	mg/L	190	12
Dissolved Oxygen	mg/L	0.84	3.74
Redox Potential	mV	-220	+261
Conductivity	µMho	817	316
Nitrate	mg/L	< 0.2	14
Ferrous Iron	mg/L	76	< 1
Dissolved Manganese	mg/L	0.81	0.059
Sulfate	mg/L	57	57
Methane	mg/L	12	0.000019
Ethene	µg/L	26	0.013
Dissolved Hydrogen	nM	2.8	0.84
Chloride	mg/L	26	12
Dissolved Organic Carbon	mg/L	6.4	< 1

All intermediate reductive dechlorination breakdown products of PCE, namely TCE, DCE, and vinyl chloride, were present in the contaminated groundwater. The final breakdown product, ethene, was present at a concentration (26 µg/L) that was more than three orders of magnitude above the background value of 0.013 µg/L. The fact that the chlorinated ethene concentrations remained essentially constant since 1999 was evidence that although reductive dechlorination occurred, the process appeared to stop with incomplete destruction of the PCE-related compounds.

In order to estimate the electron donor demand of the contaminated groundwater, lactate ($CH_3CHOHCOO^-$) was chosen as a model substrate easily metabolized by soil microbes. Balanced chemical equations were formulated with lactate as the electron donor and the following electron acceptors: oxygen, nitrate, manganese dioxide, ferric iron, sulfate, carbon dioxide, PCE, TCE, DCE, and VC. Given the measured concentrations (contaminated zone vs. background) of the electron donors, a total electron donor demand of approximately 102 mg/L of organic compounds as lactate was calculated. This result meant that at least 102 mg/L of lactate would be necessary to maintain reducing conditions and allow the chlorinated solvents to be dechlorinated.

The dissolved organic carbon (DOC) level measured in the contaminated groundwater was 6.4 mg/L, an amount insufficient to serve as an ongoing source of food and electrons to maintain reductive dechlorination. Background DOC concentrations were below the method detection limit of 1 mg/L, indicating that the only historical carbon source was residual fuel oil in the contaminated zone. Thus, it was believed that after replacement of the basement floor cut off the supply of oxygen, the PCE-related contaminants naturally degraded through anaerobic biodegradation until the existing organic carbon source/electron donor was depleted, leaving a persistent residual level of each of the chlorinated ethenes.

ENHANCED ANAEROBIC BIOREMEDIATION SYSTEM DESIGN

Overview. A series of microwells were installed by direct-push technology in a grid pattern across the treatment area. In a one-time event, carbon substrate was injected through the microwells. Substrate injection was accomplished by extracting 2.6% of the groundwater to be treated, mixing in sodium lactate and yeast extract, and re-injecting the amended groundwater.

Groundwater pumping, mixing, and re-injection was performed under controlled conditions, observing the following precautions:
- Avoid contact of the groundwater with any metal parts (except stainless steel). Such metal contact would cause slight increases in the dissolved hydrogen concentration, which could mask the biologically produced hydrogen concentration in the groundwater through the addition of abiotically generated hydrogen.
- Minimize groundwater contact with atmospheric oxygen by keeping the pumping and mixing systems as a closed loop. Exclusion of oxygen helps maintain anaerobic conditions in the aquifer. Any oxygen added by atmospheric contact will deplete a portion of the injected sodium lactate.

After carbon substrate injection, the condition of the microbial system is currently being monitored by periodic measurement of biodegradation indicator parameters from monitoring wells and microwells in the treatment zone and from the up-gradient background location. Contaminant monitoring is being performed as VOC analyses from the same wells.

Microwell Installation. The microwells were installed in a square grid pattern across the 20-by-40-ft (6-by-12-m) treatment area. A conservatively small radius of influence of 4 ft (1.2 m) for substrate injected through one of the wells was estimated. Thus, the wells were installed in six rows of 3 wells each, for a total of 18 microwells, with an inter-well spacing of 6 2/3 ft (2 m).

The microwells were installed by direct-push drilling directly through the basement floor and into the underlying soils. Boreholes were approximately 2 inches (5 cm) in diameter. Each well was advanced until the iron-cemented sandstone confining layer was reached.

Each well was finished below ground with a 1-inch (2.5-cm) inner diameter, 10-slot PVC screen. A 1-inch (2.5-cm) inner diameter PVC casing extended approximately 4 inches (10 cm) above ground and was fitted with a locking cap. As needed for groundwater re-injection, a 1-inch (2.5-cm) inner diameter PVC pipe, 8 to 10 ft (2.4 to 3 m) long, was attached with a rubber coupling to the well riser.

System Set-up and Operation. The treatment zone contained approximately 12,570 gallons (47,570 liters) of groundwater to be treated. Approximately 2.6 percent of that groundwater (330 gal or 1250 L total) was removed, mixed with carbon substrate, re-injected, and the groundwater left to react. Substrate injection was a one-time event conducted over the course of one week.

Groundwater was pumped from one row of three microwells at a time. Thus, approximately 55 gal (208 L) of groundwater (1/6 of the total to be removed) was extracted at one time. The extraction was achieved with three peristaltic pumps, one set at each well head. When any well stopped producing water, the pump was shut off until the well recovered in order to avoid pumping air. The extracted water was pumped into an empty, collapsed bladder tank, which expanded as it filled thereby contributing as little oxygen as possible to the water. The bladder tank and its twin were set approximately 7 ft (2 m) above the basement floor on scaffolding.

After 55 gal (208 L) of groundwater were pumped into one bladder tank, the following two chemicals were added: (1) sodium lactate as carbon and electron source, and (2) yeast extract for trace minerals, vitamins, and cofactors. The contaminated groundwater electron donor demand was estimated above to be 102 mg/L of lactate, which equals 127 mg/L of sodium lactate. The carbon substrate feed mixture was formulated to yield final in-aquifer concentrations after dilution of 286 mg/L of sodium lactate (more than twice the required level) and 2.8 mg/L of yeast extract (approximately one percent of the added carbon source based on a typical microbial nutrient broth: Madsen, 2000). The aqueous substrate solution was pumped into the filled groundwater tank and re-circulated to mix. The mixing was done in a closed loop in order to admit no outside air into the system.

After the added carbon substrate was thoroughly mixed, the groundwater (with additives) was pumped back into the microwells from which it came, under the force of gravity (generated by the elevation of the bladder tank). Simultaneously, 55 gal (208 L) of groundwater from a second row of three microwells was pumped into the second bladder tank, providing additional force to draw the substrate-amended water back into the aquifer.

The extraction and re-injection process took place in a systematic fashion, moving stepwise in the direction of groundwater flow. The farthest up-gradient row of microwells provided the first batch of 55 gal (208 L) of groundwater.

After the carbon substrate was added and mixed, this groundwater was re-injected while the second batch of groundwater was extracted into the second bladder tank from the next down-gradient row of microwells. The process was repeated until the final row of microwells was reached. At this point, the final batch of substrate-amended groundwater was re-injected without simultaneous extraction of further groundwater.

System Monitoring. With carbon substrate injection now completed, the following groundwater biodegradation indicator parameters are now being measured periodically to evaluate the microbial reductive dechlorination system:
- field measurements - temperature, pH, conductivity, redox potential;
- inorganic parameters - alkalinity, chloride, ferrous iron, dissolved manganese, sulfate, nitrate;
- dissolved gases - oxygen, carbon dioxide, nitrogen, methane, ethane, ethene, hydrogen;
- metabolic acids - acetic, propionic, pyruvic, butyric, lactic; and
- dissolved organic carbon.

The biodegradation parameters are being analyzed in groundwater samples from the contaminated well, the up-gradient background well, two down-gradient monitoring wells, and four microwells in different portions of the treatment area. The contaminant concentrations are also being followed during the EAB process by analysis of the same groundwater samples for volatile organic compounds (VOCs).

CONCLUSIONS

Depletion of the carbon source can bring microbially mediated natural attenuation of chlorinated ethenes in groundwater to a halt. In such cases, injection of a carbon substrate directly into the contaminated groundwater can promote EAB to complete the dechlorination process. An efficient strategy to implement EAB involves removing a fraction of the contaminated groundwater, mixing in carbon substrate, and re-injecting the water, all with minimal contact with the outside air. A sequential pattern of groundwater extraction, and re-injection across the contaminated zone can be employed to overcome the chemical delivery limitations of low aquifer permeability and low groundwater flow rates.

REFERENCES

Madsen, E.L. 2000. Professor of Microbiology, Cornell University, Ithaca, NY. Personal communication.

Wiedermeier, T.H. et al. 1998. *Technical Protocol for Evaluating Natural Attenuation of Chlorinated Solvents in Ground Water.* U.S. Environmental Protection Agency Technical Report, EPA/600/R-98/128, National Risk Management Laboratory Office of Research and Development, Cincinnati, OH.

ENHANCED BIOREMEDIATION UNDER DIFFICULT GEOLOGIC CONDITIONS – CASE STUDIES

Nichole L. Case, Susan L. Boyle, Vincent B. Dick, (Haley & Aldrich, Inc., Rochester, New York)

ABSTRACT: Enhanced in-situ bioremediation has been shown to be an effective remediation approach at numerous sites contaminated with a variety of compounds. Implementation of bioremediation enhancements is relatively straightforward in unconsolidated and permeable sediments. However, many sites have significantly more difficult field conditions, ranging from contamination existing in bedrock, low permeability silts/clays, or other difficult settings where building proximity limits access. One of the fundamental obstacles associated with implementing an enhanced bioremediation approach at sites with difficult geologic conditions is the effective delivery of nutrients or substrate to the affected subsurface. Three case studies demonstrate various approaches to effective delivery of substrate to enhance anaerobic bioremediation. The case studies include enhanced bioremediation at a site with a relatively impermeable clay (10^{-6} to 10^{-7} cm/sec); a site at which the contamination is located within a very dense glacial till; and at a site where contamination is located within bedrock.

INTRODUCTION

Many biodegradation processes, such as the breakdown of chlorinated solvents (particularly chlorinated ethenes) tend to be limited by the availability of e^- donors. Electron donors are utilized by naturally occurring bacteria to anaerobically biodegrade chlorinated solvents. A substrate containing nutrients, such as lactate, can be delivered to the subsurface and bacteria in the subsurface can metabolize the nutrients, thereby releasing e^- donors (e.g. hydrogen) which have the capability to reductively dechlorinate chlorinated ethenes (Koenigsberg and Farone, 1999).

Enhanced in-situ bioremediation has been shown to be an effective remediation approach at numerous sites contaminated with a variety of compounds (Murray, et al., 2000; Harms, et al., 2000; Lodato, et al., 2000). Implementation and application of bioremediation enhancements is relatively straightforward in unconsolidated and permeable sediments. However, many sites have significantly more difficult field conditions, ranging from contamination existing in bedrock, to impermeable clays, to other difficult geologic settings. One of the fundamental obstacles associated with implementing an enhanced bioremediation approach at sites with difficult geologic conditions is the effective delivery of the substrate to the affected areas of the subsurface. In such cases, effective use of delivery mechanics is critical. Delivery may depend on the agents (nutrients or substrates), and their capability to dissolve, diffuse, or the mechanical delivery method (drilling, injection, or other method). Three example cases illustrate these techniques.

CASE STUDY 1

Site Setting. The first case study involves remediation of a TCE-contaminated site via 2-PHASE Extraction and subsequent injection of an enhanced bioremediation substrate to facilitate bioremediation of residual concentrations. The site housed a former industrial filter manufacturer where chlorinated volatile organic compounds were utilized in degreasing operations. The predominate degreaser used was trichloroethene (TCE) which was determined to have affected the soil, groundwater, and sediments within the site.

Following performance of the RI/FS (Haley & Aldrich, 1991, 1992), a Record of Decision (ROD) was issued by the New York State Department of Environmental Conservation (NYSDEC) that selected removal and disposal of sediments with drainage controls; shallow soil removal; and an aggressive multi-phase high vacuum extraction for soil and groundwater remediation. Sediment and soil removal was successful upon completion. The 2-PHASE™ Extraction (2-PHASE) commenced in 1994. Pre-remediation concentrations at the site were >50 mg/kg in soil and >190 mg/L in groundwater. The groundwater cleanup objective was to "design and operate to the extent practicable to mitigate and control shallow source area groundwater" (NYSDEC, 1993).

The 2-PHASE system operated for approximately 4 years and almost 30,000 operational hours. Although additional extraction wells had been installed to improve mass removal rates, periodic groundwater and soil sampling and analysis concluded that the system had reached asymptotic conditions and operation was discontinued. Soil concentrations met the closure criteria. Groundwater concentrations remained elevated up to 28 mg/L in a "core" source area; still, it was determined these localized residual concentrations could fit a risk-based closure.

The localized residual groundwater concentrations provided the NYSDEC and the responsible party the opportunity to test enhanced bioremediation and potentially improve the final remedial condition for closure.

Design Criteria. The project site was attractive for enhanced bioremediation due to the relatively small, controllable size of the CVOC source core; the density of monitoring wells in and downgradient of the core area (Figure 1); and evidence of existing biodegradation. Subsurface materials consist of relatively low-permeability clayey silts, saturated to within 0.3 to 0.6m (1 to 2 ft.) of grade. Injection/treatment depth was up to 4.6± m (15± ft.) depth below grade.

Due to the need for anaerobic conditions for degradation, 2-PHASE was shut down to eliminate flow of air and relatively oxygenated surface water through the treatment zone. The NYSDEC allowed 2-PHASE shutdown prior to collection of pre-injection baseline samples.

Electron Donor Injection Strategies 283

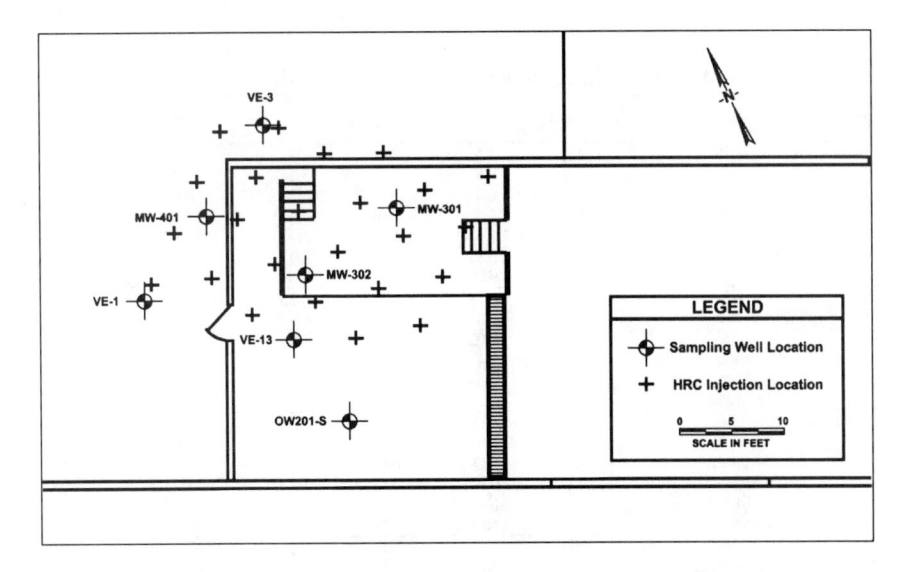

FIGURE 1: Site and HRC Injection Location Plan

Site characteristics evaluated for injection design included: dissolved oxygen (DO) and other e^- acceptor concentrations, chlorinated volatile organic compounds (CVOC) concentrations, and hydrogeologic conditions. Hydrogeologic and CVOC data existed prior to application. Samples of relatively undisturbed soil were collected for a microcosm study of indigenous bacteria and laboratory inoculation with the lactate formulation (HRC was used in this case). The initial injection grid was then fine-tuned and the amount of HRC to be injected, injection point spacing, and monitoring parameters were determined.

The grid was laid out in the field and the nodes were adjusted to permit rig access. HRC was injected with a Geoprobe rig using the rig pump, retrofitted so as to inhibit O_2 introduction. Final node spacing was targeted at 1.5±m (5± ft.) intervals (see Figure 1) with 21 points injected with roughly 15.9 kg (35 lbs.) per hole.

Monitoring Methods & Results. Pre-injection groundwater samples were used to determine baseline conditions, the HRC was then injected, and subsequent progress of subsurface conditions was monitored on a quarterly basis. Monitoring parameters used for this site included: CVOC concentrations; HRC component concentrations (metabolic acids); biodegradation indicators (e^- acceptors, endpoint gases); microbial analyses; field parameters (DO, Eh).

Table 1 summarizes the CVOC data for six quarterly events. The table also summarizes the percentage TCE loss for each well from the pre-injection event to the 15-month sampling event. Substantial decreases in TCE concentrations were observed in all of the seven wells in the test area. Five of the six wells located within the injection grid (VE-1, VE-3, VE-13, MW-302, and MW-401) experienced TCE reductions ranging from 82% to 100% over the 15 month test period.

TABLE 1: Summary of Analytical Results

Compound	Date	VE-1	VE-3	VE-13	MW-301	MW-302
TCE	Pre-HRC	26	3.5	0.68	2.8	1.1
	3 Month	14	NS	ND	3.1	0.34
	6 Month	11	0.03	0.007	ND	ND
	9 Month	8.6	3.5	0.095	0.61	0.17
	12 Month	13	1.6	0.022	1	0.32
	15 Month	4.7	0.82	0.0081	4.3	ND
	TCE Percent Loss	82%	94%	99%	-54%	100%
DCE	Pre-HRC	ND	0.34	0.25	0.26	0.18
	3 Month	1.3	NS	1.7	6.9	6.2
	6 Month	0.58	0.021	0.034	ND	2.4
	9 Month	0.61	1.4	0.4	8.9	11
	12 Month	0.72	0.640	0.063	1.5	12
	15 Month	0.26	0.22	0.028	1.8	18
Vinyl chloride	Pre-HRC	ND	ND	ND	ND	ND
	3 Month	ND	NS	0.84	ND	ND
	6 Month	ND	ND	0.024	0.14	0.21
	9 Month	ND	0.02	0.17	7.2	3.9
	12 Month	ND	0.0069	0.048	0.96	8.8
	15 Month	ND	ND	ND	0.19	ND
Ethene	Pre-HRC	ND	ND	ND	ND	ND
	3 Month	ND	ND	ND	ND	ND
	6 Month	ND	ND	0.02	ND	0.013
	9 Month	ND	ND	0.065	0.042	0.028
	12 Month	ND	ND	ND	ND	ND
	15 Month	ND	ND	0.0731	0.405	0.0505

Note that on the basis of advective transport alone, migration of dissolved lactate from injection points could not account for the decrease in CVOC concentrations. Low permeability settings such as this site appear to provide good conditions for hydrogen, when generated by microbial breakdown of lactate, to diffuse more quickly through the aquifer than is possible by advective processes. Such transport should be carefully considered in fine-grained settings, so that enhanced bioremediation is not rejected on the basis of low permeability.

CASE STUDY 2

Site Setting. The second case study involves remediation of 1,1,1-Trichloroethane (TCA) found in the soil and groundwater at an active contact lens manufacturing facility in western New York. The facility entered into a Voluntary Cleanup Agreement (VCA) with NYSDEC and was required to investigate site conditions and remediate the site. Investigations from 1997-2000 found groundwater concentrations as high as 400-500 mg/L in the source residue area (Haley & Aldrich, 2000). Impacted soils were found to be present from ground surface to depths of 11.6m (38 feet) below ground surface (bgs).

Site soils consist of a very dense till (10^{-4} to 10^{-7} cm/sec, N values ranging from 60 to >100 blow counts) which is comprised primarily of clayey silts with some sand. A high vacuum pilot test was completed at the site and was shown to provide no meaningful subsurface response to the vacuum. Additionally, due to facility constraints (existing building layout, limited available surface area for support systems), many typical remedies such as soil vapor extraction, excavation, or soil heating, could not be applied.

Enhanced bioremediation was evaluated through microcosm studies. The studies evaluated the compatibility of lactate (in this study HRC™ was used) with site soils and groundwater at various concentrations of site compounds. The study included testing degradation of TCA in the site's soil/groundwater samples at low (25 mg/L) and high (250 mg/L) concentrations. The results of the study showed that viable populations of microorganisms capable of CVOC degradation are present in the site source area soils collected, and can be stimulated by lactic acid application. The study also indicated that the degradation of TCA in the "low" concentration samples (25 mg/L) ranged up to 79±% in 28 days. 1,2-Dichloroethane (1,2-DCA) as a daughter product was also produced in the samples. Degradation of TCA in the "high" concentration samples (250 mg/L) ranged up to 92±% in 28 days. Again, DCA as a daughter product was produced in the samples.

Design Criteria. Site conditions necessitated different injection scenarios at various locations of the site. The source zone injection was designed on a 2.1m (7-foot) grid, which results in approximately 40 injection points completed to a depth of 11.6m (38 feet) bgs. The injection spans from 11.6m (38 feet) bgs to approximately 1.5m (5 feet) bgs with approximately 2.3kg (5 pounds) of HRC injected per linear foot. Other areas of the site have similar injection designs that have been modified based on depth and concentrations of contamination. There are a total of approximately 65 injection points throughout the treatment areas, as shown on Figure 2.

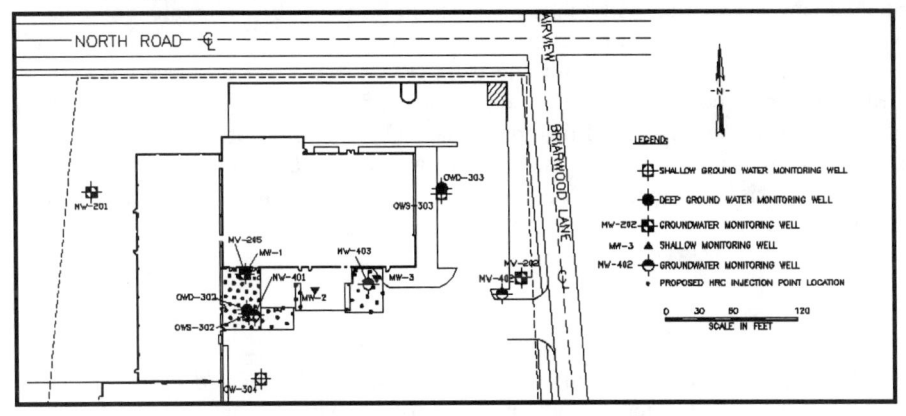

FIGURE 2: Site and HRC Injection Location Plan

Lactate or other substrates are typically introduced to the subsurface using Geoprobe-type direct push methods or wells that can allow soluble nutrients to "infuse" the affected subsurface area. However, because of the required depth of injection for this site and the high density of the soils, it was determined that a rotary injection method would likely be required to reach the desired injection depth, the deepest being approximately 11.6m (38 feet) below ground surface. A field test, with what is reported by Geoprobe to be their most powerful rig, took place at the site in July 2000. This field test was completed to determine if a more powerful direct-push rig was capable of reaching the desired injection depth. Because of the exceptionally dense soils, the rig was only able to penetrate to approximately 3m (10) feet below ground surface. Based on this field test, it was determined that a combination rotary injection-direct push technique will be required for the HRC injection.

Through discussions with various drillers, a combination rotary injection-direct push technique method was developed (Haley & Aldrich, 2001). This combination technique involves advancing a tri-cone rotary bit or narrow diameter solid-stem auger from the ground surface to the desired depth. Once the desired depth is reached the bit/auger is removed from the hole and a direct-push tool inserted to act as the injector. HRC is then injected through the tool tip under pressure using a pump as the tip is slowly removed from the borehole base. In a normal direct-push type injection, pressure is achieved and maintained in the boreholes by friction between the injection tools and the formation. Pressure is measured directly at the pump. For this project, several contingency options were identified with the driller to maintain injection pressure. A more detailed discussion of the various techniques can be found in the paper entitled "Enhanced Bioremediation of High Contaminant Concentrations in Source Residual Area" (Dick, et al., 2001).

Monitoring Methods & Results. The field injection for this site has not yet been completed, so no field data is available at the time of paper preparation. The Case Study 2 site remediation will be implemented in Spring 2001 and evaluation data will be shown at the time of paper presentation in June 2001.

Monitoring parameters for the site will include CVOC concentrations, HRC component concentrations, biodegradation indicators (e^- acceptors, endpoint gases), microbial analyses, and field parameters (DO, Eh, pH, temperature).

As with the Case Study 1 site, subsurface soil conditions should not be an automatic basis for rejection of enhanced bioremediation. The installation techniques described here may be applicable to other unconsolidated soil site settings.

CASE STUDY 3

Site Setting. The final case study involves a bedrock site located in Princeton, New Jersey. Ground water beneath one are ("Area C") of the site is contaminated with CVOCs. The primary chemicals of concern are PCE, TCE, TCA and their natural breakdown products. Ground water beneath the site occurs in a weathered and fractured bedrock aquifer in the Triassic-aged Stockton Formation. Vertical CVOC distribution is generally limited to the shallowest 30 feet of the bedrock aquifer near a former source location at Area C. Contaminated ground water is captured by one of the Area C building foundation drain systems and is not flowing off-site.

Design Criteria. To date, delivery of bioremediation substrates in bedrock has been limited to injection through wells (see, for example Sorenson, et. al., 2000) which depends on the wells' connection to the native bedrock fracture network (or other effective porosity network) to be effective. This site involves remedial design and deployment of a bedrock blast-fractured trench in two parallel alignments, and installation of in-situ treatment in each alignment. This program provides the opportunity to combine a set of three previously developed, but never combined, innovative remediation methods – bedrock blast fracturing to refractively channel flow of groundwater, reactive iron (abiotic) treatment, and enhanced bioremediation. Blast fracturing involves design and placement of an engineered alignment of relatively high permeability blast-fractured bedrock. The blast fractured trench is intended to passively (i.e. without pumping) collect contaminated groundwater flow across a significant portion of the plume; refract the collected flow into the blast fractured zone and through the installed treatment zones, reducing or destroying the contaminant concentrations; and then disperse the treated water back into the native bedrock formation. The design layout of two chevron-shaped blast fractured trenches is shown in Figure 4. The reactive iron and enhanced bioremediation (via HRC injection) will be installed separately at the apex of each blast-fractured alignment to separately evaluate their effects on hydraulics in the blast-fractured zones, treatment results, and geochemical characteristics.

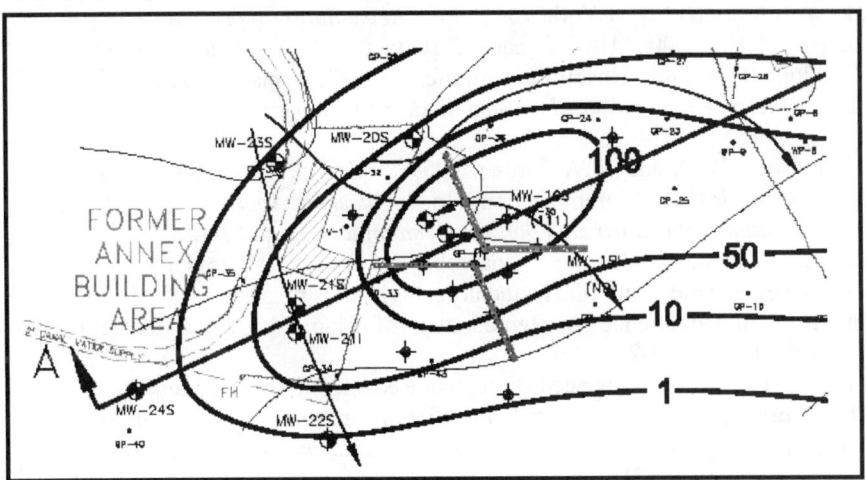

FIGURE 4: Design Layout of Two Chevron-Shaped Blast Fractured Trenches

Monitoring Methods & Results. Because the field test for this case study has not yet been completed, (Case Study 3 expected to be implemented upon approval of funding in 2001 or 2002) no field data is available to date.

CONCLUSIONS

Enhanced in-situ bioremediation is an effective remediation technology that needs increasing emphasis on deployment under difficult field conditions. Each of the case studies presented here demonstrates effective approaches to deliver substrate for enhanced anaerobic bioremediation. The delivery methods vary depending on the geological setting, ranging from a tight grid-space layout for the injection, to a combination rotary-direct push injection, to placement of an enhanced bioremediation substrate into a blast fractured bedrock zone.

REFERENCES

1. Boyle, et al., 1999. "Enhanced In-Situ Bioremediation of a Chlorinated VOC Site Using Injectable HRC", Proceedings of the Thirty-First Mid-Atlantic Industrial and Hazardous Waste Conference", June 20-23, 1999.
2. Dick, et al., 2001. "Enhanced Bioremediation of High Contaminant Concentrations in Source Residual Area", to be presented at the In Situ and On-Site Bioremediation: The Sixth International Symposium, June 4-7, 2001, San Diego, California.
3. Haley & Aldrich of New York, 1991. *Remedial Investigation Report.*
4. Haley & Aldrich of New York, 1992. *Feasibility Study Report.*
5. Haley & Aldrich of New York, 2000. *Report on VCA Investigations.*
6. Haley & Aldrich of New York, 2001. *VCA Remediation Work Plan.*
7. Harms, et al., 2000. "HRC-Enhanced Reductive Dechlorination of Source Trichloroethene in an Unconfined Aquifer." In Wickramanayake, A.R., et al. (Eds.), *Bioremediation and Phytoremediation of Chlorinated and Recalcitrant Compounds*, pp. 295-302.
8. Koenigsberg, S.S. and W.A Farone, 1999. "The Use of Hydrogen Release Compound for CAH Bioremediation." In *Engineering Approaches for In Situ Bioremediation of Chlorinated Solvent Contamination*, pp. 67-72.
9. Lee, et al., 1998. "A Combined Anaerobic and Aerobic Microbial System for Complete Degradation of Tetrachloroethylene", presented at the First International Conference on Remediation of Chlorinated and Recalcitrant Compounds, May 18-21, 1998.
10. Lodato, et al., 2000. "Enhanced Biological Reductive Dechlorination at a Dry Cleaning Facility Solvents." In Wickramanayake, A.R., et al. (Eds.), *Bioremediation and Phytoremediation of Chlorinated and Recalcitrant Compounds*, pp. 205-211.
11. Murray, et al., 2000. "HRC Enhanced Bioremediation of Chlorinated Solvents." In Wickramanayake, A.R., et al. (Eds.), *Bioremediation and Phytoremediation of Chlorinated and Recalcitrant Compounds*, pp. 287-294.
12. New York State Department of Environmental Conservation, 1993. *Record of Decision.*
13. Sorenson, et al., 2000. *Field Evaluation Report of Enhanced In Situ Bioremediation, Test Area North, Operable Unit 1-07B.*

ENHANCED MICROBIAL DECHLORINATION OF CHLOROETHENES AND CHLOROETHANES: FROM LABORATORY TESTS TO APPLICATION IN A BIOSCREEN

H. Slenders, J. ter Meer, and M. van Eekert (TNO Apeldoorn, The Netherlands)
T. Verheij (DAF, Eindhoven, The Netherlands)
P. Verhaagen and J. Theeuwen (Grontmij BV, Houten, The Netherlands)

ABSTRACT: Since the late forties DAF Trucks N.V. is producing cars at a facility along the Eindhovensch Canal. These activities resulted in soil contamination with trichloroethene (TCE). Within the framework of the Dutch research program NOBIS, the contamination and soil conditions were thoroughly investigated, and the combined dechlorination of ethenes and ethanes was demonstrated in batch experiments. A pilot system with infiltration and extraction screens was installed at a depth of 10-20 m-bgs. Currently, the microbial activity is stimulated through the infiltration of an electron donor, thus leading to reduced redox conditions and dechlorination of the target compounds.

INTRODUCTION

The workshop of a truck manufacturer in Eindhoven, the Netherlands, has been degreasing metals since 1947. As a result, large amounts of trichloroethylene (TCE) and trichloroethane (TCA) entered the soil. Source removal was considered very expensive and would cause an unacceptable disturbance of the industrial activities. As a result of low groundwater velocities in the top soil layers, the contaminant flux to the aquifer would decrease no earlier than 15 years after source removal. Therefore it was decided to create a biological treatment zone at a cross section of the plume just before the plume enters the aquifer. In this zone the initial natural conditions were not optimal for complete reductive dechlorination.

The possibilities of microbial transformation of the primary contaminants were assessed, and a biological treatment zone was designed and installed. A schematic illustration of the treatment zone is presented in Figure 1.

CONTAMINANT SITUATION AND SOIL CONDITIONS

The soil profile at the site was very complex and typical for the formations of the Nuenen group and Veghel/Sterksel deposits. The soil cover was about 30 m thick and consisted of a heterogeneous mix of sand, clay, silt and peat. The underlying aquifer (Veghel/Sterksel 30-80 m-bgs) was mainly made up of fine to coarse sands. The main spill of TCE was rapidly dechlorinated to (*cis*)1,2-dichloroethene (cDCE) in silt and peat layers between 3 and 6 m-bgs. Vinyl chloride (VC) was formed, and cDCE and VC were found in the aquifer to a depth of 50-60 m-bgs. At this site other dichloroethenes and 1,1-dichloroethane (1,1-DCA) were also found as contaminants.

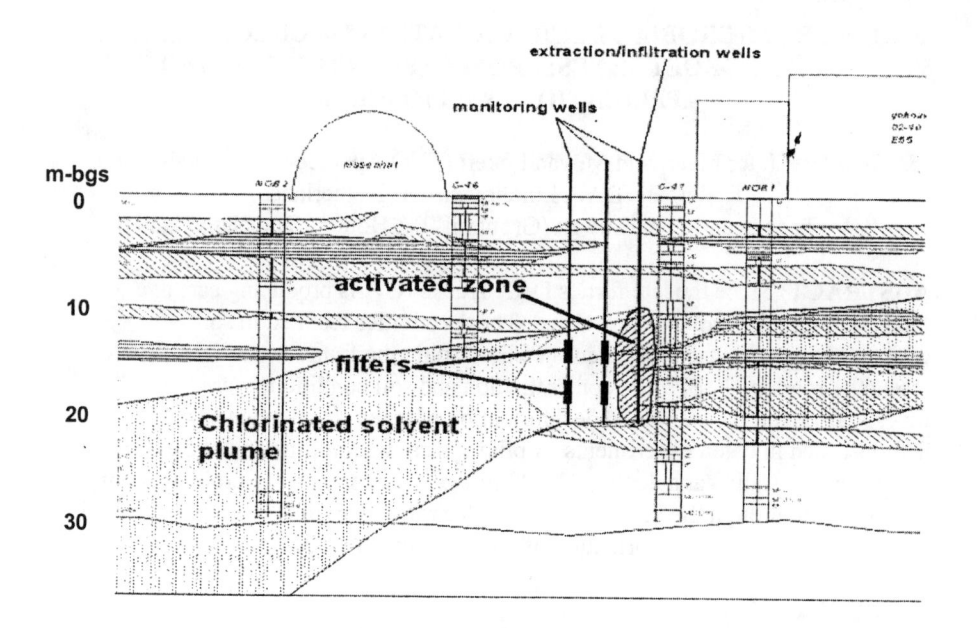

FIGURE 1: Contaminant plume in heterogeneous top layers with schematic illustration of bioscreen

The concentration of cDCE at 30 m-bgs still increased with time. While natural degradation of cDCE stagnated between 10-20 m-bgs, TCE and VC seemed to be dechlorinated along the full flowpath. Redox characterization using hydrogen measurements and thermodynamical calculations indicated that at a depth of 10-20 m iron reducing conditions prevailed. Previous laboratory experiments indicated that complete dechlorination was only possible under sulfate reducing or methanogenic conditions (Ras et al. 2000). Coincidentally, at 10-20 m-bgs the contaminant plume passed through a relatively small section. The treatment zone was planned at the zone where stagnation of cDCE degradation was found.

LABORATORY EXPERIMENTS
Objective. At an early stage of the project it was observed that a microbial dechlorinating population was present at the site. In the batch experiments described here other factors were evaluated:
- Dechlorination rates in the presence of different electron donors;
- Necessary lag phase;
- End products of dechlorination;
- Amount of electron donor needed to create a reduced zone in the soil;
- The effect of the presence of other electron acceptors.

The electron donors tested were selected according to the following factors:
- Potential for hydrogen generation;
- Applicability as a bulk electron donor;
- Unidentified stimulating compounds like trace elements;
- Ability to prevent clogging of the infiltration screens.

Finally, a combination of several donors was selected (Table 1).

TABLE 1. Electron donors used in the batch experiments

Compound	Ratio[a]	Acronym	Criteria for use[b]
Methanol		MeOH	BE
Methanol, Citrate	9:1	MeOH/Ci	BE, PC
Yeast derivative		G31	H_2, BE?, UC
Lactate, Acetate, Propionate, Butyrate	1:1:1:1	VFA	H_2, BE, UC?
Methanol, Yeast derivative, Citrate	8:1:1	MeOH/G31/Ci	H_2, BE, UC, PC

[a] Ratio based on electron equivalents
[b] H_2 = known as effective electron donor; BE = bulk electron donor; UC = presence of unidentified possibly stimulating compounds; PC = prevents clogging

Set-up. Batch experiments were carried out in 250 ml serum flasks containing 40 grams of sediment and 210 ml groundwater, contaminated mainly with cDCE (30 mg/l) and 1,1-DCA (4 mg/l). The experiments were carried out in duplicate with three different sediment layers (10-12 m-bgs, 12-14 m-bgs, and 17-18 m-bgs). The groundwater also contained around 50 mg/l sulfate. The electron donors (Table 1) were dosed in a 17 fold overdose based on the oxidation capacity of the sediment (50 to 200 mg Fe/kg sediment), the amount of electron equivalents required for the chlorinated compounds and the sulfate present in the groundwater. After 1 month of incubation, the duplicate bottles were amended with 50 mg/l NH_4Cl as a nitrogen source.

The concentration of chlorinated compounds was determined via GC. Because cDCE and 11DCA peaks could not be distinguished via the usual GC method, GC-MS measurements were carried out regularly to confirm the dechlorination of cDCE and 1,1-DCA. The VFA concentrations were analyzed via GC methods. Concentrations of anions were analysed via HPLC measurements.

Results. The laboratory experiments have shown that the reductive dechlorination of cDCE in site specific material takes place under reduced conditions. Sulfate was removed with most electron donors within 28 to 42 days. Thereafter, dechlorination of cDCE was observed. VC, and at a later stage, ethene were formed with VFA, the yeast derivative and the MeOH/G31/Ci mixture as electron donors (Figure 2). Other chlorinated compounds present, like tDCE, 1,1-DCE and 1,2-DCA, were also removed. In some cases the formation of chloroethane a possible dechlorination product of 1,1-DCA was observed after 140 days of incubation.

Transformation of the chlorinated compounds was not observed with methanol alone as the electron donor. In that case, sulfate was reduced very slowly. The application of the methanol/citrate mixture only led to the formation of VC after a relatively long lag phase of 69 to 84 days. The addition of NH$_4$Cl as the nitrogen source led to a shorter lag phase and a faster and more extensive dechlorination with VFA and yeast as the electron donors.

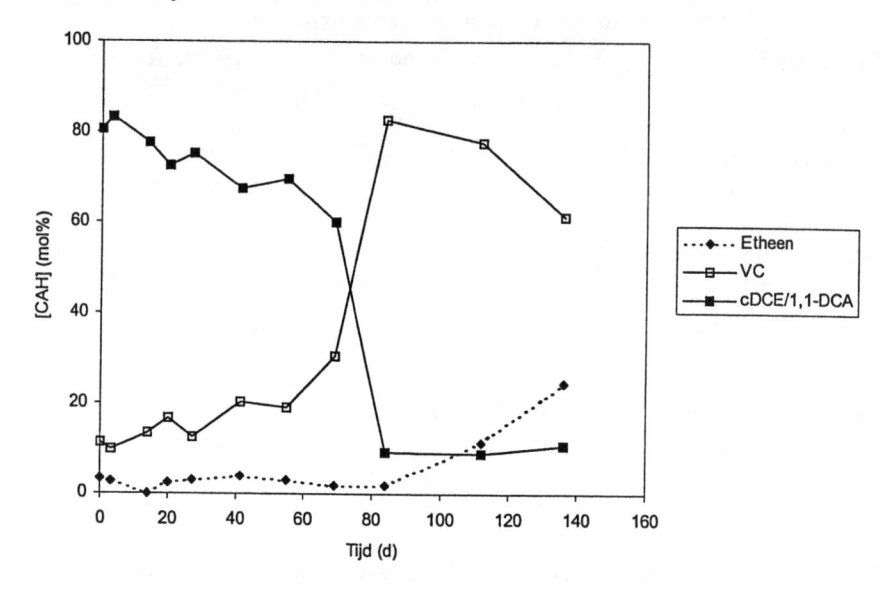

FIGURE 2: Transformation of cDCE to VC and ethene with a mixture of methanol, G31 and citrate as the e-donor and NH$_4$Cl as nitrogen source.

Conclusions for the laboratory experiments. From the results of the batch experiments it was concluded that a dechlorinating population exists at the site. A change of redox-conditions and stimulation of the microbial population were necessary to establish complete reductive dechlorination of cDCE to ethene. Reduction of sulfate in the groundwater however took place before dechlorination.

IMPLEMENTATION OF THE BIOLOGICAL TREATMENT ZONE

Design and installation: In the spring of 2000, the soil profile and the geochemistry at the selected pilot section were investigated in more detail with Cone Penetration Tests (CPTs), and soil and groundwater analyses. Silting up and clogging of the screens was investigated and two different pilot systems were designed (Slenders et al. 2000).

In both systems (Figure 3), groundwater was extracted, amended with the selected electron donor, and again infiltrated into the soil. Care was taken to maintain anaerobic conditions. In system 1 the focus was to avoid the re-extraction of stimulated groundwater or groundwater from different soil layers. Re-extraction might have lead to bioclogging, and mixing groundwater with different origins might have lead to sulfide precipitates (Beek et al. 1998). Therefore, in system 1 groundwater was extracted 13 m upstream from the infiltration zone, and was restricted to circulation in a single sandlayer at a depth of 16-19 m-bgs. This system therefore did not cover the complete cross section of the plume.

In system 2 (Figure 3) the focus was on covering and controlling the complete cross section of the plume. The bioscreen of set-up 2 consisted of three wells with each well screened from 10 to 11 m-bgs, 13 to 14 m-bgs, and 17 to 19 m-bgs. An intermittent injection scheme was used to reduce the risks of clogging.

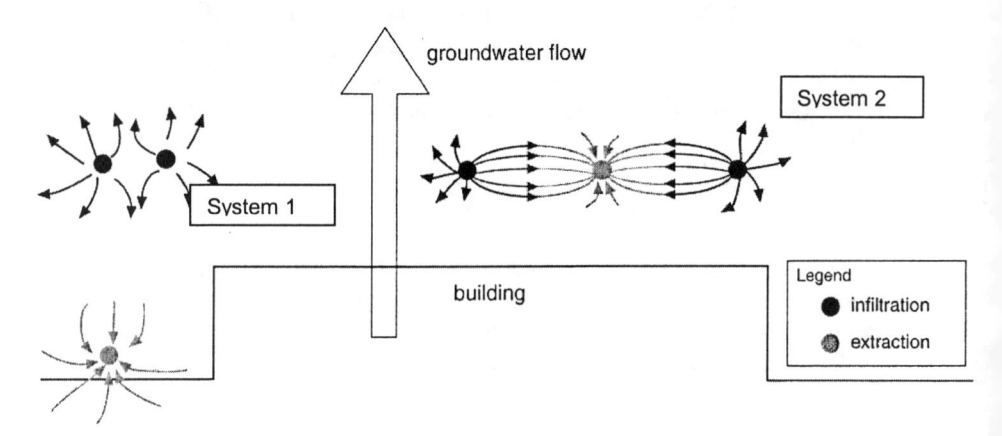

FIGURE 3. The two different infiltration systems

Installation. In June 2000 the injection and extraction wells were installed at the site. An intensive monitoring network was also completed. During the trial run, special attention was given to infiltration aspects. Flows and pressures were measured, and a tracer test was performed.

Stimulating microbiology. A mixture of methanol, lactate, and a yeast derivative (ratio 1:1:1) was chosen as the initial electron donor for the reduction of the sediment material. As a result of a pH drop, during the following infiltrations only lactate and the yeast derivative were used for the reduction of sulfate and the chloroethenes in the groundwater. This mixture was pH neutral. The applicability of hydrogen measurements as a control tool was investigated.

PRELIMINARY RESULTS OF THE PILOT SYSTEMS

Both systems could be operated at a steady flow rate of 3-4 m^3/hour, without any sign of clogging. Every injection period took 3-4 days. After these periods the systems rested for 4-6 weeks. During the tracer test with lithium, system 1 did not show a clearly delimited area of influence. The tracer dispersed widely. System 2 showed a clearly delineated area in the three sand layers it affected. The injection of electron donor also pointed out that the three zones of the bioscreen were directly influenced.

Elevated H_2 levels (>200 nM) were measured within one day after the start of injection. The background level before the start of the pilot was about 0.3 nM. This indicated that the electron donor had been converted to hydrogen, which lowered the redox potential more favorable for dechlorination (Ter Meer et al,). A large number of hydrogen measurements at different sites in the Netherlands indicated that hydrogen concentrations greater than 0.5nM correspond with favorable dechlorination conditions. Originally, the redox condition in the pilot varied from iron reducing to sulfate reducing. The latest measurements showed a fast decrease of sulfate concentrations and a favourable redox potential between –200 and –400 mV. A decreasing pH may indicate dechlorination or the formation of fatty acids. Preliminary results of chlorinated compounds seem to indicate a decrease of cDCE and an increase of VC, but more data and time are needed to proof this.

CONCLUSIONS

Enhanced natural degradation was possible through dechlorination with micro-organisms present at the site. A complex electron donor resulted in the shortest lag-phase and fastest dechlorination. Before dechlorination took place, sulfate had to be reduced first. A line of extraction and infiltration screens seemed optimal for the controlled mixing of the substrate with the original groundwater. The injection of electron donor resulted in elevated hydrogen concentrations and a decrease of pH and sulfate concentrations in all monitoring wells in the bioscreen. Data of contaminant concentrations will be available soon.

At the moment, the suitability of hydrogen measurements as a monitoring tool for the "growth" and steering of the bioscreen is being studied. Furthermore the change of contaminant flux to the aquifer will be modeled, and the effect of the bioscreen on the development of the plume in the aquifer is investigated.

ACKNOWLEDGMENTS

The authors would like to thank Andre Cinjee and Bert Hafkamp for the sampling and analysis and Arjan Krijnen and Robert Heling for their technical assistance with the field system. Furthermore SKB and DAF are acknowledged for making this research possible.

REFERENCES

Ras, N. van, T. Grotenhuis, J.L.A. Slenders, A.A.M. Langenhoff, J. Ter Meer and P.A.A. Verhaagen. 2000. "Intrinsic biorestoration within a large CAH-contamination, Phase I: Field characterization and laboratory experiments" NOBIS 95-2-09.

Slenders, H., P. Verhaagen, M. van Eekert, R.Dubbeldam, J. Ter Meer, and B. Hafkamp. 2000. "Intrinsic biorestoration within a large CAH-contamination, Phase 2a: Additional characterization, design and installation pilot system" NOBIS 95-2-09, interim report.

Ter Meer, J., J.Gerritse, C. di Mauro, M.P. Harkes and H.H.M. Rijnaarts. 1999. "Hydrogen as an indicator for in situ redox conditions and dechlorination" NOBIS 96.024.

Beek, C. van, et al. 1998. "Design and maintenance of extraction and re-infiltration systems" NOBIS 96-3-06

LINER - A NEW CONCEPT FOR THE STIMULATION OF REDUCTIVE DECHLORINATION

Emile C.L. Marnette (Tauw bv, Deventer, The Netherlands),
Haimo Tonnaer, Arne Alphenaar, (Tauw bv , Deventer, The Netherlands),
Gijsbert Jan Groenendijk (Hoek Loos bv, Amsterdam, The Netherlands)

ABSTRACT: A new technique for the distribution of substrate in the soil has been tested. The technique involves the injection of substrate into the soil to stimulate anaerobic dechlorination of chlorinated aliphatic hydrocarbons (CAH). The substrate is nebulized in a nitrogen carrier gas. The technique is called LINER - LIquid Nitrogen Enhanced Remediation.

Advantages of gas injection over infiltration of substrate in liquid form are better mixing of substrate and contaminated groundwater, no clogging of injection wells, and a larger radius of influence.

A single well test at 43 m bgs revealed that a large soil volume was treated by the injection, almost all the way up to the soil surface. In more than 50% of the monitoring wells, ethane was formed as a product of complete dechlorination. Biological degradation of PCE has been observed at a 6 m distance from the injection well.

INTRODUCTION

One of the major problems involved in soil remediation today is the treatment of deep groundwater contaminated with chlorinated hydrocarbons. Biological degradation by microorganisms will often be the best clean-up option. In practice, however, the addition of the substrate required to stimulate the biological processes in-situ is a problem. Substrate infiltrated in liquid form mixes very slowly with the contaminated groundwater and infiltration systems tend to clog easily. Furthermore, the limited radius of influence of an infiltration well requires a dense network of wells.

In cooperation with gas company Hoek Loos, engineering consultancy Tauw has developed a new remediation concept, overcoming most of the limitations inherent to the conventional in-situ biological systems for degradation of CAHs.

Remediation of soil contaminated with CAHs. The Netherlands have numerous sites contaminated with CAHs. The most common remediation approach concerns pump and treat. However, authorities frequently impose severe restrictions on groundwater extraction. In-situ air sparging based on the injection of compressed air (possibly in combination with techniques such as steam injection, electro reclamation etc.) may be an alternative for contaminants located in relatively shallow soil layers.

Over the past few years, methods to promote indigenous biological degradation of contaminations at greater depths have been developed: the bacterial population present is stimulated to biotransform CAH contamination. All methods involve introduction of a substrate (electron donor) into the subsurface.

Groundwater substrate transport. Substrate infiltrated in the subsurface as a liquid will mix very slowly with the ambient groundwater. Degradation will mainly occur at the interface of the infiltrated substrate and the contaminated groundwater. Infiltrated substrate flowing along with groundwater will may be degraded before reaching other contaminated areas downstream. Consequently, the network of infiltration points required to effectively stimulate an existing natural attenuation processes will have to be very dense. The cost of such networks makes them practically unfeasible, particularly for contaminations located at large depths. Another potential problem involved in infiltration of substrates in liquid form is clogging of the wells by biomass formation.

LINER - Gas/substrate injection. The injection of a substrate with gas as a means for distribution is a concept which has not yet been tested as a remediation technique. This method combines two concepts: stimulation of biodegradation of CAHs by the addition of a substrate, and in-situ air sparging. Due to the anaerobic nature of the targeted microbiological processes, nitrogen gas was used instead of compressed air. The flow rate at which gas is distributed both horizontally and vertically within the soil is expected to be much higher than that of water, which should make the injection of gas a much more effective procedure than infiltration of an aqueous solution. Also, introducing substrate with a carrier gas is expected to result in a much better mixing of the substrate with the groundwater than introducing the substrate as a solution. Another advantage are the relatively low costs of injecting gas at great depths.

Earlier tests indicated that due to the stirring created in the soil by the injected gas (input of energy), contaminants are more readily available for biological degradation or physical removal.which is a favorable side-effect of the method (Tonnaer et al., 2001).

Objective. The objective of this research was to demonstrate in a field pilot test that injection of a substrate with nitrogen carrier gas is a feasible alternative to the in-situ remediation methods commonly applied to remediate CAHs. The following aspects of gas injection were investigated:
- radius of influence and distribution pattern of the injected gas;
- effects of the injected substrate on the degradation rate of PCE, the original contamination.

MATERIALS AND METHODS

A pilot scale injection system consisting of a single injection well was installed at a depth of 40 m below grade. Figure 1 gives a schematical representation of the pilot setup. A number of nested monitoring wells was installed at a distance of 2,

4 and 6 m from the injection well at depths of 14-15 m, 24-25 m, 34-35 m and 43-44 m below grade.

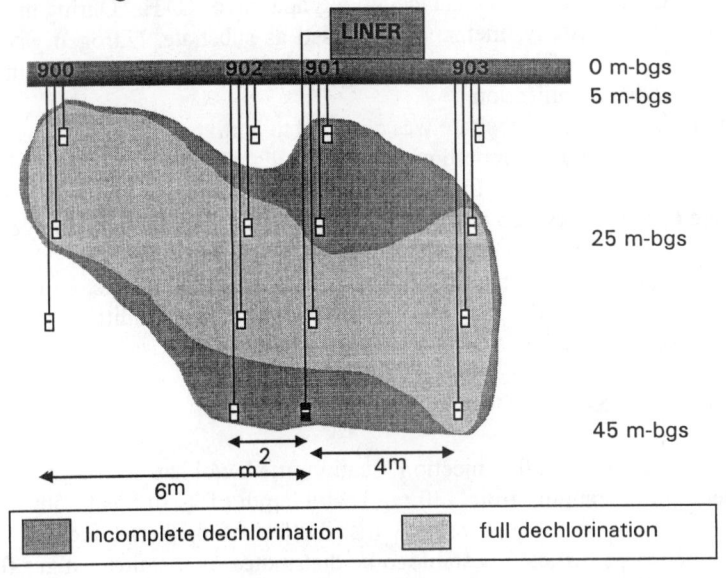

FIGURE 1. Schematical view of the pilot setup and spatial distribution of biodegradation

Figure 2 shows the LINER pilot test setup. Liquid nitrogen is vaporized and the gas pressure is reduced to the appropriate injection pressure. Methanol is nebulized into the nitrogen gas flow using nozzles.

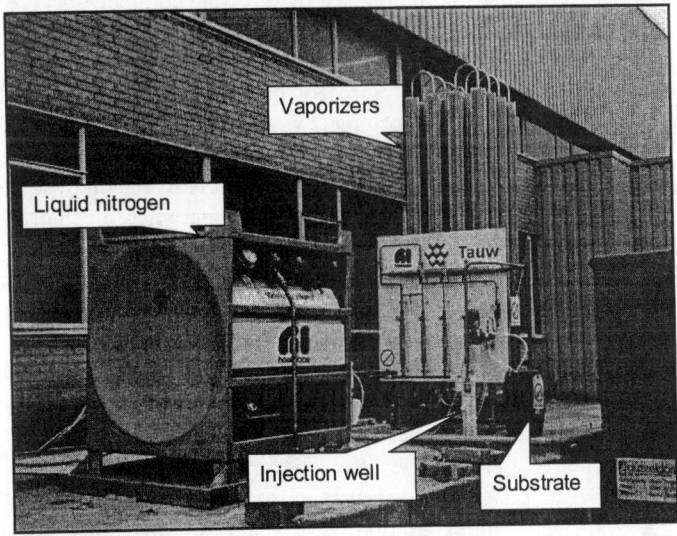

FIGURE 2. LINER pilot setup

Methanol was used because in several lab studies and field tests at other sites it proved to be a good substrate to enhance biodegradation of CAHs. During the first 12 weeks of the pilot study, methanol was used as substrate. Nitrogen gas was injected daily for 4 minutes per day. About 10 m^3 of nitrogen gas and about 1 L methanol was added per injection.

After 12 weeks, the substrate was changed to a mixture of ethyl lactate and methanol (50/50 vol%). Each month monitoring wells were sampled and analyzed for CAHs, ethene and ethane. In addition, methanol, ethyl lactate and ethanol (ethyl lactate desintegrates into ethanol and lactate) were occasionally analyzed at a selection of the monitoring wells.

Before the first substrate injection and after about 30 weeks, sulfate and methane were measured to see whether substrate addition affected electronacceptor concentrations.

RESULTS AND DISCUSSION

Substrate distribution. At the injection location, methanol has been measured in high concentrations, ranging from 130 mg/L at a depth of 39 m bgs to 500 mg/L at a depth of 14 m bgs. This vertical distribution of methanol indicates the methanol vapor to be sufficiently stable to be distributed by the nitrogen gas flow. Also ethyl lactate and ethanol were detected after injection at the location of injection over this large range in depth.

No methanol, ethyl lactate, or ethanol however, were detected in monitoring wells located laterally from the injection well. Because of the high detection limit of methanol (2 mg/L), combined with biological consumption, methanol may have reached the wells without being detected. Based on substrate measurements, no clear distribution pattern of substrate could be observed.

Stimulated biodegradation of CAHs. Since it became obvious that the distribution of substrate could not directly be assessed by substrate analyses, analyses of CAHs and degradation products was intensified. In Figure 1 the extent to which CAH degradation was stimulated is shown qualitatively. The light color represents the area where transformation of the CAHs to the end product ethene was accomplished. The dark color represents the area where dechlorination was incomplete and stopped at cis-dichloroethylene.

In Figures 3 and 4, results of two monitoring wells are shown that are representative for complete degradation to ethane (Figure 4) and incomplete degradation to cis-dichloroethylene (Figure 3). During the first 12 weeks no shift in the relative concentrations of the CAHs or their degradation products was observed in either well and therefore no significant biodegradation occurred. After 12 weeks, another substrate was used (mixture of methanol and ethyllactate) and biodegradation clearly started. The total concentration of degradation products increased compared to concentrations of PCE, the initial compound. It is not clear whether the enhanced biodegradation is a result of switching substrates, or whether the end of a lag in microbial growth was reached.

Electron Donor Injection Strategies 301

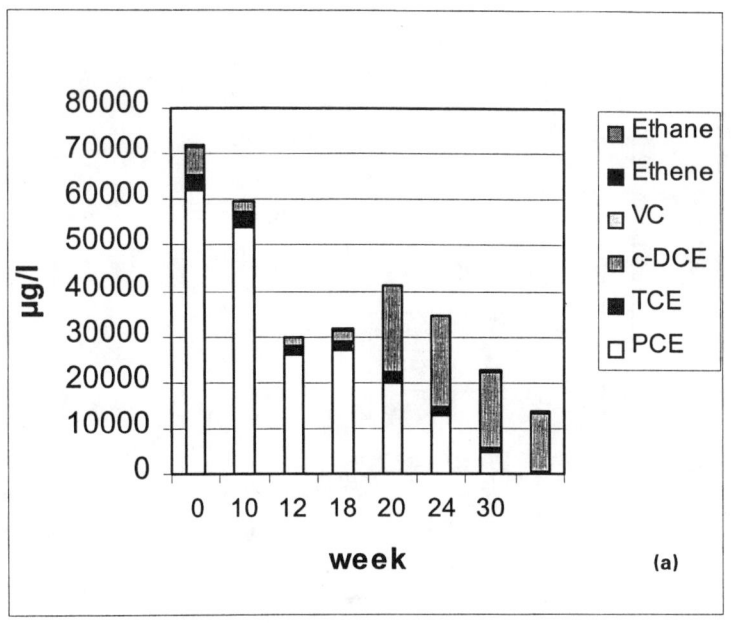

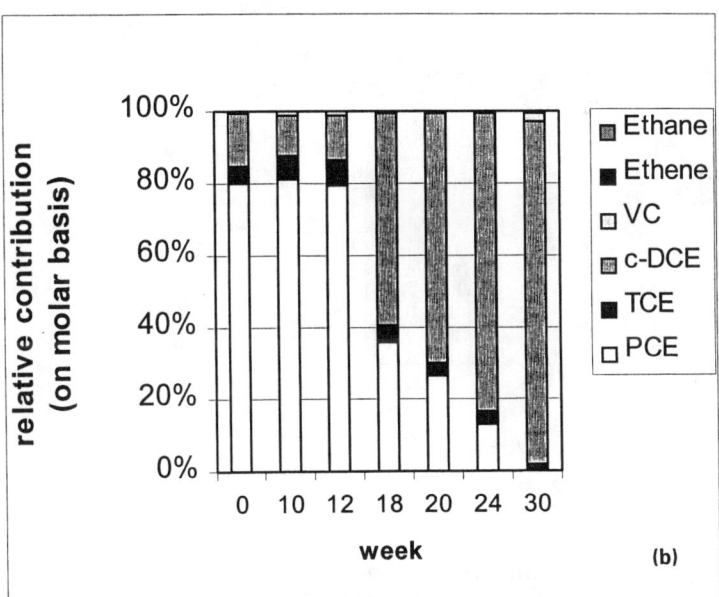

FIGURE 3. Absolute concentrations (a) and relative concentrat-ions (b) of PCE and its degradation products as a function of time at the location of Injection Well 901 at 14 m bgs.

Figure 3a shows a significant decrease of the PCE concentrations in Monitoring Well 901 (14-15 m bg). The TCE concentration remained about constant,

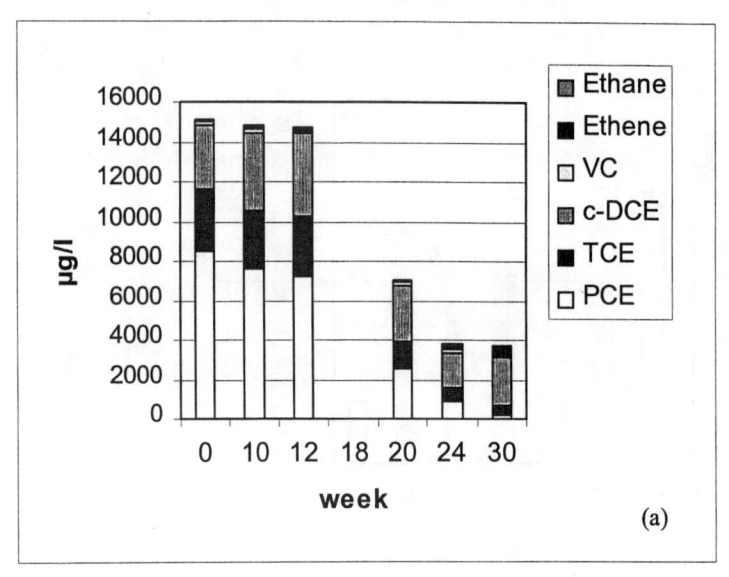

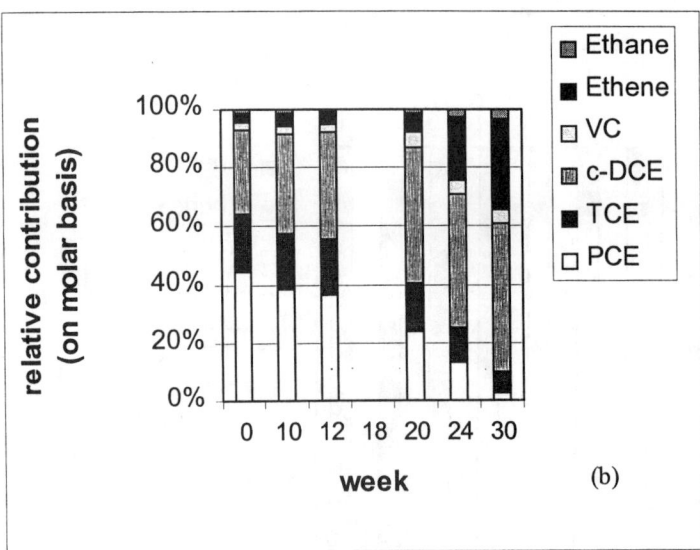

FIGURE 4. Absolute concentrations (a) and relative concentrations (b) of PCE and its degradation products as a function of time at 4 m distance from Injection Well 903 at 43 m bg.

indicating that TCE rapidly was converted to cis-dichloroethylene. The degradation seemed to stagnate at cis-dichloroethylene. A possible explanation is that there was not enough biomass yet to convert cis-dichloroethylene to vinylchloride. Figure 3b shows the relative contribution of the different compounds to the total molar concentration of ethenes. The relative contribution of c-DCE increased significantly while PCE was almost been depleted.

In Monitoring Well 903 at 4 m distance from the injection well and 43 m bgs a significant decrease in PCE and TCE concentrations was observed (Figure 4a). PCE was almost depleted and a complete transformation to ethane occurred.

Electron acceptor processes. Before the start of the pilot test, sulfate concentrations ranged from 63 to 82 mg/L and did not change significantly during the test (30 weeks). Only in well 901 (at the location of injection) did sulfate concentrations decrease to about 20 mg/L.

Methane concentrations were < 1 mg/L before the start of the test and concentrations did not increase. Methanogenesis apparently was not stimulated by the addition of the substrates.

Because the addition of the substrates resulted in chloroethene transformation without significant sulfate reduction and methanogenesis, the lack of substrate, not unfavorable electron acceptor processes, must be the cause of the absence of biological degradation under natural circumstances.

Based on the results of the pilot test, LINER appears to be a promising technique for distribution of substrates in the subsurface. In the near future, two full-scale remediation projects will be carried out using LINER.

ACKNOWLEDGEMENTS

The Dutch Soil Research Program (SKB), Philips, the province of Gelderland and the Province of South Holland are acknowlegded for their financial support.

REFERENCES

Tonnaer, H., E.C.L. Marnette, P.A.Alphenaar, C.H.J.E. Schuren, K.M.J. van den Brink. 2001. LINER-gasinjectie. Een nieuw concept voor de stimulering van de biologische CKW-afbraak. Report nr. SV-080, CUR/SKB, Gouda, The Netherlands (in Dutch).

ENHANCED CAH DECHLORINATION USING SLOW AND FAST RELEASING POLYLACTATE ESTERS

Pawan K. Sharma (Camp Dresser & McKee Inc., Walnut Creek, California)
Hoa T. Voscott (Camp Dresser & McKee Inc., Walnut Creek, California)
Benjamin M. Swann (Camp Dresser & McKee Inc., Walnut Creek, California)

ABSTRACT: This paper compares the results of two separate field tests using slow and fast releasing polylactate esters to remediate chlorinated aliphatic hydrocarbon (CAH) contamination in groundwater by enhancing reductive dechlorination. The two polylactate esters produce soluble lactic acid at different rates upon hydration in groundwater. Microbes in the groundwater metabolize the lactic acid and in the process produce hydrogen molecules that serve as electron donors for reduction-oxidation reactions, including reductive dechlorination reactions. At both sites, both the fast and slow releasing polylactate esters have been shown to effectively change groundwater conditions to enhance reductive dechlorination of CAHs by first reducing competing electron acceptors. At Site 1 only the slow releasing compound was used and at Site 2 both compounds were used in conjunction. The time required to reduce the competing electron acceptor concentrations was greatly different at the 2 sites - approximately 1 month for Site 2 compared to 7 months for Site 1. Along with quickly providing conditions favorable for reductive dechlorination at Site 2, the high levels of hydrogen produced by the fast releasing compound also induced conditions favorable for methanogenesis. A dramatic increase in methane concentration was measured at Site 2 following the injections. At Site 1, low concentrations of hydrogen produced by the slow releasing compound only slightly increased methane levels. The higher concentrations of molecular hydrogen generated by the fast releasing compound is believed to be consumed more readily in methanogenesis and may limit the acceleration of the desired reductive dechlorinating microbial activity. However, the stimulation of the methanogenic microbes did not cause a detrimental effect on the dechlorinating microbes. The high methanogenic microbial activity observed at Site 2 did waste hydrogen in the production of methane, but reductive dechlorination of the CAHs continues.

INTRODUCTION

Sites 1 and 2 are located within 10 miles of each other in Santa Clara County, California. The geology is similar at both sites, consisting of interbedded layers of clay, silty sand, and sand (clay is the predominant soil type above the water-bearing zone). The water-bearing unit at both sites is primarily composed of silty sand. At both sites, groundwater flow has been measured consistently to the north and northeast at a gradient of 0.001 to 0.002 feet per foot.

In the mid-1960s, both sites began using tetrachloroethene (PCE) and trichloroethene (TCE) for manufacturing activities. Compounds detected in soil and groundwater included PCE and TCE and their daughter (dechlorination) products: cis-1,2-dichloroethene (DCE) and vinyl chloride. The primary releases to groundwater at both sites consisted of TCE. Initial concentrations of TCE in groundwater within each site's source area were approximately 5 milligrams per liter (mg/L). In 1985, a pump and treat system was installed at Site 2 to contain the migration of contaminants to a down-gradient property. Prior to the field tests reported in this paper, TCE concentrations in groundwater at Site 2 had decreased to approximately 1 mg/L. At Site 1, no remediation activity was conducted prior to the field test and TCE levels in groundwater within the source area remained around 5 mg/L.

To date, one of the most widely used substrates to enhance anaerobic bioremediation is lactic acid. For the field tests CDM used an environmentally safe, food quality, polylactate ester produced by Regenesis Bioremediation Products that releases lactic acid upon hydration. The released lactic acid acts as a substrate for aerobic and anaerobic microbes. Anaerobic microbes ferment the lactic acid into pyruvic acid and then converted the pyruvic acid to acetic acid (Gibson and Sewell, 1992). In the conversion of lactic acid to acetic acid by acetogens, 2 moles of molecular hydrogen are given up. Dissolved hydrogen is the most reduced of all molecules under anaerobic conditions. Since hydrogen gives up an electron so readily, it is the preferred electron donor for microbes that reduce electron acceptors such as oxygen, nitrate, iron, sulfate, and CAHs (Hemond and Fechner, 1994).

The difference between the slow and fast releasing polylactate esters is that the slow releasing compound is bound in a highly viscous polymer that slowly releases the lactic acid, while the fast version is not. Because of its low viscosity (similar to water), the fast releasing compound can be injected more easily and can spread over a larger aquifer volume more rapidly than the slow releasing compound. The fast releasing polylactate ester releases lactic acid about one order of magnitude faster into the groundwater and therefore has the ability to generate higher initial concentrations of hydrogen in groundwater. Regenesis' standard HRC^{TM} product was used as the slow releasing polylactate ester and Regenesis' HRC^{TM}-primer product was used as the fast releasing polylactate ester.

Reductive dechlorination of CAHs is the process by which anaerobic microbes (dehalogenators, halorespirers) substitute a hydrogen atom for a chlorine atom on the CAH molecule. Through this process, the more chlorinated CAHs can be dechlorinated to form less chlorinated compounds (e.g., PCE to TCE to cis-1,2-DCE to vinyl chloride and finally to ethene). Reductive dechlorination occurs most readily in conditions favorable for methanogenesis. However, dechlorination of PCE and TCE also may occur in the ORP range associated with denitrification, iron (III) reduction, and sulfate reduction. In field situations, it is often observed that dechlorination of cis-1,2-DCE and vinyl chloride does not occur in the presence of levels of sulfate in excess of 50 mg/L (Weidemeier et al., 1998).

FIELD TESTS

Field tests at both sites were conducted within the contamination source areas. The polylactate esters were injected into the water-bearing zones using a pump and a truck-mounted Geoprobe™ apparatus. The injection locations were spaced approximately 10 feet apart and covered a total injection area of approximately 4,200 and 5,000 square-feet at Site 1 and 2, respectively.

For each injection location, the Geoprobe™ advanced a probe to the bottom of the water-bearing zone. Then the point on the probe was detached, allowing the polylactate ester to be pumped into the subsurface through the unit's hollow push rods. Additional polylactate ester was injected into the water-bearing zone during retraction of the probe and discontinued at the top of the water table. A few days after the injections, following dispersement of the polylactate ester into the groundwater, all the locations were grouted with neat cement.

At Site 1, the slow releasing polylactate ester was injected at a rate of approximately 6 pounds of polylactate ester per linear foot and at pressures between 500 to 1,500 pounds per square inch (psi). At Site 2, the fast and slow releasing polylactate esters were injected in adjacent injection locations. The fast releasing polylactate was injected into the groundwater first at a rate of 3 pounds per linear foot. A few hours later, subsequent to the dispersement of the fast releasing polylactate ester, the slow releasing polylactate ester was injected a few inches away at a rate of 3 pounds per linear foot.

At both sites, prior and subsequent to the injections, a monitoring well within the source area and monitoring wells adjacent to the source area (upgradient, downgradient, and crossgradient with respect to groundwater flow direction) were monitored for:

- General environmental parameters such as temperature, pH, conductivity, ORP, alkalinity, turbidity, and total organic carbon (TOC)
- Dissolved hydrogen
- Organic acids (lactic acid and its fermentation by-product acids)
- Electron acceptors including DO, nitrate/nitrite, sulfate, and total iron
- Metabolic by-products including ferrous iron, dissolved manganese, sulfide, chloride, methane, ethane, and ethene
- CAHs including chlorinated ethenes and chlorinated ethanes.

RESULTS AND DISCUSSION

Tables 1 through 5 present the analytical data from groundwater samples collected from a single monitoring well within each site's source area prior and subsequent to the injections. The tables show data for days prior to the injections as negative valued days and after the injections as positive valued days. Day 0 corresponds to the injection day: December 15, 1999 for Site 1 and October 2, 2000 for Site 2. Both sites were monitored at six to ten week intervals.

TABLE 1. General Environmental Parameters Values within the Source Area.

Day from Injection (Day 0)	Temperature (°C)	pH	ORP (mV)	TOC (mg/L)	Dissolved Hydrogen (nM)
Site 1					
-23	17.7	7.29	170	--	--
35	17.3	7.12	-235	--	0.86
103	17.1	7.11	-190	7.6	4.80
151	17.3	6.82	-303	28	4.17
209	17.8	6.41	-293	62	2.13
264	18.4	6.09	-301	87	98
328	17.1	6.72	-303	55	26
Site 2					
-7	21.0	6.92	18	<2.0	1.65
43	20.6	6.57	-341	220	240
85	19.7	6.52	-239	390	41
138	20.4	6.85	-192	180	6
195	19.6	6.66	-159	95	7.1

-- indicates not measured

TABLE 2. Organic Acid Values Within the Source Area.

Day from Injection (Day 0)	Propionic Acid (mg/L)	Butyric Acid (mg/L)	Lactic Acid (mg/L)	Pyruvic Acid (mg/L)	Acetic Acid (mg/L)
Site 1					
-23	<1.0	<1.0	<25	<10	<1.0
35	<1.0	<1.0	<25	<10	<1.0
103	2.0	<1.0	<25	<10	13
151	17	6.7	<25	<10	40
209	14	3.9	<25	<10	77
264	45	14	<25	<10	120
328	30	<1.0	<25	<10	91
Site 2					
-7	<1.0	<1.0	<25	<10	<1.0
43	180	16	<25	<10	290
85	320	77	<25	<10	400
138	120	58	<25	<10	160
195	52	7.5	<25	<10	65

Following the injection of the polylactate esters at both sites, increased levels of TOC, organic acids, and hydrogen were observed. Organic acids were not detected at either site prior to the injections. The presence of increased levels of TOC represents the release of lactic acid from the polylactate ester into the groundwater and its fermentation by-products. The presence of acetic acid indicates the occurrence of lactic acid fermentation. In addition, propionic acid and butyric acid were detected.

As expected with the higher rate of lactic acid release of the fast releasing polylactate ester, larger increases in TOC, acetic acid, and hydrogen concentrations were measured at Site 2. With the increase in hydrogen concentration to sustain reduction-oxidation reactions, ORP decreased at both

Electron Donor Injection Strategies 309

TABLE 3. Electron Acceptor Values Within the Source Area.

Day from Injection (Day 0)	Dissolved Oxygen (mg/L)	Nitrate/ Nitrite (mg/L)	Sulfate (mg/L)	Total Iron (mg/L)
Site 1				
-23	0.14	5.2	260	0.055
35	0.08	1.3	300	0.058
103	0.07	<0.05	270	0.30
151	0.35	<0.05	88	1.7
209	0.07	<0.05	51	7.3
264	0.07	<0.10	60	13
328	0.03	<0.10	6.3	7.1
Site 2				
-7	0.24	1.2	290	0.44
43	0.01	<0.1	<1.0	3.4
85	0.00	1.6	<1.0	32
138	0.00	0.98	46	24
195	0.00	<0.1	61	15

TABLE 4. Metabolic Byproduct Values Within the Source Area.

Day from Injection (Day 0)	Sulfide (mg/L)	Ferrous Iron (mg/L)	Dissolved Manganese (mg/L)	Methane (mg/L)	Ethene (mg/L)
Site 1					
-23	<1.0	<1.0	0.79	0.004	0.000026
35	<2.0	<1.0	1.4	0.004	0.000046
103	<2.0	<1.0	2.5	0.004	0.00013
151	<2.0	<1.0	4.1	0.004	0.00045
209	<2.0	1.7	6.4	0.081	0.00040
264	2.0	2.4	8.8	0.91	0.00032
328	<2.0	7.0	9.7	1.6	0.000046
Site 2					
-7	<2.0	<1.0	0.56	0.001	0.016
43	6.9	3.2	11	0.002	0.016
85	<2.0	29	1.9	0.18	0.12
138	<2.0	25	1.6	0.73	0.47
195	<2.0	15	1.8	1.4	0.45

sites to –200 to –300 mV, indicative of highly reducing conditions in groundwater. The use of the slow releasing polylactate ester at Site 1 produced a low constant concentration of hydrogen between 2 and 5 nanomolar (nM) for the first 7 months following the treatment. The constant low levels of hydrogen resulted in a decrease in nitrate level and then to the gradual decreases in sulfate concentration in groundwater. In contrast, at Site 2 the high concentration of hydrogen generated by the fast releasing polylactate ester reduced nitrate and sulfate to non-detect levels within the first monitoring event (Day 43).

Hydrogen concentrations decreased dramatically at Site 2 between the first and second monitoring events after the treatment. As the fast releasing polylactate ester degrades and hydrogen production slows, the slow releasing polylactate ester at Site 2 is expected to continue to produce hydrogen at low concentrations. By Day 138 at Site 2, hydrogen concentrations began to stabilize to levels similar

TABLE 5. CAH Results Within the Source Areas at Sites 1 and 2.

Day from Injection (Day 0)	PCE (µg/L)	TCE (µg/L)	cis-1,2-DCE (µg/L)	Vinyl Chloride (µg/L)
Site 1				
-23	<50	3,500	500	<50
35	<50	3,200	470	<50
103	<50	2,400	1,400	<50
151	<5.0	1,400	2,742	<5
209	3.4	370	3,439	<1.0
264	2.9	340	4,442	<1.0
328	<1.0	<2.0	4,952	4.4
Site 2				
-7	71	450	1,719	180
43	<1.0	11	2,315	300
85	<1.0	2.9	62.3	1,000
138	23	31	912	1,300
195	1.6	14	449	980

to those initially observed at Site 1. Hydrogen concentrations at Site 1 increased substantially after the reduction of sulfate levels (Day 264). The increase is partly attributed to the depletion of competing electron acceptors (nitrate and sulfate) in the groundwater.

Due to the reduction of nitrate and sulfate at Site 2 and the high influx of hydrogen into the groundwater, conditions became favorable for complete reductive dechlorination to ethene. TCE and cis-1,2-DCE levels at Site 2 prior to the treatment had been 450 micrograms per liter (µg/L) and 1,719 µg/L, respectively. Following reduction of competing electron acceptors, TCE concentrations at Site 2 decreased to 11 µg/L during the first monitoring event (Day 43) and 2.9 µg/L during the second monitoring event (Day 85). The increase in PCE and TCE concentrations observed at Site 2 between Day 85 and Day 138 is attributed to the desorbtion of CAHs from soil within the source area.

At Site 2, the reduction of TCE was correlated with an increase in cis-1,2-DCE concentration to 2,315 µg/L at Day 43. However with decreased concentration of TCE, cis-1,2-DCE became a favorable electron acceptor in the groundwater and concentration of cis-1,2-DCE dropped to 62.3 µg/L at Day 85. With the dechlorination of cis-1,2-DCE, increases in vinyl chloride were measured. In addition, dechlorination of vinyl chloride to ethene is inferred with the increasing concentration of ethene between Day 43 and 85, although vinyl chloride concentrations continued to increase form 300 to 1,000 µg/L. In order for complete dechlorination to occur all required microbes must be available in this delicate symbiotic relationship.

At Site 1, initial TCE levels were higher than Site 2. The TCE level at Site 1 prior to the injections was 3,500 µg/L. Under sulfate reducing conditions observed at Site 1 subsequent to the injections, concurrent reduction of TCE to cis-1,2-DCE was evident. Thermodynamically, the reduction of all CAHs is as favorable as reduction of sulfate under standard conditions (Vogel et al., 1987).

Electron Donor Injection Strategies 311

However, in field situations, the reduction of cis-1,2-DCE and vinyl chloride is observed to be inhibited under high sulfate reducing conditions. The first indication of reduction of cis-1,2-DCE to vinyl chloride did not occur until sulfate levels fell to 6.3 mg/L on Day 328 (when 4.4 µg/L of vinyl chloride was detected at Site 1). This was the first historical detection of vinyl chloride within the source area of Site 1. However, conditions at Site 1 may currently be less favorable to detoxification (complete dechlorination of CAHs to ethene) than at Site 2, possibly due to different populations of dehalogenators present at Site 2 that maybe absent from Site 1. Dehalogenators capable of dechlorinating cis-1,2-DCE and vinyl chloride may exist in smaller populations at Site 1. This is evidenced by the increase of hydrogen concentration without the dechlorination of cis-1,2-DCE to vinyl chloride after the depletion of TCE and other electron acceptors. But with the continued generation of hydrogen and current depletion of TCE (less than detection on Day 328), the population of dehalogenators capable of dechlorinating cis-1,2-DCE and vinyl chloride is expected to increase.

With the reduction of competing electron acceptors at Site 1 and continued production of hydrogen from the slow releasing polylactate ester, it is anticipated that continued reductive dechlorination will drive remaining cis-1,2-DCE to vinyl chloride and eventually to ethene, assuming that the requisite microbes will develop a sufficient population. At Site 2, the accelerated reduction of TCE and cis-1,2-DCE to vinyl chloride and ethene, is expected to continue as hydrogen continues to be generated by the slow releasing polylactate ester and the presence of all requisite microbes appear to be present to achieve complete dechlorination.

Methanogenesis (production of methane) occurs at the same reductive (ORP) conditions as reductive dechlorination. Typically levels of hydrogen above 10 nM are needed to induce methanogenesis. Following the injections, increased levels of methane were seen at both sites. This occurred only with high initial hydrogen levels. At Site 2, with the use of the fast releasing polylactate ester and high hydrogen concentrations, methane levels increased to 0.18 mg/L 85 days after the injections. At Site 1, with the low levels of hydrogen produced, methane levels increased only slightly to 0.081 mg/L 209 days after the injections. However with the higher hydrogen concentrations observed after Day 209 (98 nM), increased levels of methane (0.91 – 1.6 mg/L) were observed at Site 1.

CONCLUSIONS

Both the fast and slow releasing polylactate esters have been shown to effectively change groundwater conditions to enhance reductive dechlorination of CAHs at both sites by first reducing competing electron acceptors. Sulfate concentrations at both sites were reduced below inhibitory levels, and the continued release of hydrogen from the slow releasing polylactate ester is anticipated to facilitate the reductive dechlorination process at both sites. However, the time required to reduce the competing electron acceptor concentrations was greatly different at the two sites - approximately 1 month for Site 2 compared to 7 months for Site 1. Similar to Site 2 where the initial high hydrogen concentrations produced high methane levels, Site 1 also showed that high level of hydrogen in groundwater would induce methane production. The higher concentrations of molecular hydrogen generated by the fast releasing compound is believed to be consumed more readily in methanogenesis and may limit the acceleration of the desired reductive dechlorinating microbial activity. However, the stimulation of the methanogenic microbes did not appear to cause a detrimental effect on the dechlorinating microbes as previously reported (Fennel, et al., 1997). Even with the high production of methane observed at Site 2, reductive dechlorination of the CAHs continues.

REFERENCES

Fennel, D.E., J.M. Gosset, and S.H. Zinder. 1997. Comparison of Butyric Acid, Ethanol, Lactic Acid, and Proprionic Acid as Hydrogen Donors for the Reductive Dechlorination of Tetrachloroethene. *Environmental Science and Technology.* 32(3): 918-926.

Gibson, S.A. and G.W. Sewell. April 1992. *Applied and Environmental Microbiology.* 58(4): 1392-1393.

Hemond, H.F. and E.J. Fechner. 1994. *Chemical Fate and Transport in the Environment.*

Vogel, T.M., C.S. Criddle, P.L. McCarty. 1987. *Environmental Science Technology.* 21: 722-736.

Weidemeier, T.H., D.E. Moutoux, E.K. Gordon, J.T. Wilson, B.H. Wilson, D.H. Kampbell, P.E. Haas, R.N. Miller, J.E. Hansen, and F.H. Chapelle. 1998. *Technical Protocol for Evaluating Natural Attenuation of Chlorinated Solvents in Groundwater.*

APPLICATION OF IN-SITU REMEDIATION TECHNOLOGIES BY SUBSURFACE INJECTION

Baxter E. Duffy, **Inland Pollution Services, Inc., Elizabeth, New Jersey**

Abstract: There are many emerging remedial technologies which require the introduction of various materials to the saturated zone of a contaminated site in an effort to remediate groundwater in-situ. These remedial materials include both chemical compounds as well as proprietary products such as hydrogen release and oxygen release compounds.

A common requirement to implement an effective application of many of these materials includes not only the accurate delineation of the contaminated plume and lithology of the geologic formations in which the plume resides, but also a well thought out delivery process which considers the physical properties of the remedial material to be injected and the formation characteristics of the injection site.

Injection under pressure will provide greater distribution of materials into the formation providing a more effective remediation. Delivery processes may be developed for unconsolidated overburden formations in sands, silts, clays as well as tills. Alternate delivery processes may be developed for weathered rock and fractured rock formations. These delivery processes may include direct injection as well as re-injectable point methods. Injection pressures can vary from 90 to 700 psi. This paper describes the delivery processes in several formation types, using schematic drawings and discussion.

INTRODUCTION

The addition of materials to the saturated subsurface to remediate contaminants in situ has been performed for a number of years. As the technologies advance, many of these techniques are receiving wider acceptance and application by the consulting and regulatory community. Technologies of this type include chemical compounds targeting specific chemical reactions with contaminants in the subsurface and bio-enhancements which target initiating or enhancing natural attenuation processes. During initial applications of many of these technologies, the materials were simply poured into open boreholes or wells by gravity or low pressures, less than 70 psi. As the use of these materials has become more widely accepted, the variety of site conditions, contaminant characteristics, lithology and material characteristics encountered have yielded an equal variety of delivery methods. The use of injection techniques at pressures ranging from 90 to 700 psi to enhance delivery has evolved as a preferred method of delivery.

The observations and methods described in this paper are based upon the injection of remedial materials at over 100 sites. The locations of the subject sites are mostly in New Jersey but sites in New York, Pennsylvania, Connecticut and Puerto Rico have also been considered. The remedial materials injected include chemical compounds such as peroxides, reagents, potassium permanganate, acetic acid, sulfuric acid, sodium metabisulfite, and calcium sulfite. The majority of these methods were developed during the injection of two proprietary bio-enhancing products, Hydrogen Release Compound® (HRC) and Oxygen Release Compound® (ORC).

When installed by gravity, the distribution of the remedial material is limited to the same natural driving forces that move the contaminants present in the aquifer system. These are advection, diffusion, and dispersion. These processes may be limited by the specific lithology of a site resulting in the inhibition of the distribution of the remedial material, and in a reduction of the effectiveness of the remediation. The use of injection techniques over gravity methods provides greater physical distribution of the remedial material initially with more predictable overlapping of the area of influence of each injection point, allowing for the installation of fewer delivery points.

The effect of poor delivery of a remedial material is often seen in the "pinstripe effect" where contaminates continue to migrate in higher concentration between treatment points. Dissolved phase contaminant rebound is also observed in applications where poor distribution of the remedial materials are implemented. In these cases the remedial material may not be reaching soils which have contaminant sorbed to the soil matrix.

The process of injecting remedial materials may be utilized in unconsolidated overburden formations consisting of sands, silts, clays and even some tills by direct push drilling methods. It is in these shallow formations that the majority of injection of remedial materials has occurred, as these shallow formations are the most commonly contaminated zones. Applications have been designed and implemented for very difficult till formations as well as weathered, fractured and competent bedrock formations.

Figure 1. Direct Injection in Unconsolidated Formation

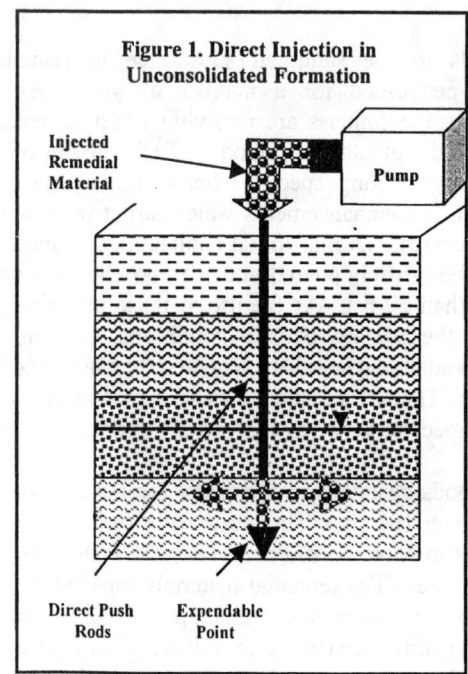

PUMPING EQUIPMENT AND INJECTION PRESSURE

The injection system will consist of the pump, mixer, hose, injection rod or casing and, in terms of required pressure, the formation into which you are injecting. The remedial material must always be evaluated to determine equipment compatibility and operator safety issues. The pressure produced by the pump may be significantly higher than the injection pressure. The pump must overcome the material characteristics and system pressure losses. Injection pumps are rated to deliver pressures as high as 2,000 psi. This pressure may never be realized within the injection system, which generally operates between 90 to 700 psi. The system components must be selected for the highest

Electron Donor Injection Strategies 315

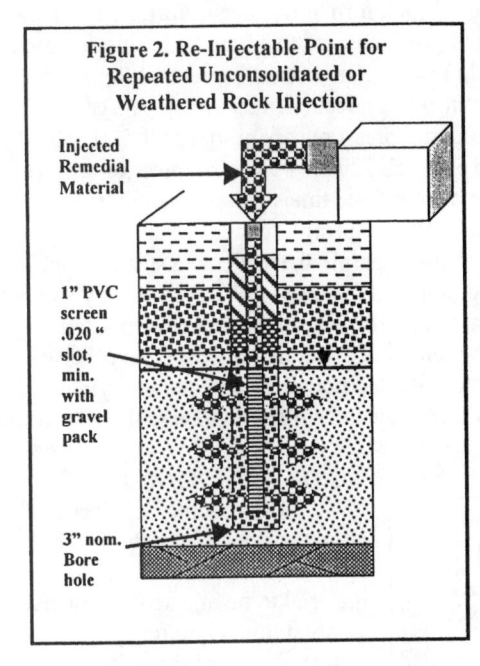

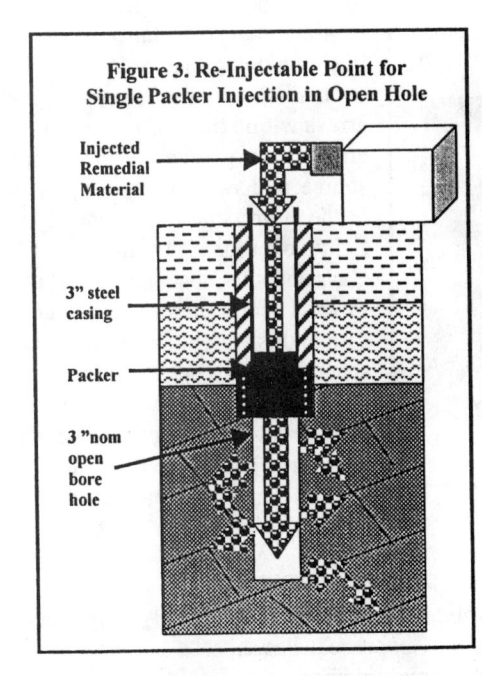

Figure 2. Re-Injectable Point for Repeated Unconsolidated or Weathered Rock Injection

- Injected Remedial Material
- 1" PVC screen .020 " slot, min. with gravel pack
- 3" nom. Bore hole

Figure 3. Re-Injectable Point for Single Packer Injection in Open Hole

- Injected Remedial Material
- 3" steel casing
- Packer
- 3" nom open bore hole

system pressure anticipated for an injection event. Where the system pressure exceeds 700 psi, it is likely that no flow pathway is present and no injection will occur before equipment failure.

The injection pressures required are, in general, material and formation dependent. The physical characteristics of the material will greatly effect the pressure demands on the pumping system. Where the material is high in suspended solids or high in viscosity, higher delivery pressures will be required. Formation characteristics will also impact injection pressures, in a well sorted beach sand, lower injection pressures will be required than in a dense, uniform clay. Another important factor in deeper injections is the hydraulic head within the system which must be overcome.

Breakthrough or Short Circuiting. "Breakthrough" or "short circuiting". occurs in shallow borings when the injection pressure within the system causes a failure in the seal between the soils and the exterior of the rod or casing string. This seal is a result of the lateral soil pressure on the rod string and may be referred to as a "smear seal". In general, the higher the clay or silt content of a formation, the stronger the smear seal. As a result, when injecting in sandy coastal plain formations, breakthrough may occur at greater depths as the smear seal is broken. When breakthrough occurs often in an injection array, the use of multiple injection borings to address deep and shallow intervals as well as very low injection pressures may be used to over come this problem.

Formation Fracturing. In tighter formations where silts, fine sands and clays and even some weathered or fractured rock formations occur, formation fracturing has been observed. The injection pressure required to create hydraulic fractures in clay soils and shale bedrock is quite low, in many cases less than 100 psi. In similar formations, pneumatic fracturing can occur at 100 to 200 psi. When utilizing high pressure pumps with pump pressure capabilities of 1,500 to 2,000 psi, the system pressure is allowed to reach 700 psi before an injection interval is terminated. This system pressure is two to three times greater than that required to perform formation fracturing. Formation fracturing is occurs when the system pressure at a given injection interval rises sharply then falls to a steady level as the fracture propagates. This phenomenon is frequently observed in clay and silt formations when injecting remedial materials. The advantage of formation fracturing is the enhancement of the remedial material delivery in an otherwise resistant formation. It is important to note that not all formations will yield and fracture, however, most unconsolidated clay and silt formations will fracture with the correct equipment.

SPACING OF INJECTION POINTS

Figure 4. Re-Injectable Point for Straddle Packer Injection in Open Hole

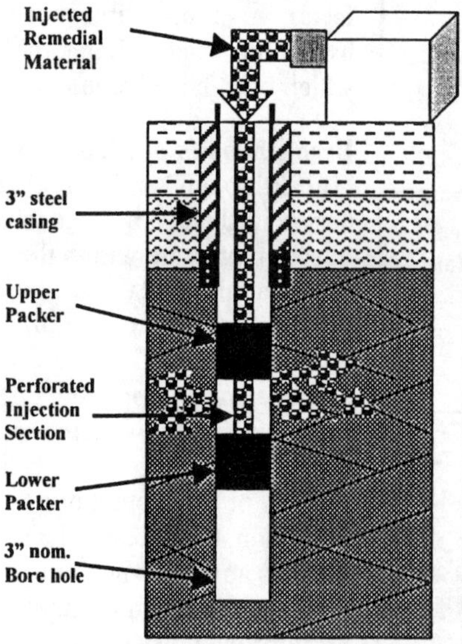

Injected Remedial Material

3" steel casing

Upper Packer

Perforated Injection Section

Lower Packer

3" nom. Bore hole

The spacing of injection points will vary with the geological formation, volume of remedial material to be injected at any given location, the migration characteristics of the remedial material and the proximity of the injection point to the plume. There are three basic orientations of an injection point relative to a contaminant plume. These are: grid arrays within the plume, barrier arrays down gradient of the plume source, and sweep arrays upgradient of the plume source. In general, in tighter overburden formations of clays and silts, spacings are set at 5' to 7' on center. In more permeable formations with higher sand content spacing may go out to 10' or 15' on center. Bedrock formations require consideration of the occurrence of weathered bedding and fracture planes to establish spacing and are typically 5' to 10' on center.

INJECTION IN UNCONSOLIDATED FORMATIONS

Injection in unconsolidated formations is generally performed by direct push drilling methods (such as Geoprobe®, Powerprobe® or Earthprobe®) although, it is important to note that other methods may be employed on a site specific basis. The injection string is advanced by driving rods (typically 1 ¼" O.D.) to the target depth. Once at depth, the pump, mixer and hose are connected to the rods. As rod diameter increases, the material volume requirement to meet the required system pressure increases and must be considered. An expendable drive point is released from the bottom of the rod string as the injection pump is initiated (Figure 1). During an initial injection, several injection locations may be completed before the injection characteristics of the site are evaluated and any breakthrough or fracturing difficulties addressed and overcome.

When utilizing the correct equipment, overburden injection difficulties are generally limited to breakthrough, attaining injection pressure required to defeat formation resistance and remedial material characteristics.

INJECTION IN RE-INJECTABLE POINTS

Where direct push injection is not practical due to formation characteristics, the application of a remedial material will occur repeatedly throughout the life of the project or repeated access with drilling equipment is an issue, the use of re-injectable points should be considered. Re-injectable points are generally constructed as micro wells. Most commonly, in unconsolidated or shallow weathered rock formations, re-injectable points are constructed of 1" diameter screen and riser, screening only the targeted injection zone. Screen lengths over 10' need to be carefully evaluated because they increase demands on system pressures. Materials of construction can be stainless steel, scd 80 PVC or scd 40 PVC depending on screen length, formation conditions, anticipated service life, and injection pressures. The re-injectable point may be installed by direct push, hollow stem auger, or air rotary drilling methods depending upon formation characteristics (Figure 2).

Re-injectable points in competent rock may be constructed as micro wells (Figure 2) in a manner similar to the unconsolidated re-injectable point or as open hole bedrock wells where the overburden is cased off (Figure 3).

Delivery of the remedial material to the re-injectable point is performed by connecting the injection pump to the top of the riser or by inserting a single packer (Figure 3) or straddle packer set (Figure 4) to target specific fracture zones or deliver material at higher pressures.

The volume of the remedial material to be delivered and the anticipated injection pressure requirements must be considered when selecting the design of the re-injectable point. Where highly weathered or fractured rock is to be addressed, care must be taken not to lodge packer assemblies into bore holes which are not competent. These bore holes may require specific construction details to avoid such a costly occurrence.

INJECTION IN TILLS AND FILLS

It is widely accepted that tills can provide for some of the most difficult drilling conditions in the overburden. Another difficult drilling condition occurs when passing through historic fills which often consist of randomly distributed construction debris. Drilling, let alone injecting in these conditions is quite difficult and may require a variety of installation techniques. These techniques may include direct push, air rotary, and packer methods to accomplish the injection goals. At sites where such conditions are encountered, it is not unusual to employ a combination of methods across a single site where varying conditions are encountered.

INJECTION APPLIACTION DESIGN CRITERIA

In consideration of the variety of drilling and injection methods available, almost any site can be evaluated and an injection plan designed and implemented for a specific remedial material. When evaluating a injection design, the key elements to consider are as follows:

Evaluate Permit Requirements. Some states will require Discharge to Groundwater (DGW) permits, Underground Injection Control (UIC) permits, or issuance of a Permit by Rule prior to the injection of remedial materials into an aquifer.

Defined Source Area. Unless you are planning to evaluate and extend the treatment area based upon monitoring results in future injections, it is not advisable to inject a remedial material in a site where the source of the contamination is not well defined. Rebound is likely to occur when critical source areas are not treated. The plume should be delineated both horizontally and vertically prior to an injection event.

Understand the Geological Conditions. It is very important to have accurate soil boring logs for unconsolidated formations and where possible a good understanding of the rock formations, including porosity, permeability, bedding plains and fractures.

Well Defined Injection Array. Prior to mobilizing the drilling and injection equipment, the desired injection array should be determined and marked out. Consideration should be given to equipment access and site restrictions including structures, process equipment, and underground and over head utilities. Adjustments to the plan can be made in the field where warranted.

Well Defined Material Handling Plan. The contractor handling the injection must understand the physical properties of the remedial material to prepare for weather, pumping, equipment compatibility, and health and safety issues.

Properly Designed Injection Point. The proper injection technique for a particular site will consider the remedial material characteristics, material volume, target depths, geologic formation, pressure requirements, and site access issues.

Proper Equipment Selection. The equipment selection must consider all the above parameters to assure that the required methods may be performed to effectively complete the injection event.

CONCLUSIONS

The methods for injecting remedial materials discussed here are a summation of methods utilized at many sites. They represent basic approaches developed with extensive field experience for a wide variety of formations. There is great value in studying the physical characteristics of the remedial material to be injected and the specific geological formations of a site where injection is to occur in order to develop a site specific injection application plan. Varying material and site conditions may warrant a field demonstration prior to a pilot or full scale application. Each individual site and each injected material will present varying obstacles to overcome in the field during the injection process.

REFERENCES

Race, S.L. and Goeke, P.M., 1999, "Study of Natural Bioremediation Projects Using Time-Released Oxygen Compounds", *Accelerated Bioremediation Using Slow Release Complounds*, (ed) S.S. Koenigsberg, Regenesis Bioremediation Products

Regenesis. 1997. *Oxygen Distribution in an Aquifer*. ORC® Technical Bulletin #2.5.1, Regenesis Bioremediation Products

Suthersan, S.S., Hydrualic and Pneumatic Fracturing, *Remediation Engineering*, Geraghty & Miller Environmental Science and Engineering Series, Ch. 9

Tact,Jr., P.D., 1999, "Vinyl Chloride Attenuation by Direct Injection of Peroxygen", *Accelerated Bioremediation Using Slow Release Complounds,* (ed) S.S. Koenigsberg, Regenesis Bioremediation Products

U.S. EPA, 2000, *Engineering Approaches to In Situ Bioremediation of Chlorinated Solvents: Fundementals and Field Applications*, EPA Report 542-R-00-008, July 2000

Wilson, D.J. and Norris, R.D., 1999, "Modeling of Remediation with ORC®: Transverse Dispersion", *Accelerated Bioremediation Using Slow Release Complounds*, (ed) S.S. Koenigsberg, Regenesis Bioremediation Products

Zahiraleslamzadeh, Z.M. and Bensch, J.C., 2000, "Enhanced Bioremediation Using Hydrogen Release Compound (HRC®) in Clay Soils", *Accelerated Bioremediation of Chlorinated Compounds in Groundwater*, (Ed) S.S. Koenigsberg, Regenesis Bioremediation Products

REHABILITATION OF A BIOFOULED RECIRCULATION WELL USING INNOVATIVE TECHNIQUES

Sarah R. Forman (URS, Linthicum, Maryland, USA)
Tim Llewellyn, Scott Morgan, and Carol Mowder (URS, Linthicum, Maryland, USA)
Dr. Suzanne Lesage, Kelly Millar, and Susan Brown (Environment Canada, Burlington, Ontario, Canada)
George DeLong (Advanced Infrastructure Management Technologies, Oak Ridge, Tennessee, USA)
Donald J. Green and Heather McIntosh (U.S. Army Garrison, APG, Maryland, USA)

ABSTRACT: A field scale Pilot Test, using recirculation well technology to deliver a mixture of vitamin B_{12} concentrate to remediate contaminant source areas at the site, has been in operation at Graces Quarters (a National Priorities List site), Aberdeen Proving Ground, Maryland since September 1999. Groundwater on site is contaminated with a mixture of chlorinated solvents, including 1,1,2,2-tetrachloroethane, carbon tetrachloride, trichloroethene, tetrachloroethene, and chloroform. This paper reports on the biofouling difficulties; assessment methods and testing; treatment methods; control/maintenance measures enacted; and numerical simulations of chemical delivery scenarios for the Pilot Test.

INTRODUCTION

A Pilot Test using recirculation well technologies to deliver a vitamin B_{12} concentrate to degrade a mix of volatile organic compounds (VOCs) in situ (Mowder et al, 2000) was shutdown in December 1999 due to biofouling of the recirculation well and its immediate vicinity. Conventional well redevelopment and chemical methods were only partially effective in the restoration of permeability. To assess the extent and type of biofouling, direct push probe (geoprobe) samples of the aquifer matrix were collected and submitted for laboratory testing of permeability, phospholipid fatty acid (PLFA), and deoxyribonucleic acid (DNA). It was determined that the affected (biofouled) area was limited in extent and the Aqua Freed® well redevelopment process (an injection of carbon dioxide under pressure) was selected as a treatment method. This technique was successful in restoring full permeability to the system. This project was developed and implemented under contract and oversight by AIMTech, a U.S. Department of Energy program managed by Lockheed Martin Energy Systems, Inc., in conjunction with the APG Installation Restoration Program. Distribution restriction statement approved for public release, distribution is unlimited, OPSEC Number 3743-A-4.

The biocide and vitamin B_{12} concentrate injection concentrations and delivery scenarios were modeled using a modified version of a previously generated MODFLOW model (Forman et al., 2000) and the MT3DMS model code. The options of pulsing the mixture into the aquifer in one day or three-day weekly injections were simulated, and relative pulse concentrations and propagation through the aquifer were evaluated. The selected delivery scenario (one day weekly pulses of vitamin B_{12} with daily biocide injection) controlled the biomass on-site and field data validated the model.

The delivery of the vitamin B_{12} concentrate was modified prior to resumption of the Pilot Test. Changes included the addition of a biocide (Tolcide®); the removal of sugar from the concentrate; reversal of flow; and pulse injection to the subsurface. Tolcide® was injected into the recirculation well prior to and during the course of the resumption of the Pilot Test with positive results. Hydrographs and a stable functioning system confirm that the biomass was controlled with the aid of the biocide. The reformulation of the vitamin B_{12} concentrate aided in the control of the biomass.

Objective. The objective of the rehabilitation techniques employed was to restore aquifer permeability in the recirculation well, its well pack, and its immediate vicinity. This was vital to the resumption of the Pilot Test on-site which was conducted to evaluate the remediation of chlorinated VOCs using a vitamin B_{12} concentrate.

Site Description. The Pilot Test site is located at Graces Quarters, a part of the Edgewood Area portion of the Aberdeen Proving Ground (APG) in Eastern Maryland. Graces Quarters is located on the Gunpowder Neck Peninsula. The Pilot Test site is located in the Primary Test Area, which is approximately 22 acres (0.089 square kilometers [km^2]) in size. This area is underlain by an unconfined surficial aquifer, which is separated from an underlying confined aquifer by a silty clay confining unit. This confining unit is absent in some areas allowing the surficial and confined aquifers to form one hydrologic unit at these locations.

Groundwater flows south-southwest in the surficial aquifer. Upon encountering the migration pathways in the confining layer, the groundwater flows into the confined aquifer and flows outward in a radial pattern that is superimposed on an overall southward flow field.

The contaminants of concern are chlorinated VOCs that include 1,1,2,2-tetrachloroethane, carbon tetrachloride, trichloroethene, tetrachloroethene, and chloroform. Repeatable total VOC concentrations are on the order of 7,000 micrograms per liter (µg/L). Concentrations as high as 185,988 µg/L (total VOCs) have been documented at (unrepeatable) geoprobe locations, suggesting the presence of residual product. These contaminants were likely introduced to the surficial aquifer, entered the confining aquifer through holes in the confining layer and spread out in the groundwater in an overall southward flowing direction.

MATERIALS AND METHODS

In September 1999, the field scale Pilot Test for reductive dechlorination of VOCs, using a vitamin B_{12}/titanium citrate mix (B_{12} concentrate) with a recirculation well, was initiated at Graces Quarters. Data indicated that a hydraulic recirculation cell was set-up and that contaminants were being successfully degraded on site. However, it became increasingly difficult to inject the water into the upper well screen of the recirculation well and system shutdown occurred in December 1999. It was thought that an increase in bacterial populations was causing the reduced permeability in the aquifer.

The collection and analyses of groundwater and geoprobe (soil) samples, and downhole video logs were used to assess the cause of reduced permeability (biofouling). Mechanical and chemical redevelopment techniques were applied to the test area to rehabilitate the recirculation well/system and allow resumption of the Pilot Test. A biocide (Tolcide®) in conjunction with the pulsed delivery of the vitamin B_{12} concentrate was used to control the bacterial growth on site. Numerical modeling was used to determine injection concentrations of the biocide and the vitamin B_{12} concentrate, and the relative distribution of the vitamin B_{12} concentrate in the aquifer resulting from these differing pulse injection scenarios.

The aforementioned field methods, laboratory analyses, and numerical modeling simulations are discussed below.

Assessment of the Extent and Type of Biofouling. In November 1999, five groundwater samples were collected from piezometers 1A, 3A, and 3C and the upper and lower screened intervals of the recirculation well (QRW1) (shown on Figure 1) for inspection of the suspected biomass. Following collection of these samples, a downhole video log was run on the upper portion of the recirculation well and several piezometers to investigate conditions in the wells. The video log allowed observation of the biomass coating the well screen and equipment in the well.

Based on the condition of the recirculation well, as revealed by the downhole video log, it was redeveloped in December 1999, using mechanical

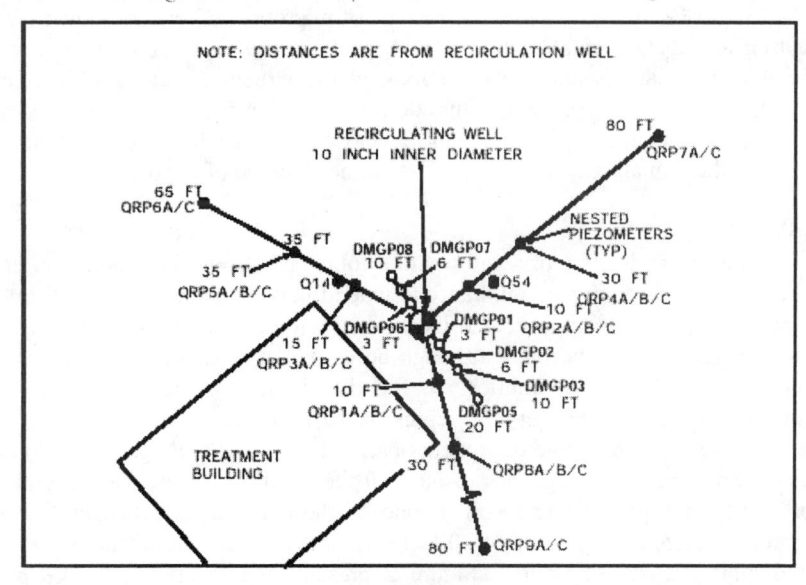

FIGURE 1
RECIRCULATION WELL
& PIEZOMETER LAYOUT

chemical (hydrogen peroxide) methods. Hydrogen peroxide was added to the recirculation well (without the downhole piping and the packer assembly) to achieve a five percent (5%) solution. The interior of the well was scrubbed with a large brush and water was evacuated to remove the biomass. This procedure was repeated to ensure the removal of biomass. In addition, the upper screened interval of the well was isolated with a packer and purged. No biomass was observed at the end of redevelopment of the recirculation well. The system was restarted, without the addition of the vitamin B_{12} concentrate.

Although this method succeeded in cleaning the bacteria from the inside of the well, it remained difficult to inject groundwater into the upper screen of the recirculation well. This seemed indicative of clogging/biofouling of the well pack and the formation. The lower screen of the well remained relatively unaffected by the biofouling.

Seven direct push probes (geoprobes—DMGP01 through DMGP03, and DMGP05 through DMGP08) were installed along two axes radiating out from the recirculation well as presented in Figure 1. Radial distances from the recirculation well ranged from 3 to 20 feet (0.91 to 6.10 meters [m]). Each geoprobe was advanced to a depth of 20 feet (6.10 m) below ground surface (bgs) and soil cores were collected at each location from 10 to 20 feet (3.05 to 6.10m) bgs. Soils were visually inspected in the field, using a hand lens, for biomass or other impact from the treatment system. Five geoprobe cores were sent to a geotechnical laboratory for permeability testing. Each sample was selected from the most transmissive zone, based on lithology, at that radial distance from the recirculation well. Six soil samples of the aquifer matrix were collected and submitted for laboratory analyses of microbial phospholipid fatty acids (PLFA), and deoxyribonucleic acid (DNA) by denaturing gradient gel electrophoresis (DGGE). PLFA analysis indicates the type of microbes present and how they are reacting to environmental factors. DNA analysis provides a closer look at the microbial populations and allows a look at the structure and identification of prominent organisms present. In addition, six samples were collected and submitted for titanium, iron, manganese, and exopolysaccharides analyses. The results of these analyses were used to determine a course of action.

RESULTS

Visual and olfactory information obtained from the geoprobe cores indicated that biofouling was unlikely to extend beyond 6 feet (1.83 m) from the recirculation well. Odor (fatty acid-like) (See Figure 2) and dark grey stained soil was observed in these borings, although no evidence of significant biomass that would impede flow in the aquifer was found. The most noticeable staining and odor were located at the vertical depths of 10 to 13 feet (3.05 to3.96 m) bgs. However, some staining and odor were observed below 13 feet bgs. A slight odor was observed at both geoprobes located 10 feet (3.05 m) from the recirculation well, with dark grey stained soil at one of these locations (DMGP08). The geoprobe located at a distance of 20 feet (6.10 m) from the recirculation well did not exhibit staining or odors. Figure 2 presents the results of the geoprobe investigation.

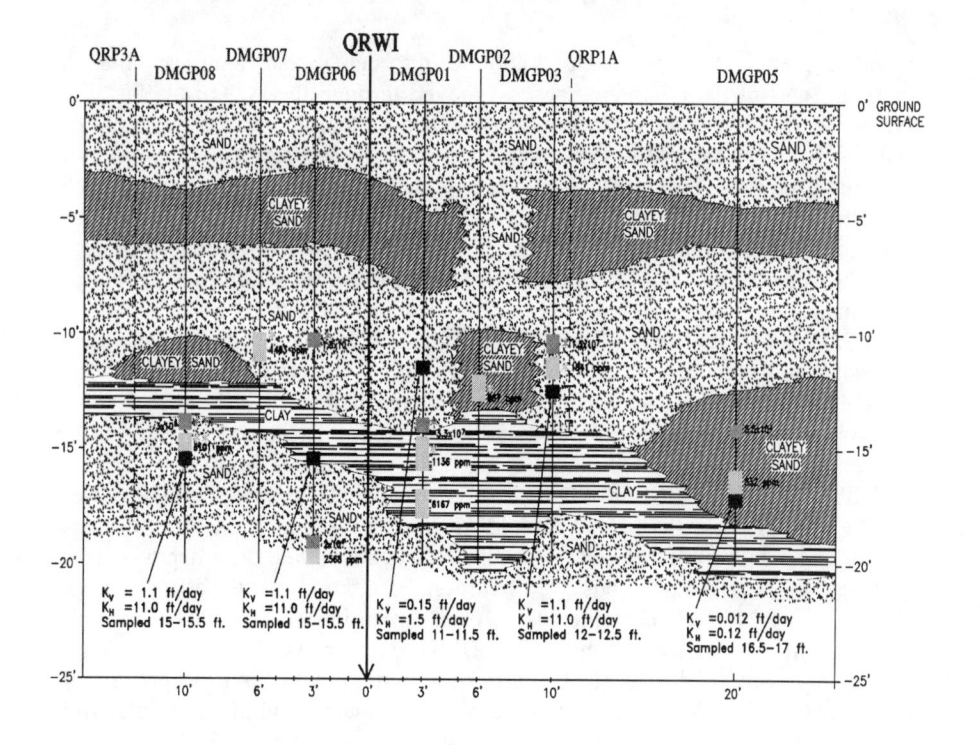

FIGURE 2
Geoprobe Investigation Results

The exopolysaccharides (EPS) detected varied from very low (μg/g) to non-detect in theses samples. Therefore, they would not be a significant cause of reduced permeability in the formation. However, a low EPS count may not mean less bacteria due to the fact that not all bacterial strains produce EPS.

Permeabilities correlated with lithology (i.e., clay/clayey sand [DMGP05] and sand [DMGP08]), rather than the amount of bacteria. However, the sample DMGP01 (sand – located 3 feet from the recirculation well [Figure 2]) had an average number of bacteria, and a reduced permeability (0.15 feet per day [0.046 meters per day {m/day}]) compared to surrounding information (1.1 feet per day [0.34 m/day]). This sample was collected at a distance of 3 feet (0.91 m) from the recirculation well at a depth of 11 feet (3.35 m) to 11.5 feet (3.51 m) bgs.

Geoprobe samples were also analyzed for titanium, iron, and manganese. The samples were analyzed for titanium because bacterial activity is associated with the degradation of citrate and the resulting deposition of titanium. The samples were analyzed for iron and manganese because the treatment may mobilize them. As presented in Figure 2, the metals results indicate that the titanium (part of the treatment process) moved through the system (as far as 20 feet [6.10 m] away), as the system did not contain titanium prior to treatment onsite. Results indicated that there was no correlation between concentration and distance from the recirculation well. The highest concentrations were associated with the presence of fines. Titanium concentrations were on the same order of magnitude as the iron concentrations. Therefore, the redistribution of titanium over the treated area would not be cause of a significant reduction in permeability. The manganese concentrations were non-detect to low within approximately 10 feet (3.05 m) of the recirculation well, with higher concentrations seen at the geoprobe collected at a distance of 20 feet (6.10 m) away (DMGP05). It is possible that manganese was dissolved and re-deposited.

Six geoprobe soil samples were analyzed for DNA and PLFA. PLFA results (Figure 2) indicated that biomass content ranged from $\sim 10^4$ cells in DMGP05 (located a distance of 20 feet [6.10 m] away from the recirculation well, at a depth of 14 14.5 feet [4.27 to 4.42 m] bgs) to $\sim 10^7$ in DMGP01 (located a distance of 3 feet [0.91 m] away from the recirculation well, at a depth of 14 to 14.5 feet [4.27 to 4.42 m] bgs). In addition, results indicated that all samples, except for DMGP05 contained a diverse microbial community, primarily composed of Gram negative bacteria. DMGP05 contained a simple microbial community composed mainly of biomarkers ubiquitous to both the prokaryotic and eukaryotic kingdoms. The Gram negative communities in these samples were in the stationary (normal) phase of growth with sample DMGP05 having the slowest turnover rate. The Gram negative communities with detectable biomarkers for decreased membrane permeability showed signs of this occurring (a bacterial response to environmental stress).

DNA analyses indicated distinct banding patterns for all samples, except for DMGP05 (See Figure 2 for sample depths). This sample did not provide enough material to produce a distinguishable pattern. Similarities were observed between samples DMGP01 (14 to 14.5 feet [4.27 to 4.42 m] bgs), DMGP06 (10 to 10.5 feet [3.05 to 3.20 m] bgs), and DMGP03 (10 to 10.5 feet [3.05 to 3.20 m] bgs). Several dominant organisms were present in all three samples. Sequences results showed several bands affiliated with the genus *Geobacter*. It is suggested that *Geobacter* is a dissimilatory iron- and sulfur-reducing bacterium (Lonergan et al, 1996).

Rehabilitation Measure(s). Based on the results of the geoprobe investigation, the Aqua Freed® well redevelopment process (an injection of carbon dioxide under pressure) was selected as a treatment method. This process is suited to clearing the well pack material and the first few feet (meters) of the formation, where the majority of clogging/biomass was now documented to occur. Following redevelopment with this process from April 26 through April 29, 2000, hydraulic testing was conducted to assess the specific capacity of the aquifer. This technique was successful in restoring full permeability to the system.

Gaseous carbon dioxide was injected into the upper and lower screened intervals of the recirculation well, as well as several nearby piezometers (QRP1A,

QRP1B, QRP2B, and QRP3B), which are located 10 to 15 feet (3.05 to 4.57 m) from the recirculation well (Figure 1). Due to the unconsolidated sediments on-site (sand with clay lenses) and shallow depth of the upper screen, the injection of the gaseous carbon dioxide (injection pressures used ranged from 25 to 30 pounds per square inch [psi] [17,577.5 to 21,093.0 kilograms per square meter {kg/m^2}]) was closely monitored to minimize channeling to the surface. The liquid phase of the Aqua Freed® process, which requires injection pressures 70 psi (49,217.0 kg/m^2) or greater, was not initiated on-site due to probable channeling. All piezometers within an 80-foot (24.38 m) radius of the recirculation well, except for QRP1A and QRP9A, exhibited effects of the gaseous carbon dioxide as a rise in water levels.

Following the injection of the gaseous carbon dioxide, surging and overpumping were used to redevelop the recirculation well. The evacuated groundwater changed from dark green to grey with a strong biological odor to light grey to clear with a slight biological odor. A total of 3,000 pounds (1,360.78 kilograms [kg]) of gaseous carbon dioxide were injected into the ground and 6,500 gallons (24,605.1 L) of groundwater were evacuated during this process. The packer, pumps, and associated downhole equipment were decontaminated with a hydrogen peroxide solution prior to reinsertion into the recirculation well.

Following redevelopment, two, 300 minute, variable rate recirculation tests were conducted at the recirculation well on May 8 and May 9, 2000 (one with the withdrawal of water from the lower screen and discharge through the upper screen, and one with the withdrawal of water from the upper screen and discharge through the lower screen). Each test consisted of three, 100 minute, intervals. The first test, involving the withdrawal of water from the lower screen and discharge through the upper screen, was conducted at flow rates of 2, 4, and 4.7 gallons per minute (gpm) (7.57, 15.14, and 17.79 Liters per minute [L/min]). The second test involved the withdrawal of water from the upper screen and discharge through the lower screen at flow rates of 2, 4, and 5 gpm (7.57, 15.14, and 18.93 L/min). Water levels for each test stabilized at each flow rate.

Results indicated that permeability was fully restored, allowing the withdrawal and discharge through both the upper and lower screens of the recirculation well at rates similar to initial conditions determined in August 1999. Based on these results, plans were made to resume the Pilot Test.

Biofouling Control/Maintenance Measures. The delivery of the vitamin B_{12} concentrate was modified prior to resumption of the Pilot Test. Changes included the addition of a biocide (Tolcide®); the removal of sugar from the concentrate; reversal of flow; increased flow rate; and pulse injection to the subsurface. Tolcide® was injected into the recirculation well prior to, and during, the resumption of the Pilot Test with positive results. Hydrographs and a stable functioning system confirm that the biomass was controlled with the aid of the biocide. The reformulation of the vitamin B_{12} concentrate aided in the control of the biomass. Figure 3 presents water level data obtained during the resumed portion of the Pilot Test.

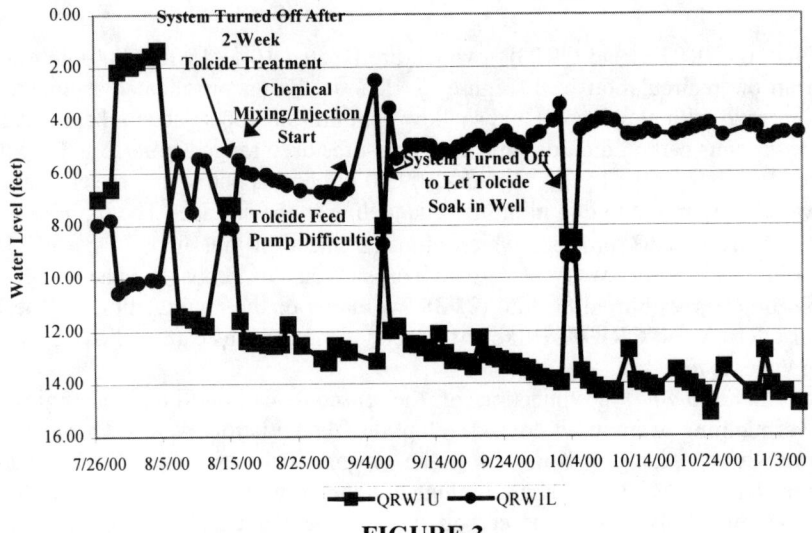

FIGURE 3
QRW1U/QRW1L Water Levels - Phase II

The Pilot Test was resumed on July 28, 2000 with the recirculation well operating at 4 gpm (15.14 L/min). Tolcide® was added to the recirculation well for 4 hours daily each day of operation. The Tolcide® pump was difficult to regulate (i.e., difficulty adjusting the injection rate observed) and injected the Tolcide® at 150 ppm (150 milligrams per Liter [mg/L]) (the upper limit of the range). As presented on Figure 2, Tolcide® feed pump difficulties (i.e., periodic shutdown) resulted in periodic increases in the water level in the lower screen of the recirculation well. Clogging of the recirculation well lower screen occurred as evidenced by odor and rising water level. Following the addition of additional Tolcide®, the water level would decrease in the lower well screen. In response to dropping water levels in the upper screen, the flow was reversed and a Tolcide® soak was performed weekly. Although periodic Tolcide® "soaks" were instituted to maintain control of the biofouling, the Pilot Test operated for the 3-months without long-term shutdowns and the Tolcide® proved very effective at controlling biofouling growth without detrimental effect to the treatment process itself. This is discussed in detail in Millar et al, 2001.

During the Pilot Test, three groundwater samples were submitted for analysis to estimate the Tolcide® concentration in the aquifer at various distances. The goal was to determine if there was a build-up of Tolcide® in the aquifer and to check on the Tolcide® dosing rate.

The vitamin B_{12} concentrate was reformulated prior to resumption of the Pilot Test. Glucose was no longer to be added to the formula, in an effort to control biofouling of the recirculation well and the formation. The Tolcide® is designed to preserve the citrate. Less vitamin B_{12} was added to the formula, although the higher injection rate would yield higher vitamin B_{12} concentrations in the recirculation well. The titanium concentration during injection was also increased to that used during the laboratory treatability test to induce faster reaction rates. Also, the vitamin B_{12} concentrate was pulse injected to inhibit bacteria.

Numerical Modeling of Biocide and Vitamin B_{12} Concentrate Delivery Scenarios. The biocide and vitamin B_{12} concentrate injection concentrations and delivery scenarios were modeled using a modified version of a previously generated MODFLOW model (Forman et al., 2000) and the MT3DMS model code. The options of pulsing the mixture into the aquifer in daily, 1-day weekly, or 3-day weekly injections were simulated, and relative pulse concentrations and propagation through the aquifer were evaluated. The selected delivery scenario (1-day weekly pulses of vitamin B_{12} with daily biocide injection) controls the biomass on-site and field data validate the model.

In addition to using Tolcide® to inhibit biofouling of the recirculation well, the vitamin B_{12} concentrate was pulse injected instead of continuously injected in the recirculation well. MT3DMS was used to simulate the vitamin B_{12} concentrate injection scenarios (i.e., 1 day per week [20 hours per day]; 3 days per week [20 hours per day]; and 7 days per week [20 hours per day]). These modeling scenarios were generated for a comparison of concentrations caused by the different delivery scenarios, not to predict absolute concentrations of analytes in the aquifer. All scenarios assumed the groundwater was recirculated at 4 gpm (15.14 L/min), Tolcide® was injected for 4 hours per day, and 600 gallons (2271.24 Liters [L]) of the vitamin B_{12} concentrate was injected weekly into the recirculation well, regardless of the duration of the injection.

For the solute transport modeling, titanium was used as the solute to evaluate the vitamin B_{12} concentrate in the aquifer, because it is a key additive in the concentrate for which analytical data are available. Based on previous titanium concentrations of 2 millimolar (mM) (96 ppm [96 mg/L]) in the recirculation well, titanium concentrations in the recirculation well of 479 ppm (479 mg/L), 160 ppm (160 mg/L), and 68.4 ppm (68.4 mg/L) were used for the 1 day, 3 days, and 7 days injection scenarios, respectively.

Modeling indicated that the 1 day per week injection scenario would yield the highest titanium concentrations at all distances from the recirculation well. These results are presented in Table 1.

Modeling simulations of these scenarios were generated with and without a reduced availability rate for titanium. This rate was determined based on Pilot Test results. Trends in the graphs are similar with and without reduced availability, although titanium concentrations are significantly lower when the reduced availability rate is taken into account. Based on laboratory treatability test results, higher concentrations of chlorinated solvents are more effectively treated with a higher concentration of vitamin B_{12} concentrate, as opposed to a slightly longer contact time at a lower vitamin B_{12} concentration. Therefore, it was determined that the Pilot Test would be resumed using a 1-day per week (20 hours per day) vitamin B_{12} concentrate injection scenario.

Although it was anticipated that the higher concentrations associated with the 1-day per week injection scenario may cause mobilization of clay and silt due to the high sodium content of the concentrate, it was also anticipated that this would equilibrate with time. Turbidity was monitored daily for the first 2 weeks and did equilibriate.

TABLE 1: Maximum Titanium Concentrations Expected at Various Distances from the Recirculation Well Based on MT3DMS Simulation

Distance from QRW1 (feet/meters)	Maximum Titanium Concentration (ppm (mg/L))		
	1 Day (20 hours) @ 479 ppm (mg/L)	3 Days (20 hours/day) @ 160 ppm (mg/L)	7 Days (20 hours/day) @ 68.4 ppm (mg/L)
Without Reduced Availability			
1/0.30	415	140	60
5/1.52	356	124	53
10/3.05	215	87	38
15/4.57	101	55	24
20/6.10	42	30	16
25/6.10	20	16	10
With Reduced Availability			
1/0.30	307	103	46
5/1.52	217	74	32
10/3.05	86	32	15
15/4.57	23	12	5
20/6.10	5	4	2
25/7.62	2	1.7	1.1

In addition, the vitamin B_{12} concentrate was added to the base of the aquifer via the lower screen of the recirculation well to treat potential residual DNAPL and to demonstrate the in-situ nature of this technology (Lesage et al, 2001).

CONCLUSIONS

Geoprobe sampling and laboratory analyses (i.e., permeability, DNA, and PLFA) are useful tools in the determination of the extent and type of bacteria present and may aid in determining the type of biocide or rehabilitation techniques to apply.

Based on field observations at the study site, it can be concluded that the Aqua Freed® process successfully rehabilitated the recirculation well and allowed completion of the Pilot Test. It can also be concluded that the use of a biocide(s) may successfully control and rehabilitate wells. The biocide (Tolcide®) was a valuable tool to maintain aquifer permeability during the Pilot Test. Numerical modeling was found to be useful in determining the concentration of vitamin B_{12} concentrate and biocide for remediation purposes.

REFERENCES

Mowder, C., T. Llewellyn, S. Forman, S. Lesage, S. Brown, K. Millar, D. Green, K. Gates, G. DeLong, F. Tenbus, 2000. "Field Demonstration of *In Situ* Vitamin B_{12}-/Catalyzed Reductive Dechlorination." In proceedings for the *Second International Conference on Remediation of Chlorinated and Recalcitrant Compounds*

Forman, S., T. Llewellyn, S. Morgan, D. Green, K. Gates, G. DeLong, 2000. "Numerical Simulation and Pilot Testing of a Recirculation Well." In proceedings for the *Second International Conference on Remediation of Chlorinated and Recalcitrant Compounds*

Lonergan, D., H. Jenter, J. Coates, E. Phillips, T. Schmidt, D. Lovley, 1996. "Phylogenetic Analysis of Dissimilatory Fe(III)-Reducing Bacteria." In Journal of Bacteriology, Apr. 1996, p.2402-2408.

Millar, K., S. Lesage, S. Brown, C. Mowder, T. Llewellyn, S. Forman, D. Peters, G. DeLong, D. Green, H. McIntosh, 2001. "Biocide Application Prevents Biofouling of a Chemical Injection/Recirculation Well." In proceedings for the *Sixth International Symposium on In Situ and On-Site Bioremediation*

Lesage, S., S. Brown, K. Millar, C. Mowder, T. Llewellyn, S. Forman, G. DeLong, D. Green, 2001. "Use of a Recirculation Well for the Delivery of Vitamin B_{12} for the In-Situ Remediation of Chlorinated Solvents in Groundwater." In proceedings for the *Sixth International Symposium on In Situ and On-Site Bioremediation*

BIOCIDE APPLICATION PREVENTS BIOFOULING OF A CHEMICAL INJECTION/RECIRCULATION WELL

Kelly Millar, Suzanne Lesage, and Susan Brown
(NWRI, Environment Canada, Burlington, Ontario, Canada),
Carol S. Mowder, Tim Llewellyn, Sarah Forman, and Dave Peters
(URS, Linthicum, Maryland, USA),
George DeLong (AIMTech, Oak Ridge, Tennessee, USA),
Donald J. Green and Heather McIntosh
(U.S. Army Garrison, APG, Maryland, USA).

ABSTRACT: A pilot test for *in situ* vitamin B_{12}-catalyzed reductive dechlorination of chlorinated solvents was conducted for 14 weeks in 1999 at Graces Quarters, Aberdeen Proving Ground, Maryland. Groundwater at the site is contaminated with 1,1,2,2-tetrachloroethane (TeCA), carbon tetrachloride (CT), trichloroethene (TCE), tetrachloroethene (PCE), and chloroform (CF). Although analytical groundwater data from the pilot test showed favorable results, biological fouling of the chemical injection/recirculation well and the near-by formation halted pilot test operations. After successful redevelopment of the injection/recirculation well, and several modifications to the chemical treatment mix and injection strategy (aimed at preventing future biofouling) the pilot test was resumed. Modifications included the use of a non-oxidizing biocide, Tolcide®, containing tetrakis(hydroxymethyl)phosphonium sulfate (THPS) as the active ingredient. The data show that daily application of Tolcide® at a 150 mg/L active concentration for 4 hours prevented biological fouling of the area surrounding the well. The inhibitory effects of Tolcide® were limited to the anaerobic treatment zone. Microbial activity was resumed in the aquifer as Tolcide® concentrations decreased. A laboratory microcosm study confirmed the inhibitory effects of Tolcide® on the site bacteria.

INTRODUCTION
A two-phase pilot test was conducted at Graces Quarters, APG, to evaluate the use of a concentrated mixture of vitamin B_{12} and Ti(III)-citrate to treat chlorinated volatile organic compounds (VOCs) in groundwater (Lesage et al. 2001). Groundwater was treated *in situ* using a recirculation well designed to optimize contact between the vitamin B_{12}/Ti(III)-citrate mixture and groundwater contaminants, yet minimize contact with the surrounding soil. A slow, continuous injection approach was used during phase one (Fall 1999), to maximize dechlorination reactions in the well. The chemical mix consisted of vitamin B_{12}, Ti(III)-citrate, and a glucose/fructose syrup. Although glucose is not a necessary component of the reactive vitamin B_{12} mixture, it was added as a preferred bacterial substrate over citrate (Millar and Lesage, 1997), to try to increase the longevity of the Ti(III)-citrate complex in the recirculation well and aquifer. Analytical results of the initial phase of treatment were favorable, with complete

degradation of CT and 50-80 % removal of TeCA occurring within the recirculation well (Mowder et al. 2000). However, excessive bacterial growth in the vicinity of the recirculation well, made treatment delivery increasingly difficult, finally shutting the system down after 14 weeks of operation.

The purpose of the second phase of the pilot test was to minimize biofouling, using a biocide and a pulsed approach to treatment injection. In addition, the vitamin B_{12}/Ti(III)-citrate mix was injected into deeper regions of the aquifer where residual DNAPL was suspected and at increased pumping rates, to deliver higher concentrations a further distance from the recirculation well.

Most chemical well treatments for biofouling, being highly oxidizing, are incompatible with the reduced redox conditions required for the vitamin B_{12}-catalyzed reaction. Therefore, the non-oxidizing biocide Tolcide® (Rhodia), containing THPS (Fig. 1), was selected. THPS was widely used as a flame retardant in fabrics before its biocidal properties were recognized. Current usage is for controlling bacteria, algae, and fungi in industrial cooling systems, oil field operations, and papermaking processes. Its benefits include low overall toxicity, low effective treatment concentrations, and no bioaccumulation. In addition, while Tolcide® is effective under anaerobic conditions, it degrades quickly under aerobic conditions; therefore its inhibitory effects do not extend outside of the treatment zone.

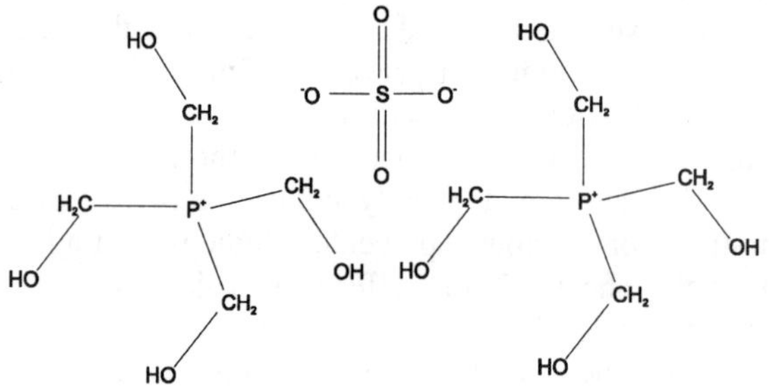

FIGURE 1. The structure of tetrakis(hydroxymethyl)phosphonium sulfate (THPS), active ingredient in Tolcide®.

The effect of the dual strategy was evaluated using water level measurements and volatile fatty acid (VFA) analytical data from the recirculation well and surrounding multi-level piezometers and monitoring wells. In addition, the results of a laboratory microcosm study to assess the inhibitory effects of Tolcide® using site water collected during the second phase of groundwater treatment, is summarized.

MATERIALS AND METHODS

Pilot Test – Phase Two. Following successful redevelopment of the recirculation well, restoration of aquifer permeability (Forman et al. 2001), and regulatory approval of the use of Tolcide® onsite, the pilot test was resumed. On July 28,

2000, system operations began with a 2½-week treatment of Tolcide® to inhibit any resident microorganisms. A 2 % active solution of Tolcide® was pumped into the recirculation well for 4 hours/day at a rate of 105 mL/min to obtain an active concentration of 150 mg/L within the well. The recirculation well was operated at a flow rate of 4 gallons per minute (gpm) in upward mode during the first week, treating and then exiting the upper well screen, and in downward mode for the final ten days, to treat the lower well screen and surrounding sand pack. For the remainder of the pilot test, the recirculation well was operated in the downward mode, opposite to the flow direction employed in phase one.

Weekly injections of the reformulated vitamin B_{12}/Ti(III)-citrate mix (without glucose) began on August 15, 2000, until November, for a total treatment time of 12 weeks. The mix was injected over a 20-hour period into the recirculation well, once per week. Tolcide® was added 4 hours/day every day to achieve a 150 mg/L active concentration within the well. Periodically, the system was backwashed by changing the direction of flow for up to 8 hours.

Water levels in the recirculation well (upper and lower screens) were recorded regularly. Monthly groundwater sampling was conducted for analysis of contaminants, dissolved gases, VFAs, metals, chloride, pH, and Eh. For a plan view of the site see Lesage et al. (2001).

Microcosms. Groundwater from the lower well screen of the recirculation well was collected on October 2, 2000. Microcosms were established in triplicate 60-mL serum vials, in an anaerobic chamber having an atmosphere of 5 % CO_2, 10 % H_2, and 85 % N_2. Four conditions were established, each consisting of site water, inoculum and the following: (A) 10 mM Ti(III)-citrate, 5 mg/L vitamin B_{12}; (B) 150 mg/L (active) Tolcide® for 4 hours, then 10 mM Ti(III)-citrate, 5 mg/L vitamin B_{12}, filter-sterilized site water; (C) 150 mg/L (active) Tolcide® for 4 hours, then filter-sterilized site water; and (D) 20 mg/L (active) Tolcide®, 1.9 mM Ti(III)-citrate, 1 mg/L vitamin B_{12}. A fifth microcosm (E) that contained site water alone was established as a control. Inocula were prepared by centrifuging groundwater from the site, then resuspending the bacterial pellets in 3 mL of site water per vial. For (B) and (C), vial contents were re-centrifuged following a 4-hour Tolcide® exposure and pellets were resuspended as above. Analyses of VFAs and methane were conducted.

RESULTS AND DISCUSSION

Water levels. For a schematic of the recirculation well and a detailed description of its operation see Mowder et al. (2000). In brief, the 10" diameter recirculation well consists of two screened intervals separated by an inflatable packer. The reference for water level measurements of the upper-screened interval is the top of the well casing while water levels of the lower-screened interval are measured relative to the packer.

Upper and lower well screen water levels during both phases of the pilot test are shown in Figure 2. During the initial phase, the system was operated in upward mode such that groundwater treatment took place in the upper part of the recirculation well. Continuous pumping of glucose-amended chemical mix resulted in significant increases in the water level of the upper well screen as early as 2 weeks into the experiment. By week six, aquifer permeability was limited (Fig. 2A) and biological growth was evident as foul odors and gelatinous masses of material were observed in the lines of the recirculating system. While aquifer conditions were originally unfavorable for bacterial growth, intrinsic microorganisms responded rapidly to the incoming carbon sources, neutral pH, and reduced redox conditions generated by the chemical treatment. Treatment delivery became increasingly difficult with frequent shutdowns due to high water levels and fouled pumps, resulting in the termination of the test after 14 weeks of operation.

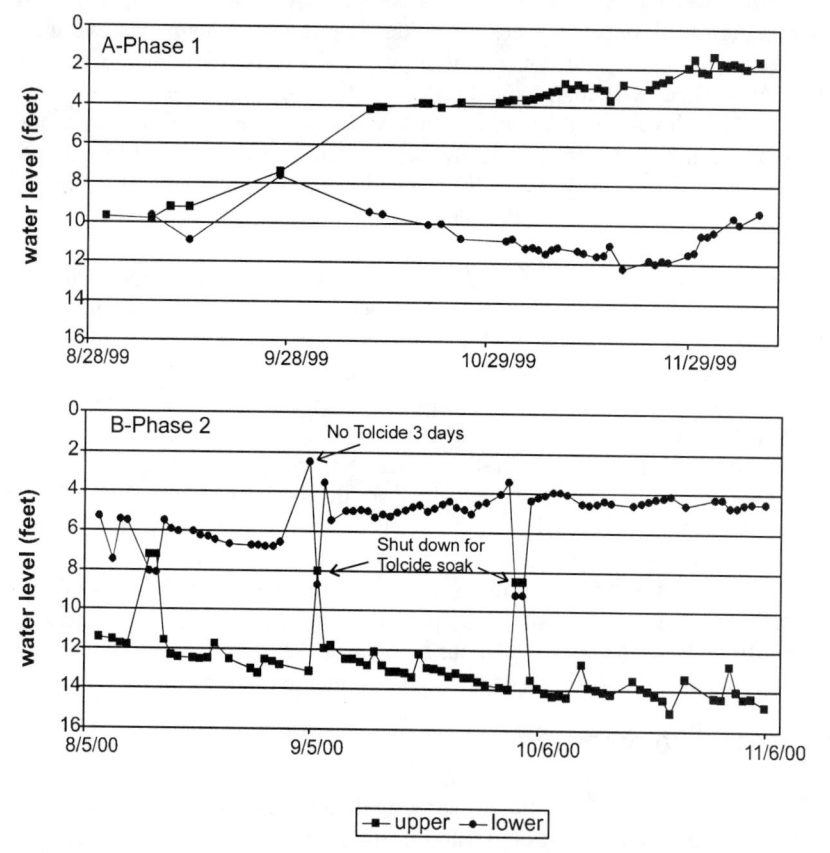

FIGURE 2. Water levels of the upper and lower well screens of the recirculation well during phase one and two of the pilot test.

During the second phase of the pilot, flow was reversed and the lower part of the aquifer received chemical treatment. Daily application of Tolcide® at 150 mg/L active concentration was effective in minimizing biological growth in the immediate vicinity of the recirculation well as was evident by the stability of the water levels of the lower well screen (Fig. 2B). The effect of Tolcide® on the bacterial community was bacteriostatic such that, biological growth was inhibited, as opposed to bactericidal, which would have resulted in cell death. The resilience of the population was evident on September 5th, 2000. The Tolcide® injection pump had failed the previous Friday, leaving the aquifer untreated for three days. This resulted in a 4-foot rise in the water level of the lower well screen. To try to restore conditions, the system was backwashed for 8 hours, by reversing flow, then shutting the system down to allow an overnight Tolcide® soak. From then on, water levels began to creep up and another flow reversal/Tolcide® soak was conducted just after the halfway point of the test. It was then decided that a more rigorous approach to controlling biological growth needed to be taken. As such, flow reversals and Tolcide® soaks were conducted weekly, the day prior to the 20-hour injection of the Vitamin B_{12}/Ti(III)-citrate mix. This proved effective, with no further increase in water levels for the remainder of the test. In contrast to the initial phase of the pilot test, with reduced aquifer permeability resulting in an 8.5 foot increase in the water level of the upper well, through regular use of Tolcide®, water levels in the lower well screen rose an average of only one foot, over the course of the 12-week treatment.

Tolcide® effects in the aquifer.

From baseline sampling in May 2000, it was evident that biological activity in the aquifer had increased since the termination of the initial pilot test. Abiotic *cis*-DCE to *trans*-DCE ratios of 2:1, observed during chemical treatment in phase 1, had increased to ratios characteristic of biological dichloroelimination of TeCA (3:1, Lorah and Olsen, 1999). Methane also began to appear at a number of the monitoring wells. Methane generation, in part, may have been stimulated by well/aquifer restorative procedures during the interim. In particular, residual CO_2 from the Aqua Freed® process (Forman et al. 2001) may have stimulated methanogenic populations that had survived the redevelopment process. Biological dechlorination was apparent at piezometer 1C (10 feet from the recirculation well) by reductions in TCE (655 µg/L to 112 µg/L) and CT (622 µg/L to 37 µg/L) that could not be accounted for by dilution, from incoming A-level waters, alone. In addition, biodegradation of CT had resulted in CF concentrations increasing from 19 µg/L in January to 40 µg/L in May 2000.

A good indicator of the distribution and effectiveness of Tolcide® is the persistence of the Ti(III)-citrate complex in the aquifer. During the second phase of treatment, piezometer 1C saw concentrations of citrate and titanium as high as 9.6 mM and 3.5 mM, respectively (Fig. 3A). MT3DMS model simulations predicted maximum titanium concentrations of 4.5 mM at 10 feet, under conditions of no degradation/precipitation. Observed fluctuations in titanium and citrate concentrations at 1C are a result of the pulsed injection and the timing of sample collection. The week 8 (Oct. 6) sampling, which occurred 2 days after injection, was the only sampling event that could have captured the concentrate. It is difficult to determine whether the peak concentrations of citrate and titanium were obtained during this event. However, acetate concentrations did not increase during the active pumping period, suggesting that citrate degradation

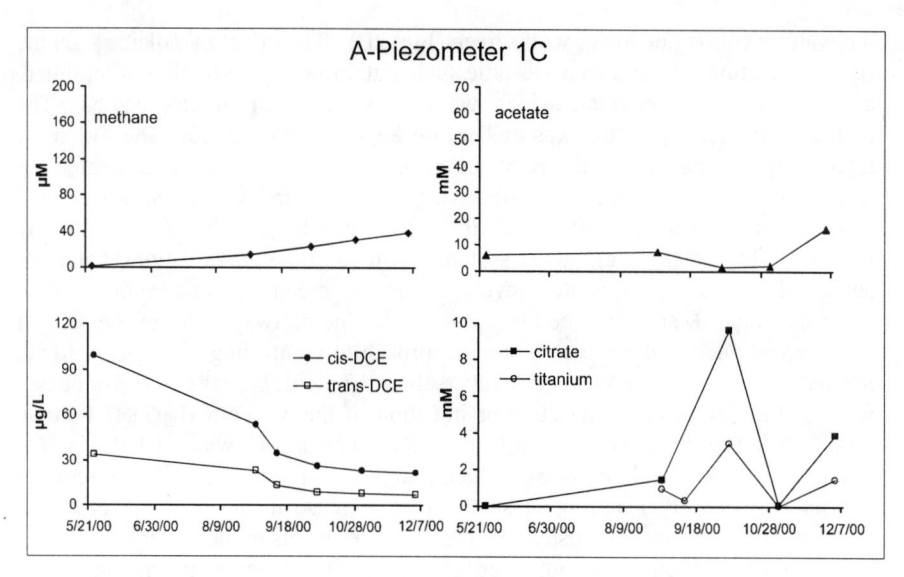

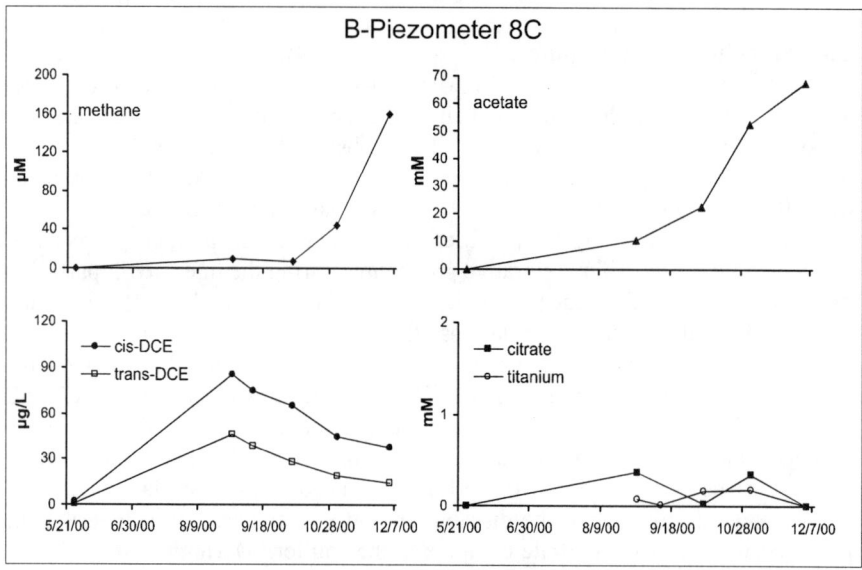

FIGURE 3. Titanium, VFAs, methane, and *cis*- and *trans*-DCE at piezometers 1C and 8C.

was minimal. In contrast, piezometer 8C, 30 feet away from the recirculation well, saw minimal concentrations of citrate, indicating that it was being metabolized along the flow path, with acetate accumulating at 8C (Fig. 3B). The increase in acetate at 1C, week 16 (Dec. 4 - four weeks post shutdown) indicates that the inhibitory effects of Tolcide® were temporary, and that microbial populations can reestablish upon its disappearance or removal from the system. In addition to acetogenic bacteria, Tolcide®, though not reportedly tested on methanogens, appeared to exhibit a bacteriostatic effect on methanogens, with little increase in 1C methane concentrations, compared to that observed at 8C. While biological degradation of *cis-* and *trans-*DCE had occurred at 1C, between baseline and week 2 (Aug. 30) sampling, further degradation appeared to be inhibited upon repeated applications of biocide. At 8C, however, where Tolcide® effects were negligible, biodegradation of *cis-* and *trans-*DCE appeared to be continuous.

Microcosms. A series of microcosms were established to assess microbial activity in the aquifer under conditions of the pilot test: (A) Vitamin B_{12} concentrate injection, (B) changeover from Tolcide® to vitamin B_{12}/Ti(III)-citrate mix, (C) 4-hour Tolcide® injection, (D) low Tolcide®, as seen 10 ft away, and (E) site water control. The effect of Tolcide® on acetogenic bacteria is presented in Table 1. In treatment (A), citrate utilization forming acetate was variable, from non-measurable (A1), to partial (A2), and complete (A3), within 47 days. Variability between replicates, and the delay in citrate utilization, was likely due to the inoculum, which, despite efforts to concentrate it by centrifugation, was very low. Citrate was also degraded in 2 of the 3 low Tolcide® treatments (D) indicating that 20 mg/L active, would not be an effective inhibitory concentration. The 4-hour, 150 mg/L Tolcide® exposure (B) was effective at preventing citrate degradation. The effect of Tolcide® on methanogens showed a similar pattern. Methane generation was greatest under conditions of the vitamin B_{12}/Ti(III)-citrate injection and lowest in treatments receiving 150 mg/L active Tolcide® for 4 hours (data not shown). Based on the results of the study, it appeared that methanogens were more resistant to Tolcide® than acetogenic strains. However, because of the dependence of methanogens on fermentative bacteria to generate H_2, formate, and/or acetate, for growth, inhibition is increased indirectly. This was apparent upon exchanging the anaerobic chamber gas mix with N_2, 7 days into the experiment. Without an external source of CO_2 and H_2, methanogens were dependent upon the activities of the slow-growing fermentative bacteria, and a subsequent drop in the rate of methane production was observed.

TABLE 1. The effects of Tolcide® on citrate utilization.

	citrate (mM) day 0	citrate (mM) day 47	acetate (mM) day 0	acetate (mM) day 47
A1	13.0	12.9	1.2	0.9
A2	-	4.8	-	20.4
A3	-	0	-	28.2
B1	12.9	14.3	1.1	0.9
B2	-	12.3	-	0.8
B3	-	12.8	-	0.7
C1	0	0	0.6	0.6
C2	-	0	-	0.6
C3	-	0	-	0.6
D1	1.3	0	0.7	3.3
D2	-	1.2	-	0.8
D4	-	0	-	3.5
E1	0	0	0.9	0.3
E2	-	0	-	0.7
E3	-	0	-	0.5

('-' indicates replicates not sampled on day 0)

CONCLUSIONS

Daily application of Tolcide® at a 150 mg/L active concentration, in combination with pulsed injection of the vitamin B_{12}/Ti(III)-citrate mix was effective for the prevention of biological fouling over a 12-week remediation treatment. Bacteriostatic activity was limited to the anaerobic treatment zone and did not interfere with ongoing biological degradation outside this region. Tolcide® was effective in preventing premature citrate utilization and in extending the area of treatment.

ACKNOWLEDGMENTS

This project was planned, implemented under contract, managed, and overseen by AIMTech, a U.S. Department of Energy program operated by Lockheed Martin Energy Systems Inc., in conjunction with the APG Installation Restoration Program.

REFERENCES

Forman, S. R., T. Llewellyn, S. Morgan, C. Mowder, S. Lesage, S. Brown, K. Millar, G. DeLong, D. J. Green, and H. McIntosh. 2001. "Rehabilitation of a Biofouled Recirculation Well Using Innovative Techniques." Battelle. This symposium.

Lesage, S., S. Brown, K. Millar, C. S. Mowder, T. Llewellyn, S. Forman, G. DeLong, and D. J. Green. 2001. "Use of a Recirculation Well for the Delivery of Vitamin B_{12} for the In-Situ Remediation of Chlorinated Solvents in Groundwater." Battelle. This symposium.

Lorah, M. M. and L. D. Olsen. 1999. "Degradation of 1,1,2,2-Tetrachloroethane in a Freshwater Tidal Wetland: Field and Laboratory Evidence." *Environ. Sci. Technol.* 33(2): 227-234.

Millar, K. and S. Lesage. 1997. "Biocompatibility of the Vitamin B_{12}-Catalyzed Reductive Dechlorination of Tetrachloroethylene." In B. C. Alleman and A. Leeson (Chs.), *In-Situ and On-Site Bioremediation: Vol.4*, pp. 471-476. Battelle, Columbus, OH.

Mowder, C. S., T. Llewellyn, S. Forman, S. Lesage, S. Brown, K. Millar, D. Green, K. Gates, G. DeLong. 2000. "Field Demonstration of *In Situ* Vitamin B_{12}-Catalyzed Reductive Dechlorination." In G. B. Wickramanayake and A. R. Gavaskar (Eds.), *Physical and Thermal Technologies: Remediation of Chlorinated and Recalcitrant Compounds*, Vol. C2-5, pp. 261-268. Battelle, Columbus, OH.

USE OF A RECIRCULATION WELL FOR THE DELIVERY OF VITAMIN B_{12} FOR THE IN-SITU REMEDIATION OF CHLORINATED SOLVENTS IN GROUNDWATER.

Suzanne Lesage, Susan Brown, and Kelly Millar
(NWRI, Environment Canada, Burlington, Ontario, Canada),
Carol S. Mowder, Tim Llewellyn and Sarah Forman
(URS, Linthicum, Maryland, USA),
George DeLong (AIMTech, Oak Ridge, Tennessee, USA),
Donald J. Green (U.S. Army Garrison, APG, Maryland, USA).

ABSTRACT: One of the advantages of the vitamin B12/titanium citrate mixture is its ability to degrade mixtures of chlorinated methanes, ethanes and ethenes. This technology, developed by Environment Canada, is being evaluated in the field at Graces Quarters, Aberdeen Proving Grounds in Maryland. The major contaminants of concern at the site are carbon tetrachloride, 1,1,2,2-tetrachloroethane (TeCA), and trichloroethene. One of the major difficulties in the implementation of chemical treatment in groundwater is ensuring proper mixing. For this reason, instead of using a traditional injection well, it was decided to use a recirculation well to do in-ground mixing of the water and the treatment mixture. The project was conducted in two three-month phases. In the initial phase (Fall 1999), the emphasis was placed on maximizing the amount of reaction occurring in the well itself while minimizing chemical addition. The purpose of the second phase (Fall 2000) was to deliver a concentrated treatment as directly as possible to areas of higher contaminant concentrations, indicative of a dense non-aqueous phase residual. The treatment mixture was added in weekly concentrated pulses of 20 hours. The pH and Eh, as surrogate parameters for the movement of the active ingredients, were monitored daily for the first two weeks and weekly thereafter, in the 24 monitoring points installed radially up to 80 ft away from the recirculation well. After three months of uninterrupted pulsed treatment the contaminant concentrations decreased to below guidelines in a 20-ft radius around the recirculation well.

INTRODUCTION:

Groundwater at Graces Quarters, Aberdeen Proving Grounds (APG), is contaminated with chlorinated volatile organic compounds (VOCs). The compounds detected at the highest concentrations at the site are 1,1,2,2-tetrachloroethane (TeCA) and carbon tetrachloride (CT), with peak dissolved phase concentrations on the order of 2,000 to 4,000 μg/L. Concentrations of trichloroethene (TCE) are below 500 μg/L, and chloroform (CF) and tetrachloroethene (PCE) concentrations are less than 100 μg/L.

A pilot test was designed to evaluate the use of vitamin B_{12} and titanium citrate to treat groundwater at the Graces Quarter site. The design parameters were based on the results of a laboratory-scale treatability study (Mowder et al. 1999), which included the use of various microcosms and column studies. The optimal combination included titanium citrate, vitamin B_{12} and glucose, which is referred to as the vitamin B_{12} concentrate in this paper. The project was conducted in two three-month phases. In the initial phase (Fall 1999), the emphasis was placed on maximizing the amount of reaction occurring in the well itself while minimizing chemical addition. The mixture was supplemented with a glucose-fructose syrup as a preferential carbon source to citrate and as a source of organic acids known to be a source of H_2 to support dechlorinating bacteria. Changes in the ratio of *cis*- to *trans*-DCE in groundwater samples over time were indicative of bacterial dechlorination activity. While this strategy was successful in allowing titanium citrate to be transported at least 20 feet from the well, it caused excessive bacterial growth resulting in the clogging of the well and the adjacent aquifer.

After using various methods to clean the well screens, it was decided to change the rate of delivery of the chemical treatment and to add Tolcide™ as a bacteriostatic agent. (Forman et al. 2001; Millar et al. 2001). The purpose of the second phase (Fall 2000) was to deliver a concentrated treatment as directly as possible to areas of higher contaminant concentrations, indicative of a dense non-aqueous phase residual. The direction of flow in the well was reversed and the vitamin B12 concentrate, without the glucose/fructose syrup, was added in weekly concentrated pulses of 20 hours, followed by four hours of Tolcide™.

MATERIALS AND METHODS

The vitamin B_{12} concentrate was generated on site in weekly batches. This process (Lesage et al. 1997) involved preparing titanium oxalate from titanium sponge and oxalic acid, followed by the addition of sodium citrate. The oxalate in solution was precipitated as calcium oxalate after the addition of calcium carbonate and calcium citrate. Sodium carbonate was added to increase the pH to above 8.0 and decrease the Eh to approximately –900 millivolts (mV). Vitamin B_{12} and glucose were then added. The

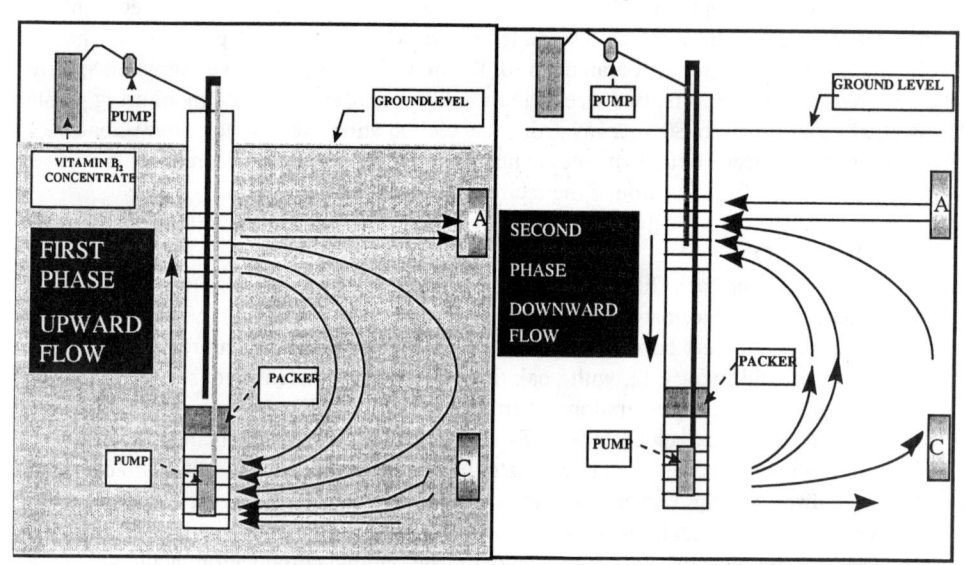

Figure 1 a and b. Schematic view of the recirculation well in downward (a) and upward (b) flow configuration. The rectangles marked A and C respectively represent the screened intervals of the monitoring wells.

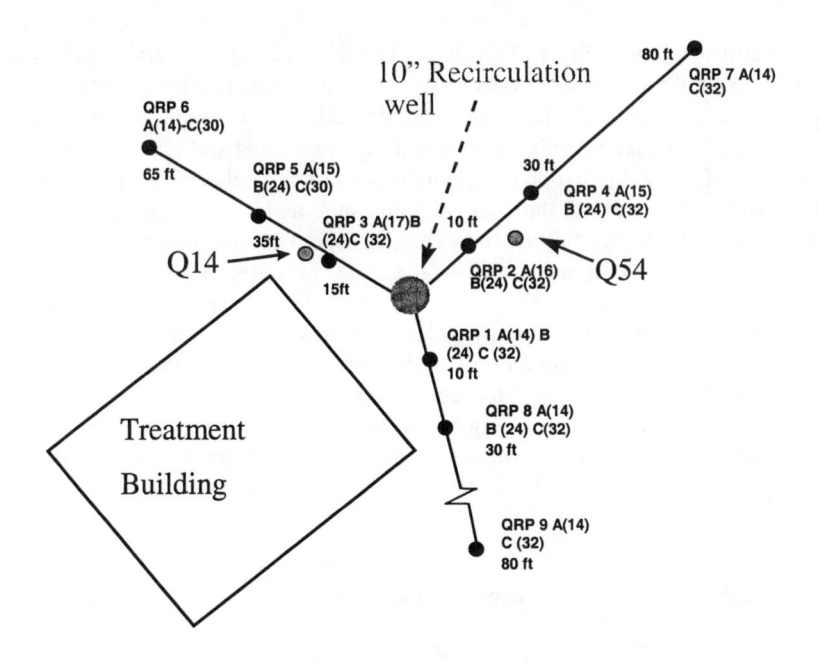

Figure 2. Schematic plan view of the site indicating the location of the monitoring wells, the depth, in feet, of the middle point of the two-feet screened intervals is indicated in brackets after each letter.

vitamin B_{12} concentrate was transferred to a collapsible pillow tank (ATL Flex-Tanks, Aero-Tec Laboratories Inc., Ramsey, NJ) to ensure anaerobic conditions were maintained and was continuously metered into the recirculation well. Approximately 500 gallons of vitamin B_{12} concentrate were prepared each week. In the first phase, glucose was added as a preferential carbon source to prevent citrate degradation by bacteria, which was found in the laboratory study to cause titanium precipitation. In the second phase, the same amount of concentrate was prepared, but glucose was omitted to reduce biofouling.

A recirculation well was used to facilitate *in situ* mixing of the vitamin B_{12} concentrate and contaminated groundwater during the pilot test. A schematic of the recirculation well and vitamin B_{12} feed system is shown in Figure 1. In the first phase (Figure 1a), contaminated groundwater from the site was drawn into the lower well screen of the 10-inch diameter recirculation well and pumped above the inflatable packer which separates the upper and lower well screens. Groundwater was pumped at 2 gallons per minute (gpm) and mixed with the vitamin B_{12} concentrate, which was added at approximately 0.05 gpm. This system allowed an approximate residence/reaction time of 30 minutes in the well before discharging the groundwater/vitamin B_{12} concentrate through the upper well screen and into the aquifer. In the second phase (Figure 1b), the direction of flow was reversed and the pumping rate was increased to 4 gpm resulting in a 10 min in-well residence time. The concentrate was added weekly to the well over a 20 hour period, followed by a Tolcide™ (Tetrakis(hydroxymethyl)phosphonium sulfate; supplier Albright and Wilson) concentrate for four hours, at a final concentration of 150 ppm, active ingredient. The four-hour Tolcide™ addition was repeated every day.

The recirculation well and surrounding monitoring wells are shown in plan view in Figure 2. During the first phase, groundwater samples collected every 2 weeks from the lower and upper part of the recirculation well, and the surrounding multiple-depth monitoring wells, were analyzed for VOCs, dissolved gases (methane, ethane, ethene, and acetylene), volatile fatty acids (VFAs - acetate, butyrate, citrate, formate, lactate, and propionate), titanium, iron, manganese, and chloride. All analyses were performed in a fixed-base laboratory. Additionally, before system start-up and at the end of the first and second phase, all of the monitoring wells were sampled and analyzed in the field for redox couples (nitrate/nitrite, ferric iron/ferrous iron, sulfate/sulfide) methane and hydrogen to evaluate the changing redox conditions of the aquifer. Eh and pH were monitored in the field using a flow-through cell. In the second phase, daily sampling of

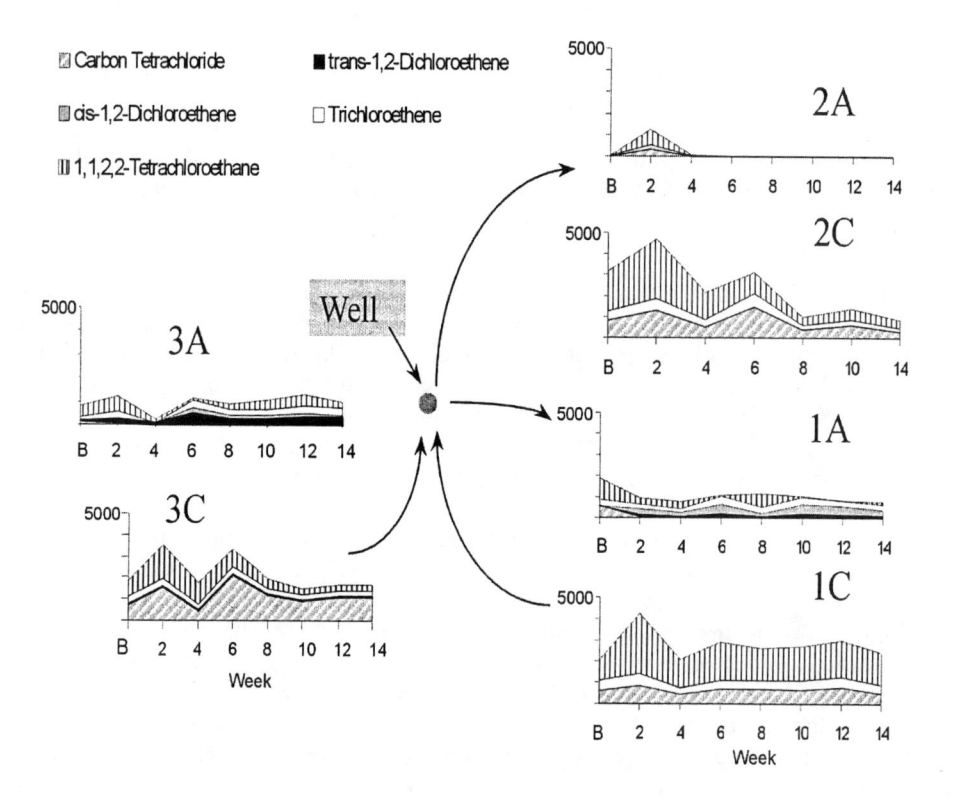

Figure 3. First phase. Water was drawn in from the C-wells level, treated in the recirculation well, then redistributed in the upper zone (A). Note the absence of CT in the A wells and the appearance of cis and trans DCE. Vertical scale, 0-1000µg/L.

Eh and pH was conducted, then reduced to monthly. The change in Eh (±100 mV) was used to select the wells for a complete set of analytical parameters.

RESULTS AND DISCUSSION

The results shown here illustrate the effectiveness of the recirculation well as an in-situ reactor, for the first phase, and as a means to distribute active ingredients in a contaminated aquifer for both the first and second phases. More details on biological activitity and on the hydraulics of the systems can be found in the references by Mowder, Millar and Forman.

First phase: In-well reaction. During system start-up and balancing of the vitamin B_{12} concentrate addition rate, VOC samples were collected from the upper and lower well screens of the recirculation well. The degradation of carbon tetrachloride was always complete within the well without the accumulation of chloroform or methylene chloride, even when the turbidity increased within the well because of the release of fines from the sodium addition. However, the percent degradation of 1,1,2,2-tetrachloroethane was extensive but seldom complete within the recirculation well (from week 6, TeCA removal ranged from 48.1 to 99.2%), but the reaction continued in the aquifer where, at 10 ft, the removal ranged from 42.6 to 99.5%. In all but one sample the concentration of TeCA was lower at ten feet than in the upper well. The reaction rate varied throughout the first phase of the pilot test. There were several reasons why the rate of reaction achieved was lower than the design value based on microsoms. The titanium concentrations were lower in the

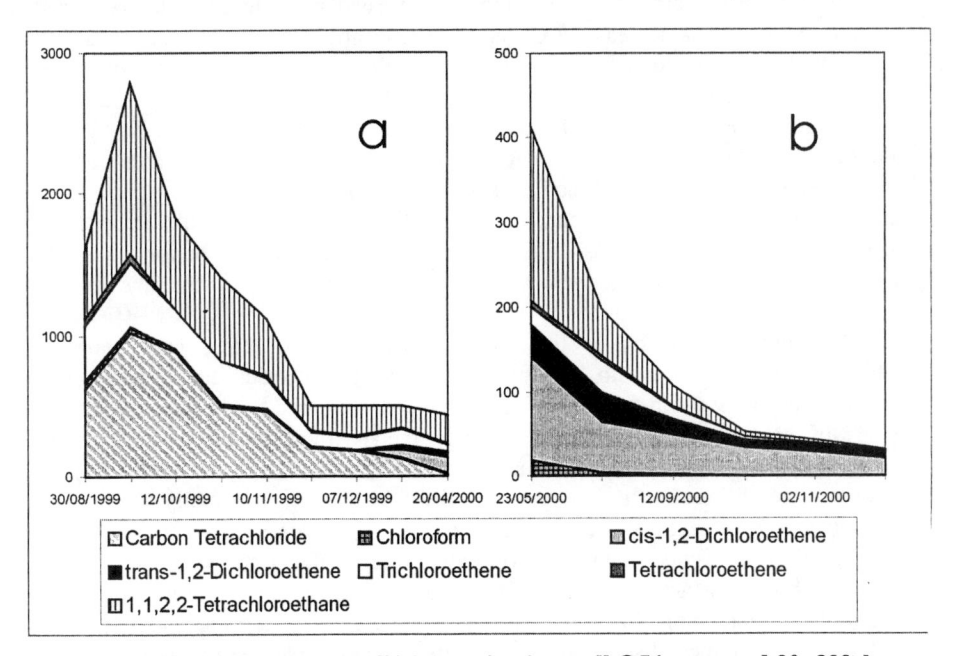

**Figure 4. Concentrations (μg/L) in monitoring well Q54, screened 20- 32ft bgs.
a 1999; b 2000**

field than in the laboratory study (2 mM titanium in the field vs. 4 mM titanium in the laboratory) because of early difficulties encountered during mixing. There was a lot of silt in the early phase, from the interaction of the mixture with surrounding clay lenses. However, after approximately 4 weeks of operation, the turbidity of the groundwater decreased. Severe weather (Hurricane Floyd) at week 3 led to power failures. Subsequent variances in 1,1,2,2-tetrachloroethane degradation in the recirculation well were probably due to fluctuations in groundwater ORP and pH.

First phase: Distribution in the aquifer. The results of the first phase are summarized in Figure 3. Each graph within the figure shows the distribution of the major contaminants and degradation products in the three monitoring wells located closest to the recirculation well. The C level monitoring wells contained only the initial contaminants with their concentrations changing very little over the whole period. It is interesting to note that in Well 1C, TeCA was the predominant contaminant, whereas Well 3C, 25 ft away from Well 1, contained mostly CT. Since CT and CF (not shown) were completely degraded within the recirculation well, a rapid decrease in the concentrations of these constituents was observed immediately in near-by shallow Wells 1A and 3A; Well 2A did not show a similar trend because it is located in a lower permeability zone. The concentration of TCE did not change significantly during that period. This may be because most of the vitamin B_{12} was reacting with the more abundant and reactive substrates (CT, TeCA, and CF) and because the residual concentration of titanium was not sufficiently high to reduce the vitamin B_{12} to the lower Eh needed for its degradation. As the vitamin B_{12} concentrate and treated groundwater extended through the aquifer with time, a decrease in CT and CF concentrations was also observed in the deeper "B" monitoring well and those further from the well. An indication of the overall performance can be seen in the trends at Q54, a well screened between 24.5 and 34.5 ft, corresponding to the B and C level monitoring wells. Figure 4a shows the data gathered in the fall of 1999 up to and including April 2000. During the period of active treatment, the concentrations of CT, TeCA and TCE decreased, but there was no evidence of any degradation products. This is not surprising since these were mostly in the upper portion of the treated area. Cis- and trans-DCE appeared in the spring after a period where some pumping occurred, but no treatment was added, while trying to restore the hydraulic conductivity of the recirculation well.

Although degradation of the *cis*- and *trans*-DCE occurred under anaerobic conditions, there has been no accumulation of vinyl chloride: the highest amount measured was 24.1 μg/L in 4A at Week 8, which had decreased to 5.3 μg/L by Week 18. The lack of accumulation of vinyl chloride under the anaerobic conditions at the site may indicate that it is degraded in the presence of vitamin B_{12}. Before injection of the vitamin B_{12} concentrate, the aquifer in the vicinity of the pilot test was essentially aerobic based on dissolved oxygen, Eh, redox, and dissolved hydrogen data. However, as expected, following injection of the vitamin B_{12} concentrate the aquifer in the vicinity of the pilot test quickly became anaerobic (Eh < -200 mV and odour of sulfide gases, indicative of sulfate-reducing bacteria). Eh and pH measurements were used as an indicator of the distribution of the remedial solution in the aquifer. For example, the Eh became negative within two weeks in monitoring Wells 1A and 3A, but remained positive in 2A. The Eh became negative in 4A after 8 weeks and within 12 weeks at 4B and C. The changes in redox conditions correlate well with the observed VOC degradation patterns.

Second Phase: Distribution in the aquifer. Because the reaction was found to continue for some distance in the aquifer, the strategy in the second phase was designed to have the treatment reach as far as possible towards the areas potentially having residual DNAPL. The flow was reversed (Figure 1b) and the pumping rate doubled to 4gpm. Before the second phase began a new baseline round of sampling was conducted. The concentration of TCE decreased significantly in the interim period (at 1C: 516 μg/L in August 1999, 655 μg/L in January 2000 and 112 μg/L in May 2000) presumably because of biodegradation. The graphs in Figure 5 show the data at all C level wells from the new baseline. The results for monitoring wells 1-3 are shown on a lower scale because the concentrations of TeCA, CT and TCE dropped so rapidly. By the end of the 12 weeks the only compounds left (at > 5 μg/L) were *cis* and *trans*-DCE at concentrations averaging 25 and 7 μg/L respectively.

Eh, pH values and titanium concentrations indicated that the treatment had reached Well 8C within 2 weeks. The large difference in concentrations between 1C and

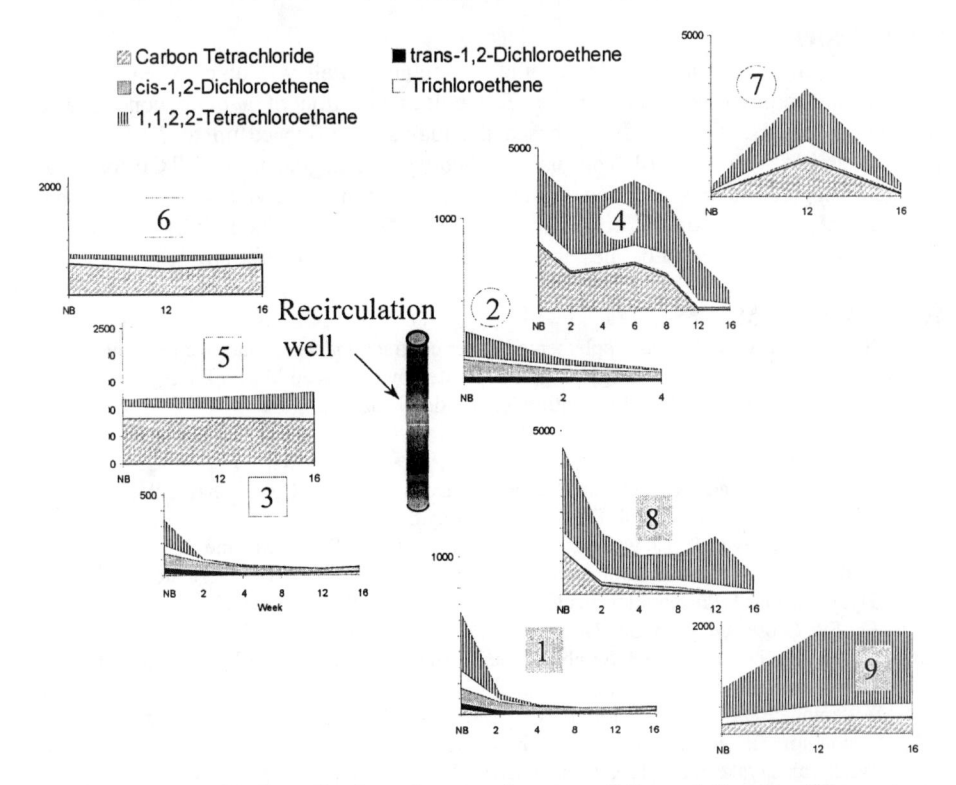

Figure 5. Fall 2000. Distribution of contaminants and degradation products over the pilot area. All C level monitoring wells. Y scale 0-1000 μg/L for wells 1-3, 0-5000 for the others. NB: New Baseline

8C indicates the probable presence of residual contamination in the vicinity of 8C. CT was completely removed by the end of 12 weeks, whereas the final TeCA and TCE concentrations were 20% and 10% of their initial concentrations, respectively. Vinyl chloride did not exceed 5 μg/L during the entire period. Well 4C was at the same distance as Well 8C, but in a lower permeability zone. Even though the treatment did not reach it until the very last week, all the CT was degraded, and TeCA and TCE were reduced to 20 and 16% of their initial concentrations. No evidence of treatment could be seen in Well 5C. The furthest Wells 9C, 7C and 6C did not receive any treatment and the concentration of the contaminants increased compared to baseline. In a longer-term treatment, this water would be recaptured and returned to the recirculation well for treatment. Modeling of the recirculation well using MODFlow indicated that this would take approximately 3 years.

The overall performance could be again assessed in Well Q54 screened between 24.5 and 34.5 ft (Figure 4b) where concentrations continued to decrease steadily. The final contaminant concentrations were identical to the 1C, 2C, and 3C averages, well below the 80 μg/L regulatory limit for cis-DCE. The situation was similar at Q14, except that 10 μg/L TCE and chloromethane remained as well.

CONCLUSIONS

The pilot test results show that the recirculation well was able to reach a 15ft radius in a continuous injection mode. The pulsed injection of a more concentrated treatment in reverse flow mode increased the radius of influence up to 30 ft in two transects, and 15 ft in the third. This also resulted in more degradation of TCE than was achieved using continuous injection at a lower concentration. The vitamin B12 treatment reduced the concentration of contaminants from >5000 μg/L total VOCs to below regulatory drinking water guidelines, in a total of six months of operation.

ACKNOWLEDGMENTS

This project was planned, implemented under contract, managed, and overseen by AIMTech, a US Department of Energy program operated by Lockheed Martin Energy Systems, Inc., in conjunction with the APG Installation Restoration Program.

REFERENCES

Lesage, S., S. J. Brown and K.R. Millar."Method for Dehalogenating Contaminated Water and Soil" U.S. Patent 5,645,374. Date of issue: July 8, 1997.

Mowder, C., S. Lesage, T. Llewellyn, D. Green, K. Millar, S. Brown and H. Steer. 1999. "Combined Anaerobic Biological and Abiotic Treatment of Chlorinated Aliphatic Hydrocarbons". Presentation made at the Fifth International Symposium on In Situ and On-Site Bioremediation, San Diego, 1999.

Mowder, Carol S., Tim Llewellyn, Sarah Forman, Suzanne Lesage, Susan Brown, Kelly Millar, Don Green, Kimberly Gates, George DeLong 2000. Field Demonstration of *In Situ* Vitamin B_{12}-Catalyzed Reductive Dechlorination. In Physical and Thermal Technologies: Remediation of Chlorinated and Recalcitrant Compounds. G.B. Wickramanayake and A.R. Gavaskar. Battelle Press. Vol. C2-5, pages 261-268.

Millar, K.R. et al. 2001 this symposium.
Forman, S. et al. 2001 this symposium.

2001 AUTHOR INDEX

This index contains names, affiliations, and volume/page citations for all authors who contributed to the ten-volume proceedings of the Sixth International In Situ and On-Site Bioremediation Symposium (San Diego, California, June 4-7, 2001). Ordering information is provided on the back cover of this book.
The citations reference the ten volumes as follows:

6(1): Magar, V.S., J.T. Gibbs, K.T. O'Reilly, M.R. Hyman, and A. Leeson (Eds.), *Bioremediation of MTBE, Alcohols, and Ethers*. Battelle Press, Columbus, OH, 2001. 249 pp.

6(2): Leeson, A., M.E. Kelley, H.S. Rifai, and V.S. Magar (Eds.), *Natural Attenuation of Environmental Contaminants*. Battelle Press, Columbus, OH, 2001. 307 pp.

6(3): Magar, V.S., G. Johnson, S.K. Ong, and A. Leeson (Eds.), *Bioremediation of Energetics, Phenolics, and Polycyclic Aromatic Hydrocarbons*. Battelle Press, Columbus, OH, 2001. 313 pp.

6(4): Magar, V.S., T.M. Vogel, C.M. Aelion, and A. Leeson (Eds.), *Innovative Methods in Support of Bioremediation*. Battelle Press, Columbus, OH, 2001. 197 pp.

6(5): Leeson, A., E.A. Foote, M.K. Banks, and V.S. Magar (Eds.), *Phytoremediation, Wetlands, and Sediments*. Battelle Press, Columbus, OH, 2001. 383 pp.

6(6): Magar, V.S., F.M. von Fahnestock, and A. Leeson (Eds.), *Ex Situ Biological Treatment Technologies*. Battelle Press, Columbus, OH, 2001. 423 pp.

6(7): Magar, V.S., D.E. Fennell, J.J. Morse, B.C. Alleman, and A. Leeson (Eds.), *Anaerobic Degradation of Chlorinated Solvents*. Battelle Press, Columbus, OH, 2001. 387 pp.

6(8): Leeson, A., B.C. Alleman, P.J. Alvarez, and V.S. Magar (Eds.), *Bioaugmentation, Biobarriers, and Biogeochemistry*. Battelle Press, Columbus, OH, 2001. 255 pp.

6(9): Leeson, A., B.M. Peyton, J.L. Means, and V.S. Magar (Eds.), *Bioremediation of Inorganic Compounds*. Battelle Press, Columbus, OH, 2001. 377 pp.

6(10): Leeson, A., P.C. Johnson, R.E. Hinchee, L. Semprini, and V.S. Magar (Eds.), *In Situ Aeration and Aerobic Remediation*. Battelle Press, Columbus, OH, 2001. 391 pp.

Aagaard, Per (University of Oslo/NORWAY) 6(2):181
Aarnink, Pedro J.P. (Tauw BV/THE NETHERLANDS) 6(10):253
Abbott, James E. (Battelle/USA) 6(5):231, 237
Accashian, John V. (Camp Dresser & McKee, Inc./USA) 6(7):133
Adams, Daniel J. (Camp Dresser & McKee, Inc./USA) 6(8):53
Adams, Jack (Applied Biosciences Corporation/USA) 6(9):331
Adriaens, Peter (University of Michigan/USA) 6(8):19, 193
Adrian, Neal R. (U.S. Army Corps of Engineers/USA) 6(6):133
Agrawal, Abinash (Wright State University/USA) 6(5):95
Aiken, Brian S. (Parsons Engineering Science/USA) 6(2): 65, 189

Aitchison, Eric (Ecolotree, Inc./USA) 6(5):121
Al-Awadhi, Nader (Kuwait Institute for Scientific Research/KUWAIT) 6(6):249
Alblas, B. (Logisticon Water Treatment/THE NETHERLANDS) 6(8):11
Albores, A. (CINVESTAV-IPN/MEXICO) 6(6):219
Al-Daher, Reyad (Kuwait Institute for Scientific Research/KUWAIT) 6(6):249
Al-Fayyomi, Ihsan A. (Metcalf & Eddy, Inc./USA) 6(7):173
Al-Hakak, A. (McGill University/CANADA) 6(9):139
Allen, Harry L. (U.S. EPA/USA) 6(3):259
Allen, Jeffrey (University of Cincinnati/USA) 6(9):9
Allen, Mark H. (Dames & Moore/USA) 6(10):95
Allende, J.L. (Universidad Complutense/SPAIN) 6(4):29
Alonso, R. (Universidad Politecnica/SPAIN) 6(6):377
Alphenaar, Arne (TAUW bv/THE NETHERLANDS) 6(7):297
Alvarez, Pedro J. J. (University of Iowa/USA) 6(1):195; 6(3):1; 6(8):147, 175
Alvestad, Kimberly R. (Earth Tech/USA) 6(3):17
Ambert, Jack (Battelle Europe/SWITZERLAND) 6(6):241
Amezcua-Vega, Claudia (CINVESTAV-IPN/MEXICO) 6(3):243
Amy, Penny (University of Nevada Las Vegas/USA) 6(9):257
Andersen, Peter F. (GeoTrans, Inc./USA) 6(10):163
Anderson, Bruce (Plan Real AG/AUSTRALIA) 6(2):223
Anderson, Jack W. (RMT, Inc./USA) 6(10):201
Andersoiu, Todd (Texas Tech University/USA) 6(9):273
Andreotti, Giorgio (ENI Sop.A.) 6(5):41

Andretta, Massimo (Centro Ricerche Ambientali Montecatini/ITALY) 6(4):131
Andrews, Eric (Environmental Management, Inc./USA) 6(10):23
Andrews, John (SHN Consulting Engineers & Geologists, Inc./USA) 6(3):83
Archibald, Brent B. (Exxon Mobil Environmental Remediation/USA) 6(8):87
Archibold, Errol (Spelman College/USA) 6(9):53
Aresta, Michele (Universita di Catania/ITALY) 6(3):149
Arias, Marianela (PDVSA Intevep/VENEZUELA) 6(6):257
Atagana, Harrison I. (Mangosuthu Technikon/REP OF SOUTH AFRICA) 6(6):101
Atta, Amena (U.S. Air Force/USA) 6(2):73
Ausma, Sandra (University of Guelph/CANADA) 6(6):185
Autenrieth, Robin L. (Texas A&M University/USA) 6(5): 17, 25
Aziz, Carol E. (Groundwater Services, Inc./USA) 6(7):19; 6(8):73
Azizian, Mohammad (Oregon State University/USA) 6(10): 145, 155

Babel, Wolfgang (UFZ Center for Environmental Research/GERMANY) 6(4):81
Bae, Bumhan (Kyungwon University/REPUBLIC OF KOREA) 6(6):51
Baek, Seung S. (Kyonggi University/REPUBLIC OF KOREA) 6(1):161
Bagchi, Rajesh (University of Cincinnati/USA) 6(5):243, 253, 261
Baiden, Laurin (Clemson University/USA) 6(7):109
Bakker, C. (IWACO/THE NETHERLANDS) 6(7):141
Balasoiu, Cristina (École Polytechnique de Montreal/CANADA) 6(9):129
Balba, M. Talaat (Conestoga-Rovers & Associates/USA) 6(1):99; 6(6):249; 6(10):131

Banerjee, Pinaki (Harza Engineering Company, Inc./USA) 6(7):157
Bankston, Jamie L. (Camp Dresser and McKee Inc./USA) 6(5):33
Barbé, Pascal (Centre National de Recherche sur les Sites et Sols Pollués/FRANCE) 6(2):129
Barcelona, Michael J. (University of Michigan/USA) 6(8):19, 193
Barczewski, Baldur (Universitat Stuttgart/GERMANY) 6(2):137
Barker, James F. (University of Waterloo/CANADA) 6(8):95
Barnes, Paul W. (Earth Tech, Inc./USA) 6(3): 17, 25
Basel, Michael D. (Montgomery Watson Harza/USA) 6(10):41
Baskunov, Boris B. (Russian Academy of Sciences/RUSSIA) 6(3):75
Bastiaens, Leen (VITO/BELGIUM) 6(4):35; 6(9):87
Batista, Jacimaria (University of Nevada Las Vegas/USA) 6(9): 257, 265
Bautista-Margulis, Raul G. (Centro de Investigacion en Materiales Avanzados/MEXICO) 6(6):361
Becker, Paul W. (Exxon Mobil Refining & Supply/USA) 6(8):87
Beckett, Ronald (Monash University/AUSTRALIA) 6(4):1
Beckwith, Walt (Solutions Industrial & Environmental Services/USA) 6(7):249
Beguin, Pierre (Institut Pasteur/FRANCE) 6(1):153
Behera, N. (Sambalpur University/INDIA) 6(9):173
Bell, Nigel (Imperial College London/UK) 6(10):123
Bell, Mike (Coats North America/USA) 6(7):213
Beller, Harry R. (Lawrence Livermore National Laboratory/USA) 6(1):195
Belloso, Claudio (Facultad Catolica de Quimica e Ingenieria/ARGENTINA) 6(6): 235, 303
Benner, S. G. (Stanford University/USA) 6(9):71
Bensch, Jeffrey C. (GeoTrans, Inc/USA) 6(7):221

Béron, Patrick (Université du Québec à Montréal/CANADA) 6(3):165
Berry, Duane F. (Virginia Polytechnic Institute & State University/USA) 6(2):105
Betts, W. Bernard (Cell Analysis Ltd./UK) 6(6):27
Billings, Bradford G. (Brad) (Billings & Associates, Inc./USA) 6(1):115
Bingler, Linda (Battelle Sequim/USA) 6(5):231, 237
Birkle, M. (Fraunhofer Institute/GERMANY) 6(2):137
Bitter, Paul (URS Corporation./USA) 6(2):261
Bittoni, A. (EniTecnologie/ITALY) 6(6):173
Bjerg, Poul L (Technical University of Denmark/DENMARK) 6(2):11
Blanchet, Denis (Institut Français du Pétrole/FRANCE) 6(3):227
Bleckmann, Charles A. (Air Force Institute of Technology/USA) 6(2):173
Blokzijl, R. (DHV Environment and Infrastructure/THE NETHERLANDS) 6(8):11
Blowes, David (University of Waterloo/CANADA) 6(9):71
Bluestone, Simon (Montgomery Watson/ITALY) 6(10):41
Boben, Carolyn (Williams/USA) 6(1):175
Böckle, Karin (Technologiezentrum Wasser/GERMANY) 6(8):105
Boender, H. (Logisticon Water Treatment/THE NETHERLANDS) 6(8):11
Böhler, Anja (BioPlanta GmbH/GERMANY) 6(3):67
Bonner, James S. (Texas A&M University/USA) 6(5):17, 25
Bononi, Vera Lucia Ramos (Instituto de Botânica/BRAZIL) 6(3):99
Bonsack, Laurence T. (Aerojet/USA) 6(9):297
Borazjani, Abdolhamid (Mississippi State University/USA) 6(5):329; 6(6):279

Borden, Robert C. (Solutions Industrial & Environmental Services/USA) 6(7):249
Bornholm, Jon (U.S. EPA/USA) 6(6):81
Bosco, Francesca (Politecnico di Torino/ITALY) 6(3):211
Bosma, Tom N.P. (TNO Environment/THE NETHERLANDS) 6(7):61
Bourquin, Al W. (Camp Dresser & McKee Inc./USA) 6(5):33; 6(6):81; 6(7):133,
Bouwer, Edward J. (Johns Hopkins University/USA) 6(2):19
Bowman, Robert S. (New Mexico Institute of Mining & Technology/USA) 6(8):131
Boyd, Sian (CEFAS Laboratory/UK) 6(10):337
Boyd-Kaygi, Patricia (Harding ESE/USA) 6(10):231
Boyle, Susan L. (Haley & Aldrich, Inc./USA) 6(7):27, 281
Brady, Warren D. (IT Corporation/USA) 6(9):215
Breedveld, Gijs (University of Oslo/NORWAY) 6(2):181
Bregante, M. (Istituto di Cibernetica e Biofisica/ITALY) 6(5):157
Brenner, Richard C. (U.S. EPA/USA) 6(5):231, 237
Breteler, Hans (Oostwaardhoeve Co./THE NETHERLANDS) 6(6):59
Bricka, Mark R. (U.S. Army Corps of Engineers/USA) 6(9):241
Brickell, James L. (Earth Tech, Inc./USA) 6(10):65
Brigmon, Robin L. (Westinghouse Savannah River Co/USA) 6(7):109
Britto, Ronnie (EnSafe, Inc./USA) 6(9):315
Brossmer, Christoph (Degussa Corporation/USA) 6(10):73
Brown, Bill (Dunham Environmental Services/USA) 6(6):35
Brown, Kandi L. (IT Corporation/USA) 6(1):51
Brown, Richard A. (ERM, Inc./USA) 6(7):45, 213
Brown, Stephen (Queen's University/CANADA) 6(2):121

Brown, Susan (National Water Research Institute/CANADA) 6(7):321, 333, 341
Brubaker, Gaylen (ThermoRetec North Carolina Corp./USA) 6(7):1
Bruce, Cristin (Arizona State University/USA) 6(8):61
Bruce, Neil C. (University of Cambridge/UK) 6(5):69
Buchanan, Gregory (Tait Environmental Management, Inc./USA) 6(10):267
Bucke, Christopher (University of Westminster/UK) 6(3):75
Bulloch, Gordon (BAE Systems Properties Ltd./UK) 6(6):119
Burckle, John (U.S. EPA/USA) 6(9):9
Burden, David S. (U.S. EPA/USA) 6(2):163
Burdick, Jeffrey S. (ARCADIS Geraghty & Mills/USA) 6(7):53
Burgos, William (The Pennsylvania State University/USA) 6(8):201
Burken, Joel G. (University of Missouri-Rolla/USA) 6(5):113, 199
Burkett, Sharon E. (ENVIRON International Corp./USA) 6(7):189
Burnell, Daniel K. (GeoTrans, Inc./USA) 6(2):163
Burns, David A. (ERM, Inc./USA) 6(7):213
Burton, Christy D. (Battelle/USA) 6(1):137; 6(10):193
Buscheck, Timothy E. (Chevron Research & Technology Co/USA) 6(1): 35, 203
Buss, James A. (RMT, Inc./USA) 6(2):97
Butler, Adrian P. (Imperial College London/UK) 6(10):123
Butler, Jenny (Battelle/USA) 6(7):13
Büyüksönmez, Fatih (San Diego State University/USA) 6(10):301

Caccavo, Frank (Whitworth College/USA) 6(8):1
Callender, James S. (Rockwell Automation/USA) 6(7):133
Calva-Calva, G. (CINVESTAV-IPN/MEXICO) 6(6):219
Camper, Anne K. (Montana State University/USA) 6(7):117

Author Index

Camrud, Doug (Terracon/USA) 6(10):15
Canty, Marietta C. (MSE Technology Applications/USA) 6(9):35
Carman, Kevin R. (Louisiana State University/USA) 6(5):305
Carrera, Paolo (Ambiente S.p.A./ITALY) 6(6):227
Carson, David A. (U.S. EPA/USA) 6(2):247
Carvalho, Cristina (Clemson University/USA) 6(7):109
Case, Nichole L. (Haley & Aldrich, Inc./USA) 6(7):27, 281
Castelli, Francesco (Universita di Catania/ITALY) 6(3):149
Cha, Daniel K. (University of Delaware/USA) 6(6):149
Chaney, Rufus L. (U.S. Department of Agriculture/USA) 6(5):77
Chang, Ching-Chia (National Chung Hsing University/TAIWAN) 6(10):217
Chang, Soon-Woong (Kyonggi University/REPUBLIC OF KOREA) 6(1):161
Chang, Wook (University of Maryland/USA) 6(3):205
Chapuis, R. P. (École Polytechnique de Montréal/CANADA) 6(4):139
Charrois, Jeffrey W.A. (Komex International, Ltd./CANADA) 6(4):7
Chatham, James (BP Exploration/USA) 6(2):261
Chekol, Tesema (University of Maryland/USA) 6(5):77
Chen, Abraham S.C. (Battelle/USA) 6(10):245
Chen, Chi-Ruey (Florida International University/USA) 6(10):187
Chen, Zhu (The University of New Mexico/USA) 6(9):155
Cherry, Jonathan C. (Kennecott Utah Copper Corp/USA) 6(9):323
Child, Peter (Investigative Science Inc./CANADA) 6(2):27
Chino, Hiroyuki (Obayashi Corporation/JAPAN) 6(6):249
Chirnside, Anastasia E.M. (University of Delaware/USA) 6(6):9

Chiu, Pei C. (University of Delaware/USA) 6(6):149
Cho, Kyung-Suk (Ewha University/REPUBLIC OF KOREA) 6(6):51
Choung, Youn-kyoo (Yonsei University/REPUBLIC OF KOREA) 6(6):51
Clement, Bernard (École Polytechnique de Montréal/CANADA) 6(9):27
Clemons, Gary (CDM Federal Programs Corp./USA) 6(6):81
Cocos, Ioana A. (École Polytechnique de Montréal/CANADA) 6(9):27
Cocucci, M. (Universita' degli Studi di Milano/ITALY) 6(5):157
Coelho, Rodrigo O. (CSD-GEOLOCK/BRAZIL) 6(1):27
Collet, Berto (TAUW bv/THE NETHERLANDS) 6(10):253
Compton, Joanne C. (REACT Environmental Engineers/USA) 6(3):25
Connell, Doug (Barr Engineering Company/USA) 6(5):105
Connor, Michael A. (University of Melbourne/AUSTRALIA) 6(10):329
Cook, Jim (Beazer East, Inc./USA) 6(2):239
Cooke, Larry (NOVA Chemicals Corporation/USA) 6(4):117
Coons, Darlene (Conestoga-Rovers & Associates/USA) 6(1):99; 6(10):131
Costley, Shauna C. (University of Natal/REP OF SOUTH AFRICA) 6(9):79
Cota, Jennine L. (ARCADIS Geraghty & Miller, Inc./USA) 6(7):149
Covell, James R. (EG&G Technical Services, Inc./USA) 6(10):49
Cowan, James D. (Ensafe Inc./USA) 6(9):315
Cox, Evan E. (GeoSyntec Consultants/CANADA) 6(8):27, 6(9):297
Cox, Jennifer (Clemson University/USA) 6(7):109
Craig, Shannon (Beazer East, Inc./USA) 6(2):239
Crawford, Donald L. (University of Idaho/USA) 6(3):91; 6(9):147

Crecelius, Eric (Battelle/USA) 6(5): 231, 237
Crotwell, Terry (Solutions Industrial & Environmental Services/USA) 6(7):249
Cui, Yanshan (Chinese Academy of Sciences/CHINA) 6(9):113
Cunningham, Al B. (Montana State University/USA) 6(7):117; 6(8):1
Cunningham, Jeffrey A. (Stanford University/USA) 6(7):95
Cutright, Teresa J. (The University of Akron/USA) 6(3):235

da Silva, Marcio Luis Busi (University of Iowa/USA) 6(1):195
Daly, Daniel J. (Energy & Environmental Research Center/USA) 6(5):129
Daniel, Fabien (AEA Technology Environment/UK) 6(10):337
Daniels, Gary (GeoTrans/USA) 6(8):19
Das, K.C. (University of Georgia/USA) 6(9):289
Davel, Jan L. (University of Cincinnati/USA) 6(6):133
Davis, Gregory A. (Microbial Insights Inc./USA) 6(2):97
Davis, Jeffrey L. (U.S. Army/USA) 6(3): 43, 51
Davis, John W. (The Dow Chemical Company/USA) 6(2):89
Davis-Hoover, Wendy J. (U.S. EPA/USA) 6(2):247
De'Ath, Anna M. (Cranfield University/UK) 6(6):329
Dean, Sean (Camp Dresser & McKee. Inc/USA) 6(7):133
DeBacker, Dennis (Battelle/USA) 6(10):145
DeHghi, Benny (Honeywell International Inc./USA) 6(2):39;6(10):283
de Jong, Jentsje (TAUW BV/THE NETHERLANDS) 6(10):253
Del Vecchio, Michael (Envirogen, Inc./USA) 6(9):281
Delille, Daniel (CNRS/FRANCE) 6(2):57
DeLong, George (AIMTech/USA) 6(7):321, 333, 341
Demers, Gregg (ERM/USA) 6(7):45
De Mot, Rene (Catholic University of Leuven/BELGIUM) 6(4):35

Deobald, Lee A. (University of Idaho/USA) 6(9):147
Deschênes, Louise (École Polytechnique de Montréal/CANADA) 6(3):115; 6(9):129
Dey, William S. (Illinois State Geological Survey/USA) 6(9):179
Díaz-Cervantes, Dolores (CINVESTAV-IPN/MEXICO) 6(6):369
Dick, Vincent B. (Haley & Aldrich, Inc./USA) 6(7):27, 281
Diehl, Danielle (The University of New Mexico/USA) 6(9):155
Diehl, Susan V. (Mississippi State University/USA) 6(5):329
Diels, Ludo (VITO/BELGIUM) 6(9):87
DiGregorio, Salvatore (University della Calabria/ITALY) 6(4):131
Di Gregorio, Simona (Universita degli Studi di Verona/ITALY) 6(3):267
Dijkhuis, Edwin (Bioclear/THE NETHERLANDS) 6(5):289
Di Leo, Cristina (EniTecnologie/ITALY) 6(6):173
Dimitriou-Christidis, Petros (Texas A&M University) 6(5):17
Dixon, Robert (Montgomery Watson/ITALY) 6(10):41
Dobbs, Gregory M. (United Technologies Research Center/USA) 6(7):69
Doherty, Amy T. (GZA GeoEnvironmental, Inc./USA) 6(7):165
Dolan, Mark E. (Oregon State University/USA) 6(10):145, 155, 179
Dollhopf, Michael (Michigan State University/USA) 6(8):19
Dondi, Giovanni (Water & Soil Remediation S.r.l./ITALY) 6(6):179
Dong, Yiting (Chinese Academy of Sciences/CHINA) 6(9):113
Dooley, Maureen A. (Regenesis/USA) 6(7):197
Dottridge, Jane (Komex Europe Ltd./UK) 6(4):17
Dowd, John (University of Georgia/USA) 6(9):289
Doughty, Herb (U.S. Navy/USA) 6(10):1

Doze, Jacco (RIZA/THE NETHERLANDS) *6*(5):289
Dragich, Brian (California Polytechnic State University/USA) *6*(2):1
Drake, John T. (Camp Dresser & McKee Inc./USA) *6*(7):273
Dries, Victor (Flemish Public Waste Agency/BELGIUM) *6*(7):87
Du, Yan-Hung (National Chung Hsing University/TAIWAN) *6*(6):353
Dudal, Yves (École Polytechnique de Montréal/CANADA) *6*(3):115
Duffey, J. Tom (Camp Dresser & McKee Inc./USA) *6*(5):33
Duffy, Baxter E. (Inland Pollution Services, Inc./USA) *6*(7):313
Duijn, Rik (Oostwaardhoeve Co./THE NETHERLANDS) *6*(6):59
Durant, Neal D. (GeoTrans, Inc./USA) *6*(2):19, 163
Durell, Gregory (Battelle Ocean Sciences/USA) *6*(5):231
Dworatzek, S. (University of Toronto/CANADA) *6*(8):27
Dwyer, Daryl F. (University of Minnesota/USA) *6*(3):219
Dzantor, E. K. (University of Maryland/USA) *6*(5):77

Ebner, R. (GMF/GERMANY) *6*(2):137
Ederer, Martina (University of Idaho/USA) *6*(9):147
Edgar, Michael (Camp Dresser & McKee Inc./USA) *6*(7):133
Edwards, Elizabeth A. (University of Toronto/CANADA) *6*(8):27
Edwards, Grant C. (University of Guelph/CANADA) *6*(6):185
Eggen, Trine (Jordforsk Centre for Soil and Environmental Research/NORWAY) *6*(6):157
Eggert, Tim (CDM Federal Programs Corp./USA) *6*(6):81
Elberson, Margaret A. (DuPont Co./USA) *6*(8):43
Elliott, Mark (Virginia Polytechnic Institute & State University/USA) *6*(5):1
Ellis, David E. (Dupont Company/USA) *6*(8):43

Ellwood, Derek C. (University of Southampton/UK) *6*(9):61
Else, Terri (University of Nevada Las Vegas/USA) *6*(9):257
Elväng, Annelie M. (Stockholm University/SWEDEN) *6*(3):133
England, Kevin P. (USA) *6*(5):105
Ertas, Tuba Turan (San Diego State University/USA) *6*(10):301
Escalon, Lynn (U.S. Army Corps of Engineers/USA) *6*(3):51
Esparza-Garcia, Fernando (CINVESTAV-IPN/MEXICO) *6*(6):219
Evans, Christine S. (University of Westminster/UK) *6*(3):75
Evans, Patrick J. (Camp Dresser & McKee, Inc./USA) *6*(2):113, 199; *6*(8):209

Fabiani, Fabio (EniTecnologie S.p.A./ITALY) *6*(6):173
Fadullon, Frances Steinacker (CH2M Hill/USA) *6*(3):107
Fang, Min (University of Massachusetts/USA) *6*(6):73
Faris, Bart (New Mexico Environmental Department/USA) *6*(9):223
Farone, William A. (Applied Power Concepts, Inc./USA) *6*(7):103
Fathepure, Babu Z. (Oklahoma State University/USA) *6*(8):19
Faust, Charles (GeoTrans, Inc./USA) *6*(2):163
Fayolle, Françoise (Institut Français du Pétrole/FRANCE) *6*(1):153
Feldhake, David (University of Cincinnati/USA) *6*(2):247
Felt, Deborah (Applied Research Associates, Inc./USA) *6*(7):125
Feng, Terry H. (Parsons Engineering Science, Inc./USA) *6*(2):39; *6*(10):283
Fenwick, Caroline (Aberdeen University/UK) *6*(2):223
Fernandez, Jose M. (University of Iowa/USA) *6*(1):195
Fernández-Sanchez, J. Manuel (CINVESTAV-IPN/MEXICO) *6*(6):369

Ferrer, E. (Universidad Complutense de Madrid/SPAIN) 6(4):29
Ferrera-Cerrato, Ronald (Colegio de Postgraduados/MEXICO) 6(6):219
Fiacco, R. Joseph (Environmental Resources Management) 6(7):45
Fields, Jim (University of Georgia/USA) 6(9):289
Fields, Keith A. (Battelle/USA) 6(10):1
Fikac, Paul J. (Jacobs Engineering Group, Inc./USA) 6(6):35
Fischer, Nick M. (Aquifer Technology/USA) 6(8):157, 6(10):15
Fisher, Angela (The Pennsylvania State University/USA) 6(8):201
Fisher, Jonathan (Environment Agency/UK) 6(4):17
Fitch, Mark W. (University of Missouri-Rolla/USA) 6(5):199
Fleckenstein, Janice V. (USA) 6(6):89
Fleischmann, Paul (ZEBRA Environmental Corp./USA) 6(10):139
Fletcher, John S. (University of Oklahoma/USA) 6(5):61
Foget, Michael K. (SHN Consulting Engineers & Geologists, Inc./USA) 6(3):83
Foley, K.L. (U.S. Army Engineer Research & Development Center/USA) 6(5):9
Follner, Christina G. (University of Leipzig/GERMANY) 6(4):81
Fontenot, Martin M. (Syngenta Crop Protection, Inc./USA) 6(6):35
Foote, Eric A. (Battelle/USA) 6(1):137; 6(7):13
Ford, James (Investigative Science Inc./CANADA) 6(2):27
Forman, Sarah R. (URS Corporation/USA) 6(7):321, 333, 341
Fortman, Tim J. (Battelle Marine Sciences Laboratory/USA) 6(3):157
Francendese, Leo (U.S. EPA/USA) 6(3):259
Francis, M. McD. (NOVA Research & Technology Center/CANADA) 6(4):117; 6(5):53,
François, Alan (Institut Français du Pétrole/FRANCE) 6(1):153

Frankenberger, William T. (University of California/USA) 6(9):249
Freedman, David L. (Clemson University/USA) 6(7):109
French, Christopher E. (University of Cambridge/UK) 6(5):69
Friese, Kurt (UFZ Center for Environmental Research/GERMANY) 6(9):43
Frisbie, Andrew J. (Purdue University/USA) 6(3):125
Frisch, Sam (Envirogen Inc./USA) 6(9):281
Frömmichen, René (UFZ Centre for Environmental Research/GERMANY) 6(9):43
Fuierer, Alana M. (New Mexico Institute of Mining & Technology/USA) 6(8):131
Fujii, Kensuke (Obayashi Corporation/JAPAN) 6(10):239
Fujii, Shigeo (Kyoto University/JAPAN) 6(4):149
Furuki, Masakazu (Hyogo Prefectural Institute of Environmental Science/JAPAN) 6(5):321

Gallagher, John R. (University of North Dakota/USA) 6(5):129; 6(6):141
Gambale, Franco (Istituto di Cibernetica e Biofisica/ITALY) 6(5):157
Gambrell, Robert P. (Louisiana State University/USA) 6(5):305
Gandhi, Sumeet (University of Iowa/USA) 6(8):147
Garbi, C. (Universidad Complutense de Madrid/SPAIN) 6(4):29; 6(6):377
García-Arrazola, Roeb (CINVESTAV-IPN/MEXICO) 6(6):369
García-Barajas, Rubén Joel (ESIQIE-IPN/MEXICO) 6(6):369
Garrett, Kevin (Harding ESE/USA) 6(7):205
Garry, Erica (Spelman College/USA) 6(9):53
Gavaskar, Arun R. (Battelle/USA) 6(7):13
Gavinelli, Marco (Ambiente S.p.A./ITALY) 6(6):227
Gebhard, Michael (GeoTrans/USA) 6(8):19

Author Index

Gec, Bob (Degussa Canada Ltd./CANADA) 6(10):73
Gehre, Matthias (UFZ - Centre for Environmental Research/GERMANY) 6(4):99
Gemoets, Johan (VITO/BELGIUM) 6(4):35; 6(9):87
Gent, David B. (U.S. Army Corps of Engineers/USA) 6(9):241
Gentry, E. E. (Science Applications International Corporation/USA) 6(8):27
Georgiev, Plamen S. (University of Mining & Geology/BULGARIA) 6(9):97
Gerday, Charles (Université de Liège/BELGIUM) 6(2):57
Gerlach, Robin (Montana State University/USA) 6(8):1
Gerritse, Jan (TNO Environmental Sciences/THE NETHERLANDS) 6(2):231; 6(7):61
Gerth, André (BioPlanta GmbH/GERMANY) 6(3):67; 6(5):173
Ghosh, Upal (Stanford University/USA) 6(3):189; 6(6):89
Ghoshal, Subhasis (McGill University/CANADA) 6(9):139
Gibbs, James T. (Battelle/USA) 6(1):137
Gibello, A. (Universidad Complutense/SPAIN) 6(4):29
Giblin, Tara (University of California/USA) 6(9):249
Gilbertson, Amanda W. (University of Missouri-Rolla/USA) 6(5):199
Gillespie, Rick D. (Regenesis/USA) 6(1):107
Gillespie, Terry J. (University of Guelph/CANADA) 6(6):185
Glover, L. Anne (Aberdeen University /UK) 6(2):223
Goedbloed, Peter (Oostwaardhoeve Co./THE NETHERLANDS) 6(6):59
Golovleva, Ludmila A. (Russian Academy of Sciences/RUSSIA) 6(3):75
Goltz, Mark N. (Air Force Institute of Technology/USA) 6(2):173

Gong, Weiliang (The University of New Mexico/USA) 6(9):155
Gossett, James M. (Cornell University/USA) 6(4):125
Govind, Rakesh (University of Cincinnati/USA) 6(5):269; 6(8):35; 6(9):1, 9, 17
Gozan, Misri (Water Technology Center/GERMANY) 6(8):105
Grainger, David (IT Corporation/USA) 6(1):51; 6(2):73
Grandi, Beatrice (Water & Soil Remediation S.r.l./ITALY) 6(6):179
Granley, Brad A. (Leggette, Brashears, & Graham/USA) 6(10):259
Grant, Russell J. (University of York/UK) 6(6):27
Graves, Duane (IT Corporation/USA) 6(2):253; 6(4):109; 6(9):215
Green, Chad E. (University of California/USA) 6(10):311
Green, Donald J. (USAG Aberdeen Proving Ground/USA) 6(7):321, 333, 341
Green, Robert (Alcoa/USA) 6(6):89
Green, Roger B. (Waste Management, Inc./USA) 6(2):247; 6(6):127
Gregory, Kelvin B. (University of Iowa/USA) 6(3):1
Griswold, Jim (Construction Analysis & Management, Inc./USA) 6(1):115
Groen, Jacobus (Vrije Universiteit/THE NETHERLANDS) 6(4):91
Groenendijk, Gijsbert Jan (Hoek Loos bv/THE NETHERLANDS) 6(7):297
Grotenhuis, Tim (Wageningen Agricultural University/THE NETHERLANDS) 6(5):289
Groudev, Stoyan N. (University of Mining & Geology/BULGARIA) 6(9):97
Guarini, William J. (Envirogen, Inc./USA) 6(9):281
Guieysse, Benoît (Lund University/SWEDEN) 6(3):181
Guiot, Serge R. (Biotechnology Research Institute/CANADA) 6(3):165
Gunsch, Claudia (Clemson University/USA) 6(7):109
Gurol, Mirat (San Diego State University/USA) 6(10):301

Ha, Jeonghyub (University of Maryland/USA) 6(10):57
Haak, Daniel (RMT, Inc./USA) 6(10):201
Haas, Patrick E. (Mitretek Systems/USA) 6(7):19, 241, 249; 6(8):73
Haasnoot, C. (Logisticon Water Treatment/THE NETHERLANDS) 6(8):11
Habe, Hiroshi (The University of Tokyo/JAPAN) 6(4):51; 6(6):111
Haeseler, Frank (Institut Français du Pétrole/FRANCE) 6(3):227
Haff, James (Meritor Automotive, Inc./USA) 6(7):173
Haines, John R. (U.S. EPA/USA) 6(9):17
Håkansson, Torbjörn (Lund University/SWEDEN) 6(9):123
Halfpenny-Mitchell, Laurie (University of Guelph/CANADA) 6(6):185
Hall, Billy (Newfields, Inc./USA) 6(5):189
Hampton, Mark M. (Groundwater Services/USA) 6(8):73
Hannick, Nerissa K. (University of Cambridge/UK) 6(5):69
Hannigan, Mary (Mississippi State University) 6(5):329; 6(6):279
Hannon, LaToya (Spelman College/USA) 6(9):53
Hansen, Hans C. L. (Hedeselskabet /DENMARK) 6(2):11
Hansen, Lance D. (U.S. Army Corps of Engineers/USA) 6(3):9, 43, 51; 6(4):59; 6(6):43; 6(7):125; 6(10):115
Haraguchi, Makoo (Sumitomo Marine Research Institute/JAPAN) 6(10):345
Hardisty, Paul E. (Komex Europe, Ltd./ENGLAND) 6(4):17
Harmon, Stephen M. (U.S. EPA/USA) 6(9):17
Harms, Hauke (Swiss Federal Institute of Technology/SWITZERLAND) 6(3):251
Harmsen, Joop (Alterra, Wageningen University and Research Center/THE NETHERLANDS) 6(5):137, 279; 6(6):1, 59

Harper, Greg (TetraTech EM Inc./USA) 6(3):259
Harrington-Baker, Mary Ann (MSE, Inc./USA) 6(9):35
Harris, Benjamin Cord (Texas A&M University/USA) 6(5):17, 25
Harris, James C. (U.S. EPA/USA) 6(6):287, 295
Harris, Todd (Mason and Hanger Corporation/USA) 6(3):35
Harrison, Patton B. (American Airlines/USA) 6(1):121
Harrison, Susan T.L. (University of Cape Town/REP OF SOUTH AFRICA) 6(6):339
Hart, Barry (Monash University/AUSTRALIA) 6(4):1
Hartzell, Kristen E. (Battelle/USA) 6(1):137; 6(10):193
Harwood, Christine L. (Michael Baker Corporation/USA) 6(2):155
Hassett, David J. (Energy & Environmental Research Center/USA) 6(5):129
Hater, Gary R. (Waste Management Inc./USA) 6(2):247
Hausmann, Tom S. (Battelle Marine Sciences Laboratory/USA) 6(3):157
Hawari, Jalal (National Research Council of Canada/CANADA) 6(9):139
Hayes, Adam J. (Triple Point Engineers, Inc./USA) 6(1):183
Hayes, Dawn M. (U.S. Navy/USA) 6(3):107
Hayes, Kim F. (University of Michigan/USA) 6(8):193
Haynes, R.J. (University of Natal/REP OF SOUTH AFRICA) 6(6):101
Heaston, Mark S. (Earth Tech/USA) 6(3):17, 25
Hecox, Gary R. (University of Kansas/USA) 6(4):109
Heebink, Loreal V. (Energy & Environmental Research Center/USA) 6(5):129
Heine, Robert (EFX Systems, Inc./USA) 6(8):19
Heintz, Caryl (Texas Tech University/USA) 6(3):9

Author Index

Hendrickson, Edwin R. (DuPont Co./USA) 6(8):27, 43
Hendriks, Willem (Witteveen+Bos Consulting Engineers/THE NETHERLANDS) 6(5):289
Henkler, Rolf D. (ICI Paints/UK) 6(2):223
Henny, Cynthia (University of Maine/USA) 6(8):139
Henry, Bruce M. (Parsons Engineering Science, Inc/USA) 6(7):241
Henssen, Maurice J.C. (Bioclear Environmental Biotechnology/THE NETHERLANDS) 6(8):11
Herson, Diane S. (University of Delaware/USA) 6(6):9
Hesnawi, Rafik M. (University of Manitoba/CANADA) 6(6):165
Hetland, Melanie D. (Energy & Environmental Research Center/USA) 6(5):129
Hickey, Robert F. (EFX Systems, Inc./USA) 6(8):19
Hicks, Patrick H. (ARCADIS/USA) 6(1):107
Hiebert, Randy (MSE Technology Applications, Inc./USA) 6(8):79
Higashi, Teruo (University of Tsukuba/JAPAN) 6(9):187
Higgins, Mathew J. (Bucknell University/USA) 6(2):105
Higinbotham, James H. (ExxonMobil Environmental Remediation/USA) 6(8):87
Hines, April (Spelman College/USA) 6(9):53
Hinshalwood, Gordon (Delta Environmental Consultants, Inc./USA) 6(1):43
Hirano, Hiroyuki (The University of Tokyo/JAPAN) 6(6):111
Hirashima, Shouji (Yakult Pharmaceutical Industry/JAPAN) 6(10):345
Hirsch, Steve (Environmental Protection Agency/USA) 6(5):207
Hiwatari, Takehiko (National Institute for Environmental Studies/JAPAN) 6(5):321
Hoag, Rob (Conestoga-Rovers & Associates/USA) 6(1):99

Hoelen, Thomas P. (Stanford University/USA) 6(7):95
Hoeppel, Ronald E. (U.S. Navy/USA) 6(10):245
Hoffmann, Johannes (Hochtief Umwelt GmbH/GERMANY) 6(6):227
Hoffmann, Robert E. (Chevron Canada Resources/CANADA) 6(6):193
Höfte, Monica (Ghent University/BELGIUM) 6(5):223
Holder, Edith L. (University of Cincinnati/USA) 6(2):247
Holm, Thomas R. (Illinois State Water Survey/USA) 6(9):179
Holman, Hoi-Ying (Lawrence Berkeley National Laboratory/USA) 6(4):67
Holoman, Tracey R. Pulliam (University of Maryland/USA) 6(3):205
Hopper, Troy (URS Corporation/USA) 6(2):239
Hornett, Ryan (NOVA Chemicals Corporation/USA) 6(4):117
Hosangadi, Vitthal S. (Foster Wheeler Environmental Corp./USA) 6(9):249
Hough, Benjamin (Tetra Tech EM, Inc./USA) 6(10):293
Hozumi, Toyoharu (Oppenheimer Biotechnology/JAPAN) 6(10):345
Huang, Chin-I (National Chung Hsing University/TAIWAN) 6(10):217
Huang, Chin-Pao (University of Delaware/USA) 6(6):9, 149
Huang, Hui-Bin (DuPont Co./USA) 6(8):43
Huang, Junqi (Air Force Institute of Technology/USA) 6(2):173
Huang, Wei (University of Sheffield/UK) 6(2):207
Hubach, Cor (DHV Noord Nederland/THE NETHERLANDS) 6(8):11
Huesemann, Michael H. (Battelle/USA) 6(3):157
Hughes, Joseph B. (Rice University/USA) 6(5):85; 6(7):19
Hulsen, Kris (University of Ghent/BELGIUM) 6(5):223
Hunt, Jonathan (Clemson University/USA) 6(7):109

Hunter, William J. (U.S. Dept of Agriculture/USA) 6(9):209, 309
Hwang, Sangchul (University of Akron/USA) 6(3):235
Hyman, Michael R. (North Carolina State University/USA) 6(1): 83, 145

Ibeanusi, Victor M. (Spelman College/USA) 6(9):53
Ickes, Jennifer (Battelle/USA) 6(5):231, 237
Ide, Kazuki (Obayashi Corporation Ltd./JAPAN) 6(6):111; 6(10):239
Igarashi, Tsuyoshi (Nippon Institute of Technology/JAPAN) 6(5):321
Infante, Carmen (PDVSA Intevep/VENEZUELA) 6(6):257
Ingram, Sherry (IT Corporation/USA) 6(4):109
Ishikawa, Yoji (Obayashi Corporation/JAPAN) 6(6):249; 6(10):239

Jackson, W. Andrew (Texas Tech University/USA) 6(5):207, 313; 6(9):273
Jacobs, Alan K. (EnSafe, Inc./USA) 6(9):315
Jacques, Margaret E. (Rowan University/USA) 6(5):215
Jahan, Kauser (Rowan University/USA) 6(5):215
James, Garth (MSE Inc./USA) 6(8):79
Jansson, Janet K. (Södertörn University College/SWEDEN) 6(3):133
Japenga, Jan (Alterra/THE NETHERLANDS) 6(5):137
Jauregui, Juan (Universidad Nacional Autonoma de Mexico/MEXICO) 6(6):17
Jensen, James N. (State University of New York at Buffalo/USA) 6(6):89
eon, Mi-Ae (Texas Tech University/USA) 6(9):273
Jerger, Douglas E. (IT Corporation/USA) 6(3):35
Jernberg, Cecilia (Södertörn University College/SWEDEN) 6(3):133
Jindal, Ranjna (Suranaree University of Technology/THAILAND) 6(4):149

Johnson, Dimitra (Southern University at New Orleans/USA) 6(5):151
Johnson, Glenn (University of Utah/USA) 6(5):231
Johnson, Paul C. (Arizona State University/USA) 6(1):11; 6(8):61
Johnson, Richard L. (Oregon Graduate Institute/USA) 6(10):293
Jones, Antony (Komex H_2O Science, Inc./USA) 6(2):223; 6(3):173; 6(10):123
Jones, Clay (University of New Mexico/USA) 6(9):223
Jones, Triana N. (University of Maryland/USA) 6(3):205
Jonker, Hendrikus (Vrije Universiteit/THE NETHERLANDS) 6(4):91
Ju, Lu-Kwang (The University of Akron/USA) 6(6):319

Kaludjerski, Milica (San Diego State University/USA) 6(10):301
Kamashwaran, S. Ramanathen (University of Idaho/USA) 6(3):91
Kambhampati, Murty S. (Southern University at New Orleans/USA) 6(5):145, 151
Kamimura, Daisuke (Gunma University/JAPAN) 6(8):113
Kang, James J. (URS Corporation/USA) 6(1):121; 6(10):223
Kappelmeyer, Uwe (UFZ Centre for Environmental Research/GERMANY) 6(5):337
Karamanev, Dimitre G. (University of Western Ontario/CANADA) 6(10):171
Karlson, Ulrich (National Environmental Research Institute) 6(3):141
Kastner, James R. (University of Georgia/USA) 6(9):289
Kästner, Matthias (UFZ Centre for Environmental Research/GERMANY) 6(4):99; 6(5):337
Katz, Lynn E. (University of Texas/USA) 6(8):139
Kavanaugh, Rathi G. (University of Cincinnati/USA) 6(2):247

Kawahara, Fred (U.S. EPA/USA) 6(9):9
Kawakami, Tsuyoshi (University of Tsukuba/JAPAN) 6(9):187
Keefer, Donald A. (Illinois State Geological Survey/USA) 6(9):179
Keith, Nathaniel (Texas A&M University/USA) 6(5):25
Kelly, Laureen S. (Montana Department of Environmental Quality/USA) 6(6):287
Kempisty, David M. (U.S. Air Force/USA) 6(10):145, 155
Kerfoot, William B. (K-V Associates, Inc./USA) 6(10):33
Keuning, S. (Bioclear Environmental Technology/THE NETHERLANDS) 6(8):11
Khan, Tariq A. (Groundwater Services, Inc./USA) 6(7):19
Khodadoust, Amid P. (University of Cincinnati/USA) 6(5):243, 253, 261
Kieft, Thomas L. (New Mexico Institute of Mining and Technology/USA) 6(8):131
Kiessig, Gunter (WISMUT GmbH/GERMANY) 6(5):173; 6(9):155
Kilbride, Rebecca (CEFAS Laboratory/UK) 6(10):337
Kim, Jae Young (Seoul National University/REPUBLIC OF KOREA) 6(9):195
Kim, Jay (University of Cincinnati/USA) 6(6):133
Kim, Kijung (The Pennsylvania State University/USA) 6(9):303
Kim, Tae Young (Ewha University/REPUBLIC OF KOREA) 6(6):51
Kinsall, Barry L. (Oak Ridge National Laboratory/USA) 6(4):73
Kirschenmann, Kyle (IT Corp/USA) 6(4):109
Klaas, Norbert (University of Stuttgart/GERMANY) 6(2):137
Klecka, Gary M. (The Dow Chemical Company/USA) 6(2):89
Klein, Katrina (GeoTrans, Inc./USA) 6(2):163

Klens, Julia L. (IT Corporation/USA) 6(2):253; 6(9):215
Knotek-Smith, Heather M. (University of Idaho/USA) 6(9):147
Koch, Stacey A. (RMT, Inc./USA) 6(7):181
Koenen, Brent A. (U.S. Army Engineer Research & Development Center/USA) 6(5):9
Koenigsberg, Stephen S. (Regenesis Bioremediation Products/USA) 6(7):197, 257; 6(8):209; 6(10):9, 87
Kohata, Kunio (National Institute for Environmental Studies/JAPAN) 6(5):321
Kohler, Keisha (ThermoRetec Corporation/USA) 6(7):1
Kolhatkar, Ravindra V. (BP Corporation/USA) 6(1):35, 43
Komlos, John (Montana State University/USA) 6(7):117
Komnitsas, Kostas (National Technical University of Athens/GREECE) 6(9):97
Kono, Masakazu (Oppenheimer Biotechnology/JAPAN) 6(10):345
Koons, Brad W. (Leggette, Brashears & Graham, Inc./USA) 6(1):175
Koschal, Gerard (PNG Environmental/USA) 6(1):203
Koschorreck, Matthias (UFZ Centre for Environmental Research/GERMANY) 6(9):43
Koshikawa, Hiroshi (National Institute for Environmental Studies/JAPAN) 6(5):321
Kramers, Jan D. (University of Bern/SWITZERLAND) 6(4):91
Krooneman, Jannneke (Bioclear Environmental Biotechnology/THE NETHERLANDS) 6(7):141
Kruk, Taras B. (URS Corporation/USA) 6(10):223
Kuhwald, Jerry (NOVA Chemicals Corporation/CANADA) 6(5):53
Kuschk, Peter (UFZ Centre for Environmental Research Leipzig/GERMANY) 6(5):337

Laboudigue, Agnes (Centre National de Recherche sur les Sites et Sols Pollués/FRANCE) 6(2):129
LaFlamme, Brian (Engineering Management Support, Inc./USA) 6(10):231
Lafontaine, Chantal (École Polytechnique de Montréal/CANADA) 6(10):171
Laha, Shonali (Florida International University/USA) 6(10):187
Laing, M.D. (University of Natal/REP OF SOUTH AFRICA) 6(9):79
Lamar, Richard (EarthFax Development Corp/USA) 6(6):263
Lamarche, Philippe (Royal Military College of Canada/CANADA) 6(8):95
Lamb, Steven R. (GZA GeoEnvironmental, Inc./USA) 6(7):165
Landis, Richard C. (E.I. du Pont de Nemours & Company/USA) 6(8):185
Lang, Beth (United Technologies Corp./USA) 6(10):41
Langenhoff, Alette (TNO Institute of Environmental Science/THE NETHERLANDS) 6(7):141
LaPat-Polasko, Laurie T. (Parsons Engineering Science, Inc./USA) 6(2):65, 189
Lapus, Kevin (Regenesis/USA) 6(7):257; 6(10):9
LaRiviere, Daniel (Texas A&M University/USA) 6(5):17, 25
Larsen, Lars C. (Hedeselskabet/DENMARK) 6(2):11
Larson, John R. (TranSystems Corporation/USA) 6(7):229
Larson, Richard A. (University of Illinois at Urbana-Champaign/USA) 6(5):181
Lauzon, Francois (Dept of National Defence/CANADA) 6(8):95
Leavitt, Maureen E. (Newfields Inc./USA) 6(1):51; 6(5):189
Lebron, Carmen A. (U.S. Navy/USA) 6(7):95
Lee, B. J. (Science Applications International Corporation) 6(8):27

Lee, Brady D. (Idaho National Engineering & Environmental Laboratory/USA) 6(7):77
Lee, Chi Mei (National Chung Hsing University/TAIWAN) 6(6):353
Lee, Eun-Ju (Louisiana State University/USA) 6(5):313
Lee, Kenneth (Fisheries & Oceans Canada/CANADA) 6(10):337
Lee, Michael D. (Terra Systems, Inc./USA) 6(7):213, 249
Lee, Ming-Kuo (Auburn University/USA) 6(9):105
Lee, Patrick (Queen's University/CANADA) 6(2):121
Lee, Seung-Bong (University of Washington/USA) 6(10):211
Lee, Si-Jin (Kyonggi University/REPUBLIC OF KOREA) 6(1):161
Lee, Sung-Jae (ChoongAng University/REPUBLIC OF KOREA) 6(6):51
Leeson, Andrea (Battelle/USA) 6(10):1, 145, 155, 193
Lehman, Stewart E. (California Polytechnic State University/USA) 6(2):1
Lei, Li (University of Cincinnati/USA) 6(5):243, 261
Leigh, Daniel P. (IT Corporation/USA) 6(3):35
Leigh, Mary Beth (University of Oklahoma/USA) 6(5):61
Lendvay, John (University of San Francisco/USA) 6(8):19
Lenzo, Frank C. (ARCADIS Geraghty & Miller/USA) 6(7):53
Leon, Nidya (PDVSA Intevep/VENEZUELA) 6(6):257
Leong, Sylvia (Crescent Heights High School/CANADA) 6(5):53
Leontievsky, Alexey A. (Russian Academy of Sciences/RUSSIA) 6(3):75
Lerner, David N. (University of Sheffield/UK) 6(1):59; 6(2):207
Lesage, Suzanne (National Water Research Institute/CANADA) 6(7):321, 333, 341

Leslie, Jolyn C. (Camp Dresser & McKee, Inc./USA) 6(2):113
Lewis, Ronald F. (U.S. EPA/USA) 6(5):253, 261
Li, Dong X. (USA) 6(7):205
Li, Guanghe (Tsinghua University/CHINA) 6(7):61
Li, Tong (Tetra Tech EM Inc./USA) 6(10):293
Librando, Vito (Universita di Catania/ITALY) 6(3):149
Lieberman, M. Tony (Solutions Industrial & Environmental Services/USA) 6(7):249
Lin, Cindy (Conestoga-Rovers & Associates/USA) 6(1):99; 6(10):131
Lipson, David S. (Blasland, Bouck & Lee, Inc./USA) 6(10):319
Liu, Jian (University of Nevada Las Vegas/USA) 6(9):265
Liu, Xiumei (Shandong Agricultural University/ CHINA) 6(9):113
Livingstone, Stephen (Franz Environmental Inc./CANADA) 6(6):211
Lizzari, Daniela (Universita degli Studi di Verona/ITALY) 6(3):267
Llewellyn, Tim (URS/USA) 6(7):321, 333, 341
Lobo, C. (El Encin IMIA/SPAIN) 6(4):29
Loeffler, Frank E. (Georgia Institute of Technology/USA) 6(8):19
Logan, Bruce E. (The Pennsylvania State University/USA) 6(9):303
Long, Gilbert M. (Camp Dresser & McKee Inc./USA) 6(6):287
Longoni, Giovanni (Montgomery Watson/ITALY) 6(10):41
Lorbeer, Helmut (Technical University of Dresden/GERMANY) 6(8):105
Lors, Christine (Centre National de Recherche sur les Sites et Sols Pollués /FRANCE) 6(2):129
Lorton, Diane M. (King's College London/UK) 6(2):223; 6(3):173
Losi, Mark E. (Foster Wheeler Environ. Corp./USA) 6(9):249
Loucks, Mark (U.S. Air Force/USA) 6(2):261

Lu, Chih-Jen (National Chung Hsing University/TAIWAN) 6(6):353; 6(10):217
Lu, Xiaoxia (Tsinghua University/CHINA) 6(7):61
Lubenow, Brian (University of Delaware/USA) 6(6):149
Lucas, Mary (Parsons Engineering Science, Inc./USA) 6(10):283
Lundgren, Tommy S. (Sydkraft SAKAB AB/SWEDEN) 6(6):127
Lundstedt, Staffan (Umeå University/SWEDEN) 6(3):181
Luo, Xiaohong (NRC Research Associate/USA) 6(8):167
Luthy, Richard G. (Stanford University/USA) 6(3):189
Lutze, Werner (University of New Mexico/USA) 6(9):155
Luu, Y.-S. (Queen's University/CANADA) 6(2):121
Lynch, Regina M. (Battelle/USA) 6(10):155

Macek, Thomáš (Institute of Chemical Technology/Czech Republic) 6(5):61
MacEwen, Scott J. (CH2M Hill/USA) 6(3):107
Machado, Kátia M. G. (Fund. Centro Tecnológico de Minas Gerais/BRAZIL) 6(3):99
Maciel, Helena Alves (Aberdeen University/UK) 6(1):1
Mack, E. Erin (E.I. du Pont de Nemours & Co./USA) 6(2):81; 6(8):43
Macková, Martina (Institute of Chemical Technology/Czech Republic) 6(5):61
Macnaughton, Sarah J. (AEA Technology/UK) 6(5):305; 6(10):337
Macomber, Jeff R. (University of Cincinnati/USA) 6(6):133
Macrae, Jean (University of Maine/USA) 6(8):139
Madden, Patrick C. (Engineering Consultant/USA) 6(8):87
Madsen, Clint (Terracon/USA) 6(8):157; 6(10):15
Magar, Victor S. (Battelle/USA) 6(1):137; 6(5):231, 237; 6(10):145, 155

Mage, Roland (Battelle Europe/SWITZERLAND) 6(6):241; 6(10):109
Magistrelli, P. (Istituto di Cibernetica e Biofisica/ITALY) 6(5):157
Maierle, Michael S. (ARCADIS Geraghty & Miller, Inc./USA) 6(7):149
Major, C. Lee (Jr.) (University of Michigan/USA) 6(8):19
Major, David W. (GeoSyntec Consultants/CANADA) 6(8):27
Maki, Hideaki (National Institute for Environmental Studies/JAPAN) 6(5):321
Makkar, Randhir S. (University of Illinois-Chicago/USA) 6(5):297
Malcolm, Dave (BAE Systems Properties Ltd./UK) 6(6):119
Manabe, Takehiko (Hyogo Prefectural Fisheries Research Institute/JAPAN) 6(10):345
Maner, P.M. (Equilon Enterprises, LLC/USA) 6(1):11
Maner, Paul (Shell Development Company/USA) 6(8):61
Manrique-Ramírez, Emilio Javier (SYMCA, S.A. de C.V./MEXICO) 6(6):369
Marchal, Rémy (Institut Français du Pétrole/FRANCE) 6(1):153
Maresco, Vincent (Groundwater & Environmental Srvcs/USA) 6(10):101
Marnette, Emile C. (TAUW BV/THE NETHERLANDS) 6(7):297
Marshall, Timothy R. (URS Corporation/USA) 6(2):49
Martella, L. (Istituto di Cibernetica e Biofisica/ITALY) 6(5):157
Martin, C. (Universidad Politecnica/SPAIN) 6(4):29
Martin, Jennifer P. (Idaho National Engineering & Environmental Laboratory/USA) 6(7):265
Martin, John F. (U.S. EPA/USA) 6(2):247
Martin, Margarita (Universidad Complutense de Madrid/SPAIN) 6(4):29; 6(6):377

Martinez-Inigo, M.J. (El Encin IMIA/SPAIN) 6(4):29
Martino, Lou (Argonne National Laboratory/USA) 6(5):207
Mascarenas, Tom (Environmental Chemistry/USA) 6(8):157
Mason, Jeremy (King's College London/UK) 6(2):223; 6(3):173; 6(10):123
Massella, Oscar (Universita degli Studi di Verona/ITALY) 6(3):267
Matheus, Dacio R. (Instituto de Botânica/BRAZIL) 6(3):99
Matos, Tania (University of Puerto Rico at Rio Piedras/USA) 6(9):179
Matsubara, Takashi (Obayashi Corporation/JAPAN) 6(6):249
Mattiasson, Bo (Lund University/SWEDEN) 6(3):181; 6(6):65; 6(9):123
McCall, Sarah (Battelle/USA) 6(10):155, 245
McCarthy, Kevin (Battelle Duxbury Operations/USA) 6(5):9
McCartney, Daryl M. (University of Manitoba/CANADA) 6(6):165
McCormick, Michael L. (The University of Michigan/USA) 6(8):193
McDonald, Thomas J. (Texas A&M University) 6(5):17
McElligott, Mike (U.S. Air Force/USA) 6(1):51
McGill, William B. (University of Northern British Columbia/CANADA) 6(4):7
McIntosh, Heather (U.S. Army/USA) 6(7):321, 333
McLinn, Eugene L. (RMT, Inc./USA) 6(5):121
McLoughlin, Patrick W. (Microseeps Inc./USA) 6(1):35
McMaster, Michaye (GeoSyntec Consultants/CANADA) 6(8):27, 43, 6(9):297
McMillen, Sara J. (Chevron Research & Technology Company/USA) 6(6):193
Meckenstock, Rainer U. (University of Tübingen/GERMANY) 6(4):99
Mehnert, Edward (Illinois State Geological Survey/USA) 6(9):179

Meigio, Jodette L. (Idaho National Engineering & Environmental Laboratory/USA) 6(7):77
Meijer, Harro A.J. (University of Groningen/THE NETHERLANDS) 6(4):91
Meijerink, E. (Province of Drenthe/THE NETHERLANDS) 6(8):11
Merino-Castro, Glicina (Inst Technol y de Estudios Superiores/MEXICO) 6(6):377
Messier, J.P. (U.S. Coast Guard/USA) 6(1):107
Meyer, Michael (Environmental Resources Management/BELGIUM) 6(7):87
Meylan, S. (Queen's University/CANADA) 6(2):121
Miles, Victor (Duracell Inc./USA) 6(7):87
Millar, Kelly (National Water Research Institute/CANADA) 6(7):321, 333, 341
Miller, Michael E. (Camp Dresser & McKee, Inc./USA) 6(7):273
Miller, Thomas Ferrell (Lockheed Martin/USA) 6(3):259
Mills, Heath J. (Georgia Institute of Technology/USA) 6(9):165
Millward, Rod N. (Louisiana State University/USA) 6(5):305
Mishra, Pramod Chandra (Sambalpur University/INDIA) 6(9):173
Mitchell, David (AEA Technology Environment/UK) 6(10):337
Mitraka, Maria (Serres/GREECE) 6(6):89
Mocciaro, PierFilippo (Ambiente S.p.A./ITALY) 6(6):227
Moeri, Ernesto N. (CSD-GEOKLOCK/BRAZIL) 6(1):27
Moir, Michael (Chevron Research & Technology Co./USA) 6(1):83
Molinari, Mauro (AgipPetroli S.p.A/ITALY) 6(6):173
Mollea, C. (Politecnico di Torino/ITALY) 6(3):211
Mollhagen, Tony (Texas Tech University/USA) 6(3):9
Monot, Frédéric (Institut Français du Pétrole/FRANCE) 6(1):153

Moon, Hee Sun (Seoul National University/REPUBLIC OF KOREA) 6(9):195
Moosa, Shehnaaz (University of Cape Town/REP OF SOUTH AFRICA) 6(6):339
Morasch, Barbara (University Konstanz/GERMANY) 6(4):99
Moreno, Joanna (URS Corporation/USA) 6(2):239
Morgan, Scott (URS - Dames & Moore/USA) 6(7):321
Morrill, Pamela J. (Camp, Dresser, & McKee, Inc./USA) 6(2):113
Morris, Damon (ThermoRetec Corporation/USA) 6(7):1
Mortimer, Marylove (Mississippi State University/USA) 6(5):329
Mortimer, Wendy (Bell Canada/CANADA) 6(2):27; 6(6):185, 203, 211,
Mossing, Christian (Hedeselskabet/DENMARK) 6(2):11
Mossmann, Jean-Remi (Centre National de Recherche sur les Sites et Sols Pollués/FRANCE) 6(2):129
Moteleb, Moustafa A. (University of Cincinnati/USA) 6(6):133
Mowder, Carol S. (URS/USA) 6(7):321, 333, 341
Moyer, Ellen E. (ENSR International./USA) 6(1):75
Mravik, Susan C. (U.S. EPA/USA) 6(1):167
Mueller, James G. (URS Corporation/USA) 6(2):239
Müller, Axel (Water Technology Center/GERMANY) 6(8):105
Müller, Beate (Umweltschutz Nord GmbH/GERMANY) 6(4):131
Müller, Klaus (Battelle Europe/SWITZERLAND) 6(5):41; 6(6):241
Muniz, Herminio (Hart Crowser Inc./USA) 6(10):9
Murphy, Sean M. (Komex International Ltd./CANADA) 6(4):7
Murray, Cliff (United States Army Corps of Engineers/USA) 6(9):281
Murray, Gordon Bruce (Stella-Jones Inc./CANADA) 6(3):197

Murray, Willard A. (Harding ESE/USA) 6(7):197
Mutch, Robert D. (Brown and Caldwell/USA) 6(2):145
Mutti, Francois (Water & Soil Remediation S.r.l./ITALY) 6(6):179
Myasoedova, Nina M. (Russian Academy of Sciences/RUSSIA) 6(3):75

Nadolishny, Alex (Nedatek, Inc./USA) 6(10):139
Nagle, David P. (University of Oklahoma/USA) 6(5):61
Nam, Kyoungphile (Seoul National University/REPUBLIC OF KOREA) 6(9):195
Narayanaswamy, Karthik (Parsons Engineering Science/USA) 6(2):65
Nelson, Mark D. (Delta Environmental Consultants, Inc./USA) 6(1):175
Nelson, Yarrow (California Polytechnic State University/USA) 6(10):311
Nemati, M. (University of Cape Town/REP OF SOUTH AFRICA) 6(6):339
Nestler, Catherine C. (Applied Research Associates, Inc./USA) 6(4):59, 6(6):43
Nevárez-Moorillón, G.V. (UACH/MEXICO) 6(6):361
Neville, Scott L. (Aerojet General Corp./USA) 6(9):297
Newell, Charles J. (Groundwater Services, Inc./USA) 6(7):19
Nieman, Karl (Utah State University/USA) 6(4):67
Niemeyer, Thomas (Hochtief Umwelt Gmbh/GERMANY) 6(6):227
Nies, Loring (Purdue University/USA) 6(3):125
Nipshagen, Adri A.M. (IWACO/THE NETHERLANDS) 6(7):141
Nishino, Shirley (U.S. Air Force/USA) 6(3):59
Nivens, David E. (University of Tennessee/USA) 6(4):45
Noffsinger, David (Westinghouse Savannah River Company/USA) 6(10):163

Noguchi, Takuya (Nippon Institute of Technology/JAPAN) 6(5):321
Nojiri, Hideaki (The University of Tokyo/JAPAN) 6(4):51; 6(6):111
Noland, Scott (NESCO Inc./USA) 6(10):73
Nolen, C. Hunter (Camp Dresser & McKee/USA) 6(6):287
Norris, Robert D. (Eckenfelder/Brown and Caldwell/USA) 6(2):145; 6(7):35
North, Robert W. (Environ Corporation./USA) 6(7):189
Novak, John T. (Virginia Polytechnic Institute & State University/USA) 6(2):105; 6(5):1
Novick, Norman (Exxon/Mobil Oil Corp/USA) 6(1):35
Nuttall, H. Eric (The University of New Mexico/USA) 6(9): 155, 223
Nuyens, Dirk (Environmental Resources Management/BELGIUM) 6(7):87; 6(9):87
Nzengung, Valentine A. (University of Georgia/USA) 6(9):289

Ochs, L. Donald (Regenesis/USA) 6(10):139
O'Connell, Joseph E. (Environmental Resolutions, Inc./USA) 6(1):91
Odle, Bill (Newfields, Inc./USA) 6(5):189
O'Donnell, Ingrid (BAE Systems Properties, Ltd./UK) 6(6):119
Ogden, Richard (BAE Systems Properties Ltd./UK) 6(6):119
Oh, Byung-Taek (The University of Iowa/USA) 6(8):147, 175
Oh, Seok-Young (University of Delaware/USA) 6(6):149
Omori, Toshio (The University of Tokyo/JAPAN) 6(4):51; 6(6):111
O'Neal, Brenda (ARA/USA) 6(3):43
Oppenheimer, Carl H. (Oppenheimer Biotechnology/USA) 6(10):345
O'Regan, Gerald (Chevron Products Company/USA) 6(1):203
O'Reilly, Kirk T. (Chevron Research & Technology Co/USA) 6(1):83, 145, 203
Oshio, Takahiro (University of Tsukuba/JAPAN) 6(9):187

Ozdemiroglu, Ece (EFTEC Ltd./UK) 6(4):17

Padovani, Marco (Centro Ricerche Ambientali/ITALY) 6(4):131
Paganetto, A. (Istituto di Cibernetica e Biofisica/ITALY) 6(5):157
Pahr, Michelle R. (ARCADIS Geraghty & Miller/USA) 6(1):107
Pal, Nirupam (California Polytechnic State University/USA) 6(2):1
Palmer, Tracy (Applied Power Concepts, Inc./USA) 6(7):103
Palumbo, Anthony V. (Oak Ridge National Laboratory/USA) 6(4):73; 6(9):165
Panciera, Matthew A. (University of Connecticut/USA) 6(7):69
Pancras, Tessa (Wageningen University/THE NETHERLANDS) 6(5):289
Pardue, John H. (Louisiana State University/USA) 6(5): 207, 313; 6(9):273
Park, Kyoohong (ChoongAng University/REPUBLIC OF KOREA) 6(6):51
Parkin, Gene F. (University of Iowa/USA) 6(3):1
Paspaliaris, Ioannis (National Technical University of Athens/GREECE) 6(9):97
Paton, Graeme I. (Aberdeen University/UK) 6(1):1
Patrick, John (University of Reading/UK) 6(10):337
Payne, Frederick C. (ARCADIS Geraghty & Miller/USA) 6(7):53
Payne, Jo Ann (DuPont Co./USA) 6(8):43
Peabody, Jack G. (Regenesis/USA) 6(10):95
Peacock, Aaron D. (University of Tennessee/USA) 6(4):73; 6(5):305
Peargin, Tom R. (Chevron Research & Technology Co/USA) 6(1):67
Peeples, James A. (Metcalf & Eddy, Inc./USA) 6(7):173
Pehlivan, Mehmet (Tait Environmental Management, Inc./USA) 6(10):267, 275

Pelletier, Emilien (ISMER/CANADA) 6(2):57
Pennie, Kimberley A. (Stella-Jones, Inc./CANADA) 6(3):197
Peramaki, Matthew P. (Leggette, Brashears, & Graham, Inc./USA) 6(10):259
Perey, Jennie R. (University of Delaware/USA) 6(6):149
Perez-Vargas, Josefina (CINVESTAV-IPN/MEXICO) 6(6):219
Perina, Tomas (IT Corporation/USA) 6(1):51; 6(2):73
Perlis, Shira R. (Rowan University/USA) 6(5):215
Perlmutter, Michael W. (EnSafe, Inc./USA) 6(9):315
Perrier, Michel (École Polytechnique de Montréal/CANADA) 6(4):139
Perry, L.B. (U.S. Army Engineer Research & Development Center/USA) 6(5):9
Persico, John L. (Blasland, Bouck & Lee, Inc./USA) 6(10):319
Peschong, Bradley J. (Leggette, Brashears & Graham, Inc./USA) 6(1):175
Peters, Dave (URS/USA) 6(7):333
Peterson, Lance N. (North Wind Environmental, Inc./USA) 6(7):265
Petrovskis, Erik A. (Geotrans Inc./USA) 6(8):19
Peven-McCarthy, Carole (Battelle Ocean Sciences/USA) 6(5):231
Pfiffner, Susan M. (University of Tennessee/USA) 6(4):73
Phelps, Tommy J. (Oak Ridge National Laboratory/USA) 6(4):73
Pickett, Tim M. (Applied Biosciences Corporation/USA) 6(9):331
Pickle, D.W. (Equilon Enterprises LLC/USA) 6(8):61
Pierre, Stephane (École Polytechnique de Montréal/CANADA) 6(10):171
Pijis, Charles G.J.M. (TAUW BV/THE NETHERLANDS) 6(10):253
Pirkle, Robert J. (Microseeps, Inc./USA) 6(1):35
Pisarik, Michael F. (New Fields/USA) 6(1):121

Piveteau, Pascal (Institut Français du Pétrole/FRANCE) 6(1):153
Place, Matthew (Battelle/USA) 6(10):245
Plata, Nadia (Battelle Europe/SWITZERLAND) 6(5):41
Poggi-Varaldo, Hector M. (CINVESTAV-IPN/MEXICO) 6(3):243; 6(6):219
Pohlmann, Dirk C. (IT Corporation/USA) 6(2):253
Pokethitiyook, Prayad (Mahidol University/THAILAND) 6(10):329
Polk, Jonna (U.S. Army Corps of Engineers/USA) 6(9):281
Pope, Daniel F. (Dynamac Corp/USA) 6(1):129
Porta, Augusto (Battelle Europe/SWITZERLAND) 6(5):41; 6(6):241; 6(10):109
Portier, Ralph J. (Louisiana State University/USA) 6(5):305
Powers, Leigh (Georgia Institute of Technology/USA) 6(9):165
Prandi, Alberto (Water & Soil Remediation S.r.l/ITALY) 6(6):179
Prasad, M.N.V. (University of Hyderabad/INDIA) 6(5):165
Price, Steven (Camp Dresser & McKee, Inc./USA) 6(9):303
Priester, Lamar E. (Priester & Associates/USA) 6(10):65
Pritchard, P. H. (Hap) (U.S. Navy/USA) 6(7):125
Profit, Michael D. (CDM Federal Programs Corporation/USA) 6(6):81
Prosnansky, Michal (Gunma University/JAPAN) 6(9):201
Pruden, Amy (University of Cincinnati/USA) 6(1):19
Ptacek, Carol J. (University of Waterloo/CANADA) 6(9):71

Radosevich, Mark (University of Delaware/USA) 6(6):9
Radtke, Corey (INEEL/USA) 6(3):9
Raetz, Richard M. (Global Remediation Technologies, Inc./USA) 6(6):311
Rainwater, Ken (Texas Tech University/USA) 6(3):9

Ramani, Mukundan (University of Cincinnati/USA) 6(5):269
Raming, Julie B. (Georgia-Pacific Corp./USA) 6(1):183
Ramírez, N. E. (ECOPETROL-ICP/COLOMBIA) 6(6):319
Ramsay, Bruce A. (Polyferm Canada Inc./CANADA) 6(2):121; 6(10):171
Ramsay, Juliana A. (Queen's University/CANADA) 6(2):121; 6(10):171
Rao, Prasanna (University of Cincinnati/USA) 6(9):1
Ratzke, Hans-Peter (Umweltschutz Nord GMBH/GERMANY) 6(4):131
Reardon, Kenneth F. (Colorado State University/USA) 6(8):53
Rectanus, Heather V. (Virginia Polytechnic Institute & State University/USA) 6(2):105
Reed, Thomas A. (URS Corporation/USA) 6(8):157; 6(10):15, 95
Rees, Hubert (CEFAS Laboratory/UK) 6(10):337
Rehm, Bernd W. (RMT, Inc./USA) 6(2):97; 6(10):201
Reinecke, Stefan (Franz Environmental Inc./CANADA) 6(6):211
Reinhard, Martin (Stanford University/USA) 6(7):95
Reisinger, H. James (Integrated Science & Technology Inc/USA) 6(1):183
Rek, Dorota (IT Corporation/USA) 6(2):73
Reynolds, Charles M. (U.S. Army Engineer Research & Development Center/USA) 6(5):9
Reynolds, Daniel E. (Air Force Institute of Technology/USA) 6(2):173
Rice, John M. (RMT, Inc./USA) 6(7):181
Richard, Don E. (Barr Engineering Company/USA) 6(3):219; 6(5):105
Richardson, Ian (Conestoga-Rovers & Associates/USA) 6(10):131
Richnow, Hans H. (UFZ-Centre for Environmental Research/GERMANY) 6(4):99

Rijnaarts, Huub H.M. (TNO Institute of Environmental Science/THE NETHERLANDS) 6(2):231
Ringelberg, David B. (U.S. Army Corps of Engineers/USA) 6(5):9; 6(6):43; 6(10):115
Ríos-Leal, E. (CINVESTAV-IPN/MEXICO) 6(3):243
Ripp, Steven (University of Tennessee/USA) 6(4):45
Ritter, Michael (URS Corporation/USA) 6(2):239
Ritter, William F. (University of Delaware/USA) 6(6):9
Riva, Vanessa (Parsons Engineering Science, Inc./USA) 6(2):39
Rivas-Lucero, B.A. (Centro de Investigacion en Materiales Avanzados/MEXICO) 6(6):361
Rivetta, A. (Universita degli Studi di Milano/ITALY) 6(5):157
Robb, Joseph (ENSR International/USA) 6(1):75
Robertiello, Andrea (EniTecnologie S.p.A./ITALY) 6(6):173
Robertson, K. (Queen's University/CANADA) 6(2):121
Robinson, David (ERM, Inc./USA) 6(7):45
Robinson, Sandra L. (Virginia Polytechnic Institute & State University/USA) 6(5):1
Rockne, Karl J. (University of Illinois-Chicago/USA) 6(5):297
Rodríguez-Vázquez, Refugio (CINVESTAV-IPN/MEXICO) 6(3):243; 6(6):219, 369
Römkens, Paul (Alterra/THE NETHERLANDS) 6(5):137
Rongo, Rocco (University della Calabria/ITALY) 6(4):131
Roorda, Marcus L. (Rowan University/USA) 6(5):215
Rosser, Susan J. (University of Cambridge/UK) 6(5):69
Rowland, Martin A. (Lockheed-Martin Michoud Space Systems/USA) 6(7):1
Royer, Richard (The Pennsylvania State University/USA) 6(8):201

Ruggeri, Bernardo (Politecnico di Torino/ITALY) 6(3):211
Ruiz, Graciela M. (University of Iowa/USA) 6(1):195
Rupassara, S. Indumathie (University of Illinois at Urbana-Champaign/USA) 6(5):181

Sacchi, G.A. (Universita degli Studi di Milano/ITALY) 6(5):157
Sahagun, Tracy (U.S. Marine Corps./USA) 6(10):1
Sakakibara, Yutaka (Waseda University/JAPAN) 6(8):113; 6(9):201
Sakamoto, T. (Queen's University/CANADA) 6(10):171
Salam, Munazza (Crescent Heights High School/CANADA) 6(5):53
Salanitro, Joseph P. (Equilon Enterprises, LLC/USA) 6(1):11; 6(8):61
Salvador, Maria Cristina (CSD-GEOKLOCK/BRAZIL) 6(1):27
Samson, Réjean (École Polytechnique de Montréal/CANADA) 6(3):115; 6(4):139; 6(9):27
San Felipe, Zenaida (Monash University/AUSTRALIA) 6(4):1
Sánchez, F.N. (ECOPETROL-ICP/COLOMBIA) 6(6):319
Sánchez, Gisela (PDVSA Intevep/VENEZUELA) 6(6):257
Sánchez, Luis (PDVSA Intevep/VENEZUELA) 6(6):257
Sanchez, M. (Universidad Complutense de Madrid/SPAIN) 6(4):29; 6(6):377
Sandefur, Craig A. (Regenesis/USA) 6(7):257; 6(10):87
Sanford, Robert A. (University of Illinois at Urbana-Champaign/USA) 6(9):179
Santangelo-Dreiling, Theresa (Colorado Dept. of Transportation/USA) 6(10):231
Saran, Jennifer (Kennecott Utah Copper Corp./USA) 6(9):323
Sarpietro, M.G. (Universita di Catania/ITALY) 6(3):149

Sartoros, Catherine (Université du Québec à Montréal/CANADA) 6(3):165
Saucedo-Terán, R.A. (Centro de Investigacion en Materiales Avanzados/MEXICO) 6(6):361
Saunders, James A. (Auburn University/USA) 6(9):105
Sayler, Gary S. (University of Tennessee/USA) 6(4):45
Scalzi, Michael M. (Innovative Environmental Technologies, Inc./USA) 6(10):23
Scarborough, Shirley (IT Corporation/USA) 6(2):253
Schaffner, I. Richard (GZA GeoEnvironmental, Inc./USA) 6(7):165
Scharp, Richard A. (U.S. EPA/USA) 6(9):9
Schell, Heico (Water Technology Center/GERMANY) 6(8):105
Scherer, Michelle M. (The University of Iowa/USA) 6(3):1
Schipper, Mark (Groundwater Services) 6(8):73
Schmelling, Stephen (U.S. EPA/USA) 6(1):129
Schnoor, Jerald L. (University of Iowa/USA) 6(8):147
Schoefs, Olivier (École Polytechnique de Montréal/CANADA) 6(4):139
Schratzberger, Michaela (CEFAS Laboratory/UK) 6(10):337
Schulze, Susanne (Water Technology Center/GERMANY) 6(2):137
Schuur, Jessica H. (Lund University/SWEDEN) 6(6):65
Scrocchi, Susan (Conestoga-Rovers & Associates/USA) 6(1):99; 6(10):131
Sczechowski, Jeff (California Polytechnic State University/USA) 6(10):311
Seagren, Eric A. (University of Maryland/USA) 6(10):57
Sedran, Marie A. (University of Cincinnati/USA) 6(1):19
Seifert, Dorte (Technical University of Denmark/DENMARK) 6(2):11
Semer, Robin (Harza Engineering Company, Inc./USA) 6(7):157

Semprini, Lewis (Oregon State University/USA) 6(10):145, 155, 179
Seracuse, Joe (Harding ESE/USA) 6(7):205
Serra, Roberto (Centro Ricerche Ambientali/ITALY) 6(4):131
Sewell, Guy W. (U.S. EPA/USA) 6(1):167; 6(7):125; 6(8):167
Sharma, Pawan (Camp Dresser & McKee Inc./USA) 6(7):305
Sharp, Robert R. (Manhattan College/USA) 6(7):117
Shay, Devin T. (Groundwater & Environmental Services, Inc./USA) 6(10):101
Shelley, Michael L. (Air Force Institute of Technology/USA) 6(5):95
Shen, Hai (Dynamac Corporation/USA) 6(1): 129, 167
Sherman, Neil (Louisiana-Pacific Corporation/USA) 6(3):83
Sherwood Lollar, Barbara (University of Toronto/CANADA) 6(4):91, 109
Shi, Jing (EFX Systems, Inc./USA) 6(8):19
Shields, Adrian R.G. (Komex Europe/UK) 6(10):123
Shiffer, Shawn (University of Illinois/USA) 6(9):179
Shin, Won Sik (Lousiana State University/USA) 6(5):313
Shiohara, Kei (Mississippi State University/USA) 6(6):279
Shirazi, Fatemeh R. (Stratum Engineering Inc./USA) 6(8):121
Shoemaker, Christine (Cornell University/USA) 6(4):125
Sibbett, Bruce (IT Corporation/USA) 6(2):73
Silver, Cannon F. (Parsons Engineering Science, Inc./USA) 6(10):283
Silverman, Thomas S. (RMT, Inc./USA) 6(10):201
Simon, Michelle A. (U.S. EPA/USA) 6(10):293
Sims, Gerald K. (USDA-ARS/USA) 6(5):181
Sims, Ronald C. (Utah State University/USA) 6(4):67; 6(6):1
Sincock, M. Jennifer (ENVIRON International Corp./USA) 6(7):189

Author Index

Sittler, Steven P. (Advanced Pollution Technologists, Ltd./USA) *6*(2):215
Skladany, George J. (ERM, Inc./USA) *6*(7):45, 213
Skubal, Karen L. (Case Western Reserve University/USA) *6*(8):193
Slenders, Hans (TNO-MEP/THE NETHERLANDS) *6*(7):289
Slomczynski, David J. (University of Cincinnati/USA) *6*(2):247
Slusser, Thomas J. (Wright State University/USA) *6*(5):95
Smallbeck, Donald R. (Harding Lawson/USA) *6*(10):231
Smets, Barth F. (University of Connecticut/USA) *6*(7):69
Smith, Christy (North Carolina State University/USA) *6*(1):145
Smith, Colin C. (University of Sheffield/UK) *6*(2):207
Smith, John R. (Alcoa Inc./USA) *6*(6):89
Smith, Jonathan (The Environment Agency/UK) *6*(4):17
Smith, Steve (King's College London/UK) *6*(2):223; *6*(3):173; *6*(10):123
Smyth, David J.A. (University of Waterloo/CANADA) *6*(9):71
Sobecky, Patricia (Georgia Institute of Technology/USA) *6*(9):165
Sola, Adrianna (Spelman College/USA) *6*(9):53
Sordini, E. (EniTechnologie/ITALY) *6*(6):173
Sorensen, James A. (University of North Dakota/USA) *6*(6):141
Sorenson, Kent S. (Idaho National Engineering and Environmental Laboratory./USA) *6*(7):265
South, Daniel (Harding ESE/USA) *6*(7):205
Spain, Jim (U.S. Air Force/USA) *6*(3):59; *6*(7):125
Spasova, Irena Ilieva (University of Mining & Geology/BULGARIA) *6*(9):97
Spataro, William (University della Calabria/ITALY) *6*(4):131

Spinnler, Gerard E. (Equilon Enterprises, LLC/USA) *6*(1):11; *6*(8):61
Springael, Dirk (VITO/BELGIUM) *6*(4):35
Srinivasan, P. (GeoTrans, Inc./USA) *6*(2):163
Stansbery, Anita (California Polytechnic State University/USA) *6*(10):311
Starr, Mark G. (DuPont Co./USA) *6*(8):43
Stehmeier, Lester G. (NOVA Research Technology Centre/CANADA) *6*(4):117; *6*(5):53
Stensel, H. David (University of Washington/USA) *6*(10):211
Stordahl, Darrel M. (Camp Dresser & McKee Inc./USA) *6*(6):287
Stout, Scott (Battelle/USA) *6*(5):237
Strand, Stuart E. (University of Washington/USA) *6*(10):211
Stratton, Glenn (Nova Scotia Agricultural College/CANADA) *6*(3):197
Strybel, Dan (IT Corporation/USA) *6*(9):215
Stuetz, R.M. (Cranfield University/UK) *6*(6):329
Suarez, B. (ECOPETROL-ICP/COLOMBIA) *6*(6):319
Suidan, Makram T. (University of Cincinnati/USA) *6*(1):19; *6*(5):243, 253, 261; *6*(6):133,
Suthersan, Suthan S. (ARCADIS Geraghty & Miller/USA) *6*(7):53
Suzuki, Masahiro (Nippon Institute of Technology/JAPAN) *6*(5):321
Sveum, Per (Deconterra AS/NORWAY) *6*(6):157
Swallow, Ian (BAE Systems Properties Ltd./UK) *6*(6):119
Swann, Benjamin M. (Camp Dresser & McKee Inc./USA) *6*(7):305
Swannell, Richard P.J. (AEA Technology Environment/UK) *6*(10):337

Tabak, Henry H. (U.S. EPA/USA) *6*(5):243, 253, 261, 269; *6*(9):1, 17
Takai, Koji (Fuji Packing/JAPAN) *6*(10):345

Talley, Jeffrey W. (University of Notre Dame/USA) 6(3):189; 6(4):59; 6(6):43; 6(7):125; 6(10):115
Tao, Shu (Peking University/CHINA) 6(7):61
Taylor, Christine D. (North Carolina State University/USA) 6(1):83
Ter Meer, Jeroen (TNO Institute of Environmental Science/THE NETHERLANDS) 6(2):231; 6(7):289
Tétreault, Michel (Royal Military College of Canada/CANADA) 6(8):95
Tharpe, D.L. (Equilon Enterprises LLC/USA) 6(8):61
Theeuwen, J. (Grontmij BV/THE NETHERLANDS) 6(7):289
Thomas, Hartmut (WASAG DECON GMbH/GERMANY) 6(3):67
Thomas, Mark (EG&G Technical Services, Inc./USA) 6(10):49
Thomas, Paul R. (Thomas Consultants, Inc./USA) 6(5):189
Thomas, Robert C. (University of Georgia/USA) 6(9):105
Thomson, Michelle M. (URS Corporation/USA) 6(2):81
Thornton, Steven F. (University of Sheffield/UK) 6(1):59, 6(2):207
Tian, C. (University of Cincinnati/USA) 6(8):35
Tiedje, James M. (Michigan State University/USA) 6(7):125; 6(8):19
Tiehm, Andreas (Water Technology Center/GERMANY) 6(2):137; 6(8):105
Tietje, David (Foster Wheeler Environmental Corportation/USA) 6(9):249
Timmins, Brian (Oregon State University/USA) 6(10):179
Togna, A. Paul (Envirogen Inc/USA) 6(9):281
Tolbert, David E.(U.S. Army/USA) 6(9):281
Tonnaer, Haimo (TAUW BV/THE NETHERLANDS) 6(7):297; 6(10):253
Toth, Brad (Harding ESE/USA) 6(10):231

Tovanabootr, Adisorn (Oregon State University/USA) 6(10):145
Travis, Bryan (Los Alamos National Laboratory/USA) 6(10):163
Trudnowski, John M. (MSE Technology Applications, Inc./USA) 6(9):35
Truax, Dennis D. (Mississippi State University/USA) 6(9):241
Trute, Mary M. (Camp Dresser & McKee, Inc./USA) 6(2):113
Tsuji, Hirokazu (Obayashi Corporation Ltd./JAPAN) 6(6):111, 249; 6(10):239
Tsutsumi, Hiroaki (Prefectural University of Kumamoto/JAPAN) 6(10):345
Turner, Tim (CDM Federal Programs Corp./USA) 6(6):81
Turner, Xandra (International Biochemicals Group/USA) 6(10):23
Tyner, Larry (IT Corporation/USA) 6(1):51; 6(2):73

Ugolini, Nick (U.S. Navy/USA) 6(10):65
Uhler, Richard (Battelle/USA) 6(5):237
Unz, Richard F. (The Pennsylvania State University/USA) 6(8):201
Utgikar, Vivek P. (U.S. EPA/USA) 6(9):17

Valderrama, Brenda (Universidad Nacional Autónoma de México/MEXICO) 6(6):17
Vallini, Giovanni (Universita degli Studi di Verona/ITALY) 6(3):267
van Bavel, Bert (Umeå University/SWEDEN) 6(3):181
van Breukelen, Boris M. (Vrije University/THE NETHERLANDS) 6(4):91
VanBroekhoven, K. (Catholic University of Leuven/BELGIUM) 6(4):35
Vandecasteele, Jean-Paul (Institut Français du Pétrole/FRANCE) 6(3):227
VanDelft, Frank (NOVA Chemicals/CANADA) 6(5):53
van der Gun, Johan (BodemBeheer bv/THE NETHERLANDS) 6(5):289

van der Werf, A. W. (Bioclear Environmental Technology/THE NETHERLANDS) 6(8):11
van Eekert, Miriam (TNO Environmental Sciences /THE NETHERLANDS) 6(2):231; 6(7):289
Van Hout, Amy H. (IT Corporation/USA) 6(3):35
Van Keulen, E. (DHV Environment and Infrastructure/THE NETHERLANDS) 6(8):11
Vargas, M.C. (ECOPETROL-ICP/COLOMBIA) 6(6):319
Vazquez-Duhalt, Rafael (Universidad Nacional Autónoma de México/MEXICO) 6(6):17
Venosa, Albert (U.S. EPA/USA) 6(1):19
Verhaagen, P. (Grontmij BV/THE NETHERLANDS) 6(7):289
Verheij, T. (DAF/THE NETHERLANDS) 6(7):289
Vidumsky, John E. (E.I. du Pont de Nemours & Company/USA) 6(2):81; 6(8):185
Villani, Marco (Centro Ricerche Ambientali/ITALY) 6(4):131
Vinnai, Louise (Investigative Science Inc./CANADA) 6(2):27
Visscher, Gerolf (Province of Groningen/THE NETHERLANDS) 6(7):141
Voegeli, Vincent (TranSystems Corporation/USA) 6(7):229
Vogt, Bob (Louisiana-Pacific Corporation/USA) 6(3):83
Volkering, Frank (TAUW bv/THE NETHERLANDS) 6(4):91
von Arb, Michelle (University of Iowa) 6(3):1
Vondracek, James E. (Ashland Inc./USA) 6(5):121
Vos, Johan (VITO/BELGIUM) 6(9):87
Voscott, Hoa T. (Camp Dresser & McKee, Inc./USA) 6(7):305
Vough, Lester R. (University of Maryland/USA) 6(5):77

Waisner, Scott A. (TA Environmental, Inc./USA) 6(4):59; 6(10):115

Walecka-Hutchison, Claudia M. (University of Arizona/USA) 6(9):231
Wall, Caroline (CEFAS Laboratory/UK) 6(10):337
Wallace, Steve (Lattice Property Holdings Plc./UK) 6(4):17
Wallis, F.M. (University of Natal/REP OF SOUTH AFRICA) 6(6):101; 6(9):79
Walton, Michelle R. (Idaho National Engineering & Environmental Laboratory/USA) 6(7):77
Walworth, James L. (University of Arizona/USA) 6(9):231
Wan, C.K. (Hong Kong Baptist University/CHINA) 6(6):73
Wang, Chuanyue (Rice University/USA) 6(5):85
Wang, Qingren (Chinese Academy of Sciences/CHINA [PRC]) 6(9):113
Wani, Altaf (Applied Research Associates, Inc./USA) 6(10):115
Wanty, Duane A. (The Gillette Company/USA) 6(7):87
Warburton, Joseph M. (Parsons Engineering Science/USA) 6(7):173
Watanabe, Masataka (National Institute for Environmental Studies/JAPAN) 6(5):321
Watson, James H.P. (University of Southampton/UK) 6(9):61
Wealthall, Gary P. (University of Sheffield/UK) 6(1):59
Weathers, Lenly J. (Tennessee Technological University/USA) 6(8):139
Weaver, Dallas E. (Scientific Hatcheries/USA) 6(1):91
Weaverling, Paul (Harding ESE/USA) 6(10):231
Weber, A. Scott (State University of New York at Buffalo/USA) 6(6):89
Weeber, Philip A. (Geotrans/USA) 6(10):163
Wendt-Potthoff, Katrin (UFZ Centre for Environmental Research/GERMANY) 6(9):43
Werner, Peter (Technical University of Dresden/GERMANY) 6(3):227; 6(8):105

West, Robert J. (The Dow Chemical Company/USA) 6(2):89
Westerberg, Karolina (Stockholm University/SWEDEN) 6(3):133
Weston, Alan F. (Conestoga-Rovers & Associates/USA) 6(1):99; 6(10):131
Westray, Mark (ThermoRetec Corp/USA) 6(7):1
Wheater, H.S. (Imperial College of Science and Technology/UK) 6(10):123
White, David C. (University of Tennessee/USA) 6(4):73; 6(5):305
White, Richard (EarthFax Engineering Inc/USA) 6(6):263
Whitmer, Jill M. (GeoSyntec Consultants/USA) 6(9):105
Wick, Lukas Y. (Swiss Federal Institute of Technology/SWITZERLAND) 6(3):251
Wickramanayake, Godage B. (Battelle/USA) 6(10):1
Widada, Jaka (The University of Tokyo/JAPAN) 6(4):51
Widdowson, Mark A. (Virginia Polytechnic Institute & State University/USA) 6(2):105; 6(5):1
Wieck, James M. (GZA GeoEnvironmental, Inc./USA) 6(7):165
Wiedemeier, Todd H. (Parsons Engineering Science, Inc./USA) 6(7):241
Wiessner, Arndt (UFZ - Centre for Environmental Research/GERMANY) 6(5):337
Wilken, Jon (Harding ESE/USA) 6(10):231
Williams, Lakesha (Southern University at New Orleans/USA) 6(5):145
Williamson, Travis (Battelle/USA) 6(10):245
Willis, Matthew B. (Cornell University/USA) 6(4):125
Willumsen, Pia Arentsen (National Environmental Research Institute/DENMARK) 6(3):141
Wilson, Barbara H. (Dynamac Corporation/USA) 6(1):129
Wilson, Gregory J. (University of Cincinnati/USA) 6(1):19

Wilson, John T. (U.S. EPA/USA) 6(1):43, 167
Wiseman, Lee (Camp Dresser & McKee Inc./USA) 6(7):133
Wisniewski, H.L. (Equilon Enterprises LLC/USA) 6(8):61
Witt, Michael E. (The Dow Chemical Company/USA) 6(2):89
Wong, Edwina K. (University of Guelph/CANADA) 6(6):185
Wong, J.W.C. (Hong Kong Baptist University/CHINA) 6(6):73
Wood, Thomas K. (University of Connecticut/USA) 6(5):199
Wrobel, John (U.S. Army/USA) 6(5):207

Xella, Claudio (Water & Soil Remediation S.r.l./ITALY) 6(6):179
Xing, Jian (Global Remediation Technologies, Inc./USA) 6(6):311

Yamamoto, Isao (Sumitomo Marine Research Institute/JAPAN) 6(10):345
Yamazaki, Fumio (Hyogo Prefectural Institute of Environmental Science/JAPAN) 6(5):321
Yang, Jeff (URS Corporation/USA) 6(2):239
Yerushalmi, Laleh (Biotechnology Research Institute/CANADA) 6(3):165
Yoon, Woong-Sang (Sam) (Battelle/USA) 6(7):13
Yoshida, Takako (The University of Tokyo/JAPAN) 6(4):51; 6(6):111
Yotsumoto, Mizuyo (Obayashi Corporation Ltd./JAPAN) 6(6):111
Young, Harold C. (Air Force Institute of Technology/USA) 6(2):173

Zagury, Gérald J. (École Polytechnique de Montréal/CANADA) 6(9): 27, 129
Zahiraleslamzadeh, Zahra (FMC Corporation/USA) 6(7):221
Zaluski, Marek H. (MSE Technology Applications/USA) 6(9):35
Zappi, Mark E. (Mississippi State University/USA) 6(9):241

Zelennikova, Olga (University of Connecticut/USA) 6(7):69

Zhang, Chuanlun L. (University of Missouri/USA) 6(9):165

Zhang, Wei (Cornell University/USA) 6(4):125

Zhang, Zhong (University of Nevada Las Vegas/USA) 6(9):257

Zheng, Zuoping (University of Oslo/NORWAY) 6(2):181

Zocca, Chiara (Universita degli Studi di Verona/ITALY) 6(3):267

Zwick, Thomas C. (Battelle/USA) 6(10):1

KEYWORD INDEX

This index contains keyword terms assigned to the articles in the ten-volume proceedings of the Sixth International In Situ and On-Site Bioremediation Symposium (San Diego, California, June 4-7, 2001). Ordering information is provided on the back cover of this book.

In assigning the terms that appear in this index, no attempt was made to reference all subjects addressed. Instead, terms were assigned to each article to reflect the primary topics covered by that article. Authors' suggestions were taken into consideration and expanded or revised as necessary. The citations reference the ten volumes as follows:

6(1):   Magar, V.S., J.T. Gibbs, K.T. O'Reilly, M.R. Hyman, and A. Leeson (Eds.), *Bioremediation of MTBE, Alcohols, and Ethers*. Battelle Press, Columbus, OH, 2001. 249 pp.
6(2):   Leeson, A., M.E. Kelley, H.S. Rifai, and V.S. Magar (Eds.), *Natural Attenuation of Environmental Contaminants*. Battelle Press, Columbus, OH, 2001. 307 pp.
6(3):   Magar, V.S., G. Johnson, S.K. Ong, and A. Leeson (Eds.), *Bioremediation of Energetics, Phenolics, and Polycyclic Aromatic Hydrocarbons*. Battelle Press, Columbus, OH, 2001. 313 pp.
6(4):   Magar, V.S., T.M. Vogel, C.M. Aelion, and A. Leeson (Eds.), *Innovative Methods in Support of Bioremediation*. Battelle Press, Columbus, OH, 2001. 197 pp.
6(5):   Leeson, A., E.A. Foote, M.K. Banks, and V.S. Magar (Eds.), *Phytoremediation, Wetlands, and Sediments*. Battelle Press, Columbus, OH, 2001. 383 pp.
6(6):   Magar, V.S., F.M. von Fahnestock, and A. Leeson (Eds.), *Ex Situ Biological Treatment Technologies*. Battelle Press, Columbus, OH, 2001. 423 pp.
6(7):   Magar, V.S., D.E. Fennell, J.J. Morse, B.C. Alleman, and A. Leeson (Eds.), *Anaerobic Degradation of Chlorinated Solvents*. Battelle Press, Columbus, OH, 2001. 387 pp.
6(8):   Leeson, A., B.C. Alleman, P.J. Alvarez, and V.S. Magar (Eds.), *Bioaugmentation, Biobarriers, and Biogeochemistry*. Battelle Press, Columbus, OH, 2001. 255 pp.
6(9):   Leeson, A., B.M. Peyton, J.L. Means, and V.S. Magar (Eds.), *Bioremediation of Inorganic Compounds*. Battelle Press, Columbus, OH, 2001. 377 pp.
6(10): Leeson, A., P.C. Johnson, R.E. Hinchee, L. Semprini, and V.S. Magar (Eds.), *In Situ Aeration and Aerobic Remediation*. Battelle Press, Columbus, OH, 2001. 391 pp.

A

abiotic/biotic dechlorination 6(8):193
acenaphthene 6(5):253
acetate as electron donor 6(3):51; 6(9):297
acetone 6(2):49
acid mine drainage, (*see also* mine tailings) 6(9):1, 9, 27, 35, 43, 53
acrylic vessel 6(5):321
actinomycetes 6(10):211
activated carbon biomass carrier 6(6):311; 6(8):113

activated carbon **6(8)**:105
adsorption **6(3)**:243; **6(5)**:253; **6(6)**:377; **6(7)**:77; **6(8)**:131; **6(9)**:86
advanced oxidation **6(1)**:121; **6(10)**:33
aerated submerged **6(10)**:329
aeration **6(6)**:203
anaerobic/aerobic treatment **6(6)**:361; **6(7)**:229
age dating **6(5)**:231, 237
air sparging **6(1)**:115, 175; **6(2)**:239; **6(9)**:215; **6(10)**:1, 9, 41, 49, 65, 101, 115, 123, 163, 223
alachlor **6(6)**:9
algae **6(5)**:181
alkaline phosphatase **6(9)**:165
alkane degradation **6(5)**:313
alkylaromatic compounds **6(6)**:173
alkylbenzene **6(2)**:19
alkylphenolethoxylate **6(5)**:215
Amaranthaceae **6(5)**:165
Ames test **6(6)**:249
ammonia **6(1)**:175; **6(5)**:337
amphipod toxicity test **6(5)**:321
anaerobic **6(1)**:35, 43; **6(3)**:91; 205; **6(5)**:17, 25, 261, 297, 313; **6(6)**:133; **6(7)**:249, 297; **6(9)**:147, 303
anaerobic biodegradation **6(1)**:137; **6(5)**:1; **6(8)**:167
anaerobic bioventing **6(3)**:9
anaerobic petroleum degradation **6(5)**:25
anaerobic sparging **6(7)**:297
aniline **6(6)**:149
Antarctica **6(2)**:57
anthracene **6(3)**:165, 251; **6(6)**:73
aquatic plants **6(5)**:181
arid-region soils **6(9)**:231
aromatic dyes **6(6)**:369
arsenic **6(2)**:239, 261; **6(5)**:173; **6(9)**: 97, 129
atrazine **6(5)**:181; **6(6)**:9
azoaromatic compounds **6(6)**:149
Azomonas **6(6)**:219

B

bacterial transport **6(8)**:1
barrier technologies **6(1)**:11; **6(3)**:165; **6(7)**:289; **6(8)**:61, 79, 87, 105, 121; **6(9)**:27, 71, 195, 209, 309
basidiomycete **6(6)**:101
benthic **6(10)**:337

benzene **6(1)**:1, 67, 75, 145, 167, 203; **6(4)**:91,117; **6(8)**:87; **6(10)**:123
benzene, toluene, ethylbenzene, and xylenes (BTEX) **6(1)**:43, 51, 59, 107, 129, 167, 195; **6(2)**:11, 19, 137, 215, 223, 270; **6(4)**:99; **6(5)**:33; **6(7)**:133; **6(8)**:105; **6(10)**: 1, 23, 49, 65, 95, 123, 131
benzo(a)pyrene **6(3)**:149; **6(6)**:101
benzo(e)pyrene **6(3)**:149
BER, *see* biofilm-electrode reactor
bioassays **6(3)**:219
bioaugmentation **6(1)**:11; **6(3)**:133; **6(4)**:59; **6(6)**:9, 43, 111; **6(7)**:125; **6(8)**:1, 11, 19, 27, 43, 53, 61, 147, 175
bioavailability **6(3)**:115, 157, 173, 189, 51; **6(4)**:7; **6(5)**:253, 279, 289; **6(6)**:1
bioavailable FeIII assay **6(8)**:209
biobarrier **6(1)**:11; **6(3)**:165; **6(7)**:289; **6(8)**:61, 79, 105, 121; **6(9)**:27, 71, 209, 309
BIOCHLOR model **6(2)**:155
biocide **6(7)**:321, 333
biodegradability **6(6)**:193
biodegradation **6(1)**:19,153; **6(3)**:165, 181, 205, 235; **6(10)**:187
biofilm **6(3)**:251; **6(4)**:149; **6(8)**:79; **6(9)**:201, 303
biofilm-electrode reactor (BER) **6(9)**:201
biofiltration **6(4)**:149
biofouling **6(7)**:321, 333
bioindicators **6(1)**:1; **6(3)**:173; **6(5)**:223
biological carbon regeneration **6(8)**:105
bioluminescence **6(1)**:1; **6(3)**:173; **6(4)**:45
biopile **6(6)**:81, 127, 141, 227, 249, 287
bioreactors **6(1)**:91; **6(6)**:361; **6(8)**:11, 35; **6(9)**:1, 265, 281, 303, 315; **6(10)**:171, 211
biorecovery of metals **6(9)**:9
bioreporters **6(4)**:45
biosensors **6(1)**:1
bioslurping **6(10)**:245, 253, 267, 275
bioslurry and bioslurry reactors **6(3)**:189; **6(6)**:51, 65
biosparging **6(10)**:115, 163
biostabilization **6(6)**:89
biostimulation **6(6)**:43
biosurfactant **6(3)**:243; **6(7)**:53
bioventing **6(10)**:109, 115, 131
biphasic reactor **6(3)**:181

Keyword Index

biological oxygen demand (BOD) 6(10):311
BTEX, *see* benzene, toluene, ethylbenzene, and xylenes
Burkholderia cepacia 6(1):153; 6(7):117; 6(8):53
butane 6(1):137, 161
butyrate 6(7):289

C

cadmium 6(3):91; 6(9):79, 147
carAa, see carbazole 1,9a-dioxygenase gene
carbazole-degrading bacterium 6(6):111
carbazole 1,9a-dioxygenase gene (*carAa*) 6(4):51
Carbokalk 6(9):43
carbon isotope 6(4):91, 99, 109, 117; 6(10):115
carbon tetrachloride (CT) 6(2):81, 89; 6(5):113; 6(7):241; 6(8):185, 193
cesium-137 6(5):231
CF, *see* chloroform
charged coupled device camera 6(2):207
chelators addition (EDGA, EDTA) 6(5):129, 137, 145, 151; 6(9):123, 147
chemical oxidation 6(7):45
chicken manure 6(9):289
chlorinated ethenes 6(7):27, 61, 69, 109; 6(10):163, 201, 231
chlorinated solvents 6(2):145; 6(7):all; 6(8):19; 6(10):231
chlorobenzene 6(8):105
chloroethane 6(2):113; 6(7):133, 249
chloroform (CF) 6(2):81; 6(8):193
chloromethanes 6(8):185
chlorophenol 6(3):75, 133
chlorophyll fluorescence 6(5):223
chromated copper arsenate 6(9):129
chromium (Cr[VI]) 6(8):139, 147; 6(9):129, 139, 315
chrysene 6(6):101
citrate and citric acid 6(5):137; 6(7):289
cleanup levels 6(6):1
coextraction method 6(4):51
Coke Facility waste 6(2):129
combined chemical toxicity (*see also* toxicity) 6(5):305
cometabolic air sparging 6(10):145, 155, 223

cometabolism 6(1):137, 145, 153, 161; 6(2):19; 6(6):81, 141; 6(7):117; 6(10):145, 155, 163, 171, 179, 193, 201, 211, 217, 223, 231; 239
competitive inhibition 6(2):19
composting 6(3):83; 6(5):129, 6(6):73, 119, 165, 257; 6(7):141
constructed wetlands 6(5):173, 329
contaminant aging 6(3):157, 197
contaminant transport 6(3):115
copper 6(9):79, 129
cosolvent effects 6(1):175, 195, 203, 243
cosolvent extraction 6(7):125
cost analyses and economics of environmental restoration 6(1):129; 6(4):17; 6(8):121; 6(9):331; 6(10):65, 211
Cr(VI), *see* chromium
creosote 6(3):259; 6(4):59; 6(5):1, 237, 329; 6(6):81, 101, 141, 295
cresols 6(10):123
crude oil 6(5):313; 6(6):193, 249; 6(10):329
CT, *see* carbon tetrachloride
cyanide 6(9):331
cytochrome P-450 6(6):17

D

2,4-DAT, *see* diaminotoluene
DCA, *see* dichloroethane
1,1-DCA, *see* 1,1-dichloroethane
1,2-DCA, *see* 1,2-dichloroethane
DCE, *see* dichloroethene
1,1-DCE, *see* 1,1-dichloroethene
1,2-DCE, *see* 1,2-dichloroethene
c-DCE, *see* cis-dichloroethene
DCM, *see* dichloromethane
DDT, *see also* dioxins *and* pesticides 6(6):157
2,4-DNT, *see* dinitrotoluene
dechlorination kinetics 6(2):105; 6(7):61
dechlorination 6(2):231; 6(3):125; 6(5):95; 6(7):13, 61, 165, 173, 333; 6(8):19, 27, 43
DEE, *see* diethyl ether
Dehalococcoides ethenogenes 6(8):19, 43
dehalogenation 6(8):167
denaturing gradient gel electrophoresis (DGGE) 6(1):19; 6(4):35

denitrification **6(2)**:19; **6(4)**:149; **6(5)**:17, 261; **6(8)**:95; **6(9)**:179, 187, 195, 201, 209, 223, 309
dense, nonaqueous-phase liquid (DNAPL) **6(7)**:13, 19, 35, 181; **6(10)**:319
depletion rate **6(1)**:67
desorption **6(3)**:235, 243; **6(5)**:253; **6(6)**:377; **6(7)**:53, 77; **6(8)**:131
DGGE, *see* denaturing gradient gel electrophoresis
DHPA, *see* dihydroxyphenylacetate
dialysis sampler **6(5)**:207
diaminotoluene (2,4-DAT) **6(6)**:149
dibenzofuran-degrading bacterium **6(6)**:111
dibenzo-p-dioxin **6(6)**:111
dibenzothiophene **6(3)**:267
dichlorodiethyl ether **6(10)**:301
dichloroethane (DCA) **6(2)**:39; **6(7)**:289
1,1-dichloroethane (1,1-DCA; 1,2-DCA) **6(2)**:113; **6(5)**:207; **6(7)**:133, 165
1,2-dichloroethane (1,2-DCA) **6(5)**:207
dichloroethene, dichloroethylene **6(2)**:97, 155; **6(4)**:125; **6(5)**:105,113; **6(7)**:157, 197
cis-dichloroethene, *cis*-dichloroethylene (*c*-DCE) **6(2)**:39, 65, 73; 105, 173; **6(5)**:33, 95, 207; **6(7)**:1, 13, 61, 133, 141, 149, 165, 173, 181, 189, 205, 213, 221, 249, 273, 281, 289, 297, 305; **6(8)**:11, 19, 27, 43, 73, 105, 157, 209; **6(10)**:41, 145, 155, 179, 201
1,1-dichloroethene, 1,1-dichloroethylene (1,1-DCE) **6(2)**:39; **6(7)**:165, 229; **6(8)**:157; **6(10)**:231
1,2-dichloroethene and 1,2-dichloroethylene (1,2-DCE) **6(2)**:113
dichloromethane (DCM) **6(2)**:81; **6(8)**:185
diesel fuel **6(1)**:175; **6(2)**:57; **6(5)**:305; **6(6)**:81, 141, 165; **6(10)**:9
diesel-range organics (DRO) **6(10)**:9
diethyl ether (DEE) **6(1)**:19
dihydroxyphenylacetate (DHPA) **6(4)**:29
diisopropyl ether (DIPE) **6(1)**:19, 161
1,3-dinitro-5-nitroso-1,3,5-triazacyclohexane (MNX) (*see also* explosives *and* energetics) **6(3)**:51; **6(8)**:175
dinitrotoluene (2,4-DNT) **6(3)**:25, 59; **6(6)**:127, 149
dioxins **6(6)**:111

DIPE, *see* diisopropyl ether
dissolved oxygen **6(2)**:189, 207
16S rDNA sequencing **6(8)**:19
DNAPL, *see* dense, nonaqueous-phase liquid
DNX, *see* explosives and energetics
DRO, *see* diesel-range organics
dual porosity aquifer **6(1)**:59
dyes **6(6)**:369

E

ecological risk assessment **6(4)**:1
ecotoxicity, (*see also* toxicity) **6(1)**:1; **6(4)**:7
ethylenedibromide (EDB) **6(10)**:65
EDGA, *see* chelate addition
EDTA, *see* chelate addition
effluent **6(4)**:1
electrokinetics **6(9)**:241, 273
electron acceptors and electron acceptor processes **6(2)**:1, 137, 163, 231; **6(5)**:17, 25, 297; **6(7)**:19
electron donor amendment **6(3)**:25, 35, 51, 125; **6(7)**:69, 103,109, 141, 181, 249, 289, 297; **6(8)**:73; **6(9)**:297, 315
electron donor delivery **6(7)**:19, 27, 133, 173, 213, 221, 265, 273, 281, 305
electron donor mass balance **6(2)**:163
electron donor transport **6(4)**:125; **6(7)**:133; **6(9)**:241
embedded carrier **6(9)**:187
encapsulated bacteria **6(5)**:269
enhanced aeration **6(10)**:57
enhanced desorption **6(7)**:197
environmental stressors **6(4)**:1
enzyme induction **6(6)**:9; **6(10)**:211
ERIC sequences **6(4)**:29
ethane **6(2)**:113; **6(7)**:149
ethanol 6(1):19,167,175, 195, 203; **6(5)**:243; **6(6)**:133; **6(9)**:289
ethene and ethylene **6(2)**:105,113; **6(5)**:95; **6(7)**:1, 95, 133, 141, 205, 281, 297, 305; **6(8)**:11, 43, 167, 175, 209
ethylene dibromide **6(10)**:193
explosives and energetics **6(3)**:9, 17, 25, 35, 43, 51, 67; **6(5)**:69; **6(6)**:119, 127, 133; **6(7)**:125

F

fatty acids **6(5)**:41
Fe(II), *see* iron
Fenton's reagent **6(6)**:157
fertilizer **6(5)**:321; **6(6)**:35; **6(10)**:337
fixed-bed and fixed-film reactors
   6(5):221, 337; **6(6)**:361; **6(9)**:303
flocculants **6(6)**:279
flow sensor **6(10)**:293
fluidized-bed reactor **6(1)**:91; **6(6)**:133, 311; **6(9)**:281
fluoranthene **6(3)**:141; **6(6)**:101
fluorogenic probes **6(4)**:51
food safety **6(9)**:113
formaldehyde **6(6)**:329
fractured shale **6(10)**:49
free-product recovery **6(6)**:211
Freon **6(2)**:49
fuel oil **6(5)**:321
fungal remediation **6(3)**:75, 99; **6(5)**:61, 279; **6(6)**:17, 101, 157, 263, 319, 329, 369
Funnel-and-Gate™ **6(8)**:95

G

gas flux **6(6)**:185
gasoline **6(1)**:35, 75, 161, 167, 195; **6(10)**:115
gasoline-range organics (GRO) **6(10)**:9
manufactured gas plants and gasworks **6(2)**:137; **6(10)**:123
GCW, *see* groundwater circulating well
gel-encapsulated biomass **6(8)**:35
GEM, see genetically engineered microorganisms
genetically engineered microorganisms (GEM) **6(4)**:45; **6(5)**:199; **6(7)**:125
genotoxicity, (*see also* toxicity) **6(3)**:227
Geobacter **6(3)**:1
geochemical characterization **6(4)**:91
geographic information system (GIS) **6(2)**:163
geologic heterogeneity **6(2)**:11
germination index 6(3):219; **6(6)**:73
GFP, *see* green fluorescent protein
GIS, *see* geographic information system
glutaric dialdehyde dehydrogenase **6(4)**:81
Gordonia terrae **6(1)**:153
green fluorescent protein (GFP) **6(5)**:199
GRO, see gasoline-range organics

groundwater **6(3)**:35; **6(8)**: 35, 87, 121; **6(10)**:231
groundwater circulating well (GCW) **6(7)**:229, 321; **6(10)**:283, 293

H

H_2 gas, *see* hydrogen
H_2S, *see* hydrogen sulfide
halogenated hydrocarbons **6(9)**:61
halorespiration **6(8)**:19
heavy metal **6(2)**:239; **6(5)**:137, 145, 157, 165, 173; **6(6)**:51; **6(9)**:53, 61, 71, 79, 86, 97, 113, 129, 147
herbicides **6(5)**:223; **6(6)**:35
hexachlorobenzene **6(3)**:99
hexane **6(3)**:181, **6(6)**:329
HMX, *see* explosives and energetics
hollow fiber membranes **6(5)**:269
hopane **6(6)**:193; **6(10)**:337
hornwort **6(5)**:181
HRC® (a proprietary hydrogen-release compound) **6(3)**:17, 25, 107; **6(7)**:27, 103, 157, 189, 197, 205, 221, 257, 305, **6(8)**:157, 209
^{2}H-tetradecane (*see also* tetradecane) **6(2)**:27
humates **6(1)**:99
hybrid treatment **6(10)**:311
hydraulic containment **6(8)**:79
hydraulically facilitated remediation **6(2)**:239
hydrocarbon **6(6)**:235; **6(10)**:329
hydrogen (H_2 gas) **6(2)**:199; **6(9)**:201
hydrogen injection, in situ **6(7)**:19
hydrogen isotope **6(4)**:91
hydrogen peroxide **6(1)**:121; **6(6)**:353; **6(10)**:33
hydrogen release compound, *see* HRC®
hydrogen sulfide (H_2S) **6(9)**:123
hydrogen **6(2)**:231, **6(7)**:61, 305
hydrolysis **6(1)**:83
hydrophobicity **6(3)**:141
hydroxyl radical **6(1)**:121
hydroxylamino TNT intermediates **6(5)**:85

I

immobilization **6(8)**:53
immobilized cells **6(8)**:121
immobilized soil bioreactor **6(10)**:171

in situ oxidation **6(7)**:1
industrial effluents **6(6)**:303, 361
inhibition **6(9)**:17
injection strategies, in situ **6(7)**:19, 133, 173, 213, 221, 265, 273, 305, 313; **6(9)**:223; **6(10)**:23, 163
insecticides **6(6)**:27
intrinsic biodegradation **6(2)**:89, 121
intrinsic remediation, *see* natural attenuation
ion migration **6(9)**:241
iron (Fe[II]) **6(5)**:1
iron barrier **6(8)**:139, 147, 157, 167
iron oxide **6(3)**:1
iron precipitation **6(3)**:211
iron-reducing processes **6(2)**:121; **6(3)**:1; **6(5)**:1, 17, 25; **6(6)**:149; **6(8)**:193, 201, 209; **6(9)**:43, 323
IR-spectroscopy **6(4)**:67
isotope analyses **6(2)**:27; **6(4)**:91; **6(8)**:27
isotope fractionation **6(4)**:99, 109, 117

J

jet fuel **6(10)**:95, 139

K

KB-1 strain **6(8)**:27
kerosene **6(6)**:219
kinetics **6(8)**:131, **6(1)**:1, 19, 27, 167; **6(2)**:11, 19, 105; **6(3)**:173; **6(4)**:131; **6(7)**:61
Klebsiella oxytoca **6(7)**:117
Kuwait **6(6)**:249

L

laccase **6(3)**:75; **6(6)**:319
lactate and lactic acid **6(7)**:103, 109, 165, 181, 213, 265, 281, 289; **6(8)**:139; **6(9)**:155, 273
lagoons **6(6)**:303
land treatment units (LTU) **6(6)**:1; **6(6)**:81, 141, 287, 295
landfarming **6(3)**:259; **6(4)**:59; **6(5)**:53, 279; **6(6)**:1, 43, 59, 179, 203, 211, 235
landfills **6(2)**:145, 247; **6(4)**:91; **6(8)**:113
leaching **6(9)**:187
lead **6(5)**:129, 145, 151, 157

lead-210 **6(5)**:231
light, nonaqueous-phase liquids (LNAPL) **6(1)**:59; **6(4)**:35; **6(10)**:57, 109, 245, 253, 275
lindane, *(see also* pesticides) **6(5)**:189
linuron *(see also* herbicides) **6(5)**:223
LNAPL, *see* light, nonaqueous-phase liquids
Lolium multiflorum **6(5)**:9
LTU, *see* land treatment units
lubricating oil **6(6)**:173
luciferase **6(3)**:133
lux **6(4)**:45

M

mackinawite **6(9)**:155
macrofauna **6(10)**:337
magnetic separation **6(9)**:61
magnetite **6(3)**:1; **6(8)**:193
manganese **6(2)**:261
manufactured gas plant (MGP) **6(2)**:19; **6(3)**:211, 227; **6(10)**:123
mass balance **6(2)**:163
mass transfer limitation **6(3)**:157
mass transfer **6(1)**:67
MC-100, *see* mixed culture
media development **6(9)**:147
Meiofauna **6(5)**:305; **6(10)**:337
membrane **6(5)**:269; **6(9)**:1, 265
metabolites **6(3)**:227
metal reduction **6(8)**:1
metal precipitation **6(9)**:9, 165
metals, biorecovery of **6(9)**:9
metals speciation **6(9)**:129
metal toxicity *(see also* toxicity) **6(9)**:17, 129
metals **6(5)**:129, 305; **6(8)**:1; **6(9)**:9, 17, 27, 105, 123, 129, 155, 165
methane oxidation **6(10)**:171, 187, 193, 201, 223, 231
methane **6(1)**:183; **6(8)**:113
methanogenesis **6(1)**:35, 43, 183; **6(3)**:205; **6(9)**:147
methanogens **6(3)**:91
methanol **6(1)**:183; **6(7)**:141, 289, 297
methanotrophs **6(10)**:171, 187, 201
methylene chloride **6(2)**:39; **6(10)**:231
Methylosinus trichosporium **6(10)**:187
methyl *tert*-butyl ether *or* methyl *tertiary*-butyl ether (MTBE) **6(1)**:1, 11, 19, 27, 35, 43, 51, 59, 67, 75, 83, 91, 107,

115, 121, 129, 137, 145, 153,161, 195, *6(2)*:215; *6(8)*:61; *6(10)*:1, 65
MGP, *see* manufactured gas plant
microbial heterogeneity *6(4)*:73
microbial isolation *6(3)*:267
microbial population dynamics *6(4)*:35
microbial regrowth *6(2)*:253; *6(7)*:1, 13; *6(10)*:319
microcosm studies *6(7)*:109; *6(10)*:179
microencapsulation *6(8)*:53
microfiltration *6(9)*:201
microporous membrane *6(9)*:265
microtox assay *6(3)*:227
mine tailings (*see also* acid mine drainage) *6(5)*:173; *6(9)*:27, 71
mineral oil *6(5)*:279, 289; *6(6)*:59
mineralization *6(2)*:121; *6(3)*:165; *6(6)*:165; *6(8)*:175; *6(9)*:139, 155
MIP, *see* membrane interface probe
mixed culture *6(8)*:61
mixed wastes *6(3)*:91; *6(7)*:133; *6(9)*:139
MNX, *see* 1,3-dinitro-5-nitroso-1,3,5-triazacyclohexane
modeling *6(1)*:51; *6(2)*:105, 155, 181, *6(4)*:125, 131, 139, 149; *6(6)*:339, 377; *6(8)*:185; *6(9)*:27, 105; *6(10)*:163
moisture content *6(2)*:247
molasses as electron donor *6(3)*:35; *6(7)*:53, 103, 149, 173; *6(9)*:315
monitored natural attenuation (*see also* natural attenuation) *6(1)*:183, *6(2)*:11, 163, 199, 223, 253, 261
monitoring techniques *6(2)*:27,189, 199, 207; *6(4)*:59
motor oil *6(5)*:53
MPE, *see* multiphase extraction
multiphase extraction (MPE) well design *6(10)*:245, 259
MTBE, *see* methyl *tert*-butyl ether
multiphase extraction *6(10)*:245, 253, 259, 267, 275
municipal solid waste *6(2)*:247
Mycobacterium sp. IFP 2012 *6(1)*:153
Mycobacterium adhesion *6(3)*:251
mycoremediation *6(6)*:263

N

naphthalene *6(1)*:1; *6(2)*:121; *6(3)*:173, 227; *6(5)*:1, 253; *6(6)*:51; *6(8)*:95; *6(9)*:139; *6(10)*:123

NAPL, *see* nonaqueous-phase liquid
natural attenuation *6(1)*:27, 35, 43, 51, 59, 75, 83, 183, 195; *6(2)*:1,39, 73, 81, 89, 97, 105, 137, 145, 173, 181, 215; *6(4)*:91, 99, 117; *6(5)*:33, 189, 321; *6(8)*:185, 209; *6(9)*:179; *6(10)*:115, 163
natural gas *6(10)*:193
natural organic carbon *6(2)*:261
natural organic matter *6(2)*:81, 97; *6(8)*:201
natural recovery *6(5)*:132, 231
nitrate contamination *6(9)*:173
nitrate reduction *6(3)*:51; *6(5)*:25; *6(9)*:331
nitrate utilization efficiency *6(6)*:353
nitrate *6(2)*:1; *6(3)*:17, 43; *6(6)*:353; *6(8)*:95, 147; *6(9)*:179, 187, 195, 209, 223, 257
nitrification *6(4)*:149; *6(5)*:337; *6(9)*:215
nitroaromatic compounds (*see also* explosives and energetics) *6(3)*:59, 67; *6(6)*:149
nitrobenzene, *see also* explosives and energetics *6(6)*:149
nitrocellulose, *see also* explosives and energetics *6(6)*:119
nitrogen fixation *6(6)*:219
nitrogen utilization *6(9)*:231
nitrogenase *6(6)*:219
nitroglycerin, *see also* explosives and energetics *6(5)*:69
nitrotoluenes, *see also* explosives and energetics *6(6)*:127
nitrous oxide *6(8)*:113
^{13}C-NMR, *see* nuclear magnetic resonance spectroscopy
nonaqueous-phase liquids (NAPLs) *6(1)*:67, 203; *6(3)*:141; *6(7)*:249
nonylphenolethoxylates *6(5)*:215
nuclear magnetic resonance spectroscopy (^{13}C-NMR) *6(4)*:67
nutrient augmentation *6(3)*:59; *6(5)*:329; *6(6)*:257; *6(7)*:313; *6(9)*:331; *6(10)*:23
nutrient injection *6(10)*:101
nutrient transport *6(9)*:241

O

oily waste *6(4)*:35; 6(6):257; *6(10)*:337, 345
oil-coated stones *6(10)*:329

optimization *6(5)*:279
ORC® (a proprietary oxygen-release compound) *6(1)*:99,107; *6(2)*:215; *6(3)*:107; *6(7)*:229; *6(10)*:9, 15, 87, 95, 139
organic acids *6(2)*:39
organophosphorus *6(6)*:17, 27
advanced oxidation *6(6)*:157, *6(10)*:311
oxygen-release compound, *see* ORC®
oxygen-release material *6(10)*:73
oxygen respiration *6(9)*:231; *6(10)*:57
oxygenation *6(1)*:107, 145
ozonation *6(1)*:121; *6(10)*:33, 149, 301

P

packed-bed reactors *6(9)*:249; *6(10)*:329
PAHs, *see* polycyclic aromatic hydrocarbons
paper mill waste *6(4)*:1
paraffins *6(3)*:141
partitioning *6(9)*:129
PCBs, *see* polychlorinated biphenyls
PCP toxicity (*see also* toxicity) *6(3)*:125
PCP, *see* pentachlorophenol
PCR analysis, *see* polymerase chain reaction
pentachlorophenol (PCP) *6(3)*:83, 91, 99, 107, 115, 125; *6(5)*:329; *6(6)*:279, 287, 295, 329
percarbonate *6(10)*:73
perchlorate *6(9)*:249, 257, 265, 273, 281, 289, 297, 303, 309, 315
perchloroethene, perchloroethylene *6(7)*:53
permeable reactive barriers *6(3)*:1; *6(8)*:73, 87, 95, 121, 139, 147, 157, 167, 175, 185; *6(9)*:71, 309, 323; *6(10)*:95
pesticides *6(5)*:189; *6(6)*:9, 17, 35
PETN reductase *6(5)*:69
petroleum hydrocarbon degradation *6(4)*:7; *6(5)*:9, 17, 25; *6(8)*:131; *6(10)*:65, 101, 245, 345
phenanthrene *6(2)*:121; *6(3)*:227, 235, 243; *6(6)*:51, 65, 73
phenol *6(6)*:303, 319, 329
phenolic waste *6(6)*:311
phenol-oxidizing cultures *6(10)*:211, 217, 239
phenyldodecane *6(2)*:27
phosphate precipitation *6(9)*:165
PHOSter *6(10)*:65

photocatalysis *6(10)*:311
physical/chemical pretreatment *6(1)*:1, 51; *6(2)*:253; *6(3)*:149; *6(5)*:9, 33, 41, 53, 61, 69, 77, 85,105, 113, 121, 129,137, 145, 151, 157, 165, 189, 199, 207, 279, 337; *6(6)*:59, 157, 241; *6(7)*:1, 13; *6(9)*:113, 173; *6(10)*:239, 311, 319
phytotoxicity (*see also* toxicity) *6(5)*:41, 223
phytotransformation *6(5)*:85
pile-turner *6(6)*:249
PLFA, *see* phospholipid fatty acid analysis
polychlorinated biphenyls (PCBs) *6(2)*:39,105,173; *6(5)*:33, 61, 95, 113, 231, 289; *6(6)*:89, *6(7)*:13, 61, 69, 95, 109, 125, 133, 141, 149, 165, 181, 189, 197, 205, 213, 241, 249, 273, 297, 305; *6(8)*:11,19, 27, 43, 157, 167, 193, 209; *6(10)*:33, 41, 231, 283
polycyclic aromatic hydrocarbons (PAHs) *6(2)*:19, 121, 129, 137; *6(3)*:141, 149, 157, 165, 173, 181, 189, 197, 205, 211, 219, 227, 235, 243; *6(4)*:35, 45, 59, 67; *6(5)*:1, 9, 17, 41, 237, 243, 251, 253, 261, 269, 279, 289, 305, 329; *6(6)*:43, 51, 59, 65, 73, 81, 89, 101, 279, 295, 297; *6(7)*:125; *6(8)*:95; *6(9)*:139; *6(10)*:33, 123
polymerase chain reaction (PCR) analysis *6(4)*:29, 35, 51; *6(8)*:43
polynuclear aromatic hydrocarbons, *see* polycyclic aromatic hydrocarbons
poplar lipid fatty acid analysis (PLFA) *6(3)*:189
poplar trees *6(5)*:113, 121, 189
potassium permanganate *6(2)*:253; *6(7)*:1
precipitation *6(9)*:105; *6(10)*:301
pressurized-bed reactor *6(6)*:311
propane utilization *6(1)*:137; *6(10)*:145, 155, 179, 193
propionate *6(7)*:265, 289
Pseudomonas fluorescens *6(3)*:173
pyrene *6(3)*:165, 235; *6(4)*:67; *6(6)*: 65, 73, 101
pyridine *6(4)*:81

R

RABITT, *see* reductive anaerobic biological in situ treatment technology
radium 6(5):173
rapeseed oil 6(6):65
RDX, *see* research development explosive
rebound 6(10):1
recirculation well 6(7):333, 341; 6(10):283
redox measurement and control 6(1):35; 6(2):11, 231; 6(5):1; 6(9):53
reductive anaerobic biological in situ treatment technology (RABITT) 6(7):109
reductive dechlorination 6(2):39, 65, 97, 105, 145, 173; 6(4):125; 6(7):45, 53, 87, 103, 109, 133, 141, 149, 157,181, 197, 205, 213, 221, 249, 257, 265, 273, 289, 297; 6(8):11, 73, 105, 157, 209
reductive dehalogenation 6(7):69
reed canary grass 6(5):181
research development explosive (RDX) 6(3):1, 9, 17, 25, 35, 43, 51; 6(6):133; 6(8):175
respiration and respiration rates 6(2):129; 6(4):59; 6(6):185, 227
respirometry 6(6):127; 6(10):217
rhizoremediation 6(5):9, 61, 199
Rhodococcus opacus 6(4):81
risk assessment 6(2):215; 6(4):1
16S rRNA sequencing 6(8):43; 6(9):147
rock-bed biofiltration 6(4):149
rotating biological contactor 6(9):79
rototiller 6(6):203
RT3D 6(10):163

S

salinity 6(9):257
salt marsh 6(5):313
SC-100, *see* single culture
Sea of Japan 6(5):321
sediments 6(3):91; 6(5):231, 237, 253, 261, 269, 279, 289, 297, 305; 6(6):51, 59; 6(9):61
selenium 6(9):323, 331
semivolatile organic carbon (SVOC) 6(2):113
sheep dip 6(6):27
Shewanella putrefaciens 6(8):201
silicon oil 6(3):141, 181
single culture 6(8):61
site characterization 6(10):139
site closure 6(2):215
slow-release fertilizer 6(2):57
sodium glycine 6(9):273
soil treatment 6(3):181; 6(6):1
soil washing 6(5):243; 6(6):241
soil-vapor extraction (SVE) 6(1):183; 6(10):1, 41, 131, 223
solids residence time 6(10):211
sorption 6(5):215, 253; 6(6):377; 6(8):131; 6(9):79, 105
source zone 6(7):13, 19, 27, 181; 6(10):267
soybean oil 6(7):213
sparging 6(10):33, 145, 155
stabilization 6(6):89
substrate delivery 6(7):281
sulfate reduction 6(1):35; 6(3):43, 91; 6(5):261, 313; 6(6):339; 6(7):69, 95; 6(8):139, 147, 193; 6(9):1, 9, 17, 27, 35, 43, 61, 71, 86, 105, 123, 147
sulfide precipitation 6(9):123
surfactants 6(5):215; 6(6):73; 6(7):213, 321, 333; 6(8):131
sustainability 6(6):1
SVE, *see* soil vapor extraction
SVOC, *see* semivolatile organic carbon
synthetic pyrethroid 6(6):27

T

TCA, *see* trichlorethane
1,1,1-TCA, *see* 1,1,1-trichloroethane
1,1,2-TCA, *see* 1,1,2-trichloroethane
2,4,6-TCP, *see* 2,4,6-trichlorophenol
1,1,1,2-TeCA,*see* tetrachloroethane
1,1,2,2-TeCA, *see* tetrachloroethane
1,3,5-TNB, *see* 1,3,5-trinitrobenzene
TAME, *see* tertiary methyl-amyl ether
TBA, *see* tertiary butyl alcohol
TBF, *see* tertiary butyl formate
TCE oxidation, *see* trichloroethene, trichloroethylene
TCE, *see* trichloroethene
TCP, *see* trichlorophenol
t-DCE, *see* trans-dichloroethene, trans-dichloroethylene
technology comparisons 6(7):45; 6(9):323
terrazyme 6(10):345

tertiary butyl alcohol (TBA) **6(1)**:19, 27, 35, 51, 59, 91, 145, 153, 161
tertiary butyl formate (TBF) **6(1)**:145, 161
tertiary methyl-amyl ether (TAME) **6(1)**:59, 161
tetrachloroethane (1,1,1,2-TeCA, 1,1,2,2-TeCA) **6(5)**:207; **6(7)**:321, 341; **6(8)**:193
tetradecane (see also ^{2}H-tetradecane) **6(3)**:181
thermal desorption **6(3)**:189, **6(6)**:35
TNB, see trinitrobenzene
TNT, see trinitrotoluene
TNX, see 1,3,5-trinitroso-1,3,5-triazacyclohexane
tobacco plant **6(5)**:69
toluene **6(1)**:145; **6(2)**:181; **6(7)**:95; **6(8)**:35, 131
total petroleum hydrocarbons (TPH) **6(2)**:1; **6(5)**:9; **6(6)**:127, 173, 179, 193, 227, 241, 249; **6(10)**:15, 73, 115, 337
toxicity **6(1)**:1; **6(3)**:67, 189, 227; **6(4)**:7; **6(5)**:41, 61, 223, 305; **6(9)**:17, 129
TPH, see total petroleum hydrocarbons
trace gas emissions **6(6)**:185
trans-dichloroethene, trans-dichloroethylene **6(5)**:95, 207; **6(7)**:165
transgenic plants **6(5)**:69
transpiration **6(5)**:189
Trecate oil spill **6(6)**:241; **6(10)**:109
trichloroethane (TCA) **6(7)**:241, 281
1,1,1-trichloroethane (1,1,1-TCA; 1,1,2-TCA) **6(2)**:39, 113, 464; **6(5)**:207; **6(7)**:87,165, 281
1,1,2-trichloroethane (1,1,2-TCA) **6(5)**:207
trichloroethene, trichloroethylene (TCE) **6(2)**:39, 65, 73, 97, 105, 113, 155, 173, 253; **6(4)**:125; **6(5)**:33, 95, 105, 113, 207; **6(7)**:1, 13, 53, 61, 69, 77, 87, 109, 117, 133, 141, 149, 157, 181, 189, 197, 205, 213, 221, 241, 249, 265, 273, 281, 297, 305; **6(8)**:11, 19, 27, 35, 43, 53, 73, 105,147, 157, 193, 209; **6(10)**:41, 131, 145, 155, 163, 171, 179, 187, 201, 211, 217, 223, 231, 239, 283, 319
2,4,6-trichlorophenol (2,4,6-TCP) **6(3)**:75; **6(8)**:121
trichlorotrifluoroethane **6(2)**:49

trinitrobenzene (TNB) **6(3)**:9, 25
1,3,5-trinitroso-1,3,5-triazacyclohexane (TNX) **6(8)**:175
trinitrotoluene (TNT) **6(3)**:35, 67; **6(5)**:69, 77, 85; **6(6)**:133

U

underground storage tank (UST) **6(1)**:67, 129
uranium **6(5)**:173; **6(7)**:77; **6(9)**:155, 165
UST, see underground storage tank

V

vacuum extraction **6(1)**:115
vadose zone **6(1)**:183; **6(2)**:39, 65, 97, 105, 113, 155, 173; **6(3)**:9; **6(5)**:33, 105; **6(7)**:1,13, 61, 133, 141, 197, 205, 213, 249, 273, 281, 305; **6(8)**:11,19, 43, 73, 157, 209; **6(10)**:41, 163
vegetable oil **6(6)**:65; **6(7)**103, 213, 241, 249
vinyl chloride **6(2)**:73; **6(4)**:109; **6(5)**:95; **6(7)**:95,149, 157, 165, 173, 289, 297, **6(10)**:231
vitamin B$_{12}$ **6(7)**:321, 333, 341
VOCs, see volatile organic carbons
volatile fatty acid **6(7)**:61
volatile organic carbons (VOCs) **6(2)**:113, 189; **6(5)**:113, 121

W

wastewater treatment **6(5)**:215; **6(6)**:149; **6(9)**:173
water potential **6(9)**:231
weathering **6(4)**:7
wetlands **6(5)**:33, 95, 105, 313, 329; **6(9)**:97
white rot fungi, (see also fungal remediation) **6(3)**:75, 99; **6(6)**:17, 157, 263
windrow **6(6)**:81, 119, 141
wood preservatives **6(3)**:83, 259; **6(4)**:59; **6(6)**:279

X

xylene **6(1)**:67

Y
yeast extract *6(7)*:181

Z
zero-valent iron *6(8)*:157, 167; *6(9)*:71
zinc *6(4)*:91; *6(9)*:79